Microbes and the Fetlar Man

W. Watson Cheyne

Sir William Watson Cheyne

Microbes and the Fetlar Man
The Life of Sir William Watson Cheyne

Jane Coutts

humming earth

Published by

Humming Earth
an imprint of
Zeticula Ltd
Unit 13
196 Rose Street
Edinburgh
EH2 4AT
Scotland

http://www.hummingearth.com
admin@hummingearth.com

First published in 2015
Reprinted 2016
Copyright © Jane Coutts 2015, 2016

ISBN 978-1-84622-061-6 hardback
ISBN 978-1-84622-055-5 paperback

All rights reserved. No part of this publication may be reproduced, stored in a retrieval system, or transmitted in any form or by any means, electronic, mechanical, photocopying, recording or otherwise, without the prior permission of the publishers.

To my mother and father

Sir Watson with his ubiquitous cigar.

Acknowledgements

Biographies are labour-intensive, and without the help and support of an enormous number of people, they would never reach completion. This book belongs to all the people, in Shetland, England, Scotland, Spain and further afield who have helped in its production, particularly to the following.

I should like to thank my husband John, himself a native of Fetlar, for his inexhaustible patience over the years of writing this book. His Fetlar stories and the memories of his family have brought the tale of Sir Watson to life. His father managed Cheyne's home farm. I am indebted to him in particular for advice on the nuances of Shetland dialect in chapter five. I am especially grateful to his eldest sister, Mrs Catherine Brown, in New Zealand who, at the time of writing is in her late nineties, and remembers Sir Watson personally. Her clear memories and first-hand knowledge of Leagarth and Sir Watson's practices at the Home Farm are, I think, unique. Thanks to her daughters and son-in-law for collating the answers to my questions and sending them on. I should also like to thank my parents, sister and brother-in-law for their long-standing support, for their unwavering faith that this biography would eventually find its way onto bookshelves, and for part-funding the research. Thanks to my son Frank for helping with the photos and my son Temmy for moral support.

In a professional capacity, I cannot sufficiently thank Professor Mary Wilson Carpenter for her detailed input into chapter five, particularly for discussing newly-discovered material with me. I should also like to thank Dr Ruth Richardson for her invaluable and comprehensive help and support on some of the medical history aspects of the book, for assistance in my research and for the brainstorming which produced part of the title. My thanks also go to Professor Anne Crowther for reviewing the chapters referring directly to Lord Lister, particularly Cheyne's student days. I am grateful to Professor Michael Worboys for his input, also to Professor Marguerite Dupree for her support at a critical time, and for agreeing to read the full manuscript. The staff at the archives of the Royal College of Surgeons of Edinburgh provided invaluable help at the beginning of the research, as did the University of Edinburgh archives, and in London, I wish to thank the staff at the Royal College of Surgeons of England, King's

College Archive, the Wellcome Museum, the National Archives at Kew, and the British Library for their patience and assistance.

I cannot sufficiently acknowledge the help of the staff at Shetland Archives. I particularly thank Blair Bruce for his swift responses to requests for information, and Brian Smith, the Shetland Archivist, for commenting on the chapters with Shetland content. Also in Shetland, I should like to thank old friends for support and accommodation during research periods, particularly the Society of Our Lady of the Isles (SOLI), Joyce Garden and the Kapushniaks. Thanks go to Pierre Cambillard for looking up plaques in the Fetlar Kirk (and for taking on Brough Lodge). Thanks also to the Trustees of Fetlar Museum Trust for permission to reproduce some of their digitised photograph collection, for allowing me access to the collections, and to revisit a building which has played a significant role in my life.

Elsewhere in Shetland, I would especially like to thank Mr and Mrs K. Hughson, formerly of Fetlar, and now living in Yell, for sharing their memories of nearly a century and for their many years of support and contribution to the museum collection in Fetlar. They were caretakers of Leagarth House for many years, and more than answered my questions about its heyday. Many thanks also to David Clark and Andrina Tulloch in Yell for finding me information on Sir Watson's Shetland boats and on Walter Shewan. Thanks to Eileen Mullay and Susan Thompson for providing photographs of Tangwick Haa at the last moment. Thanks also to Yvonne McHugh in California for finding me information on Captain Cheyne from California newspapers.

I have saved until almost last my debt to the strongest advocates of this work, the Cheyne family. Sir Watson's grandchildren, great-grandchildren and great-great grandchildren have not only placed at my disposal any papers, information and personal memories of their illustrious ancestor, they have also been the most enthusiastic supporters of the production of his biography. I should like to thank in particular Bridget Lanyon, Sir Watson's great-granddaughter, for accommodating me at Leagarth House prior to the worst storm of 2013, and for her patience as I raced to record the contents of Sir Watson's library in two days and nights (thanks also to Gregor the builder, who came to the rescue when my camera batteries ran out). I should like to thank the Cheyne family in general for its permission to digitise photographs and documents from the house, and for its support in loaning artefacts for the Cheyne exhibition at Fetlar Interpretive Centre. I am indebted to the late Andrew (Sandy) Cheyne and the late Sir Joseph Cheyne for their unique personal memories of their grandfather. A friend with whom I corresponded during the research was astounded to hear that there were still people with personal memories of Sir Watson, where in so many other respects, the Lister era is now seen as a relatively distant moment in British history.

I will end with a rather unusual acknowledgement to all the "ghosts", without whose foresight in collecting, saving or documenting what must have seemed like everyday incidents, parts of this book would not have been possible. They include, in the 19th and 20th centuries, Lawrence Williamson of Mid Yell, Shetland, J.J. (Jeemsie) Laurenson in Fetlar, Margaret Mathewson and of course, Sir William Watson Cheyne and his father, Captain Andrew Cheyne. Tribute is due in general to the indefatigable efforts of ordinary Shetlanders in their enthusiasm for preserving and discussing their history. Finally, I should like to acknowledge the late Dr Dorothy Shineberg of the Australian National University for her extensive correspondence with Sir Joseph Cheyne in order to publish a portion of Captain Cheyne's journals, and her conviction that Captain Cheyne had been misunderstood by history.

Contents

	Acknowledgements	*vii*
	Illustrations	*xiii*
	Preface: Rediscovering Sir Watson	*xvii*
	Introduction	1
	Note on names and locations	7
1	Dreams of Empire: Captain Andrew Cheyne	9
2	Fetlar: A childhood by the sea	35
3	Discovering Lister: A revolution in surgery	62
4	Edinburgh to Vienna: An introduction to microbes	86
5	Mr Cheyne enjoys a lark: A patient describes Lister's surgical ward	102
6	With Lister to London: Pioneering British bacteriology	132
7	Quite singular beauty: Illustrating microorganisms and tuberculosis	156
8	Where tubercle bacilli most do congregate: The tuberculin affair	186
9	Surgical towels in the chamber pot: Antisepsis or asepsis?	220
10	Microbes at the Cape: The Boer War	240
11	Hospitals or barracks?: Lessons from the War	276
12	Picnics and concert parties: Building Leagarth	296
13	Getting older: 1900-1914	331
14	Antiseptics at the Dardanelles: World War I	355
15	A very interesting lecture: Cheyne and politics	374
16	Coming Home: The Fetlar years	396
	Notes and References	*437*
	Select Bibliography	*510*
	Photograph Credits	*532*
	Index	*536*

Leagarth House with verandah in the 1920s.

Illustrations

Sir William Watson Cheyne	ii
Sir Watson with his ubiquitous cigar	vi
Leagarth House with verandah in the 1920s	xii
Sir William Watson Cheyne	xvi
A garden party in the grounds of Leagarth House, early 1900s	xviii
Tangwick Haa, childhood home of Captain Andrew Cheyne	11
Tangwick Haa	12
This image may be Andrew Cheyne	13
This may be Grace Watson, Cheyne's aunt, as an old lady	38
The West Manse in Fetlar	43
The West Manse, Fetlar, in 2014	46
One of the last houses in Fetlar with a fire in the centre	48
Certificate for Natural History from the University of Edinburgh	71
Certificate for Chemistry from the University of Edinburgh	73
Cheyne's certificate for Anatomy	74
Cheyne's certificates for Junior Surgery	75
Cheyne's certificates for Senior Surgery	76
Lister's carbolic hand spray	90
Registration document from the University of Vienna.	95
Vienna class attendance certificate	96
Margaret Mathewson, Lord Lister's Shetland patient	103
A Shetland woman walking down a road knitting	105
Illustration from Cheyne's *Antiseptic Surgery*	133
Drawing of Bacillus alvei, found inside one of Cheyne's books	162
Page from Cheyne's own collection of his Reports	168
Pages from Cheyne's own amendments to *The Treatment of Wounds*	207
A 'Watson Cheyne' probe	213
Basil Cheyne, Sir Watson's youngest son	218
Lister's carbolic steam spray	222
Signed photograph of Sir William Watson Cheyne	234
Probably on the march to Bloemfontein	247
What appears to be a destroyed railway bridge	251
Attending to bodies during the Boer War	252
Erecting a temporary bridge	254

Men clearly in some distress, and may be suffering from fever	255
Building probably used as a temporary hospital	256
Difficulties medical teams faced with transport	258
Lord Roberts and Boer General Cronje	261
In talks after General Cronje's surrender	262
Lord Roberts leaving Cronje and his secretary	263
Cronje's secretary (far left) speaking to Lord Roberts	264
Scenes of devastation in the Boer laager	265
Devastation, probably in Cronje's laager after his surrender	267
Showing the difficulties medical teams faced with transport	270
Royal Army Medical Corps group in South Africa	277
Sir Watson with his camera outside the Chapel	300
Brough Lodge, Fetlar, the home of the Nicolson family	302
The tower at Brough Lodge, Fetlar, a mid-19th century folly	302
Lady Annie Nicolson	304
Sir Arthur T.B.R. Nicolson of Brough Lodge	305
Flit boats coming into Brough, Fetlar	306
The flit boat transport to (or from) the Earl of Zetland	307
Golf at Brough Lodge	308
Leagarth House in its original state	311
Car being brought ashore by flit boat	313
Vera Nicolson of Brough Lodge	314
The Kirk on Sunday	316, 317
Flittin peats by boat from Lambhoga, Fetlar	318
Flittin peats by pony in Lambhoga, Fetlar	319
One of Sir Watson's boats	320
Sir Watson in one of his boats	321
Standing by the flit boat	322
A Leagarth party disembarking into the flit boat	323
A Leagarth party, early 1900s	324
Images from the guest book at Leagarth	326
Sir Watson (left) in his motor boat	327
Meta Cheyne, Basil Cheyne. Outside Leagarth House, early 1900s	328
Framed photograph of Lady Curzon	334
Sir Lister Cheyne, Sir Watson's eldest son	335
Complimentary dinner given for Cheyne	343
The notice contains signatures of the guests	344
The Order of St. Michael and St. George (K.C.M.G.)	373
J.J. ('Jeemsie') Laurenson	383

Leagarth House and gardens from the sea	396
The silver tray engraved "Beechgrove, 1920"	397
The Leagarth staff	399
Leagarth staff, 1920s, standing outside	400
The entrance to the laboratory next to Cheyne's apartments	401
Prescription written by Sir Watson for Christian Anderson	402
Frank Coutts, manager of Cheyne's Home Farm	404
Overview of Leagarth gardens, early 1920s	406
Sir Watson in the completed gardens at Leagarth	407
In the sunken alpine garden at Leagarth	408
The head gardener at Leagarth, James Young Robertson ("Jimmy Young"), and his assistant, Willie Garriock	409
Sir Watson in the verandah at Leagarth	410
The Fetlar community	411
Fetlar people dressed for a concert party, probably 1920s	411
One of the many Leagarth concert parties, 1920s	412
Sports on the lawn at Leagarth	412
A picnic on the lawn outside the verandah at Leagarth, 1920s	413
Sir Watson teaching his son Basil to fish in the Houbie Burn, Fetlar	415
The trout ladder Sir Watson built on the Houbie Burn	416
Sir Watson on one of his many journeys	418
Possibly Ella Davis, on Sir Watson's New Zealand trip in 1928	425
The illuminated copy of the welcome speech	427, 428
Probably Ella Davis holding a pickaxe outside a hut	430
Hunter Cheyne, who followed his father into medicine	431
Sir Watson's grandchildren on Tresta Beach, Fetlar	433
Sir Watson's final resting place in the Fetlar Kirkyard	434
Sir Watson and his dog, outside the verandah at Leagarth	436
... with his ubiquitous cigar	509

Sir William Watson Cheyne

Preface

Rediscovering Sir Watson

In the year 2000, at the turn of the millennium, I was manager of the museum on the remote Shetland island of Fetlar. I was fascinated by the fact that, in the early 1900s, not only had one of Britain's most eminent surgeons lived on the island, he had also spent most of his childhood and retirement there. More intriguing still was the fact that, though his early life read like a tale from Robert Louis Stevenson, and his scientific work was of considerable significance, relatively little detailed attention had been afforded them in the history of medicine. In fact, in the course of subsequent research, I was to discover that, far from being simply an assistant to his mentor Lord Lister, Cheyne was one of the most prominent founding fathers of medical bacteriology in Britain.

Sir William Watson Cheyne, or "Sir Watson" as he was almost universally known after he was made a baronet, was Lord Lister's house surgeon and close friend, and set up some of the first systematic laboratory procedures in an English hospital. He was also a pioneer of the methods which led to the momentous advances in treatment of disease at the end of the 19th century. He was highly regarded internationally, particularly by Robert Koch, the Nobel prize-winner who discovered the microorganism which causes tuberculosis. Finally, Cheyne was Lord Lister's staunchest ally and defender in the development and adaptation of the antiseptic techniques which contributed to safer hospitals. Why, then, had he not remained a more significant figure in history?

Tentatively, we took initial steps to look into the matter. With the help of the Cheyne family and the Fetlar community, we were able to bring together material from his professional and home life, and produce an exhibition which opened in 2000. To our enormous surprise, it won the Best Educational Initiative Award at Scottish Museum of the Year, and I always considered this to have been Sir Watson's achievement, not ours.

Where his story was clearly significant in the history of medicine, I also wanted to establish whether his life as a whole could be of interest to the general public, given some of the extraordinary events of his childhood. Any doubts we may have had were dispelled at the opening of the exhibition. We had invited Sir Joseph Cheyne, Sir Watson's grandson and one of the few left alive at the time who remembered him personally, to open the exhibition and give a short talk to an assembled group of school children from two islands. They were crowded into the relatively small space, with a bust of the great man watching over them. Normally, it would have been almost impossible to find something to keep so many children still for even five minutes in a museum, yet for well over half an hour, the entire group sat cross-legged on the floor, completely spellbound. Sir Joseph, who had inherited his grandfather's gift for telling a compelling story, finished the tale, and for a few moments silence still reigned. One of us had to begin the applause to signal that it was all over.

A garden party in the grounds of Leagarth House, early 1900s

As part of a number of broader projects, I digitised Cheyne's photographs from Leagarth along with important documents, one of which was his own handwritten copy of the essay he submitted for the American Boylston Prize at Harvard in 1880. Combined with other essays, it became one of his main works, *Antiseptic Surgery*. I had seen the bound, handwritten manuscript on the shelves of the library at Leagarth House many times, but it was only when I began to examine Cheyne's works in more detail that I realised its significance. It seems most discoveries of this ilk are made by chance.

The most revealing, however, was reserved for the photographs. On some occasions, they were so faded as to be almost blank to the naked eye, but the process of digitisation brought them back to life like ghosts appearing out of nowhere on the screen. One image stands out in particular. From a faded mass of sepia, a scene of a garden party emerged, local people lined up in chairs neatly around the lawn, the Leagarth staff serving tea from a long table with a calm summer sea in the background.

One final set of photographs emerged much later. One afternoon Gerald Cheyne, Sir Watson's great-grandson, who was visiting Leagarth with friends, came into the museum with a small package he had found in the house. Inside were glass lantern slides, but we could not quite see the images. Our scanner had insets to be able to scan glass slides, as there were others in the museum collection, so I worked on them as soon as I could. As they began to appear on the screen, I realised they were the photographs Sir Watson had taken during his march with Lord Roberts from Modder River to Bloemfontein during the Boer War. Four prints of the moment of the Boer General Cronje's surrender had been hanging in Leagarth since Sir Watson's day, but these glass images were quite different. They showed horses trying to pull medical supplies across contaminated rivers over precarious, temporary bridges, and medical teams scaling impossible slopes with wagons. The slides were, unfortunately, not always complete, and some had been subject to the ravages of time, but they were, nevertheless, the slides Sir Watson had given at his much-publicised talks after the war. He was one of the first medical men to use lantern slides in his lectures.

If I had ever had any doubts that this material was as important as Sir Watson himself, it became clear with the photographs, the glass slides and Sir Watson's hand-written manuscript that Leagarth House and the island of Fetlar were sitting on important contributions to the history of British medicine. We went on to take interpretive exhibitions on Cheyne's work to the teaching department of the Gilbert Bain hospital in Lerwick, and promoted the permanent display through the Scotland and Medicine project based at the Royal College of Surgeons in Edinburgh. At the Lister Centenary Conference at King's College London in 2012, I was gratified to hear his name in a substantial number of papers other than my own. Peter Thompson, Consultant Urological Surgeon at King's College

Hospital who introduced the section at the conference, said that next time he hoped it would be Cheyne's turn to be celebrated. Over eighty years after Sir Watson's death, I hope this account of his extraordinary life goes some small way to restoring him to his rightful place in history.

Jane Coutts
July 2015

Introduction

Sir William Watson Cheyne was one of the most prominent members of the inner scientific circle of Lord Lister, best remembered as the father of antiseptic surgery. Cheyne's significance among Listerians of the late 19th century lies partly in the fact that he remained with Lister for a good part of his professional life, but more importantly because he was regarded as Lister's bacteriologist. He conducted and informed many of the laboratory experiments which allowed Lister to adapt and progress his antiseptic theory and method on the basis of a causal relationship between bacteria and sepsis.

Lister attracted not only a loyal circle of advocates but also a battalion of detractors. Cheyne conquered his early shyness to become the foremost of Lister's intellectual bodyguards, and was so instrumental in maintaining the profile of chemical antiseptics in surgery that he has been referred to as "Lister's bulldog"[1]. However, beyond his close working relationship with Lister, Cheyne was an important figure in introducing from the Continent the laboratory methodologies which helped bacteriology to develop as a science in British medicine in its own right.

The name of Lord Lister is well known to anyone interested in the history of medicine. In the mid-late 19th century, he was leading steps to reduce mortality rates in hospitals in Britain, informed principally by the work of Louis Pasteur in France. Underpinning all this work, however, was the birth of the science of bacteriology, and the movement to prove, first, that minute living "germs" could be the cause of infection in wounds and the deadly diseases so much feared by the 19th century public, and secondly, to show that a very specific bacterium caused these diseases. The implication for Lister was that, by preventing bacteria from entering wounds sustained either in accidents or through surgery, sepsis could be avoided. It meant there would be fewer amputations, and fewer people would be at risk of dying from post-operative complications.

The aspect of Lister's work which tends to be down-played in the public mind, however, is his role as a scientist. By linking the study of microbes to surgery, and focusing on a particular branch of germ theory (a germ theory of putrefaction),[2] he was able to provide experimental evidence of the relationship of bacteria to wounds. The methods he developed to exclude bacteria from wounds took much of the guesswork out of surgery, and created a more solid foundation for systematic development. His students were thoroughly grounded in his methods, and some took his work further, adapted it, and made sure its bacteriological base was not forgotten in subsequent developments.

Contemporary descriptions of Sir William Watson Cheyne as a solid character of somewhat exotic origins are not too wide of the mark, but not necessarily for the right reasons. They dwell on the "Norse blood" of his native Shetland, but, in fact, his background and early influences are much more complex. Before he was two years old, he had travelled the globe as an incidental player in his father's colourful South Sea adventures, and then grew up in a small, remote island community with only a hazy memory of his parents. The circumstances of his birth and childhood gave him a relentless taste for innovation and a lifelong obsession with the sea. One commentator wrote that Cheyne was "never happier than when sailing a boat; he could handle a tiller to perfection."[3] On the other hand, family tragedy followed him almost from the moment of his birth, and he began his medical career rather by chance, a shy young man with little faith in his abilities.

The King's College Review described Cheyne in 1918 as "the complement of Lister ... The *Master* was a man of peace and without temper; the disciple was a clean, straight fighter, descended from ancestors of immense courage and easily wrought fury, but cheery and not bearing malice."[4] The image of 'immense courage and easily wrought fury' could apply with no difficulty to his father, but Cheyne himself rather liked to cultivate his 'Norseman' image as he grew older. He was immensely proud of his Shetland origins.

No biography would therefore be complete without a detailed assessment of this background, and I have devoted the first chapter to Cheyne's father, his mother and the fateful journey on which their son was born. Their influence, whatever he remembered of it, can only have contributed to the pioneer who heard in Lord Lister's lectures "the wonderful era that was dawning for surgery"[5]. The tale also illustrates all too well the events which more than once jeopardised his survival. His first two years were spent in the company of mariners, port-dwellers, traders and island communities in the Pacific, and the family more than once faced dangers which could have spelled disaster. The contrast with the Church of Scotland Manse where he spent the remainder of his childhood could hardly have been greater.

Chapter two explores his childhood in Fetlar, one of the remoter islands of the Shetland archipelago, under the guardianship of his aunt and uncle, and in constant contact with the sea. I examine its influences on his health and future work, and how it led him, by chance rather than design, to Lister. I also consider the effects of the spectacular death of his father, and the legacy of his adventures. In later years, Sir Watson was to remain loyal to the memory of his father, whose rather unorthodox approach to life had perhaps deprived him of the success and recognition he deserved within his own lifetime. The question of the legacy of Captain Andrew Cheyne is important for understanding the man his son became, and the competing influences on his early decisions.

In Chapters three and four, I detail the birth of Cheyne's scientific and surgical career, his introduction to antiseptic surgical methods and his early involvement in the study of microbes. His student years in Edinburgh brought an end to an early period of indecision and depression after the death of his father, and gave him access to the channels he needed for his talents to develop. A subsequent study tour of European universities gave him a grounding in the all-important laboratory methodology which would make him so valuable to Lister. These years also laid the foundation for his role as Lister's champion in the theoretical debates about the role of antiseptics, and provided him with an insight into the complexity of the arguments.

Chapter five describes day-to-day life on Lister's wards through the diary of a Shetland woman who recorded her experience as a patient in Edinburgh. Margaret Mathewson was something of a pioneer herself, embarking on hospital treatment against the advice of her family in an era where surgical intervention had a poor reputation, and she describes Cheyne and the environment in which he worked from the rare point of view of a patient. I was keen to elaborate the material from the diary in a chapter of its own, as it gives a detailed, close-up view of Cheyne as a young man. Their shared Shetland origins helped Mathewson cope with an otherwise difficult experience, and her descriptions of Cheyne provide an unusual cameo of a brief but very significant moment in his life.

Chapter six details the early developments in Cheyne's career as a scientist and surgeon. With first-hand knowledge of the infant science of bacteriology from the European institutions where it was making its debut, he was set to become Britain's most prominent demonstrator of new continental laboratory methodology which would lead to momentous advances in medicine. We are often left with the impression that he felt more at home in the laboratory than in the operating theatre, and that he fully appreciated at the time the pioneering nature of the work, thriving on the excitement of it. I consider in this chapter the importance of Lister's decision for Cheyne to accompany him to King's College Hospital in London, where they initially met with a degree of opposition to their methods, and especially the theory behind them. It was this period which particularly consolidated Cheyne's career. At a time when there were few opportunities in Britain for the formal study of the relationship of bacteria to disease, Lister's support provided Cheyne with the environment to become one of the first pioneers of the field in the country.

Chapter seven examines the public face of Cheyne's laboratory work, and how it made him a leading authority at European level as well as at home. The chapter considers his central role in introducing continental laboratory techniques to Britain, and the part they played in adapting the theory and practice of Listerism. This period of Cheyne's life best illustrates his own achievements, and I argue that this is where he deserves recognition in his own right.

Chapter eight elaborates on this with regard to tuberculosis, where Cheyne developed expertise in the study of the disease in bones and joints. His work on the subject illustrates his staggering capacity for detail. His integrity and objectivity were also put to the test when he was asked to run trials on a new treatment for tuberculosis developed by his friend Robert Koch in Germany. Tuberculin was released prematurely, and was the subject of great hopes as well as serious concerns in the international medical community. Cheyne's objective assessment established him as one of the country's most reliable investigators in medical science. He was nevertheless forced to concentrate on a career in surgery, and leave behind any ambitions he may have had in bacteriology. I argue, however, that he never entirely left the study of microbes behind, and also that his popular image as a "safe", but not particularly innovative surgeon is misleading. This is illustrated by some of his more interesting surgical innovations, including confirmation of a procedure for treating "water on the brain" in children. The mid-1880s and '90s, in which he was experiencing an increasingly heavy workload, were also when he married and entered into perhaps the most stable time in his life, both professionally and domestically.

In Chapter nine, I consider Cheyne's public persona within the context of his arguments against surgical methods which were supplanting the use of chemical antiseptics. The chapter illustrates the combination of wit and maturity which made him one of the grand old medical men of his era. Cheyne was a man of meticulous detail, with unwavering commitment and loyalty on the one hand, and a restless and pioneering imagination on the other. If there is any contradiction in these qualities, it is manifest in his witty anecdotes on opponents, whom he generally regarded as having missed the point.

Chapter ten details Cheyne's experience during the Boer war, as Lord Roberts' Consulting Surgeon on the gruelling march from Modder River to Bloemfontein. Combining Cheyne's own comments with contemporary accounts of the conditions they faced, the chapter gives a graphic illustration of the few months which gave him practical experience in the treatment of war wounds.

Chapter eleven details his response to the high-profile typhoid outbreak he witnessed at Bloemfontein, which formed the basis for an interest in the organisation of medical services in war conditions. His role in the subsequent inquiry into the military authorities' management of the outbreak brought him into the national media when he was called on to give a balanced view of what had become a public scandal.

On his return from the war, his thoughts turned increasingly to home, and he bought land in Fetlar to build a retirement home for himself. He clearly already had the vision that would bring his life full circle. Chapter twelve considers the more light-hearted side of Cheyne as he began to return to his roots and build a home for his family. In mild competition with the island's existing laird, he succeeded in creating a hub for the community, staging regular entertainment and creating employment.

Throughout the period from 1900 to 1920, Cheyne was nevertheless still resident in London, at Harley Street, and visited Fetlar in summer with large parties of colleagues, family and friends. In London, meanwhile, he consolidated his growing reputation as a surgeon, and treated some of the most prominent members of British society. I examine this period in Chapter thirteen, and illustrate how his success led to national honours, including a baronetcy. His overwhelming workload, however, gradually led to a deterioration in his health. By this time, his first wife had died, and he had remarried, probably in an attempt to manage his private life alongside his many professional commitments.

At a late stage in his career, he found himself at the forefront of naval medical services during World War I when he acted as a Consultant Surgeon to the Royal Navy. He ultimately attained the rank of Surgeon Rear-Admiral, based at Chatham Royal Naval Hospital. He was simultaneously President of the Royal College of Surgeons of England, placing him theoretically in a powerful position to influence medical policy during the war. His stance in favour of prioritising chemical antiseptics as a form of first aid, however, helped consolidate a reputation as a die-hard. In Chapter fourteen, I examine how he was delighted to find himself returning temporarily to experimental bacteriology, and ran trials to determine how antiseptics could improve early treatment of wounds at the Front. It led to an acrimonious public debate with Sir Almroth Wright over antiseptics and the treatment of wounds in conditions of war.

Before the end of the war, however, he was persuaded to stand for Parliament in 1917. In Chapter fifteen, I illustrate his political career with reference to his involvement in the medical debates of the early part of the century. He was by now seen as one of the grand old men of British medicine, but it was characteristic of him not to let himself slide quietly into retirement when he considered there were still issues to be resolved.

He was returned twice as MP, but was only able to serve just over a term before his health got the better of him, and he retired permanently to Leagarth. Even in retirement, however, he rarely sat still, and in the final Chapter I illustrate how he reacted to further family tragedy and came full circle, to where he had started out, travelling around the world. His last cruise involved an accident which, by all accounts, precipitated his final illness.

The life of Sir William Watson Cheyne spanned the flamboyant era of colonial expansion and some of the most important medical developments of the 19th century. His own role in these advances was central, and has not previously been studied in the kind of detail it deserves. I hope this biography will go some way towards giving him due prominence in future work on the period.

Note on names and locations

Cheyne is pronounced as in "chain" (as opposed to "Cheyney"), though in Fetlar it is pronounced Sheen. I have referred to Sir William Watson Cheyne in a number of different ways throughout the book. In the chapters on his childhood, I have referred to him simply as William, to avoid confusion with his father, and also because there were complications over his name during this period, as will become clear. I have largely used Cheyne, Captain Cheyne or Andrew to refer to his father.

In his professional capacity, I have referred to him simply as Cheyne, unless he is cited as Watson Cheyne in publications. However, after he was made a baronet, I have also used "Sir Watson", as this is how he was universally known to colleagues and students, and to his tenants in Fetlar. There is evidence that he specifically chose this form after the baronetcy was conferred.

Excellent maps of Scotland, including Fetlar and Mainland Shetland, are available online at maps.nls.uk.
For example, http://maps.nls.uk/view/74430916 (accessed 5 June 2015) shows Leagarth in 1882.
Maps of the Paardeberg conflict can be found, for example at http://www.warmuseum.ca/cwm/exhibitions/boer/boerwarmaps_e.shtml#paardeberg3 (accessed 5 June 2015)

1

Dreams of Empire

Captain Andrew Cheyne

It all began with sandalwood. In the 1840s, events in the Pacific were changing the dynamics of European trade in the Far East. They would also form the background to the first years of Sir William Watson Cheyne, bacteriologist, surgeon, medical pioneer and Lord Lister's most devoted disciple in the development of antiseptic surgery. Cheyne's early years read like a chapter from Robert Louis Stevenson, and his life was, in many ways, more compelling than a novel. On the one hand, a shy and unassuming man, his professional tenacity earned him the respect of the British, European and North American medical establishments, and the epithet of "Lister's bulldog".[1] Whether or not he ever recognised the full influence of the colourful characters and events of his childhood, they nevertheless left an indelible mark on the person he became. They gave him relentless drive, a quiet sense of humour and an abiding obsession with the sea.

If we are to tell Cheyne's story in its entirety, we must begin with the extraordinary tale of his father, and with sandalwood, a tree of six to eight feet producing an oil with an "exquisite perfume".[2] It had long been a commodity which commanded high prices on the Chinese markets, where it was burned by Buddhist communities as incense, and western merchants had capitalised on the trade for many years. For over a century, the British East India Company had maintained a monopoly on the China trade by Royal Charter. In 1834, however, the British government had finally yielded to the free trade lobby and opened the door to a wide range of companies, who very quickly established themselves in a variety of dealings in the Far East - legal and illegal - in order to trade with China.

An important problem lay in the fact that China was one of the few suppliers of the inexhaustible British and colonial demand for tea, but Chinese merchants were unwilling to accept goods as payment and insisted on specie (coin). British merchants attempted to circumvent the restrictions by exchanging goods directly in China where they could, an enterprise which involved finding commodities the Chinese wanted badly enough to exchange for tea. They found their magic solution in opium, and began to trade supplies from Bengal. The trade was illegal in China, but opium consumption had nevertheless led to a serious addiction problem. When the Chinese authorities decided to act definitively and confiscate opium in the ports, it led to two wars with the British in the 1840s and '50s.

There was, however, another commodity of serious interest to the Chinese, one which remained legal, if not easy to source. Buddhist communities had maintained a substantial demand for sandalwood. Western merchants had begun to access it in Fiji from around 1803 to 1816, followed by the Marquesas, Hawaii, the Austral Islands and finally the New Hebrides and New Caledonia.[3] Sydney-based merchants in particular were well placed to exploit the trade, and when supplies of sandalwood were spotted on the Isle of Pines, entrepreneurs sent ships out to investigate. The quality was generally inferior to the Indian wood and fetched a lower price in China, "but it could be cheaply bought, and adventurers were lured to seek it by the prospect, often illusory, of quick fortunes"[4].

One such adventurer was to be found in Sydney late in 1840, operating on behalf of one of the merchant companies based in the town. Andrew Cheyne first appeared in the Sydney newspapers as Master of the brig *Bee*, when he sailed to the Bay of Islands, New Zealand, on 25th September, 1840[5], but his real career began with the emerging trade in sandalwood from the new discoveries in the Pacific. He was, in fact, no stranger to islands.

Born in 1817 in Shetland, in the northernmost reaches of the British Isles, Andrew Cheyne was steeped in maritime trade in an era when the archipelago was one of Britain's main centres for dried, salted whitefish.[6] His parents, James Cheyne and Elizabeth Robertson, were not married. The Church of Scotland maintained an influential presence in Shetland, and ministers and the kirk session "kept a strict watch on the timing of the birth of the first-born child of a marriage to ensure that it had not been conceived outside wedlock."[7] Any 'offenders' were brought before the congregation for three Sundays before their child could be baptised. They ran the risk of public humiliation, but consequences for a woman in this position were often compounded by economic and practical difficulties if the father of the child did not marry or recognise her. Prior to the amendment of the Poor Law in Scotland in 1845, the kirk session and the heritors[8] made decisions about who could receive financial help, and their criteria often included whether the person had some alternative means of supporting themselves, particularly if they were able-bodied.[9] Given these criteria and the moral code, an able-bodied single mother was unlikely to be high on the list of recipients.

Andrew Cheyne's mother was lucky enough to escape the system, as her son was taken in from the start by his father's family. The boy's father was a member of a landed family of merchants in Northmavine in the north of the Shetland Mainland, the biggest land mass in the archipelago. They owned their own fleet of three vessels [10], and exported dried cod and ling to mainland Europe [11]. Andrew's father was James Cheyne, a son of the laird of Tangwick. When Andrew was born, James' father was still in residence at Tangwick Haa, the family home, and is likely to have been the one to have made the decision to bring the boy up there. He died within a few years of Andrew's birth, however, and the boy is likely only to have remembered the new laird, John Cheyne who, to all intents and purposes, brought him up. The new laird had a son of his own, Henry, who was already 13 or so when Andrew was born.[12]

Tangwick Haa, Shetland, childhood home of Captain Andrew Cheyne

It created a complex situation, but not a unique one in Shetland, where it was not unknown for families of means to bring up their illegitimate children, though the mother was rarely included in the arrangement. Another instance involved Thomas Irvine of Midbrake, a landowner on the neighbouring island of Yell during almost exactly the same period, whose illegitimate daughter was cared for in his own household until she married, but no further mention is made of her mother.[13] She is likely to have been a servant at Midbrake, and was probably sent away. Andrew's mother, Elizabeth, was probably a servant at Tangwick Haa.

Microbes and the Fetlar Man

Tangwick Haa, Shetland

Local tradition says Andrew Cheyne was born on a stormy night by a raging sea, and the family clearly took responsibility for the child from the first. He was to be born at the Haa, but there is some suggestion the building went on fire that night, probably a consequence of the storm, and his mother was evacuated to another building, the East Booth. When the sea began to come into the Booth, however, she had to be evacuated again to a place called Hogaland, where it is likely her son was finally born. The anecdote, which ends by saying that local people considered Andrew a child "born between wind and water"[14], is not without its relevance. He became a man characterised by both of these elements and, it may equally be said, fire.

Yet his mother was relatively lucky. Her son grew up in his father's family home, in reasonable comfort. Neither does she appear to have been denied contact with him. He remained in touch with her for the rest of his life, and sent her gifts and money when he had it.[15] In fact, when his biological father died, and Andrew inherited "three and one half merk land twelve pennies the merk in the Room of Nounsburgh [modern spelling Nounsbrough] in the Parish of Aithsting"[16], he turned one of the two farms over to his mother. She had since married a man called John Forbes, and the couple lived out their lives on this land.[17] Andrew's generosity on this occasion is also indicative of where his ambitions lay. He had no interest in a settled existence in Shetland, or at least saw no future in it, but at the same time he battled for the status and respectability that would persistently elude him.

Dreams of Empire

This image, originally an ambrotype, was found amongst papers connected with Andrew Cheyne and the Watson family. It may be Andrew Cheyne.

There is evidence Andrew had a half sister, Elizabeth, his father's daughter by another relationship with a woman called Janet Anderson. She was born two years after Andrew, in 1819, lived in the area until she died in

1888, and was known as Betsy.[18] She may have been fostered, and did not necessarily grow up with Andrew at the Haa, but she certainly spent a good deal of time with him later.[19] Captain and Miss Cheyne are regularly noted together in the minister's diary.[20] These complex relationships were to have consequences for Andrew's family well into the future. There is no evidence to suggest he was treated unfairly, but the circumstances of his birth were nevertheless a factor. The Cheynes were known locally as relatively fair landlords, and have been described as having "a good reputation among their fishing tenants … in various ways not typical of the landlords of the day."[21] Tradition suggests that Andrew learned his trade on one of the family's three ships [22], and when his uncle died in 1840, he was left the sum of £15, with the implication that he was expected to use it to make an independent career for himself.[23] The family house and business concerns were inherited by his cousin Henry (Harry) Cheyne, whose five sons were all educated in Edinburgh and became successful and established members of the law profession and the Church.[24]

Andrew's fate - and character – were a world apart. It is unlikely he ever expected to inherit a significant role in the family business, and whether this, or his own leanings, sealed his fate as a rather flamboyant adventurer is open to speculation. He was still at the local Tangwick school in 1832, at the age of 15, where he probably studied reading, writing, grammar, arithmetic, bookkeeping and navigation.[25] Local tradition also suggests he was at least partly tutored by the Reverend William Watson, the Church of Scotland minister in Northmavine at the time.[26] Whether or not this was the case, he was well enough acquainted with the minister to remain in contact with him when the latter moved to the island of Fetlar, particularly as Watson continued to visit Northmavine after he had left.[27]

Any instruction he may have received from the church helped shape at least part of Andrew Cheyne's character. The moral contradictions he faced later in life made him into an enigmatic adventurer on the one hand, but also someone who seemed determine to carve out a respectable reputation for himself on the other. He was described in later life as "hardy, humourless, scrupulous to a fault, unforgiving in his judgments on the foibles of others"[28] and with puritanical standards of morality he himself transgressed on more than one occasion. Any stigma he experienced in the community or family as a child may have contributed to his rather outspoken morals. He was probably trying to compensate.

It was an age of adventure and discovery where Europeans of modest means were increasingly making names for themselves, and the South Seas were the new El Dorado. The sea had long been a source of opportunity.

Dreams of Empire

In the previous century, Captain James Cook had begun life as the son of a large, farm-labouring family, and his illustrious maritime career had secured him a respectable place in history. The scientific expeditions at the turn of the 19th century had begun to make a significant contribution to knowledge of the Pacific, and after the eventual removal of the East India Company's monopoly, merchant opportunities in the South Seas offered more and more ways for ordinary men with wit and ambition to establish themselves in business.

If, indeed, Andrew Cheyne received tuition from the Manse at Hillswick, it may be the Manse library as much as any training in seamanship that led him away from the comforts of home. Alongside the inevitable sermons and essays listed in the library during Watson's time (and even a *History of the Popes*), we find Homer's *Odyssey*, the *Arabian Nights*, *Chinese Tales*, *Don Quixote*, a publication called "The Adventurer" and – most significantly of all – *Robinson Crusoe*.[29] They gave him a taste for the exotic, and he was by no means alone. Only a generation earlier, Matthew Flinders, in charge of an important scientific expedition to Australia arranged by Sir Joseph Banks, and the first to denote Australia as a continent, had carried a copy of *Robinson Crusoe* with him since he was a boy.[30] It had become an essential and iconic text for the budding adventurer.

For better or worse, by 1840, Andrew Cheyne had taken his small legacy and set off to seek his fortune in the emerging Pacific trades. He may have felt that, given his limited means, his best opportunities lay in doing a good job well and behaving with as much integrity as possible. After his first position as master of the *Bee*, which sailed from Sydney to New Zealand in September 1840, he repeated the voyage in February 1841 on the *Diana*.[31] He was only 24 years old, but was held in high regard by the Sydney agent Ranulph Dacre, who sent him as supercargo on the *Diana*, accompanied by another brig, the *Orwell*, to negotiate the first shipment of sandalwood to China from the Isle of Pines in New Caledonia. As supercargo, his job was to coordinate the commercial business of the two brigs *Orwell* and *Diana*, under the command of Captains Hughes and Watson.

In his logs and publications, Cheyne gives an honest and, at times, vividly negative account of Sydney merchants during a period when the rush for sandalwood encouraged the same arbitrary behaviour as the gold rushes would do a few years later. Time was of the essence. First-comers to the islands where the wood was discovered stood the best chance of establishing a monopoly on trade and local labour. Ship owners were exacting, often unscrupulous, and urgency took precedence over human considerations.

Microbes and the Fetlar Man

Old vessels, often barely seaworthy, were used for the trade, as they could be bought cheaply and used for little else.[32] This only changed as the trade progressed, and discerning merchants realised their return cargos from China could not be insured in old vessels. Those in charge of the ships battled with constant repairs and worked in dangerous conditions.[33]

On islands where contact with western traders had been limited or was, in some cases, an almost completely new experience [34], skilled negotiators were at a premium, particularly as "the slightest false step could transform a scene of cheerful co-operation into one of deadly hostility."[35] This is where Andrew Cheyne came into his own. Traders did not carry their own workforce with them on board, and needed to negotiate terms with local headmen, whose priorities did not always coincide with those of western merchants. Cheyne was a keen observer, and noted cultural nuances which made him successful in negotiation where others had failed.

Against a background of ruthless competition, Cheyne's journal describes the dangers of the trade, and characterises the Sydney merchants as callous and overbearing, even towards those they put in charge of their operations.[36] His first negotiations on the Isle of Pines enabled him to fill the *Orwell*'s hold with cargo, which was despatched directly to China with Captain Watson while Cheyne, his crew and a shore party remained behind to cut and prepare the second load. Cheyne was to take this to Sydney, where he would order more ships to be sent, while the shore party prepared the next load. It was a race against time before any other vessels discovered there was a good supply of wood there. Furthermore, because it was yam planting time and the island leaders withdrew the supply of local labour, Cheyne had a reduced workforce, and was himself forced to join the crew in cutting and preparing the wood before they raced back with it to Sydney. On their arrival at Port Jackson, they lay offshore for a while and sent a previously arranged private signal to the owners ("The Portuguese flag at the fore"[37]) so that they could come on board without alerting anyone's attention to the cargo.

Cheyne had done everything right and fulfilled his contract. He had successfully negotiated with a people whose daily routines he had needed to learn. He had established a trade monopoly and a workforce on the island, despatched a cargo of sandalwood to China, returned to Sydney with another load and set in motion a system for the ship owners to exploit the resources with no immediate competition. Though he was a young man, the effort had exhausted him, and he had succumbed to fever and illness. He said he was "actually so weak at the time [of arriving in

Sydney], as to require frequently to be assisted on deck."[38] The owners of the ship, however, "Messrs. Dacre, Jones and Elgar"[39] refused his request for a doctor, or to be allowed ashore. He was to proceed directly to Manila. Cheyne wrote, "This was very unkind and I may say inhuman on their part, as I risked my life and contracted disease by over anxiety for their benefit, but I am sorry to say that the character of the Sydney merchants at that time was far from being respectable or honourable."[40] They managed to wriggle out of paying him his promised bonus, and he was not confident enough to fight the matter. He was a young man in a cut-throat world, but he would learn, for better or worse.

His visit to the Isle of Pines had given him his first experience of negotiation and leadership and, importantly, he had become one of the first Europeans to have sustained experience of the culture of the island. A London Missionary Society vessel had landed there briefly only a few months earlier, and had left Samoan missionaries behind [41], but Cheyne was the first European to have the opportunity to make detailed observations of island life to add to the growing public and scientific interest in exotic cultures. He had an inquiring mind and an interest far beyond the information required for mastering the art of negotiation, so he took detailed notes which, from the start, he intended for publication.

His first book was published by J.D. Potter of London in 1852, and was immediately successful. It prompted him to consider writing a broader work, to include an exhaustive description of life on the islands, aimed at a more general audience. In 1856 he published a number of articles in *Nautical Magazine* on his Pacific island voyages from 1841 to 1847 [42], and J.D. Potter followed with an edited version of the work.[43] Sadly for Cheyne, however, his unfinished, unabridged notes, containing more of the cultural information he himself wished to publish, had to wait more than a century, until 1971, when Dorothy Shineberg recognised the importance of his observations and published his journal as *The Trading Voyages of Andrew Cheyne, 1841-1844*.[44]

It was not in Cheyne's temperament to restrict his publications to observations alone. He was a keen observer on the one hand, and had a voracious moral outlook on the other. The latter makes his works lively reading, but sometimes overwhelms the objective information he is trying to convey. Although he was generally considered a fair man, Cheyne was nevertheless intolerant of what he saw as transgressions of strict moral standards, and noted of the Isle of Pines that "No one can visit this Island without feeling deep regret, that so lovely a spot of God's creation should be inhabited and daily sullied by the deeds of those depraved Wretches whom I have given an account of."[45] Years later he was gratified to note that they had been "civilized"[46] by the influence of the sandalwood traders, though public opinion of these latter individuals tended to be rather less complimentary.

Cheyne's approach was influenced by his many narrow escapes, almost losing his crews and ships to attacks by islanders, and being introduced abruptly to concepts of property and trust which were different to the ones he had been brought up to believe were inalienable. However, some of his negative judgements of islanders may also have been an attempt to paint a black picture to potential rival traders in order to frighten them away. Seen in the context of his day, it is probably more useful to imagine that Cheyne considered it customary, and even obligatory, to distinguish between the behaviour of Europeans and indigenous populations. He was a young man with a reputation to make in European colonial society, no matter how much his sense of adventure led him along other paths, and he may have been trying to overcompensate. On the other hand, though he was much given to discussing the concept of treachery at European hands as well as by natives, he reserved his most negative terms for the latter. His language suggests he had been reading rather too many adventure stories, and remarks such as "the natives are extremely cruel, void of affection, and are truly wretches in every sense of the Word, degraded beyond the power of conception"[47] give the impression he was more judgemental than was strictly necessary, even by 19th century standards.

Andrew Cheyne's ambitions, on the other hand, went far beyond running sandalwood for unscrupulous traders, or publishing the observations from his voyages. He intended to establish his own trading post, which he considered the cornerstone of an eventual empire. At the end of 1842, as he was returning from China, he stopped off at Ponape in Micronesia, and immediately saw the potential. The waters were rich in tortoiseshell and *bêche-de-mer*, or sea cucumber, a marine creature considered a delicacy on the Chinese market. Cheyne also intended to provide a strategic watering stop for whaling vessels, and stayed for five months.

His plans were sound on paper, but – as would happen many times in the future – he failed to take proper account of the human element. Trade with vessels calling at the island was already controlled by a motley group of mutineers, deserters from naval vessels and runaway convicts who maintained a tightly-guarded monopoly. Cheyne's courage, or perhaps foolhardiness in taking on this "rogues' ring"[48] on Ponape was underpinned by an absolute faith in "armchair imperialism"[49], a belief that the British government would protect its traders overseas just for the asking. He asked the British authorities in China to intervene, and was genuinely nonplussed when they refused. In her introduction to Cheyne's account, Shineberg commented:

> ... it is strange to see some of the more fanciful of the fallacies of armchair imperialism being taken so seriously by a practical empire-builder. The notion, for example, that the British Crown would gladly lend itself to the aid of loyal sons in a just cause in remote and unprofitable islands he assumed as a matter of course, and it seems also to have survived the polite indifference of the consular officials in China whose aid he sought.[50]

Only a few months later, James Paddon became the first trader to establish a successful colony on the strength of sandalwood, on the small island of Aneityum. His first batch of colonists, from Auckland, was attacked and killed, but others followed, and the settlement thrived by branching out into a number of ventures, acting as a water stop for whaling vessels. It had a library (which was unfortunately damaged by fire), and Paddon himself seems to have traded in a reasonably fair way with his local workforce.[51] A combination of misjudgement and naivety had brought Cheyne's vision to nothing. Paddon's settlement became sufficiently successful to attract Presbyterian missionaries, whose societies not only saw opportunities for attracting donations to convert new populations, but were concerned about reports they had heard of the degenerate behaviour of sandalwooders and the dubious morals engendered by their way of life.[52]

By May 1843, Cheyne had to admit temporary defeat for his embryonic empire when supplies sent by merchants in China turned out to be totally unsuitable for local trade. Any profit he had made from his previous voyage was wiped out, but Cheyne, unwilling to accept failure completely, embarked on another voyage, this time in the *Naiad*, a vessel in which he had bought a quarter share.[53] He attempted to set up a network of trading posts on islands in the western Pacific, wherever he saw possibilities. Cheyne was always ready to maximise the potential of his ventures, and was reluctant to see any of his acquisitions standing idle, but could not expand his network if he were forced to remain in one place. At his depot at Koror in the Palau islands, he therefore left his enterprises in the hands of a team of fugitive criminals and escapees from the rigidity of colonial life, who were not known for their dependability. Within a very short time, they had destroyed the venture and left Cheyne with debts. For the next year, he wandered the Pacific in the *Naiad*, taking in what cargo he could find until, in November 1845, an earth tremor at Paddon's station in Aneityum damaged the *Naiad* so severely that he had to abandon her. In September 1846 he bought land at Port Resolution in the New Hebrides, and purchased the island of Wassau.[54]

Cheyne had not quite given up on his vision. Despite his catastrophic ventures, he had more in mind than financial success in the region, and still saw it as his opportunity to set up a personal empire where he could enjoy the lifestyle and status expected by Europeans in colonial society. In fact, he not only envisioned an empire, he fancied himself as its emperor. It is likely that Cheyne had heard of the success of another Shetlander, John Clunies-Ross, in carving out a kingdom for himself on the Cocos Islands in the Indian Ocean. Ross, the son of a schoolmaster, had developed plantations on the island in the 1820s, basing his wealth on copra, the dried flesh of the coconut. He adapted aspects of the economic system prevalent in the Shetland fishing industry at the time, and had absolute control over his workforce, largely Malay workers he had inherited from a rather unscrupulous character called Hare. He was finally forced to make concessions to the workers when they organised a strike in 1837. Ross, however, had already taken his successes one stage further by declaring himself 'King'. His family was even granted the title in perpetuity by Queen Victoria. It lasted until the 1970s, when they were forced to cede ownership to Australia amid concerns from the United Nations that the system was an unacceptable anachronism.[55]

Charles Darwin, when he visited the islands in the Beagle in 1836, was less than complimentary, indicating that, while traders themselves may have considered this type of labour system acceptable and essential for success, it was not necessarily considered morally permissible by all sectors in Britain:

> The Malays are now nominally in a state of freedom, and certainly are so, as far as regards their personal treatment, but in most other points they are considered as slaves. From their discontented state, from the repeated removals from islet to islet … things are not very prosperous.[56]

This form of economic bondage was not dissimilar to the one prevalent on Shetland estates at the time, where the average islander owned no land of his own, but lived in tied accommodation from which he or his family could be evicted at a moment's notice when the landowner's economic plans changed. Fishermen rented their boats and equipment from the landowner at the beginning of the season, and borrowed from him, or the merchant, in order to do so. The price of fish was pre-set in the landlord's favour. Even if it was a good catch, they could only obtain goods in the landowner's or merchant's shop and had to borrow again from him the following season. This kept them in a perpetual economic relationship with the merchant-landowner, from which they could not easily extricate themselves.

Even where landowners acted with fairness, workers remained in constant debt and an endless cycle of dependency. For the landowner, it secured a self-perpetuating monopoly and made sure he had a ready workforce. Cash rarely changed hands, as its value would have been worth more to the worker than to the merchant. It was not too unlike the logic of the British opium traders in China who were reluctant to dilute their profits by paying cash for their goods, as a direct exchange in tea resulted in greater returns at the end of the day. The system Ross had known in Shetland was referred to as Truck, and he had transposed aspects of it to his plantations in the Cocos Islands. He paid his workers in his own currency, the *cocos rupee*[57], which could only be redeemed at his own store. Andrew Cheyne may have had something similar in mind, a combination of this type of exchange system and a successful colony like Paddon's at Aneityum.

By the late 1840s he had experienced some success with *bêche-de-mer*, destined for China or Manila, but just as he was appearing to break new ground, he had word that his father, James Cheyne, had died, and he sailed for home to regulate his affairs. He was in Edinburgh on 28th April 1849 to register the inheritance of two farms at Nounsbrough in Shetland, one of which he turned over to his mother and step-father.[58] At this stage, he was "wishing to put his first real profits to the best possible advantage"[59] and clearly saw his future in the Pacific, not as a landowner or merchant in Shetland. He returned to Australia in May 1849 in the *Elizabeth*. He conducted business there until 1851 when he returned to London to collect his Master's certificate, and to negotiate more work. It may have been during this sojourn in London that he arranged for the publication of his first work, *A Description of Islands in the Western Pacific Ocean*. With a bright future looming, and a good reputation behind him, Cheyne was finally on the way to success. Just as he was beginning to consolidate it, however, he added a complication to his life which was to change its course for ever. He decided to marry.

He had known his future wife's family since he was a young man in Northmavine, and was well acquainted with her father, the Reverend William Watson. Eliza Watson was much younger than Cheyne, however, and her father had moved to Fetlar in June 1830 when she was only two years old. Fetlar, a much smaller island than the Shetland Mainland where Andrew had grown up, was also heavily involved in the whitefish industry. The family remained in contact with Andrew and the Cheynes at Tangwick Haa, and Eliza, the Minister's youngest daughter, became engaged to Captain Cheyne.

It is difficult to say who initiated the match. On the one hand, Watson was already over 80 [60], and had married off one of his three remaining daughters to the man who was set to succeed him at the Manse, David Webster. This was a wise move which ensured that she and her sisters, Grace and Eliza, could go on living at the Manse after Watson's death or retirement, should he be in a position to retire. On the face of it, Captain Cheyne had a bright future ahead of him. He was still only 34, and was beginning to make some headway towards consolidating his projects in the Pacific. He may also have exaggerated accounts of his success, and Eliza, an impressionable young woman, may well have been happy to go along with them.

In the light of subsequent events, another issue may have been discussed with the Watsons and Websters. Andrew had certainly not given up his plans of personal empire, and may have interested them in the idea, particularly if the scheme included a place for missionaries. Cheyne liked the idea of a settlement furnished with all the respectable trappings of a British colony. I have found no evidence that the Watsons or Websters contributed financially to his ventures, but they may have entertained his colony project, and Eliza, with her considerable experience of the Church, may have seemed an excellent choice of wife. There is an indication, however, that the Watsons were at one point becoming anxious about Andrew's apparent procrastination where the marriage was concerned. The couple had been engaged "for some time"[61] by December 1851 when the Reverend William Watson, by this time Minister of the Parish of Fetlar and North Yell, noted in his diary that "Captain Cheyne" seemed to be "wishing to shuffle off his engagement to Eliza"[62]. Cheyne was in London, negotiating a number of commercial possibilities, and if Watson was correct in his assessment of the situation, may have been having difficulty deciding whether he was quite ready to marry. On the other hand, Watson later noted that Captain Cheyne had obligations regarding his command of the *Elizabeth*, and had not been free to come back to Fetlar to marry until this point.[63]

Cheyne was at a time in his career when it might have been logical to marry if he were to consolidate his successes, and his ambitions were clear. He had already spent a number of years perfecting his trade, and was on the verge of securing a reasonable position and income. He had his Master's Certificate of Competency, a good reputation and a wealth of experience as a mariner and negotiator in one of the most opportunistic and dangerous areas of commerce in the world. Taking on a young wife, just as he was about to break through all the challenges, was nevertheless an enormous decision. If, indeed, he had any thoughts of postponing his marriage, and Eliza's father was not simply misreading the situation, it may be an indication that the Watsons themselves were keen that the wedding should go ahead.

Eliza Watson was the youngest of a large family of girls, with only one brother, though two sisters and her brother had died before she was born. Another, Isabella, died before Eliza was five. Their mother died giving birth to Eliza herself, and the girl may have been named after a first-born daughter who had only survived to the age of 20.[64] She was much younger than her surviving sisters, and they were protective of her. Her older sister Christian had married their father's assistant and successor, David Webster, in August 1851.[65] If Andrew did, indeed, intend to return to the Pacific in order to persevere with his plans for a trading colony, it is unlikely the Watsons were unaware of it, and must, at least in part, have been conscious of the dangers. It is also possible that Eliza was determined to accompany him, and that her father, on the basis of the plans Cheyne presented, had no objections.

On the other hand, Captain Cheyne had a whole variety of potential schemes in the pipeline, and one may even have been an attempt to give himself more secure employment closer to home. Assisted emigrant passages had become a lucrative alternative to shipping convicts. Transportation of convicts had been abolished in New South Wales in 1850, but continued in Van Diemens Land (Tasmania) until 1853, and there was pressure on the British government to discontinue shipments.[66] Emigration, however, was becoming popular, particularly where passages were funded. While he was in Shetland in 1849, Cheyne had gathered 180 signatures from Shetland workers who wished to emigrate to Australia but who could not afford the passage to join emigrant ships in London. He used the signatures to present a case to the Colonial Land and Emigration Commissions, through the Orkney and Shetland MP Arthur Anderson, to fund assisted passages from Shetland to London.[67] The British government had been making land grants and free passages to poor settlers to Australia since the 1830s, but Shetlanders, Cheyne argued, faced unfair disadvantages in the initial costs of reaching the assisted ships.[68]

According to correspondence in the House of Commons, he was suggesting a "ship should be sent for them, free of expense, as they had not the means of paying their passage to London".[69] When the issue was finally raised in the House as part of general discussions on emigration several years later (in 1854), it was clear he had been offering to take on this responsibility himself, and thereby secure himself a government contract. By 1853, his proposals were supported by John D. Loch, the Immigration Agent at Hobart, who noted that female servants were in great demand in Van Diemen's Land (Tasmania). He said that emigration for respectable women with few means was obstructed by the fact that they could afford neither the obligatory "outfit",[69] including one pound towards the passage, nor the fare to the port of departure. Loch indicates, however, that a system was in place in Ireland to fund "the cost of transit to the vessel, in which they [the women] are accompanied by a guardian"[70], so there was clearly a precedent. Cheyne had furnished Loch with a glowing recommendation when it came to single Shetland women:

> The servant girls of Shetland receive about 2d or 21.10s. a year, with board and lodging. They are the daughters of small crafters [sic] and fishermen; accustomed to cows; clean, neat, active, and good-tempered girls, willing to learn, and particularly steady and moral; Presbyterians; likely to make good domestic servants; and speaking a language rather English than Scotch. The community being small, the character of every one is known. The rules of their church are strict, and they are all communicants; and Captain Cheyne states that the creditable appearance the women make at church is, considering the smallness of their means, surprising. The land is so poor that it does not support the population above half the year, and they then depend upon meal from Scotland, obtained by the produce of their fishing. The young men, finding no employment at home, go away in ships or to England; and hence there are in Shetland many more females than men, and many girls are left unmarried.[71]

Loch indicates that Cheyne still had hopes for the scheme in April 1852, when he was approached by "a great many people"[72] in Shetland, wishing to find out how they could obtain assisted passages. In other words, Cheyne was considering this a very viable option for his future living in the period leading up to his marriage. It would have kept him

Dreams of Empire

closer to home, sailing between Lerwick and London. Moreover, Mr Loch later noted that if the proposals were granted, they were thinking in terms of long-term contracts, which would have given Cheyne a level of security. He said

> If the same ships were regularly employed, the captains, matrons and surgeons would be most proper and efficient persons for communicating information and procuring a proper class of emigrants.[73]

He considered Captain Cheyne eminently suitable for the role, not only because he was a Shetlander who could locate suitable candidates, but also because he had presented his connection with clergymen (Watson and Webster) as testimony to his good character.[74] These proposals were not, unfortunately for Andrew, heard in the Commons until July 1854, but if they are indicative of his plans when he was already gathering signatures in 1849, he may have been placing considerable hopes in the scheme.

Behind the idea was also a recent demographic imbalance in Shetland, and Watson's island of Fetlar had suffered heavily from the reorganisation of land from the 1820s onwards. This had led to large-scale eviction of tenants, a movement which became known as the Clearances, where lairds converted their tenants' smallholdings into enclosed grazing land for imported cross-bred sheep. The remaining tenanted land could no longer sustain the previous population levels, and emigration had become a way of life. It also meant that any sons born into remaining families who could not be accommodated on the land had to seek work elsewhere, largely at sea. By the 1850s, the Greenland whaling industry was at its height, and ships, based mainly in Dundee, passed through Shetland in March each year to sign on Shetland men, who were sought after for their maritime skills. In 1850, at least 50 whaling ships arrived in Shetland and signed on 20-30 local men each.[75] The work was during the summer months only, but the general exodus to sea, including men leaving for work on merchant ships, left more women than men in many Shetland communities. Fetlar, where Watson and Webster were based, had been a major focus for these experimental land reforms.[76]

Although I have found no evidence that Cheyne ever succeeded in his venture to obtain subsidised transport for Shetland women to the emigrant ships, the effort he put into his proposals may indicate he was genuinely trying to find steadier employment closer to home before he became a married man. There is nevertheless no indication he wished to give up his plans and holdings in the Pacific either, and was probably looking for a compromise between the two. The correspondence from Mr Loch later indicated that, although the scheme was clearly aimed at single women, Captain Cheyne noted that he could, if desired, bring out a small number of families at the same time [77], and may even have been thinking of the possibilities for a successful colony of his own somewhere down the line.

After he returned to Britain in November 1851, Cheyne gave up his command of the *Elizabeth*, and his future father-in-law's diary indicates that a new ship was being built for him in Sunderland.[78] The wedding eventually went ahead in time for him to reach the ship, which was to be ready in April. Eliza must have been fully aware when she married that she was either to go with her husband on his next voyage, or risk not seeing him again for some time. Her father wrote to Mr Inkster on 14th February 1852 "to proclaim Capt. Cheyne & Eliza Watson thrice in Church tomorrow"[79], and the banns were read the following day. They were married by Eliza's father at the Fetlar Kirk on Thursday 19th February. Heavy snow had fallen in the night, and there was a strong wind all morning which drove the snow into drifts. The few guests nevertheless arrived safely: Anne Henderson, Mrs Edmondston from the neighbouring island of Unst, Andrew's half sister Betsy Cheyne, and Basil Spence. Eliza's sister and brother-in-law, Mr and Mrs Webster, were also there.[80]

After the wedding, Mr and Mrs C., as Eliza's father liked to call them in his diary, to distinguish from Mr and Mrs W. (Webster), remained in Shetland for over a month longer, visiting friends and relatives. They finally left Fetlar at 10 o'clock on a calm Monday morning, the 29th March, and the snow had abated sufficiently for Eliza's sisters, Christian and Grace, to accompany them some of the way to the boat which would take them away. Eliza's father simply noted in his diary, "Mr & Mrs Cheyne left at 10 am for Hillswick, Ler[k] (Lerwick), Leith, Lon(don), Mauritius. May God Almighty conduct, guide, protect them in their journey thru life & bring them safe at last to his heavenly kingdom."[81]

Eliza was clearly excited about the whole adventure. She wrote to her sister Christian from Edinburgh on April 8th that "old Campbell put up all his colours in honour of *us* and gave us a salute when we came on board. Was not that grand(?)."[82] They were not in Edinburgh long, and left only a short while after she wrote the letter. She complained she had not had time to buy clothes, and the black dress made for her in Lerwick would have to serve her until she reached London.[83] Her family wrote back on April 20th.

It is not clear what became of the ship Watson had indicated was being built for Captain Cheyne in Sunderland. Nor had anything transpired over Andrew's plans to transport emigrants between London and Lerwick. The couple next appear in London, living in reasonable lodgings at Number 11 Albert Square, Commercial Road East, in Limehouse [84], close to the River Thames. By this time, they had news to write home about. Eliza was expecting a child. Around this time too, it became clear that Andrew was resorting to other plans, presumably while he awaited the protracted outcome of the emigrant scheme. In July 1852, he signed a contract to take a vessel called the *Lady Montague* to Van Diemen's Land (Tasmania).[85] The ship would not be carrying settlers. It would be carrying 290 male convicts and military personnel, bound for Hobart Town.[86] It was to be one of the last convict ships to land there.

Had it been possible when all these plans unfolded, the news may have made Eliza's family more reluctant to allow her to travel. It was not particularly unusual for wives to travel with their husbands on this type of voyage, but by the time the ship finally set sail, Eliza was five months pregnant, and Captain Cheyne could very easily have sent her back to her family in Shetland had he considered the impending voyage too dangerous. We are left with the speculation that either Eliza herself was determined to continue, that Andrew simply saw nothing difficult in the idea, or alternatively, that he fully intended at this stage to come straight home after he had delivered his cargo.

Eliza Cheyne had been brought up surrounded by the sea, but in the relatively safe atmosphere of a Church of Scotland Manse. She was no stranger to travelling, and had accompanied her father on his visits to the Church of Scotland Assembly [87], but she had suggested in her letter to Christian from Edinburgh that she had been sick, most likely seasick, for a good part of the journey from Shetland to Edinburgh.[88] If she were indeed the one to insist, she cannot have been fully aware of the rigours of the journey awaiting her.

Assuming the ship's owner, Mr Vaux, did not simultaneously own two vessels with the same name, which seems unlikely, the *Lady Montague* was already at the centre of an enquiry into her previous voyage, which had been inauspicious to say the least. She had been shipping coals to Aden when the Master died en route. The second mate took over and disobeyed orders to return to London, chartering the ship instead to India and China, where he proceeded to run between Mumbai (Bombay) and Guangzhou (Canton) in particular for several months. This was an infamous route for shipping opium, and although the cargo is not specifically stated, the possibility that it included opium must remain a strong one.

Eventually, the vessel took on a group of Chinese emigrants bound for California via Callao in Peru, setting off in February 1850, but by the time they reached Hobart, 193 emigrants and crew had died of fever. Despite the efforts of a doctor sent on board by the Governor of Hobart, the fever broke out again, and by the time the ship reached Callao, 274 people were dead, 171 of whom were emigrants, and the vessel had to be fumigated. The *Lady Montague* was ultimately returned to London, and John Vaux, the owner, was forced to appear before the Board of Trade after he deposited an unusually large sum in the Merchant Seamen's Fund as 'dead money' in accordance with the Mercantile Marine Act. The only conclusion to be drawn was that the food supplies taken on board had been so substandard as to be unusable.[89] However, a subsequent enquiry in 1854, which attempted to trace any crew members who could serve as witnesses, noted the possibility that some of the Chinese had been poisoned.[90] The vessel had returned to London sometime around June 1852.

The next voyage of Mr Vaux's ship was commanded by Captain Andrew Cheyne, who had on board his new wife, five months pregnant. On 27th July 1852 in Plymouth, Captain Cheyne signed his government contract "to transport the offenders named in the list to 'the Island of Van Diemen's Land'"[91], and they finally set sail on 9th August 1852.[92] The first part of the journey seemed to be relatively free of difficulties, and Captain Cheyne had "the security of monthly wages and the leisure of navigational duties."[93] He was probably even able to work on the notes from his previous voyages, and no doubt intended to publish them as a work more appealing to the general public than his specialised 1852 publication on marine directions. Eliza had the help of a servant [94], and was by no means the only female cabin passenger. The wives and families of some of the military personnel kept her company through the latter stages of her pregnancy. Captain Hawkins' wife and two children were on board, and Lieutenant Gray of the 99th Regiment was also accompanied by his wife. Two other women, Mary McLelland and Ann Ayscough [95], were no doubt equally glad of the company. The ship's surgeon, Samuel Donnelly, records no incidents or ill health on Eliza's part, despite her delicate condition.[96]

If the following suspicions at the Foreign Office were ever confirmed, the tranquillity ended sometime in November. Word was sent to the Home Secretary, Viscount Palmerston, that the Foreign Office had received a communication from Count Walewski, the French Ambassador, to say that someone on an English ship had fired illegally at the *Minos*, a French vessel, "on the occasion of the two Vessels having fallen in with each other at sea".[97] The Board of Trade had conducted enquiries and had come to the conclusion that there was "considerable probability that the outrage in question was committed by some persons on board the *Lady Montague*, Captain Cheyne, Master."[98] The matter was referred to Viscount Palmerston "to take such steps in this matter as may be in his power with a view of discovering and punishing the perpetrators of the outrage."[99] I have been unable to ascertain the outcome, but the key to the matter may lie with the "50 Soldiers of the 99th Regiment"[100] who were on board to guard the convicts. Whether they fired in error or whether someone was out of control, it had caused an international incident, and Captain Cheyne's name had reached Viscount Palmerston in the least auspicious of ways. He, not the military personnel, was ultimately in charge of the vessel and had to answer for it.

The *Lady Montague* proceeded to Hobart, it appears, without further incident, though conditions below deck were probably not particularly good for the convicts. The surgeon, Samuel Donnelly, recorded the deaths of nine convicts from September onwards, beginning with a 27 year-old called Diver, who succumbed to pthisis (tuberculosis). Others followed in October, November and early December, mostly from respiratory conditions such as pneumonia and bronchitis, or from diarrhoea and dysentery.[101] At 6 pm on 9th December, 12 miles off Hobart Town in Van Diemen's Land, the vessel was boarded by a pilot named William Lawrence, who registered it on behalf of the port authorities and summarily confirmed the nine deaths.[102] The ship was allowed into Hobart on 10th, where the cargo was unloaded and the three remaining sick convicts were sent to the hospital in the town.[103] A sick passenger, Lieutenant Gray of the 99th Regiment[104], no doubt also had to seek medical care in Hobart.

The following day, an angry letter of protest appeared in the *Colonial Times*, addressed to the Secretary of State for the Colonies, and signed by members of the Anti-Transportation League in Hobart:

> Sir,
> Since the Council's Protest against the introduction of the prisoners per *Martin Luther*, the *Lady Montague*, with 280 [sic] male convicts, has arrived from Plymouth, furnishing a proof that the Earl of Derby's administration is as insensible to the claims of national justice and honor as his predecessor's ... From henceforth the colonists will look to their own resources for terminating a system which is a disgrace to England ...[105]

Not all colonists, however, were opposed to the landing of convicts, and the larger landowners snapped up the new labour supplies, especially when their existing labour force began to drift off to the more lucrative prospects offered by the emerging goldfields.[106]

On 14th December, only a few days after the *Lady Montague* reached Hobart, Eliza went into labour and gave birth to her son on board. He was named William Watson Cheyne after his maternal grandfather, and on 10th January the Reverend John Lillie of St. Andrew's Scottish Church in Hobart Town came on board to christen the child.[107] So began the life of a man who was to become one of Britain's earliest pioneers of medical bacteriology, and perhaps the extraordinary circumstances of his birth and early years were part of what made him into the most tenacious of Lord Lister's supporters. He began his life on a ship, and would never be free of an obsession with them for the rest of his life. The fact that he survived at all was nothing short of a miracle, given the events of their journey home.

Shineberg's introduction to Andrew Cheyne's *Voyages* notes: "Captain Cheyne apparently decided that, since he had to return his wife and infant son to Shetland, he would do so by way of the Pacific Ocean,"[108] but we are offered only sketchy clues as to why. In Hobart, he discussed his plans for Shetland emigration with the Immigration Agent, John Loch, whose account was ultimately presented to the House of Commons in 1854. It indicated that, after Cheyne left Hobart, he intended to be back in England in August or September 1853, when he could travel to Lerwick in the *Lady Montague* to collect emigrants bound for London.[109] This indicates very clearly that he was still pinning his hopes on the emigration scheme, and that he expected to be back in London quite soon.

The itinerary he adopted in reality, however, shows that he spent considerable time wandering around the area before heading home via China, Japan and America, the opposite route to the one they had come, and quite a distance out of their way. It has been tentatively suggested that there may have been "some trouble with the Department which committed him to choose his route"[110] and this may allude to the incident with the French ship. Alternatively, he may have wanted to take the time to finish his manuscript on sailing directions, some of which was written on the return journey.[111] It is likely to have been a combination of factors, but something clearly made him deviate from his original plans, as they were nowhere near London by August 1853. His contract had been to deliver the convicts to Hobart. As Shineberg shows, however, Captain Cheyne's motives were at least partly business-oriented, and he clearly took time to attend to some affairs in the Pacific. In reality, we have no record of Eliza's health at this time, as there is no mention of her in the surgeon's log [112], which means there can have been no particular cause for concern about her or William. There was probably no indication she required any more than the normal convalescence of a nursing mother. In fact, they had managed to communicate details of the birth to her father in Fetlar, but he did not mention it in his diary until April 13th. After noting other letters and details of the weather, he remarked simply "Eliza confined on 14th Nov[emb]er last – a Son – recovered very soon."[113]

Captain Cheyne's motives for remaining in the Pacific, and for taking such a circuitous route home, have been the source of a great deal of speculation. They are important, as are the details of their journey, because they had tremendous consequences for his son's early years and the fate of his wife. Hobart Town shipping records indicate that Captain Cheyne left there on 19th January 1853 "for Guam in ballast",[114] though he had indicated to Loch in Hobart that he was ultimately heading for China.[115] He had, in fact, no intention of going to Guam. 'Clearing for Guam' was a "favourite euphemism"[116] of sandalwooders when they left Australian ports for indeterminate South Sea islands. In the earlier days, they had hoped to find a cargo from the known sandalwood locations, or, even better, discover a new stand of sandalwood which could be the foundation for establishing a trading post, a monopoly and an overnight fortune. The real sandalwood boom was over by this time, but there were still opportunities, especially if traders combined other marine commodities such as tortoiseshell and *bêche-de-mer*.

Captain Cheyne's attempts to establish his trading empire on Ponape had fallen foul of runaway convicts and mutineers, and his hopes of involving the British government had come to nothing, but he was nevertheless a man with a very strong dose of imagination, if a little naïve about its application at times. If he 'cleared for Guam' out of Hobart, he was almost certainly intending a renewed attempt to secure his base, to check on his property and to consolidate his affairs. He may also have wished to collect cargo for trade on the way home, especially as he returned via Hong Kong and China, to help make the voyage as profitable as possible. He equally needed accurate information on precisely this route for his book. Captain Cheyne was nothing if not determined. He may have been trying to ensure the best of all worlds and cover all objectives.

The family spent 13 days in the New Hebrides, and reached Hong Kong on 31st March, where letters were waiting from Eliza's family. Her sister

Grace had written to her, and her father the same "to wait her in Hong Kong"[117], which means this was an intended stop on their journey. They stayed for two months before they continued their cruise to Manila.[118] Grace clearly knew they were to be in Hong Kong, but neither she nor her father seem to have been aware they had left again by June.

All the evidence points to the fact that Captain Cheyne intended to spend some time consolidating his business interests, particularly as they called in at the Isle of Pines. If indeed the *Lady Montague* were to blame for the incident with the French, he may have been hoping it would sort itself out before he returned, and was lying low. It could theoretically have jeopardised his reputation in terms of the emigrant scheme. On the other hand, when his next book appeared in print a few years later, it emerged that he had spent considerable time and effort documenting the shipping route from Australia to the trading nations of the Far East. The book was entitled *Sailing Directions from New South Wales to China and Japan*, and in it he claimed to have discovered "a couple of islands just south of Japan on the *Lady Montague* cruise"[119], probably the ones he named Becher Island and Volcano Island [120] in the Amakirrima group. This goes some way to explaining how they found themselves sailing across the Pacific, but they were, by this time, very late if they intended to arrive in London by August. Had he foreseen the consequences of all this procrastination, it is hard to know whether he would have chosen differently.

They arrived in San Francisco on 16th September 1853 and stayed until the end of the year.[121] We have no clear evidence of Eliza's health, but if Captain Cheyne had needed to get his wife and child home in a hurry, he had by now taken a considerable detour. In theory, he could have found her a passage home on another vessel had her health been sufficient cause for concern, but we should also ask ourselves whether Eliza herself – at least at this point – was reluctant to be sent home. Whatever the reason, they remained in San Francisco for some time, and this may have involved Cheyne's relations with his crew.

A report in the *Daily Alta* (California) on 6th October 1853 indicates that a certain George Williams, a seaman on board the *Lady Montague*, brought proceedings against Captain Cheyne for maltreatment and abuse.[122] The charges were dismissed, but not until he had appeared at a hearing. It was not uncommon for crews to hold their captains to ransom, knowing that it left the latter open to court proceedings and costly delays. Shineberg noted: "All the logs available have a number of entries relating to the abuse of officers, the refusal of seamen to do duty, the inability through drunkenness to do duty and desertions from the ship. Masters and merchants dreaded court actions in Sydney which, whatever the outcome, would cost them a great deal of money by holding up the ship. Knowing this, the crew had the advantage of threatening action against the master … and of being aware that they themselves ran little risk of prosecution."[123]

On the other hand, Andrew's temper was legendary, and he was not known for his tolerance, particularly where an uncooperative workforce was concerned.[124] A smallpox epidemic was sweeping the Pacific, making it difficult to find crews. Only months before, a ship called the *Cynosure* was carrying smallpox as it left San Francisco, and decimated the Makah, a North-West Coast people on the Olympic Peninsula.[125] If Cheyne's remaining crew was trying to bargain for higher wages than he was prepared to pay, they could have been holding him to ransom over it, and he genuinely lost his temper on them. Either way, his court appearance is likely to be related. Their stay in San Francisco was prolonged to around four months, but although the charges against him were dismissed, Andrew did not leave immediately after the hearing, and may have had difficulty competing for a crew. His relations with his crews were to remain problematic most of his life. As Master of the *Acis*, in the mid-1860s, he was to receive a letter from a Mr Davey, who was upset at having been accused of taking chisels to trade with the natives. Davey did not wish to work for Captain Cheyne "to be suspected a second time."[126]

While the family was waiting in San Francisco, on November 22nd, 1853, a severe earthquake hit the city, leaving a half mile wide fissure and allowing the waters of Lake Merced to flow out to the sea.[127] None of this could have been particularly beneficial to Eliza's health, and we know nothing of how her baby William was faring. She wrote to her father at some point during this period, but he did not receive it until 22nd December, and his diary does not suggest anything untoward.[128] They finally arrived in Callao, Peru, on 22nd January 1854, and were still there as late as June.[129] Their initial delay was probably a result of further crew shortages in the wake of the smallpox epidemics, but then major disaster struck. Captain Cheyne discovered that the *Lady Montague* was unfit for the remaining journey to England, and he, Eliza and William were forced to remain in Callao until she was sold. This is the vessel Andrew had indicated he would be using to transport Shetland emigrants to London if his proposals for subsidy were successful,[130] so if he was still setting any store by this for his return, the sale of the ship would complicate these plans too.

Only 19 years earlier, in his account of the voyage of the *Beagle* in 1835, Charles Darwin had described Callao as "a filthy, ill-built seaport,"[131] where the inhabitants appeared to be "a depraved, drunken set of people"[132] and the atmosphere was "loaded with foul smells"[133]. Unless the port had improved significantly since then, this is the environment in which William Watson Cheyne began the second year of his life. His mother had also begun to show signs of worsening health. Once again, it is a miracle he survived at all.

We have no evidence to suggest that Captain Cheyne did not try and send Eliza and William home ahead of him, which leaves us with the possibility that Eliza herself refused to go. She may not have felt confident enough to travel alone with her son, though with so many ships passing along the route it is hard to imagine Andrew could not have engaged someone trustworthy to accompany her. Alternatively, Eliza may have regained enough strength during the wait in Callao to think her health was improving. She knew her husband well by this time, and could have been forgiven for thinking that, had she returned to Fetlar without him, there was nothing to stop him tidying up the business with the *Lady Montague* and heading straight back to his affairs in the Pacific. She might never have seen him again. It would clearly have relieved him of a daily responsibility at a time when he was poised to take advantage of major commercial opportunities.

Whatever the circumstances, William and his mother remained with Captain Cheyne until the business was completed. There is no further record of how they made their way back to Britain, though they are unlikely to have travelled overland, and probably eventually sailed as passengers on another vessel around Cape Horn. Andrew had business in London with the *Lady Montague*'s owners, and the family did not arrive back in Fetlar until November 1854.[134] On 19th, they registered the birth of their son, William Watson Cheyne, at his grandfather's church.[135] He was nearly two years old, and until now his family had consisted entirely of his mother and father, his society a motley collection of seamen and his culture the cut-throat trades of South Sea merchant adventurers in the dubious glamour of the Pacific ports.

2

Fetlar

A childhood by the sea

When he had returned his wife and son to Eliza's father in Fetlar, Captain Cheyne remained in Britain for almost a year[1], and may have been concerned enough about her health to have postponed an immediate return to the Pacific. Relations with her family, however, were deteriorating rapidly. It was becoming clear that Eliza's condition was serious, and may not just have been the result of fatigue from the prolonged journey. She was suffering from advanced tuberculosis, and her return threw the household into turmoil. As far as the Websters and her father were concerned, Captain Cheyne was unequivocally to blame for the dangerous state of her health.[2]

By late 1855, however, Captain Cheyne had to address his business concerns, and could not afford to do otherwise. The voyage with his wife and child had been fraught with difficulties, and he had not succeeded in consolidating any of his plans. His proposals to obtain subsidies for Shetland emigrants had been presented to the House of Commons in 1854[3], but had not come to anything as far as I can ascertain. On the Isle of Pines, he had discovered that the headman he had dealt with there had died, and that the new one had formed an economic alliance with two sandalwood establishments and a moral one with a group of French Catholic missionaries[4]. Neither of these revelations can have been welcome news at the Fetlar Manse.

There were other reasons why Captain Cheyne deemed it unwise to remain too long in Shetland, or even in Britain. The rift with his in-laws had reached terminal proportions. Eliza's condition, and the subsequent disputes in the household, had also left her father in weak health. The Reverend William Watson was by now in his mid-80s, and felt he had misplaced his trust, a mistake he had neither the means nor the energy to rectify. He nevertheless made sure his son-in-law felt the sharp end of his wrath. Indeed, by all accounts, the entire Fetlar community was outraged at the poor health in which Eliza had arrived home.[5] The journey had

35

involved hardships above and beyond the norm, and they considered that Captain Cheyne had postponed their return far longer than was strictly necessary. It was these circumstances, rather than his taking her away in the first place, which had so enraged them all. No one could see any reason why the Captain should have chosen to return via America if Eliza were ill, and this alone was interpreted as deliberately putting the lives of his wife and son in danger.

The feud brewed away while Captain Cheyne remained in the country, but there is no evidence to determine whether Eliza herself was in favour of being separated from her husband. She may have resisted being sent home alone with William, even if someone could have been found to accompany her. We should also bear in mind Captain Cheyne's obligations. At the very least, he had no choice but to ensure the *Lady Montague*'s disposal on the owner's behalf. He could not simply abandon it. The owner too, may have been aware of the route, and happy to go along with Cheyne's plans, grateful they were not as dubious as the ship's previous voyage where the second mate had hijacked it, leaving Mr Vaux to answer to the Board of Trade for the deaths of 254 crew and Chinese emigrants.[6] At the Manse, Captain Cheyne may not have dwelt too long on the fact that he had been anxious to finish his latest publication, which would deal with the shipping route from Australia to China and Japan. His in-laws must have been fully aware of it, as it had been published, and he had given a signed copy to his relative, Henry Cheyne [7], so had probably done the same for his wife's family. In all events, the Websters and Watson were unwilling to listen to any mitigating circumstances, and Captain Cheyne was no longer welcome in the household.

When he finally left in October 1855, he set off for the Pacific in the *Wild Wave* [8], leaving his wife and son in the care of Eliza's father, her sister Christian and her brother-in-law, David Webster. It is unlikely he saw any of them again. During the short period William spent with his parents, he no doubt retained only a vague recollection of them both. He was nearly three before his father left again for the South Seas, but given Captain Cheyne's disposition and general concerns about his livelihood, he may even have spent only short periods in Fetlar during this time, if at all, and the boy's only real memories of him must have been some hazy recollections of the long voyage. Andrew spent at least some of the period in London, arranging for the publication of his second book, *Sailing Directions from New South Wales to China and Japan*.[9] It received a glowing review in *Nautical Magazine*, commending him for taking the time to make notes of his observations and publish them, which so few other "mercantile captains"[10] had bothered to do. It must have lifted his spirits, given the gloomy conditions at home. The book proved popular, "quoted for many years as an authority"[11], and there were second and third editions in 1859 and 1862.[12]

After his departure, Eliza continued to weaken, and eventually died on 25th July, 1856.[13] She was buried by her father in a ceremony at the local churchyard, in a windswept spot close to the sea, and Fetlar tradition recalls William Watson's words as he buried his daughter: "Oh, that he, who was glad to eat the crumbs at my table, should treat my daughter so."[14] Captain Cheyne was in China.[15] Watson had buried his wife and all but two of his eight children. The strain was too much for him, and he followed Eliza to the grave less than two months later on September 23rd.[16] He was succeeded at the Manse by his son-in-law David Webster. William was less than four years old, and in the space of a few months, had lost his mother and grandfather, and to all intents and purposes his father too.

It remains unclear whether anyone tried to inform Captain Cheyne of his wife's death, but there is no evidence he tried to claim his son. It would have been strange had he not been informed at some point. He still had relatives in Shetland who could have passed on the news, even if Eliza's family did not. However, a trust disposition signed on 16th August 1856 by the Reverend William Watson, only a month before his death, ten months after Andrew had left, suggests he expected his own family to remain responsible for William. The document made ample provision for the child's "maintenance and education"[17]. He was, after all, Watson's only grandson, but the implications may go far deeper than family obligation.

He bequeathed a third of his estate to each of his remaining daughters, Christian and Grace, and the final third to his grandson, in trust. In making the arrangements, he was effectively handing over responsibility for William to his two daughters and son-in-law, David Webster, and ensuring they had funds to maintain him [18], in the full knowledge that this would be how William was to be brought up. It suggests some sort of arrangement which cut Captain Cheyne out of any responsibility for his son, of which Andrew may have had full knowledge, and to which he may even have agreed, willingly or otherwise. The only other explanation is that they intended to keep the news of Eliza's death from Andrew, hoping he would never return, not a particularly realistic possibility, given that Captain Cheyne had any number of ways of finding out.

Christian had been around 40 when she married, and was 44 by this time.[19] She was unlikely to have children of her own. Grace, his only other surviving child, was 42 and unmarried.[20] The responsibility for William clearly weighed heavily on Watson's mind. The third share was to be held in trust for the boy until he reached the age of 21, but the annual interest was, in the meantime, to be used to provide for his education. Christian and Grace were to receive his share if he were to die before he reached the age of 21 [21]. Nowhere in the document is there any mention of William's father, and the boy is referred to as the "only child of Eliza Watson, my youngest daughter".[22]

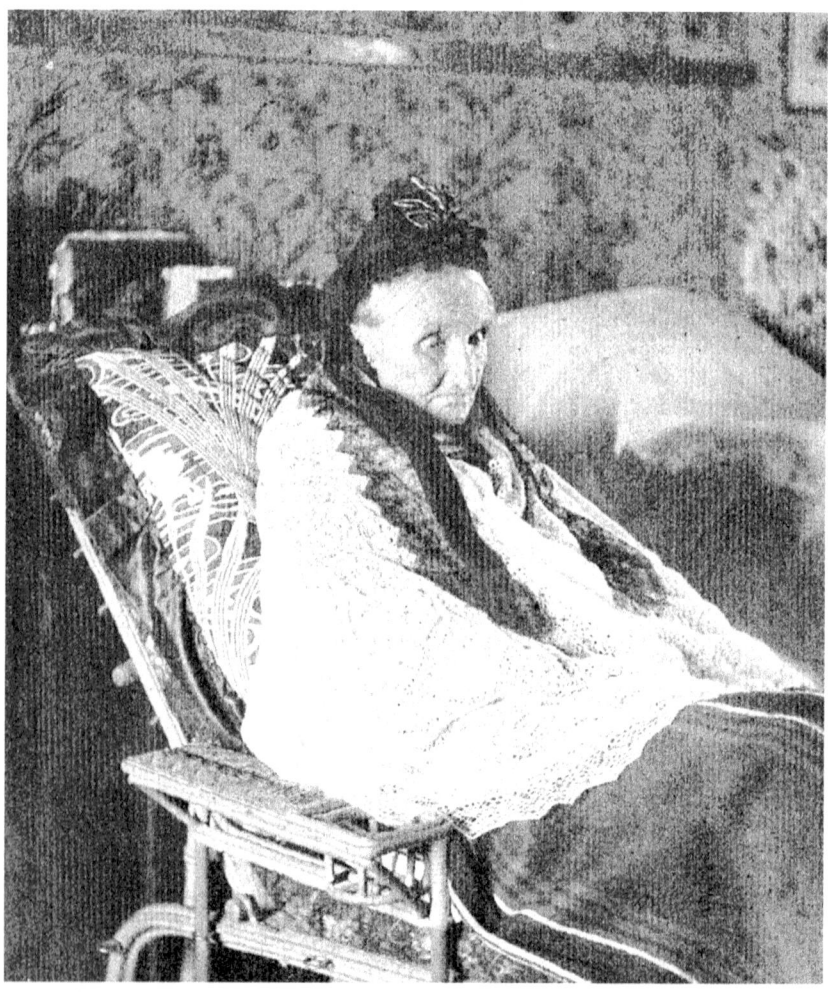

This may be Grace Watson, Cheyne's aunt, as an old lady

The wording of the document is a subtle indication of events to come. The arrangements with Captain Cheyne, whatever the details, went much further than the practical considerations of William's upbringing, and were the beginning of a campaign to erase all memory of him from the family record. It is possible that the arrangement the family had negotiated with him included a promise that, if they assumed William as their charge, and provided for his future, Andrew would never attempt to communicate with his wife or son again. When Captain Cheyne left, his wife was still alive, and if the family still considered her survival a possibility at this point, any arrangement about William is likely to have included Eliza. Andrew clearly accepted the arrangement, and we can only assume that he either admitted the consequences of his mistakes, or he realised he could not possibly pursue his colonial aims with the responsibility for Eliza and William, and took the opportunity presented to him.

During the brief period Watson was still alive, the arrangements seem to have been confined to practicalities, but the full weight of the ban came into force when David Webster took over at the Manse. Andrew's name was never mentioned again, and his flamboyant career was hidden from his son to prevent the boy from any misplaced youthful admiration.[23] In a family where sons were destined unequivocally for the Ministry, nothing was left to chance.

William became "William Watson", and his guardians dropped Cheyne from his name, at least in his presence.[24] He grew up with, at best, a hazy memory of both parents, and the absence of his father from the family record was to have a lasting effect on him. In the immediate term, he was perhaps too young to know much about it, but losing both his parents cannot but have left him confused. Interestingly, his grandfather's trust document still referred to him by his full name, William Watson Cheyne [25], and though this may have been for legal reasons, and to avoid any subsequent problems over his entitlement, it may also indicate that Andrew's memory was not banished entirely from the household until the Websters took over.

Was Andrew as dubious and heartless a character as he has been portrayed? He was certainly given to unpredictable rages, and, whether through naivety or design, attracted enemies. Contemporary reports reached Shetland that he had a reputation as a "blackbirder"[26], a trader who hoodwinked workers into slavery by trickery as opposed to force. More recently, it has been argued that this accusation was little more than gossip, and that Andrew was tarnished by the generally poor reputation of sandalwooders [27]. According to Shineberg, he treated South Sea islanders as fairly as he could, at least by his own standards, and within the context of the day. He said at one point in his journals,

> It has ever been my aim to treat all savages with whom I have had intercourse, kindly and humanely, ever bearing in mind that we sought them, not they us; indeed I have often punished members of my own crew for taking things from them without payment, and consequently made many enemies thereby. Finally, I challenge anyone that ever sailed with me - if they adhered to the truth - to say, that I defrauded a savage to the value of a cocoanut.[28]

In her extensive research on Captain Cheyne, Shineberg "found no good cause to doubt"[29] this statement. He was certainly a gifted dreamer, and his initial failure to consolidate gains may have been a result, at least in part, of bad luck and naivety. He has been described as behaving more like a character from the "*Boy's Own Paper* instead of a hard-headed sea captain."[30]

His determination nevertheless led to the manifestation of a more ruthless side of his character, and the account of a contemporary European, Alfred Tetens, who was initially in Andrew Cheyne's service in Palau, puts him in a much less favourable light. Tetens, from North Germany, may have had his own reasons for painting Cheyne so blackly, as he ultimately threw in his lot with Edward Woodin, Cheyne's partner turned rival.[31] Tetens' work nevertheless includes details that cannot be ignored. He furnishes an account of how Cheyne and Woodin carved out a trade monopoly between themselves in Palau, how Cheyne gradually drove Woodin out, and even encouraged disputes between groups of islanders in order to attack Woodin's new trading operations in the north of the island.[32]

Tetens describes Andrew Cheyne as capable of "devilish designs"[33]. He suggests that he "understood how to direct the jealousies between north and south so cleverly that Koror decided to begin a campaign against its northern rivals. Woodin should be cut off from his supplies and his ship, which would be plundered and destroyed."[34] Shineberg interprets the incident quite differently. In the generally cut-throat world of trading in the islands, she considers Tetens to have had ulterior motives, and Captain Cheyne to have had negotiating powers above the average, regularly awakening professional jealousies.[35] Captain Stevens, who was later sent to investigate Captain Cheyne's death, could find no substantial charges against him, but noted that he "appears to have evinced a spirit of dictation, combined with a somewhat unpleasant determination to overthrow every obstacle which blocked his path ..."[36]

In Andrew Cheyne's defence, his naivety in some matters does not correspond with Tetens' ruthless image of him, and he was, by and large, fair in his dealings with both natives and Europeans. He was nevertheless a hard man, and his capricious behaviour and frustrating lack of success led to well-documented and spectacular fits of temper.[37] His alleged treatment of Eliza, however, remains a mystery, and we have no way of knowing the full circumstances. The fact remains that, had she already been as ill as the Fetlar tradition implies, he was either a hard man, or faced a dilemma in deciding whether to engage someone to take her back to her family while he finished his business in the Pacific. The contradictions in his character suggest that he was not always capable of making difficult decisions when it came to people.

His character should also be seen in the context of the environment in which he was attempting to make his living. He was a sharp observer of cultural nuances and knew how to exploit them. His considerable ability to chronicle information on Pacific island life was clearly frustrated at times by the necessities of commerce. However, he never mastered, or saw the need to master, his intolerance. He disapproved as much of what he saw as moral transgressions in his fellow countrymen as he did in Pacific islanders, even if he did not always follow his own codes of conduct to the letter. His moral attitudes were "indistinguishable from those of contemporary Protestant missionaries in the South Seas"[38], harking back to the Northmavine Manse. On the other hand, Captain Stevens discovered Cheyne's relations with island women to have been "of a somewhat unlimited nature".[39] Perhaps he was not alone in Victorian society in thinking that his morals need only apply to others.

If his opinions were considered outspokenly puritanical by other traders, they were clearly not stringent enough for the Fetlar Manse. For the Websters, his erratic movements were irresponsible, jumping from one venture to another with no real success. They may even have noted the fact that his experience as a trader and merchant could theoretically have served him as well in Shetland, and he had no need to go off to the Pacific or anywhere else to make his fortune. He could have built a more solid reputation for himself at home, and gradually acquired a respectable income. Captain Cheyne was, above all else, a determined adventurer, who was convinced he stood to make not only a fortune but also a personal empire, whereas he would have remained nothing out of the ordinary at home.

There is no evidence that he ever tried to break the family's ban and communicate with his son. He may even have accepted responsibility for Eliza's death and not wanted the burden of supporting their son at a time when he had difficulty supporting himself. The boy's guardians became, to all intents and purposes, the only parents he ever really knew. It is unlikely that any of them were aware of the new liaisons Andrew formed with Palauans. John Davey, who had at one time worked for Cheyne, said his "intercourse with women was notorious, but he had never known him to abduct or take possession of a woman."[40] It emerged in the mid-20th century that Captain Cheyne had at least one grandson in Palau, called Umang (or Otto by the German missionaries who educated him), who was living there in 1906/7.[41] William had a half-sister or brother he knew nothing of as a child, and as an adult would have at least one nephew. As an older man, he may even have heard about them. I can find no proof that they ever met, and can only speculate.

We should also consider the evidence that Eliza herself may have wished to go with her husband on the journey. Andrew had his own business to conduct in the South Seas, and his stops at old haunts on the return journey raise questions about whether he ever had any intentions of settling his family there, or at least in the region as a whole. The families of other sandalwooders and South Sea traders had set up house in similar circumstances, and missionaries and their wives flocked to the settlements in search of souls. The small islands of the Pacific were considered a sort of free-for-all for European traders, and attracted the cross-section of society we are accustomed to hearing about in tales of the Wild West. On the other hand, there was nothing romantic about life in these settlements; it was brutal and unpredictable. If Cheyne had made plans to keep his wife and son there with him for any length of time, Eliza may have stood no better chance of survival than at the Fetlar Manse, and would almost certainly have experienced the kind of hardship faced by the Henrys, a Scots settler family who "feared God and subscribed to the Presbyterian Missionary Society."[42]

Andrew Henry had stations on Eromanga, Aneityum and (later) Espiritu Santo. He had a large family of children, was in perpetual debt and frequently entirely without food. His wife wrote on at least one occasion to Robert Towns, Paddon's principle rival, making an emotional plea to have the interest payments reduced on their debts.

> Just think, Dear Sir, what we must have suffered on Lifue when we were for six months half starved, where all my children had

Measles. We were wrecked in December and it was June before the *Ariel* came for us, during that time we were often without food.[43]

When Elizabeth Henry was not giving birth to more children, she worked alongside her husband supplying sandalwood, while her eldest daughter Mary took care of the household and the younger children. On one occasion, when her husband was away, Elizabeth was forced to defend their station against an Eromangan attack. They never made the fortune they had been seeking, and Elizabeth died after giving birth to 12 children.[44] How long would Eliza Cheyne have survived under similar circumstances, especially if she was already succumbing to tuberculosis?

The West Manse in Fetlar, where William Watson Cheyne grew up. Cheyne's family had had to vacate the Manse in the 1880s when his uncle died and a new Minister, Mr Campbell, took up residence.

Microbes and the Fetlar Man

David Webster may have had another cause for complaint against his brother-in-law. When Andrew's case for subsidised emigration from Shetland came before Parliament in 1854 [45], he had presented his relationship with Webster as evidence of his own good character. The new minister may have regretted consenting to the use of his name, if indeed he had consented in the first place, especially if Captain Cheyne's reputation was under scrutiny by the authorities in any way over the incident of the French ship. Webster was noted locally as a particularly zealous minister.[46] It is even feasible that he had initially shown interest in Cheyne's proposed trading settlements in the Pacific if they had included a place for missionaries or solved a problem of pressure on parish funds when it came to single women without means. Webster may have considered Cheyne's subsequent behaviour a betrayal of trust.

The most compelling evidence to suggest that Eliza's death was the result of an unfortunate tangle of issues, however, involves the conditions in which she grew up. Sir Joseph Cheyne, Captain Cheyne's great-grandson, was of the opinion that Eliza's condition was merely exacerbated by the voyage, but had already been engendered by the damp conditions at the Fetlar Manse.[47] His impressions are more than supported by the considerable correspondence on the matter, first from her father, and later Webster.[48] Four of Eliza's sisters and a brother lived no longer than their twenties, and it is possible that the conditions of the Hillswick Manse had not been much better than those in the Fetlar one. As early as August 1839, when Eliza was only eleven, Watson had written to Sir Arthur Nicolson, the principle heritor responsible for the upkeep of the Fetlar Manse, imploring him to make necessary repairs:

> I would beg of you to get repaired in proper time some damages done to the Manse & Offices by the winter storms. The Manse has become very damp from want of harling, & if it could be Roman cemented, new lead ridged and harled while the good weather continues it would save you and the other Heritors much future trouble and expense.
> Were you to take a look at the Manse any time in passing, I shall be happy to point out to you its deficiencies.[49]

The conditions were to be no better for William. He had become plain William Watson, and it is difficult to know whether he had any recollection of his change of identity, but given the later importance he attached to reinstating his name when he discovered the truth, it is likely it only served to fire his imagination about his father, the more the truth was kept from him. Whatever psychological damage the Websters may have done by hiding Andrew's identity from his son, it is fairly clear that, without the provisions Watson made for them to take care of him, he may never have reached adulthood, and would almost certainly not have had access to the education which led him into medicine.

We have little information about William's childhood, except what he himself revealed in later years. We can, however, deduce a little about life on the island. Fetlar is only around seven miles by four in area. Today, it is considered relatively remote, perched on the outer north-east edge of the Shetland archipelago, but in the 1850s it was a major centre for the haaf, or deep-sea fishing industry [50], which sent dried, salted whitefish to European markets such as Spain. Haaf fishing stations tended to be close to the deep sea fishing grounds, to facilitate landing and processing, so places which seem remote in today's world were much less so then. Where an Orcadian, in the archipelago to the south of Shetland, has always been regarded as a farmer with a boat, a Shetlander has always been unequivocally "a fisherman with a croft"[51]. In other words, farming was given a low priority, and fishing was the main industry, operated largely by merchants and landowners.

*

In the early 1800s, the main landowner in Fetlar, Sir Arthur Nicolson, had threatened to evict the families of any tenants who attempted to pay their rent with cash earned through employment on whaling ships. In the economic system known as Truck, their labour as fishermen was worth more to the landowner than cash. For perhaps the first and only time in the island's history, the local men allegedly marched on the laird's residence and demanded a change in the system. Unusually for Nicolson, he conceded on that occasion [52], but before long introduced sweeping changes to the estate the population had not anticipated, and from which it would never recover.

From the 1820s onwards, he began to reorganise his land, evicting a large proportion of his tenants. The result was massive depopulation [53], which continued well into the century, and any remaining tenants had little choice but to continue within the tied system, in constant debt. One of the few tenants who amassed enough money to buy out his land in the island in the 1850s was a man called Williamson, who had some success at the California gold diggings,⁵⁴ another potential route to freedom, though not many returned with any more than they had taken out. The established system of tenancies remained in Fetlar, and as no one was in much of a position to question the system openly, the minister was often called on to speak on the community's behalf.

Ministers were also approached to contribute local statistical and descriptive information for national surveys. William's grandfather had compiled the Fetlar contribution to the Second Statistical Account (of Scotland). He made no attempt to disguise his contempt for Nicolson, describing the latter's new mansion as architecturally "nondescript"⁵⁵ and blaming him for having introduced sheep scab to the island by importing cross breeds, which were not as hardy or adaptable as the local ones. This type of outspokenness presented the minister in a good light with islanders, and Watson seems to have been respected in Fetlar.⁵⁶ Parishioners also seem to have been fond of Watson's daughters ⁵⁷, who played an active role and took on parish duties.

The West Manse, Fetlar, in 2014

The Reverend William Watson's main problem seems to have been the incursion of the Methodists and the Free Kirk, who had been stealing his congregation. The Methodists, having moved into the island, had the apparent audacity to plant their chapel as close to the established church as they could. After the Disruption in 1843, dissenters from the Church of Scotland had formed the Free Kirk, which also established itself in Fetlar, patronised by an important merchant family on the east side, the Smiths.

On one occasion, a temporary band of Quakers joined the throng, and even Watson's own household staff trooped off to hear them preach.[58] The competition for souls could be unremitting at times, but not everyone was prepared to be swayed. At one point a man called Gardiner went around the island preaching against the Established Church, and one woman responded by saying that Watson was a good Christian and his daughters were kind to the poor.[59]

It seems William's mother, aunts and grandfather were well thought-of. William probably enjoyed similar respect, and identified himself very early on with a culture that was almost entirely dependent on the sea. Given the nature of small communities, it must have been a particularly successful network of tangled webs which kept from him any information about the identity of his father. On the other hand, people may have been slightly afraid of William's uncle, David Webster. He was known as a particularly zealous minister, and was prepared to go to some length to make sure his flock was kept in what he considered to be good moral order. He apparently initiated a crusade against fiddles, which he termed the "devil's instrument"[60], and when he came across one in a household, simply broke it to pieces. On one celebrated occasion, an old man was sitting by the fire when Webster came in and noticed the fiddle hanging on the wall. He watched the minister advancing towards the fiddle and then took the poker in his hand, stood up and said, "Dat's far enoch! If du lays wan hond on dat fiddle, du'll be afore dee makker a lok shöner dan du lipp'nd."[61] ("That's far enough. If you lay one hand on that fiddle you'll be meeting your maker a lot sooner than you expected").

Given the general expectations of ministers in the islands at the time, Webster was not particularly unusual in his enthusiasm, but was rather noteworthy in the way he went about it. We can only suppose he was as strict with William, particularly if he considered the boy in danger of emulating his father's behaviour. It is likely that William was more than a little in awe of him. On the other hand, Webster was as active in community matters as Watson had been. One of his first petitions as Minister was to secure a post boat for the island. Unlike its neighbouring island, Yell, which had five Receiving Officers for the mail, Fetlar had none. Its other neighbour, Unst, had three. The Post Office authorities offered an annuity for a boatman, but not yet a boat. Webster wrote to Nicolson asking if he would petition them, and bemoaned the fact that, once again, Fetlar did not receive the same concessions as its neighbours. He urged Nicolson to write to the authorities on the grounds that "when they are so very liberal to these two Islands, I think they might well afford to give us one."[62] They also had to take up a collection in the island, and Webster contributed one pound. The merchant Gilbert Smith was expected to contribute money too. The young William Watson can hardly have failed to be aware of the activity around the issue, especially after Webster organised men to go out and collect the signatures for a petition from each household. It was signed by 118 male islanders, all over the age of 14.[63]

William was not the only child in the household, however. He grew up alongside Webster's son John, who was a few years older than him. When Christian Webster had married her husband in 1851, he had been a widower for three years, and had brought a young son of four with him to the marriage. The boy's mother had died a few days after he was born, and his parents had hardly been married a year. On 20th January 1850 Webster had moved from Aberdeenshire to become a missionary in the Parish of North Yell and Fetlar, and became assistant to the Reverend William Watson. Within just over a year he had also married Watson's daughter.[64] They had no children of their own, but brought up Eliza's child and Webster's son side by side at the Manse[65].

One of the last houses in Fetlar to have the fire in the centre of the room, probably at Aith. This photograph is likely to have been taken in the early 1900s, but the inside of the house could also be typical of what Cheyne saw as a boy

John was seven before William even met him, and there is no evidence that the boys were particularly close. In 1861, when William was eight, both children were resident at the Manse, and described as scholars [66]. They may have been educated partly at the local school and partly at the Manse. There had been schools on the island, in one form or another, for some time. In the early 1800s, a summer school at the Booth of Urie, a fishing station in Fetlar, had been taught by the East Yell schoolmaster Andrew Dishington Mathewson, the father of a woman who was later to play an unexpected role in Cheyne's career. A.D. Mathewson was only 14 years old himself when he began to teach at the school, and his family had moved into Fetlar when his father had been evicted from their croft on the neighbouring island of Yell[67]. By the 1850s and '60s, during William's school years, the parish school was on Mathewson's native island of Yell, but a school at Still in Fetlar, operated since 1740 by the Society for the Propagation of Christian Knowledge, was of longer standing.[68] It taught rudimentary reading, so that children would grow up being able to read the Bible. There was even a library run by the Parish and open on the first Monday of every month from 12 noon to 2 pm. Anyone paying an annual subscription of one shilling could borrow twelve books a year.[69] Along with the ubiquitous religious texts, including an alarming body of information on death, were books on seamanship and a lone copy of *Robinson Crusoe*.[70] By William's time, the available reading matter may have broadened slightly, but we may, I think, safely assume that he was exposed at least to the same faraway influences as his father if he was allowed access to Defoe's novel.

At the school, the two boys are likely to have been taught reading, writing, arithmetic, book-keeping, and navigation. Cheyne's entry in the *Lives of the Fellows of the Royal College of Surgeons of England* implies he went to the local school [71], but as it calls it a grammar school, which it was not, the writer was no doubt referring to Cheyne's later education at Aberdeen Grammar School. The likelihood is that the two boys' early education was at least supplemented at the Manse.[72]

The family fully intended that both their charges would enter the ministry, and their schooling needed to prepare them for the studies they would have to undertake further afield. John complied with the family's wishes with no apparent protest. He studied at Aberdeen [73] and then Edinburgh University, and eventually followed his father into the Ministry, serving in a number of parishes on the Scottish mainland. It must have seemed to the Websters the most natural thing in the world that the children in their care would continue in the tradition of the Watsons and Websters and find a secure position in the Ministry. To his aunt and uncle's horror, however, William was bent on a career at sea.

The West Manse in Fetlar occupies one of the most extraordinarily beautiful sites in Britain, alongside a sandy beach which separates a freshwater loch from the sea, and where enormous breakers crash to land during the frequent stormy weather. If he had looked out of the windows at the Manse, towards the beach, William could not have avoided seeing the churchyard where his mother and grandfather were buried. The Manse garden had some of the only substantial trees on the island, planted by a minister in the previous century as an experiment, a challenge to the Shetland winds. It was by this time a veritable woodland in comparison with the surrounding landscape, a paradise for a small boy, running in and out of the undergrowth. Apparently, on one occasion, William was caught breaking a live chicken's leg so that he could experiment with setting it, rather an extreme prelude to his future career.[74] Most significantly of all, however, every morning when he awoke and every night before he went to bed, William could see and hear the waves.

He watched the men preparing the boats to go out to sea, the wooden sixereens, each manned by six men. He watched them returning home with their catches. At some time during an otherwise sheltered childhood, William learned to handle a boat himself, no mean skill on the seas around Fetlar. He was not a strong child. His health always showed signs of weakness, and his guardians may not have encouraged him to learn any maritime skills, but this did not deter him in the slightest. He saw the trappings of the sea everywhere he looked, and he lived and breathed it. Even a visit to another part of the same island often involved a boat journey, and travel by sea was more common than by land. Supplies were brought into the island by boat, and people and livestock left by boat. William's grandfather had once described in a letter a gruelling journey to his old parish in Northmavine in appalling weather conditions, involving several boats and pony. The journey took seven days, and was dangerous in places, given the wind and the snow.[75] Today, the journey from the Manse in Fetlar to Hillswick by ferry and road takes just over two hours. In Watson's day, there were so many unknowns it was difficult to predict how long a journey would take. In all events, the young William Watson could hardly have avoided boats, and the most striking legacy of his childhood became his obsessive love of the sea and its ways of life. Oddly enough, it would even indirectly lead him to his career in medicine.

He apparently knew nothing of his father and his maritime history, so this clearly had no direct bearing on his determination to go to sea. He was more likely to have been influenced by the island men he saw leaving for the merchant navy, a common enough occurrence in large families where younger sons in particular had to seek work elsewhere when they were old enough. They came home with much coveted cash for their families, and with tales of the world beyond the islands. Shetland men in general were often some of the best travelled in Britain, and had amassed a wealth of knowledge about the world, a fact often overlooked today when the islands seem remote and cut off. Captain Cheyne did not need to be there to fire his son's imagination. He could have heard the same exotic tales in almost any household in Fetlar where there were sons home from the sea.

Even if his guardians had not been completely opposed to the idea, however, another problem stood in the way of William's maritime career. His health may already have been delicate, as he was to show signs of tuberculosis as a young man [76], the problem which had plagued so many of his mother's family. This was not the first time fate would lead him away from his first choice of career. His aunt and uncle were quite determined he would not work at sea, fearing any life which might lead him down the same precarious path as his father, but at times they must have been close to despair. His heart was set on it despite any early signs of ill health. Even more alarmingly, the damp at the Manse had not improved in the slightest since his grandfather's day.

The moment William's grandfather died, his uncle David Webster renewed the battle with those responsible for the upkeep of the Manse. The damp was made worse by the damage of each winter storm, and in 1857 [77], Webster took matters into his own hands and lobbied the heritors (the local landowners who were responsible for maintaining the Manse) to make sweeping alterations which would stabilise the building for the foreseeable future. He was careful to note that he was not, after all, asking the earth. He did "not wish a fine house and did not wish to put them to one sixpence of unnecessary expense"[78], but he hoped they would see the advantages of proper, long-term repairs.

He began a gruelling correspondence with the main heritor, Sir Arthur Nicolson, the principle landowner in Fetlar, and at first received frustrating replies about how a "properly qualified tradesman"[79] would have to assess the situation. The Manse had been this way since Watson's time, and Sir Arthur stalled by asking why he had not been informed of it sooner. Webster replied, "I could not well have done so as it might have been an undue interference on my part, as I had no claim to the Manse whatever, while the last incumbent lived & inhabited it, and although it has been in a very uncomfortable state for a number of years past, & the health of his family has suffered in consequence, yet at his time of life he was unable to undergo the trouble & fatigue which must necessarily result from a thorough repair being put upon the house."[80] Webster was clearly unaware of Watson's letter of 1839 [81], and Nicolson had conveniently forgotten about it.

Attributing the poor health of Watson's family to the condition of the Manse was a dramatic understatement, but to be fair, many of them died before they even reached Fetlar, and the Hillswick Manse may not have been in much better condition. Eliza's health, however, was almost certainly a result of the Fetlar house in which she had spent most of her life, and by the time her father died, most of his children had preceded him to the grave. Had Christian and Grace succumbed too, there is no indication what would have happened to William.

By the time the boy was four years old, the situation was already intolerable. Webster wrote "… [the Manse] has really become uninhabitable and injurious for the health to live in. During the last winter it was frequently in such a state that no person could go in slippers from the dining room to the kitchen with dry feet."[82] Webster asked if they would consider raising the floor.[83] He also asked for the chimney tops to be replaced, "else we will not be able to live in the house for smoke during winter."[84] Webster himself had needed to "put in boxes"[85], but they were already "completely done" [86] by this time, and the atmosphere in the rooms can hardly have been conducive to the health of a family with a history of tuberculosis.

The heritors continued to have objections to the alterations, particularly as raising the ground floor would involve gutting the house, and Webster only really began to make any progress when he threatened to ask the Presbytery to conduct an inspection, which would have cost them money.[87] Webster persevered, making alternative suggestions involving the addition of two rooms, a new staircase and larger windows downstairs, to get around their objection to raising the floors.[88]

Part of the problem was that very few of the heritors were resident in Fetlar and had not actually seen the state of the Manse. Gilbert Smith, a merchant living on the other side of the island, was really the only one to whom Webster had an opportunity to point out the defects in person.[89] Nicolson, though he owned a mansion house on the island, was rarely resident, and most correspondence with him was directed to Edinburgh.

The Nicolson estate was considerable, and extended to areas around the Scottish capital, as well as land in Shetland.

Webster's perseverance paid off, and the heritors finally appeared to have agreed to his modified suggestions by September. However, this delayed the work for at least another six months, as little external building work can be done in the Shetland winter, even today. He agreed to remain in the house while the repairs were taking place, "so as not to put the Heritors to any unnecessary expense."[90]. Apart from the inconvenience, as these repairs would take some time, the dust and musty atmosphere increased the chances of William's health taking a turn for the worse, not to mention everyone else's.

The heritors also only agreed to foot the bill for the Manse repairs if Webster agreed to it becoming a "free manse"[91], maintaining it himself for the remainder of his incumbency, and freeing them of any further responsibility or expense. He only agreed to this, rather sensibly, if the repaired house were inspected by tradesmen of his own choosing, and provided any maintenance only involved "the repairs a house may annually or occasionally be expected to require"[92]. Given the unpredictable storm damage during the Shetland winters, he was making a wise decision. He also asked them not to request the Presbytery to free the Manse from the heritors until all his conditions had been satisfied.[93] He clearly did not trust them to see the process through.

In March 1858, work had still not started, and Webster was forced to renew the correspondence amid fears that "the best workmen" would "be all employed"[94]. In April he was still negotiating the plans, and offered the possibility that, with two Lerwick masons and the majority of the labourers from the local population, Nicolson would save money, because Fetlar men could be got "at one half the wages"[95]. This almost certainly referred to the rate of the wages themselves, and not the expenses incurred in taking in labourers from outside. The estate was interested in keeping local labour costs - and expectations - down. On 29th April, Nicolson had still taken no steps to authorise the work, even though the other heritors had given their approval.[96] William was too young to be aware of the efforts his uncle was making to improve their living conditions, but he was almost certainly affected by the ongoing dampness, and by the dust and mustiness of the house while it was being repaired. It was to establish a breeding ground for an ever-present threat of tuberculosis. He can also hardly have failed to be aware, as he grew older, of the ongoing tussle between his uncle and the Nicolson estate, and he cannot have had a particularly favourable impression of Nicolson or his managers.

In May 1859, Webster insisted the plans be forwarded to the Presbytery, which he warned was likely to take action if work were not started immediately.[97] Nicolson carried on other, rather lengthy correspondence with Webster at the time, but chose for some reason to delay the matter of the Manse repairs interminably. He finally agreed to the suggestions on 11th May 1859, at the eleventh hour, just before a Presbytery meeting where the matter was to be taken in hand.[98] This may have been a ruse to fend off the Presbytery, however, because by 13th May Nicolson still had not authorised estimates from tradesmen. Again, Webster was afraid that the best tradesmen would have gone, and "men from the south would not come for such a small job."[99] Nicolson had successfully fended off the wrath of the Presbytery, but still managed to delay the work as long as he could before it eventually went ahead.

While William was paddling through the hallway at the Manse, however, his father's life had finally taken a turn for the better. Captain Cheyne had remained in the Pacific after Eliza's death, and the Websters are likely to have had some idea of his activities through contacts in Northmavine. William's guardians were nevertheless meticulous in ensuring he heard nothing of his father. Within a few years, Andrew had earned enough capital working for others in the China coastal trade to buy his own vessels, and was trading *bêche-de-mer*.[100] He began, at this stage, to consolidate his dreams of empire by purchasing 10,000 acres of land in Palau and developing plantations of sugar, coffee and tobacco. He entered into trade agreements with a local leader, the Ibedul of Koror, drawing up rather presumptuous documents which became known as the *Treaty of Commerce* and *Constitution of the Pelew Islands*".[101] Despite its grandiose title, the *constitution* is misleading. It had little to do with social agreements, but was more of an attempt by Captain Cheyne to formalise a monopoly on trade.

Traders were operating in dangerous circumstances to unknown rules, which they were learning as they went along. Shineberg maintains that islanders were as agile as the traders themselves in manipulating agreements and negotiating the best deals for themselves [102], and Andrew was in competition with other European traders. Where he most misread the situation, however, was in believing he could impose European style treaties on local politics, or even that he was in any way authorised to do so. His biggest mistake of all was to assume they would be binding. Unlike Ross in the Cocos Islands, he did not have an automatic monopoly, and had to fight for it by all means at his disposal.

A consistent workforce had been a constant frustration to Captain Cheyne, from his early sandalwood-trading days to his *bêche-de-mer* enterprises. He was reliant on local labour, which was not always available when he required it, despite his attempts to close written agreements. When the yam harvest was due, for example, all local hands were suddenly unavailable [103], and Cheyne never quite learned to incorporate local routines into his own. His answer was to persist with European economic models, and try to hold his workforce to agreements which had no binding value, or to bring in a tied workforce from outside. He initially advertised for Chinese labourers to work the plantations.[104]

Most significantly, behind Cheyne's dreams of a trade monopoly were his ambitions to recreate a sort of principality, and to elevate himself to a status equivalent to monarch. Though on fairly shaky ground, he eventually succeeded in becoming "Cheyne of Palau instead of Cheyne of Tangwick, with his own farm and fishing tenants, his own curing establishment, his own boats and two sea-going ships."[105] He drew up neat contracts and covenants, "duly signed, with copies [in] pseudo-legalistic language."[106] He had even persuaded the local leader to agree to a petition to ask the British Government to set up a protectorate in the islands, which the leader then regretted, but which would have consolidated Cheyne's position even further.[107] Tetens said, with a note of alarm, that, "had Cheyne's *Constitution*, which was lodged with the Consul in Manila, ever been taken seriously, he would have been more or less the King of Palau."[108]

The Koror chiefs did not deliver their part of the 'contract', however, and broke Cheyne's monopoly. The plantations 'were allowed to run to ruin and his property robbed as soon as he left the island'[109]. Captain Cheyne had created a kingdom on quicksand. He tried to transpose formal European business habits and a good deal of romanticism onto a well-established cultural framework he never fully understood, and with which, if we believe Tetens' account, he was playing dangerous games. He had based his trading "on due payment of promised wages and on contracts and convenants"[110], but "lacked the warmth and tact which would have been more useful to him than a mountain of paper."[111] His workforce was quite up to the challenge and happily traded with his competitors. There were plenty of traders ready to step into his place.

*

At around the age of 11, and with no apparent knowledge of his father, William Watson left the security of the Manse to begin a secondary education. Shetland's first secondary school, a fee-paying section of the

Anderson Educational Institute, had first opened in 1862, the gift of Arthur Anderson, a Shetland philanthropist and partner in the London firm which founded P&O shipping.[112] The Institute was new, however, and the Websters, with traditional views of a good education, were not about to jeopardise William's future in an unfamiliar setting.[113] They also had ambitions for him beyond the opportunities Shetland could prepare him for. He was taken to Aberdeen, where he would attend one of the oldest grammar schools in Britain, an establishment which counted Lord Byron amongst its illustrious alumni.[114] John Webster was already in the city.

The legacy left by Watson to cater for William's education was not enormous, but a reasonable sum for a parish minister who had lived into his eighties. In fact, he was described as having had an above average income for a parish clergyman, and was "passing rich on 200 pounds a year".[115] He had income from land which was not tied to the Church, and which was part of the general inheritance that went to William and his two aunts. The interest was to be used to support William and his education until he reached the age of 21, and the disposable legacy amounted to £779/11/6½.[116] A third of this would have been the equivalent of around £20,000 in 2013.[117]

Today, Aberdeen is a 12-14 hour sea journey from␣Lerwick, the capital of Shetland, which is, in turn, a good two and a half hours' journey by sea and road from Fetlar, even with modern ferries. In the mid-19th century, it took much longer, and William and his aunts could easily have taken several days to reach the Scottish mainland. They travelled from Fetlar to the Shetland Mainland by packet, calling at different islands along the way to load and unload goods. When they reached a place in the islands where a packet was leaving for Aberdeen, they may have been lucky and travelled the following day, but could have waited for a vessel for some time. Despite this, Shetlanders were well used to travelling back and forth by sea to centres of population on the Scottish mainland and further afield, and it was unlikely to have been the first time William had made the journey. At least one of his aunts remained with him in lodgings in Aberdeen, probably both of them on a rota basis.[118]

An extension to Aberdeen Grammar School in 1863 had permitted a broader curriculum, and William had the opportunity to study Latin, Greek, Ancient Geography, English, mathematics, modern languages, art and gymnastics.[119] His delicate health probably precluded any enjoyment of the last subject on the list. In fact, his obituary in the *British Medical Journal (BMJ)* would later note that "he was never strong enough to indulge in the violent sports and games of youth" and that "this deprivation of social opportunities accounted for the extreme shyness which always was his greatest handicap in life. He felt this handicap keenly, realising that his diffidence and shyness had often been mistaken for exclusiveness or bearishness."[120] It was presumably at Aberdeen that he received the instruction in French and German which would serve him so well later on.

William only stayed two seasons at the Grammar School, but there is some suggestion he may have stayed part of a third, which does not show up on school records.[121] If indeed he left the school in his third year, his sudden decision coincides with one of the most momentous events of his youth. He was notified of his father's death, and finally had to be made aware of his real identity. As if this were not sufficient shock for the boy, who had either been told his father was already dead, or had come to see the Websters as his parents, he had to learn of the gruesome circumstances in which the events had transpired.

Captain Cheyne's attempts to maintain his monopoly of trade in Palau had failed. The local leader on whom he relied for labour, the Ibedul of Koror, had been trading with other merchants for some time, and Cheyne considered this a clear breach of his written agreement. He was given to uncontrollable outbursts of rage, and when these failed to bring about any change in the Ibedul's habits, Cheyne retaliated in kind by trading with the chief's rivals. He spent most of his time travelling between his business concerns, but when he was next in Koror, around January or February 1866, he was "lured from his house near the beach, and there strangled and beaten on the head and breast with a stone until he died."[122] Shetland tales of his fate as a meal for the islanders are apocryphal, but were perhaps tinged with an element of wishful thinking, and possibly even talk of divine justice at the Fetlar Manse.

A year later, Captain Charles Stevens was sent in the British warship *HMS Perseus* to investigate Andrew Cheyne's murder. He landed in Palau on April 6th, 1867 and stayed at Koror for a week. He found next to nothing remaining of Captain Cheyne's house, close to the beach and the site of his death. He examined Cheyne's "tomb"[123], composed of the pieces of coral the local people used for building. When the party visited the Ibedul's house, their European interpreter recognised a number of items which had belonged to Captain Cheyne, including a bedstead and chairs. He questioned a variety of people, including John Davey, who had sent Captain Cheyne a letter saying he could not travel with him if he were to be accused of taking stores to trade with the natives. Davey nevertheless assured Captain Stephens that, "in trading with the natives [Cheyne] always did so fairly, giving an equivalent on all occasions for their produce".[124] As Davey had his own axe to grind with Cheyne, his statement is likely to be accurate. He also thought the chiefs had made the decision to murder Cheyne in council, as "the ordinary natives wouldn't dare to commit a murder of the sort without authority."[125] One of the leaders ultimately confirmed Davey's suspicions about the decision having been made in council. Captain Stevens determined that the Ibedul had authorised the murder, so ordered that he be executed, not by the British authorities, but by one of the Ibedul's own men.[126] This he somehow or other managed to effect, and according to the translator of Tetens' account, the Ibedul was buried at Koror alongside Captain Cheyne.[127] Stevens noted in his report that the morning after the murder, the women and some of the chiefs had removed Cheyne's body and buried it at Koror, negating once and for all any speculation that he had been cannibalised. The *Perseus* left on April 13th, and Andrew Cheyne had finally managed to bring British authority to bear on the Palauans, a little too late for his own purposes.

If Cheyne's expectations of European-style honour are to be taken as his chief weakness, he had paid a high price. He had been a gifted observer and meticulous recorder of little-known cultures, and like so many others in his profession at the time, a somewhat naive adventurer with a few too many pretensions of grandeur beyond his means. He had an understanding of European legal and economic systems in a theoretical way, but was unable to turn them to use in a culture where they were inappropriate. He was by no means alone. His Palau scheme "evinced that passion for improvement which was the mainspring of a generation of British political, industrial and religious empire builders."[128] He "appeared to be personally disturbed to see land capable of development and people capable of reform being allowed to run, as it were, on their own ruinous course."[129]

Exactly when William learned the truth about his father is unclear. It has been suggested he read his own registration records at the Grammar School [130], and the authorities may not have been warned he should not have access to them, perhaps a careless oversight on the part of the Websters. Correspondence began trickling through to Shetland, and by the summer of 1867 [131], Captain Cheyne's affairs were being taken in hand. William's sudden exit therefore coincides almost exactly with the point at which the circumstances of Andrew's death became known at home. It must also have became clear he would inherit the remaining share of the Cheyne family land at Nounsbrough, at which point William's own trustees may have felt they no longer had any choice but to tell him the truth, especially if he had already been asking difficult questions on seeing his school records.

The most remarkable fact is not that he did not know his father's real identity, but that the Websters had been so successful in withholding the information for so long. When he found out is significant, as it may help to explain why William began to drift. He left Aberdeen Grammar School suddenly, at a point when the enfolding saga of his father was clearly becoming too much for him. At the age of 14, he had been jolted out of his childhood and forced to come to terms with the fact that, not only had his father been alive until this point, he had been a rather flamboyant adventurer with a colourful, if somewhat dubious maritime history. In a boy already deeply unhappy with the professional path he was being encouraged to take, this can only have caused confusion at best, and the beginnings of a period of depression at worst. Quite independently of knowing anything of his father, William had been drawn to a career at sea. Now he must have felt he had a family precedent, where until now he had seen only a dynasty of Presbyterian ministers.

Nevertheless, he enrolled shortly afterwards, in November 1868, at King's College Aberdeen to study for a Master of Arts degree.[132] His guardian was an alumnus of King's College [133], so on the one hand, this may have been a concession to his guardians, who were still determined he should enter the Ministry. On the other hand, if he was already entertaining a career in medicine, this may have been a first step. Crowther and Dupree have noted that medical students in Cheyne's era

> ... were frequently advised to study Arts, and if possible to take an Arts degree, before attempting medicine. It was argued that medical students needed a firm grounding in the classics in order to understand the terminology of their profession, but by the mid-nineteenth century the Arts curriculum also included general science and modern languages, regarded as a desirable foundation for modern medicine.[134]

On the other hand, the way in which his father had died may have made him reconsider a purely maritime career, and choose instead a broader education until he could be sure. In a conspicuous act of defiance, he nevertheless assumed his father's name, and began to refer to himself thereafter as William Watson Cheyne.

Microbes and the Fetlar Man

His new studies turned out to be only a temporary solution. He was restless, and could not settle. He himself later suggested that it was because he was "always at the bottom of the class"[135] and was "inherently lazy",[136] but this may have been an attempt to liven up the talk he was giving, and he may not have cared to recall the period in too much detail. It is more likely that he was "depressed and bored,"[137] and not particularly interested in his studies. He did not complete his degree, and left after less than two years in the summer of 1870.[138]

Shortly afterwards, he made the momentous decision to enrol at Edinburgh University to study medicine. He reasoned that, if he qualified in medicine, he could become a ship's surgeon, but it is unlikely he employed this reasoning openly with his guardians. If they suspected it, they probably hoped another few years of study would help him get over his obsession. On the other hand, they would probably have been swayed by the idea of medicine as a preparation for missionary work. It was not uncommon for budding missionaries to take medical courses, though their priorities tended to lie with their studies for the Ministry and medicine was a useful added qualification.[139]

This time, however, William Watson Cheyne was determined. He returned to Shetland to wait to enrol at Edinburgh, and seems to have spent the summer in Unst, the neighbouring island to Fetlar. He stayed with the Edmondston family at Buness, whom he knew well. Jessie Saxby, who was born an Edmondston and was later to become an author, was older than William, and recalled his staying with them to prepare for his university course. It was probably at this point rather than when he enrolled at Aberdeen University, as he had fallen so far behind as to need help in reaching the required level. She recalled,

> I often saw Watson Cheyne as a boy, as he used to come over from Fetlar to visit at Buness. One year he spent some months there that he might be "coached up" before starting on a University course. Our American cousin (Tom Edmondston, who eventually became Laird of Buness) was a very clever and highly educated young man, and he had a gift of imparting knowledge to boys. He told us that Willie Watson (as the great man was spoken of in those days), "gobbled up languages, and all the terrible isms, like sweetmeats!" It did not seem to matter how profound, or how hard to understand, were the books put in the boy's hands; he mastered them all without - seemingly - much trouble.[140]

When the time came to enrol at Edinburgh in the autumn of 1870, 'Willie Watson' found lodgings alongside John Webster, in the house of Mrs Elizabeth Pollock and her daughter at 4, Montague Street.[141] He was about to embark on studies at one of the most prestigious medical schools in the world, with an impressive array of alumni and staff, including Charles Darwin [142]. James Young Simpson [143], the first to use chloroform successfully as an anaesthetic, had studied there, and Arthur Conan Doyle would be a student there in the mid-1870s, just after Cheyne. Part of the inspiration for Sherlock Holmes is said to have come from Joseph Bell, also a lecturer in Medicine at Edinburgh, well known for picking out people at random and correctly guessing their profession using deductive techniques.[144] Edinburgh University was a place where legends were born, though it had attracted ill-founded criticism from the English universities that its medical examinations were too lenient.[145]

Most significantly of all, Joseph Lister had trained there under James Syme, and after a period as Professor of Surgery at Glasgow, had returned to the capital in 1869 as Syme's successor to the Chair of Clinical Surgery.[146] By the time Cheyne enrolled at Edinburgh University in 1870, Lister was attracting a considerable following.

3

Discovering Lister

A revolution in surgery

Prior to the mid-19th century, surgeons did not attempt particularly complicated operations on a routine basis, and areas such as abdominal surgery were considered only in emergencies.[1] Prior to the development of anaesthesia in the late 1840s, the patient had been conscious during an operation, and we have become familiar with the rather sordid image of operators racing against time. It tends to be portrayed as a sort of theatrical performance, with celebrity surgeons competing for the best times to the cheers of an audience in the wings. Robert Liston in London, hailing originally from Scotland, was an acknowledged champion in terms of speed.[2] Important advances in surgery in the mid-19th century, however, heralded a period of change. The steps to safer surgery are generally cited as anaesthesia, antiseptics and advances in nursing and public health. More recently, however, a simplistic "before and after" image has been questioned in favour of a more contextualised assessment, taking into account the social background and a series of complex, interlocking advances in both theory and practice.[3] In fact, even 19th century surgeons who supported the same basic methods could differ considerably in their theoretical understanding and approach.

The world into which Cheyne stepped in 1870 when he began his studies at Edinburgh University was riddled with competing arguments. Part of the problem lay in the fact that many surgeons and general practitioners were interested in *how* things should be done, but not necessarily *why*. Without universally-accepted evidence for theory, surgeons tended to base their practice on experience of the techniques that seemed to give them the best results, and on the teaching they received during their training. It was not that these practices lacked a theoretical background or framework, but this did not concern them as much as the apparent efficacy of the treatment. They simply wanted their patients to get well. Their reputation - and living - depended on it. At university level, professors differed widely in their acceptance or rejection of the theories underlying the different forms of treatment, even within the same establishment. Students had to choose their loyalties, sometimes on the basis of genuine thought and evaluation, and sometimes on the approach more likely to give them a pass mark in exams.

Where patients were concerned, people who could afford private health care were generally treated at home, and hospitals were more likely to be full of charity patients and the less well-off. In an ever more industrialised society, surgeons in hospitals spent a good part of their time operating on wounds from accidents, as the general public was exposed to greater opportunities for injury at work or in the streets. There can be little doubt that, by modern standards, hospitals harboured considerable risk of infection in the mid-19th century, particularly when it came to surgery. Patients were often victims of sepsis in wounds, for which there was no universally recognised explanation.

On the one hand, a series of breakthroughs laid the foundation for improvements and greater patient confidence. On the other, the portrayal of these improvements as a combined revolution, before which all was darkness and after which all became light, belies their complexity and is apt to distort the real achievements. There is even evidence that the situation was later exaggerated in order to extol the improvements, particularly by those who had been involved in them [4], and matters were not helped by the competing, and sometimes acrimonious camps of opinion which formed in medicine, surgery and public health. William Watson Cheyne was to become a major protagonist in these debates, and the way in which he later recalled his experiences was very firmly rooted in his loyalties and the dilemmas they sometimes presented.

He later painted a bleak picture of his first experience of a hospital ward, calling it "a place in which most of the patients were visibly ill, with flushed faces, parched lips, delirium, severe pain, etc., and many of them were evidently on the verge of death; the wards were pervaded with a peculiar mawkish odour which was very trying to newcomers."[5] If this were not sufficiently alarming to a modern audience, his description of surgical procedure could hardly fail to raise a few eyebrows today:

> ... the surgeon kept an old frock-coat hanging up outside the door of the operating theatre, which he put on on his arrival at the hospital; it was covered with dried blood and dust and dirt, and the bloodier it was the prouder was the surgeon. He turned up the sleeves and the coat-collar, and was ready to operate. The

instruments were laid out on a table; no doubt they were washed before they were put away, but there was no special preparation of them before being used again.[6]

Patients sometimes stayed in hospital for months, so they were exposed to considerable opportunities for infection [7], especially as it was not unusual for two or three people to share a bed.[8] Even when Cheyne began his studies, not all surgeons routinely washed their hands in disinfecting solution before operating on a patient, or before moving from one infected wound to another, though some were aware of the benefits of cleanliness. John Rudd Leeson, a contemporary of Cheyne, who began his studies at St. Thomas' Hospital in London, but ultimately moved to Edinburgh, recalled,

> I remember the house surgeon in the theatre with his bevy of threaded needles dangling from the front flap of his coat, half a dozen of them ready for use, the silken threads sweeping the well-worn cloth which had grown old in the presence of sepsis.[9]

He went on to say that students went straight from the post-mortem room to attend midwifery cases, and although there was a sign saying they should wash their hands in chlorinated solution in between, Leeson "never knew of anyone doing it, nor was there any intimation as to where or how it was to be obtained".[10] Significantly, the students did not have "the least idea of the meaning and reason of the notice."[11] In short, Leeson's early training did not include an explanation of why disinfection was important.

The first of the major breakthroughs in surgery was the incorporation of anaesthesia. Before this, the pain was unbearable. There had been a general recognition that one of the biggest drawbacks of surgery was the need to control patients during an operation, and subject them to medical "discipline".[12] A patient crying out and writhing in agony did not make it easy for a surgeon to focus on the task in hand. After a brief flirtation with a trance technique known as "mesmerism", ultimately condemned by the medical establishment as unscientific [13], a method using ether was imported from the United States, where it had been successful in dentistry.

In December 1846, Robert Liston, Professor of Surgery at University College Hospital in London, gave a public demonstration of the application of anaesthetic by ether in surgery. He hoped to knock the "mesmeric quackery"[14] off its pedestal. His operation of the thigh under anaesthetic reported identical results to those of the American dental operation. The patient "moaned and stirred restlessly, but did not cry out".[15] Liston's response, in a letter to James Miller of Edinburgh, was a jubilant "Hurrah! Rejoice! Mesmerism, and its professors, have met with a "heavy blow and great discouragement.""[16]

Liston's work was adopted with enthusiasm by the medical journals, which thereafter effectively shut out mesmerism altogether. The anaesthesia debate centred instead on how to restrict ether to its scientific use in the operating theatre, and to stop it spreading to the general populace, who had become entranced by the idea for less professional reasons.[17] Only shortly afterwards, however, in November 1847, Sir James Young Simpson, Professor of Midwifery in Edinburgh, developed chloroform as an alternative to ether, and within weeks of the publication of his paper, it had almost entirely replaced the former methods in operating theatres. When Queen Victoria was administered a chloroform anaesthetic as she gave birth to Prince Leopold, any remaining doubts the profession and the public may have had were effectively put to rest, and it was universally employed.[18]

Effective anaesthesia was a turning point. It gave the surgeon time to develop more careful procedures, and it laid an important foundation for advancing the understanding of anatomy and disease. However, it did not reduce the number of surgical deaths. Firstly, as many surgeons were prepared to operate on cases hitherto considered too dangerous, more operations were taking place. Surgeons were experimenting more, not always with successful results, because anaesthesia did not address the underlying causes of post-operative infection. However, it did help to focus medical attention more on the thorny question of why wounds so often turned septic.

In the same year as Simpson first used chloroform, a Hungarian surgeon in Vienna, Ignaz Semmelweis, recognised that medical practitioners were transferring infection between women in childbirth by not washing their hands between patients, and he dramatically reduced mortality rates by washing his hands in a chloride of lime solution.[19] Semmelweis' colleagues largely rejected his findings, and he himself was slow to publish a response. Tragically, he was forced into a mental asylum as a result of his frustrations, and died of septicaemia at the (unwashed) hands of a surgeon who was either one of his many detractors or who had never heard of the debate in the first place.[20] Semmelweis' ideas were interpreted in Britain simply as confirmation of what British surgeons had presumed all along - that infections in childbirth were contagious.

Microbes and the Fetlar Man

The most vehement attack on Semmelweis came from Simpson, who had himself recognised that puerperal fever (infection at childbirth) spread between patients, and insisted his staff and students use chlorine disinfectant to prevent it.[21] He was so angered by a letter from one of Semmelweis' advocates implying no knowledge of Simpson's own work, that he was stimulated to compile statistical evidence to show that infections in childbirth were essentially the same as the post-surgical infections he called "surgical fever",[22] and that, if so, something might be done to reduce the unacceptably high post-surgical mortality rates. He coined the phrase "hospitalism"[23] to refer in general to hospital-related infection. Simpson's conclusion was that existing hospitals needed to be torn down and redesigned to include areas where patients could be isolated in contagious cases. This particularly drastic solution was apparently carried out to the letter in Lincoln.[24] St. Thomas' Hospital in London, however, was the first to be rebuilt on the "pavilion system"[25], with air circulating more freely around the wards.

Florence Nightingale was an ardent supporter of this movement. Like Simpson, she was active in changing the underlying concepts of hospital design, to which she added systematic training for nurses. She founded the School of Nursing at St. Thomas' Hospital in London, where young women, mostly from the middle classes, were given a solid grounding in efficiency and cleanliness. From here, they were sent out to hospitals all over the country, where they passed on the practices to existing staff. Nightingale had read Simpson's statistical work [26], and even repeated his methods, producing some statistics which were not particularly accurate [27]. She nevertheless recognised the importance of this type of evidence in changing practice.

Most of her post-Crimea working life was devoted to hospital design, and this is a key factor in assessing how she understood disease. She agreed with Simpson that hospitals needed to be completely redesigned, and considered that air and light were the most important elements in preventing infection. She remained uneasy with the concept that germs could be the root cause of disease, and her views were informed by the theoretical background of anticontagionism, in which "miasms" in the air were thought to spread infection. In her *Notes on Nursing* (1860), she explained:

> Go into a room where the shutters are always shut, (in a sick room or a bedroom there should never be shutters shut), and though the

room be uninhabited, though the air has never been polluted by the breathing of human beings, you will observe a close, musty smell of corrupt air, of air, i.e. unpurified by the effect of the sun's rays. The mustiness of dark rooms and corners, indeed, is proverbial. The cheerfulness of a room, the usefulness of light in treating disease, is all important.[28]

This external cleanliness also improved patient morale and went a long way towards excluding some of the risks of infection so prevalent in the old hospitals. New hospitals were designed to higher standards, and were more purpose-built to encourage a healthy environment for patients. A "Nightingale Ward" was characterised by well-spaced beds with windows between, and heaters in the centre of the ward, which created up-currents of hot air, to help displace impurities and circulate fresh air.[29] Simultaneously, an architect, George Godwin, took a strong interest in hospital design, and Florence Nightingale quotes a number of his ideas in her *Notes on Nursing*.[30]

The underlying causes of infection, however, persisted, and patients continued to die in hospitals as a result of infection sustained during surgery and post-operative care. A paradigm shift was nevertheless gathering momentum which would affect how diseases were understood and which, can "be seen as moving from defining diseases by their symptoms and results to defining them in terms of processes and causes."[31] What caused disease, and why did it affect some and not others?

The responses to this often depended on the issues in question. In the growing public health sector, the debate centred largely on outbreaks of disease, such as cholera, and how to control their spread. A body of opinion known as contagionism supported the view that disease spread from infected individuals - animal or human - and contagionists were therefore in favour of quarantine to combat outbreaks. Anticontagionists, on the other hand, promoted the idea that "miasms" in unhealthy places could penetrate the body and cause disease, but that a healthy body was more likely to be able to resist. This was often mixed with more than an element of morality. A physically healthy body depended in part on moral health, temperance and good diet, as well as a clean living environment. In other words, many of the disease control measures focused on improving the environment in which diseases could arise, alongside persuading people to lead healthy and morally upright lives.[32]

For surgeons, who were interested mainly in wound management, the debate centred on how to prevent surgical and accident wounds turning septic. Because there was no consensus on what caused septic diseases, methods of avoiding them varied, with inconsistent rates of success. Some methods were "highly individualistic"[33] and if the wound healed by first intention, with no inflammation, the process was considered successful and could be repeated. Though many surgeons avoided theory where they could, there were nevertheless theoretical assumptions behind their methods. Many worked on the principle that septic conditions arose *inside* the body, and were the result of spontaneous chemical processes in damaged tissues. They assumed that healthy tissues could not be affected, but were vulnerable if they were damaged by adverse factors such as abnormalities or bad nutrition. On the other hand, unhealed, open wounds were subject to inflammation and putrefaction because they involved dead tissue.

To reduce inflammation and prevent sepsis, surgeons tried to reduce "unhealthy" inflammation and encourage "healthy inflammation" and the natural healing processes of the body. They treated the wound with a variety of poultices, cold water dressings and ointments, and promoted good diet and the strengthening of the body in general. When wounds turned septic, however, they sometimes turned to chemical antidotes, or antiseptics, which were already in use in the mid-19th century to attempt to counteract sepsis when it had already set in. In 1853, for example, Golding Bird, a Fellow of the Royal College of Physicians, advocated the use of "undiluted nitric acid or caustic potash".[34] This, then, was the part of the general background to British medicine when Cheyne embarked on his studies.

When he finally decided to enrol at Edinburgh University, he was committing himself to around four years of study. Medical degrees had been standardised considerably as part of the process which had regulated the profession as a whole in the mid-19th century, and students followed "specific patterns of study laid down by the General Medical Council."[35] The 1858 Medical Act required doctors to qualify and enter themselves on a national register, a move aimed partly at raising standards in the profession, and partly at making it easier for qualified doctors to attract patients without having to compete with unqualified practitioners.[36]

Presumably some of Watson's settlement was still available to help finance Cheyne's studies, but the interest on the capital sum was not going to be enough. Cheyne did not have access to the capital itself until he was 21, and when he began in Edinburgh, he was barely 18. The cost of a full medical education in Edinburgh in the early 1870s could vary, but was estimated at around £300 over a minimum of four years [37], especially if students were not able to live at home and had lodgings, as was the case with Cheyne.[38] Divided between the four years, this was probably around £75 a year. He had to buy "a minimum of eight books, and an array of medical equipment, for example, including a microscope. The cost of these necessities was around £15."[39] The combined legacy of his grandfather amounted to just under £780, including unrealised assets and debts owed him. Divided between the three beneficiaries, this was around £260 each.

Nor had Cheyne's father left him any money. There was none to leave. After his last vessel, the *Acis*, was sold at auction to cover debts and the crew's wages, the total legacy amounted to a mere 17 dollars and 78 cents. The Acting Consul in Manila sent this on to the Board of Trade either late in 1866 or early 1867.[40] The information handed down to members of the Cheyne family suggests that the Websters asked for an enquiry [41] to ascertain whether any of the land in Palau could be redeemed. The correspondence between the Consul in Manila and Andrew's family in Northmavine, however, suggests that Henry Cheyne, Andrew's half brother with whom he had grown up, maintained the quest for information from early 1867. He wrote to the Consul in Manila on 5th April asking about the estate and effects, but was told that, although lengthy deeds to land which belonged to Cheyne in Palau were available at the Consulate, the property would be "very difficult to realize".[42] The Consul, Mr Ricketts, considered that, "The natives of the Pellew with whom land is of little or no value would I fancy hardly repurchase it, and no Englishman would, I should think, deem it advisable either to purchase the property or to reside there."[43]

Moreover, there is some indication there were legal issues concerning Andrew's half sister, Elizabeth, who was still resident in Ollaberry on the Shetland Mainland. It has been suggested that Captain Cheyne's papers were divided and that titles regarding lands in Shetland were sent to his son's trustees, whereas commercial papers may have been sent to his half-sister.[44] In any case, it appears there was little either the Websters or the Cheynes could do about it, and Cheyne received no real material legacy from his father except some land at Nounsbrough, which would have to be sold.

Cheyne had, however, inherited from his grandfather a small share in other land in Shetland, along with his two aunts, and it realised some income in rents from tenants. The Reverend William Watson had been considered reasonably well off for a minister, on "200 pounds a year"[45], so if Webster in any way emulated this sum, at least during this period, they had a reasonable living. They employed at the Manse a housemaid, a general domestic servant, a dairy maid and a farm servant.[46] They were nevertheless still financing the education of their other charge, John Webster, who had graduated with an MA from Aberdeen University in 1869 [47], but continued his studies and was not licensed by the Church until 1874.[48] He was boarding at the same Edinburgh residence as Cheyne in 1871, so the family had to find the money for both sets of fees and living expenses. The time would come when land had to be sold to help finance it all.

Nor was it necessarily a secure investment from the point of the view of the Websters. When Cheyne began his studies in Edinburgh, he was initially prevented from taking the full medical degree because of symptoms of "incipient tuberculosis".[49] It was a dangerous profession, and students were open to a variety of sources of infection. Graduation in medicine required practical experience as well as lecture attendance and exam passes, and many young doctors, even students, died in the course of their work. Crowther and Dupree have noted that "[m]ortality among young practitioners was excessive, with tuberculosis the main cause ..."[50]

Cheyne himself counted his actual medical studies from May 1871 [51], but his certificates show that he took courses in botany and junior chemistry in the winter session of 1870, beginning in October.[52] Students at Edinburgh were recommended to start in May, for the summer session [53], but Cheyne decided to begin in October, with preliminary courses. The idea of starting in May was to avoid a bottleneck of exams over the course of time.[54] To obtain their qualifications, medical students at Edinburgh were required to attend 100 lectures in each of the main subjects and 50 lectures in botany, natural history and medical jurisprudence, adding hospital and dispensary experience, and attendance at births.[55] They were also required to sit three 'professional exams', incorporating the group of subjects they had studied to that point. For example, the first professional exam incorporated botany, natural history and chemistry, and was taken after the second year of study, before proceeding to clinical subjects.[56]

George Skelton Stephenson from Grimsby, a contemporary of Cheyne's at Edinburgh, described his entrance exams in March 1871 in "the fine library of the university"[57], and highlighted the general nature of the education students were expected to bring with them to their studies. These preliminary examinations lasted three days, and included English, German, Algebra, Latin, Greek/Logic and an examination on Mechanics, Hydrostatics and Pneumatics. The English exam involved an essay on either 'Ascent in a Balloon' or 'The Siege of Paris', and the German test was a translation from Schiller's *Wilhelm Tell*.[58] Success in these exams allowed the students to matriculate.

Discovering Lister

Thereafter, presence at lectures was compulsory, as well as good performance in exams, and if students persistently failed to attend, they would not obtain their "class ticket".[59] Certificates included a comment on attendance alongside the exam mark and specification. Cheyne's first certificate in junior chemistry notes that he "acquitted himself with distinction"[60] as a member of the class, and obtained an exam mark of 84%. In botany, he had attended the lectures "with diligence"[61], attained 78.6% in the exam and walked away with First Class Honours and the University medal. He had started as he meant to go on, whether medicine was a means to an end or not.

Botany was taught by Professor John Hutton Balfour, a venerable old man with a flowing beard who accompanied his students on "botanical excursions"[62] into "remote districts"[63] on Saturdays. He was popularly known as "Old Woody Fibre".[64] Stephenson recalled that Balfour showed them "the principle geological features of the country"[65] as they walked and that the old professor imparted philosophy, "deep wisdom"[66] and "sound religious enthusiasm"[67] along the way. The students took sandwiches and a "spud"[68] for lunch. He could be strict with his students when it came to exams [69], however, so Cheyne's 78.6% was clearly merited.

Certificate for Natural History from the University of Edinburgh

Cheyne's health may have improved by May 1871, when he registered for natural history. He was once again described as having "acquitted himself with distinction"[70] and came away with 75%. His teacher was Professor Wyville-Thomson, a "courteous, genial and kindly gentleman"[71] whose lectures, according to Stephenson, "were beautifully delivered in the most choice English".[72] He was nevertheless, like most of the lecturers, distant with his students, who had little direct contact with him and generally only saw him on lecture days. Shortly after Cheyne had completed his natural history course, Wyville-Thomson sailed in December 1872 as instigator and chief scientist on the first major expedition to study the deep sea on *HMS Challenger*[73], borrowed from the Royal Navy and refurbished on behalf of the Royal Society. Cheyne must have been especially impressed by this on two counts. Firstly, the privilege of going to sea as a scientist, and secondly, the fact that the ship was equipped with natural history and chemical laboratories, as well as a darkroom for developing photographs.[74] All these things were to become a focus of interest for Cheyne himself in his studies and in later life.

Cheyne's notes survive from Professor Wyville-Thomson's 1871 lectures. They define the terminology for the study of the animal kingdom, and give us an idea of the content of the exams he had to pass in his first year of study. For the first class exam in the Junior Division on June 8th 1871, he had two and a half hours to answer eight questions. After defining a number of basic terms, presumably word-for-word as Wyville-Thomson had dictated in his lectures, candidates had to discuss, among other things, pseudopodia, coelenterate and medusidae, and finish with a question on the structure, zoological relations & history of the tape worm "in the condition in which we find it in the intestine of Man."[75] He then had a general class exam on Thursday 6th July, in which he had two and a half hours to answer six more questions in the Junior Division and two and a half hours to answer six in the Senior Division, by which time he had progressed to discussing the "mode and origin of the placenta."[76] The notebooks contain drawings, sometimes on the back of card he had clearly acquired for other purposes. One of these cards shows his early interest in music, with spaces for him to note operas and their composers.[77]

Chemistry at Edinburgh was taught by Alexander Crum Brown, who apparently had difficulties with class discipline.[78] Cheyne, on the other hand, was clearly not part of the general merriment. He was apparently so interested in the subject, he elected to take it again at the beginning of his second year. The *British Medical Journal* would later pick up on this point, saying "he was so devoted to chemistry that not only did he get the first prize amongst the many students at the University of Edinburgh, but … he worked for a second year at chemical science, and again won the prize."[79] In fact, he managed 98% in the senior chemistry exam, taking the first medal.[80] Cheyne's interests already lay in the scientific side of medicine, and his 'devotion' to chemistry would not go amiss, but it also indicates that he was beginning to find his feet, and a focus for his life.

University of Edinburgh.
CLASS CERTIFICATE OF MERIT.
FACULTY OF MEDICINE

I Certify that Mr William Watson Cheyne acquitted himself with distinction as a Member of the Class of Chemistry during the Winter Session 1871-72 and obtained 98 per cent of the available marks at the class examinations and a medal, in the senior division of the Class

Alex. Crum Brown
Professor

Cheyne's certificate for Chemistry from the University of Edinburgh.

Alongside chemistry in his second year, he registered for anatomy, taught by William Turner, who had more success with discipline in his classes than Crum Brown. He would simply stop speaking and stare down any interruptions until the offender was shamed into submission.[81] His main interests were in aspects of evolution: comparative mammalian anatomy, anthropology and craniology.[82] In his second year subjects, Cheyne outdid himself, and almost everyone else, with 83% in anatomy [83] to go with his spectacular chemistry result.

We should not assume results like these were easily obtained. "Teachers, examiners and the GMC (General Medical Council) expected a high failure rate."[84] During the period in which Cheyne studied, the drop-out rate for male medical students at Edinburgh (and most were male), even after counting those who died, was over 30%,[85] and failure in the professional exams was one of the most likely causes. Over half the students registering for medical study in England and Scotland in 1871 (the same cohort as Cheyne), took longer than the minimum four years to qualify.[86] Some chose to take extramural courses, or to take a course twice for practice, but many of them were forced to repeat exams because of failing to reach the required pass levels.

Teachers were not given to leniency. English critics had suggested that Scottish medical qualifications were inferior and that examiners were lax, prompting a defence of the system from Scotland. William Turner argued that Scottish qualifications were "based on a more extended system of education, scientific, practical and clinical"[87] with "a wider range of subjects"[88]. With standards to maintain in the face of criticism, teachers were unlikely to be too forgiving.

University of Edinburgh.
CLASS CERTIFICATE OF MERIT.
FACULTY OF ARTS.

I Certify that Mr. Wm. Watson Cheyne acquitted himself with distinction as a Member of the Class of Anatomy during the Winter Session 1873-4 and obtained by acting as one of the Class Prosectors

Wm. Turner, Professor

Cheyne's certificate for Anatomy.

In October 1872, Cheyne finally began to study physiology and surgery.[89] James Spence was Professor of Surgery. He was popularly known to his students as "Dismal Jimmie"[90], and even Cheyne, a model student, referred to the lectures on surgery as "very dreary performances full of curious theories about the reactions of the body and inflammation, quite unintelligible to me, evidently something which had to be memorised but nothing to think about."⁹¹ This did not prevent Cheyne from achieving 96% in the exam in the junior division. Spence was exonerated by George Stephenson who apparently "often heard that he had a very keen sense of humour, and that he could so apply himself, after dinner, greatly to the enjoyment of his guests who were always prominent scientific men."⁹²

Cheyne's certificate for Junior Surgery.

Social life for students in Edinburgh in the 1870s has been described in a number of memoirs, though, as Crowther and Dupree note, these tended to provide "a filtered view of student life"⁹³, avoiding mention of the many temptations of the city. Conan Doyle, on the other hand, suggested that students living in lodgings were given a "preparation for life"⁹⁴, by which he meant it "ruins some and makes strong men of many."⁹⁵ George Stephenson, whose account of his student days is very much a 'filtered' one, recalled rapturous visits to historic sites and regular bathing at Granton.⁹⁶ There were some organised sports at the University by this time, notably rugby and cricket clubs ⁹⁷ but, given his delicate health, Cheyne is unlikely to have been an active member. He is more likely to have been interested in a university boat club ⁹⁸, though he would have found it very tame after Shetland waters. Stephenson and two friends had a "covered-in boat, the *Norna*"⁹⁹ and took "pleasant sails up and down the Forth"¹⁰⁰, but he never mentioned Cheyne as one of his circle. Stephenson came from Grimsby, so was also familiar with the choppy waters of the North Sea.

Cheyne was a shy student, as he admitted himself [101], which may have helped him to study diligently and remain largely undistracted by the temptations of the city. On the other hand, many of his contemporary students complained there was nothing to do in Edinburgh but study. Cheyne said that, in the absence of a students' club until the mid-1870s, he initially tried to spend periods between lectures in the library to escape the weather.[102] In the evenings and at weekends, for those who could not afford the annual two-guinea fee to join the Royal Medical Society and listen to the weekly papers and debates, walks and visits to places of interest seem to have been two of the main pastimes.[103] Cheyne left few records of social acquaintances he may have made during this period, and few of the names of his fellow students resonate in his later writings unless they became professional colleagues.

University of Edinburgh.

FIRST CLASS CERTIFICATE OF MERIT.

FACULTY OF MEDICINE

I Certify that Mr W. Watson Cheyne acquitted himself with high distinction as a Member of the Class of Surgery (Senior Division) during the Winter Session 1874-5 and obtained first Class Honours. 99 per Cent – University Medal & First Pr

James Spence   Professor

Cheyne's certificate for Senior Surgery.

Life was lived very much off campus in the 1870s, and students withdrew to their separate lodgings after lectures.[104] Stephenson recalled that they had to "cast about for lodgings"[105] and that finding good ones was a matter of luck. Cheyne found accommodation with his guardian's son, John Webster, in the house of a Mrs Elizabeth Pollock at 4, Montague

Street.[106] There is no evidence that the two young men were particularly close or had much in common. Webster was studying for the ministry at Divinity Hall.[107] Perhaps because of his own shyness, Cheyne dedicated himself entirely to his studies. This is how he was living in April 1871, when the census was taken, but later on, when he wrote out his notes for Professor Wyville-Thomson's lectures, possibly by the end of the year, he gave his address as 22 W. Preston St.[108] There may have been financial constraints by this time, and these may also have resulted in pressure for him to complete his degree as soon as possible, meaning he would have to study all the harder. If any courses had to be repeated, they incurred the costs of an extra term. He referred to himself during this period as "a shy, uncouth youth without any influence or money."[109]

Whatever the reason, Cheyne seems suddenly to have left behind his early waywardness and lack of focus, and had clearly made a decision that, since he must make his way in the world by some means, he would put his mind as best he could to the study of medicine. He was probably still encouraged by the thought of using his degree to take him to sea, but was nevertheless almost resigned to his studies, giving them his best as a means to an end. Some of the classes clearly fascinated him, but in others, he was frustrated by the lack of stimulation.

All this was to change dramatically through a chance encounter which would divert the course of his life for ever. The vivid tale of how he first became acquainted with the work of Joseph Lister, Professor of Clinical Surgery, is by now so well known to medical history that I would be doing Cheyne an injustice if I were not to allow him to tell it in his own words. He was, above all else, a gripping storyteller, perhaps a legacy of all those tales he had heard in Fetlar from the men home from sea. Even his systematic scientific arguments were peppered with illustrative anecdotes and explanatory sidelines. He never lost this extraordinary ability to bring facts to life. His talk for the first Lister Memorial lecture, many years later, would plot the history of the development of antiseptics as a tale of wonder: "The story that I have to tell to-day is one of the most fascinating tales that I know, and far exceeds in interest the most exciting detective story ever written."[110] Here then, is his account of the first of the moments which would change the direction of his professional life.

> ... in October 1872, when I was in my second year, I took out the classes of surgery, physiology and anatomy. There was an interval between the classes from twelve to one o'clock, and I did not know

what to do with myself during this interval. My lodgings were a considerable distance away, and there was no object in walking there and back; what I wanted was shelter from the biting east winds and the wet. At that time there was no Students' Club at Edinburgh, as there is now, and the only thing I could think of was to go and sit in the general library. But it was small and crowded with students of all faculties, and there was no chance of a shy youth like myself getting a seat. So it occurred to me that I might as well take out my hospital ticket in October 1872 as in the following May, which would be the usual time; in going round the hospital I might learn something, and anyway I would get shelter from the wintry blasts between noon and one o'clock.

I shall never forget the first day that I went to the hospital. At twelve o'clock I crossed the road to the infirmary and found crowds of students hurrying along the lobbies, a good many, indeed, running. I joined the company, full of curiosity, and found that we were going to Lister's clinical lecture. We ultimately reached the large operating theatre and found it filled with men crowding the gallery and standing at the back and in the gangways: there must have been over 200 there, and they were senior men, for the clinical surgery class was not usually attended till the fourth year. I should think that I was the only one of my year there. Presently a man walked into the arena, sat down on a chair placed on the floor of the theatre, and, crossing his legs and putting his right hand on his thigh, he began his lecture. This man was Lister. He had not spoken many minutes before I became fascinated with the subject and could not help seeing the wonderful era that was dawning for surgery. I had my notebooks for the other classes in my hand and I started to take down what he said, and, as he spoke very slowly and clearly and I scribbled very quickly, I was able to take down practically the whole lecture, though, not having learnt any surgery, I did not understand it at all clearly at first. But what a contrast with the lectures on systematic surgery! ... And then at twelve o'clock, simply seeking shelter from the "cauld blasts o' Auld Reekie" and not knowing that I had met my fate, I became entranced by the wonderful vision laid before us by Lister.

Is it any wonder that I left the theatre an enthusiast for the profession that I had chosen, and that on lecture days I was one of those who ran? An enthusiast I still remain. On my way home that first afternoon I bought a superior notebook into which, as soon as I got home, I copied out my notes of Lister's lecture and all that I could remember, looking up the words I could not understand in my text-book on surgery. This I continued to do after every lecture, and this notebook, containing Lister's lectures for the winter session 1872-73, and part of the following summer session, I have presented to this College [The Royal College of Surgeons of England], and it will be found with the Lister relics if any one cares to look at them.[111]

It may be useful at this point to summarise the background to Joseph Lister's lectures, and what it was about them that made Cheyne into such an instant convert. Lister has been widely regarded as the British equivalent in scientific stature of his contemporaries Louis Pasteur in France and Robert Koch in Germany, and it was almost certainly his experimental work and its implications for surgery that excited Cheyne most. Lister had been brought up in an enquiring household. His father was involved in optics and microscopy, and pioneered improvements in achromatic lenses.[112] They were Quakers, pacifists, and as a child, Lister was encouraged in academic and scientific pursuits. At Grove House School in Tottenham, he learned French and German, which gave him access to two of the leading languages of scientific research in the 19th century.[113] He also became an accomplished scientific artist.[114]

Lister graduated from the University of London in 1847, and had been present at Robert Liston's first demonstration of the use of ether.[115] He resumed his studies in 1849 after a period of stress and breakdown following an attack of smallpox, but returned to his studies too soon.[116] He nevertheless won a number of prizes, and ultimately became a dresser at University College Hospital. He was interested in and - as his disciples would later eulogise - moved by cases where a patient came successfully through an operation, only to die from one of a variety of forms of infection in the wound. His interest in the subject led his professor, William Sharpey, to advise him to visit James Syme [117], one of Edinburgh's leading surgeons. Syme had achieved considerable success with amputations and advocated washing his hands on a clean towel between operations.

Lister eventually became Syme's assistant or "supernumerary clerk"[118], and married his daughter Agnes, forming one of the most effective working partnerships in the history of science. Lister never recovered from Agnes' death in 1893, and her immense contribution has now been recognised. Throughout their married life, she documented the results of her husband's experiments, effectively allowing him to conduct research alongside his hospital work. From Edinburgh, Lister moved on to take up the post of Professor of Surgery at Glasgow University, where he was able to turn his attention seriously to the subject of septic diseases in wounds.[119] He had already conducted important scientific work on inflammation and coagulation of the blood, for which he had been made a member of the Royal Society in 1860, when he was only 33.[120]

Inflammation was considered to be an important element in sepsis, though, as I have noted, "healthy"[121] forms of inflammation were considered beneficial to healing, and were encouraged. Lister believed that "sepsis was closely bound to putrefaction of the blood and discharges in the wound"[122] and that, if he could prevent putrefaction, he could most probably address sepsis. His initial attempts to do so involved making the surgical environment as clean as possible and conducting fairly random trials with antiseptic chemicals on dressings. Cheyne later noted that Lister insisted on "piles of clean towels … in the wards, and plenty of water and basins, and those engaged in dressing the cases had to wash their hands frequently. He also used kettles of boiled water, often mixed with Condy's Fluid [a permanganate disinfectant manufactured in London], with which he tried to wash out the putrefying discharges, and … administered sulphite of soda as recommended by Polli."[123] The methods, however, met with no immediate success. It was all trial and error.

The most fundamental theoretical debate on putrefaction was whether it was caused spontaneously by chemical reaction, a stance known as spontaneous generation, or whether microscopic life forms (germs) were generated from pre-existing life. The debate had been important in France in the 1860s, particularly in the work of Louis Pasteur on fermentation. He maintained that pure organic fluids (milk, for example) only decomposed in flasks if they were contaminated by the air, and that they remained pure if the flasks were sealed and excluded air. Importantly, the basis for Pasteur's thinking was that it was not the air itself which contaminated the liquid, but microorganisms living in it. His opponents insisted that, if life forms were found in these liquids, they were the *result* of chemical decomposition, not the cause of it.[124] In other words, they believed these organisms arose from a chemical reaction, not from other organisms.

Around 1864, according to Cheyne, as Lister was walking home with his colleague Thomas Anderson, Professor of Chemistry at Glasgow, the two men began to discuss the causes of putrefaction.[125] Anderson referred him to some recently published papers on fermentation and spontaneous generation by Pasteur. It was not usual for surgeons to be familiar with theoretical developments in chemistry. Cheyne later suggested that Lister had not even been following the spontaneous generation arguments at the time.[126] However, when he read Pasteur's papers, he discovered that

> ... it was not the gases of the air [miasms] that the surgeon had to fear, but minute living particles floating in the air and settling on surrounding objects in the form of dust. And he learned further that so long as these particles were excluded from the putrescible material after it had been boiled, there was no sign of living organisms or putrefaction in the material. On the other hand, as the protection was removed and the dust was allowed free admission, minute bodies belonging to the lowest class of vegetable life could be seen swarming in the fluid, and putrefaction very quickly set in.[127]

Lister was not the first to make a connection between Pasteur's work and the use of chemical antiseptics to exclude harmful microorganisms from wounds. However, for a number of reasons, he had the most impact. In an address to the British Medical Association in August 1864, a surgeon named Thomas Spencer Wells had linked Pasteur's published work on "germs, fermentative change and disease"[128] to sepsis in wounds, indicating that living organisms played a role. Earlier still, in France, in 1861-2, Trousseau had linked microorganisms to puerperal fever, which was killing women in childbirth.[129] Wells advocated the use of antiseptic substances on wounds to counteract the effects of these organisms, in order to make the tissues of the body more resistant to them. However, according to Cheyne, he did not "follow up the logical conclusions which he drew from Pasteur's works"[130]. He dwelt instead on the implications for hospitalism and the visible "cleanliness and purity" of hospitals.[131]

Lister, on the other hand, set about testing the relationship of microorganisms to fermentation and putrefaction through laboratory experimentation, first on urine which is naturally sterile. He kept the liquid in flasks in his house and excluded the air. Leeson recalled seeing an array of test tubes on a table in the window of Lister's dining room, stopped with cotton wool and containing liquids in varying states of purity or decay.[132] According to Leeson, these flasks of "aged urine"[133] later accompanied Lister, perched on his own and his wife's knees, on the train to London.

In the notes Cheyne made at Lister's lectures, it is clear that Lister imparted his methodology and its theoretical background to his students, and it is most likely to have been this, as much as the surgical content of the lectures per se, which so stimulated Cheyne's imagination. Lister introduced them to the work of Pasteur, and its implications for surgery.

Significantly, Lister's lecture of February 24th 1873 explained to his students some recent experiments he had made on fermentation in milk. This entry is revealing in the wealth of information it provides about how advanced Lister was in his methodology and inference, and that he was imparting this to his students. He took them through the methodology, his logical conclusions and their implications:

> During the Xmas holidays I took 5 flasks and heated each with a spirit lamp and also heated their glass caps and allowed them to cool below a shade. So with 6 test tubes. Then before the milk was shed I purified the milkman's arms, the cow's udder & milk ducts with carbolic acid. Unfortunately the cow did not yield her milk readily. Each test tube was also filled with milk by one squirt and the caps quickly replaced. The flasks went wrong but the flasks in cooling sucked in air. No 2 flasks are alike. Two contained bacteria one of them large bacteria, the other small ones. The others contained fungi of various kinds each one growing from one spot. One has Aspergillus glaucus another Penicillium glaucum &c ...
>
> ... The test tubes had a better chance. Development occurred in most. Two days after the Experiment the milk in one was perfectly natural but a mould proceeded from one spot on the side of the tube. I examined it after a time and found the milk quite sweet and that the organism had just touched the milk. This was after a month and afterwards it produced various changes in the milk. In one test tube the milk has curdled & the solid part has fallen to the bottom while the light goes to the top. In another the growth has produced translucency but little alteration in the quality of the milk. Hence each organism is associated with particular change.[134]

The significance of this work and the inferences Lister drew from it will become clear from developments detailed in later chapters, but it already shows how he was giving his students an introduction not only to experimental methodology, but also how to draw logical inferences from the results and apply them to surgery. Cheyne considered himself privileged, and though he seems to have been well above the average student in absorbing it, he was not alone. Many years later, other students felt the need to write accounts of their time with Lister, showing that they too saw something extraordinary in what he was saying; this comes out in the memoirs of Leeson and Stephenson for example. Compared to Cheyne's other lectures, Lister's teaching indeed represented something exciting, particularly in terms of the fact that they were being taught it at all.

From the moment of Cheyne's first attendance at Lister's lectures, he followed them almost religiously. He later said he considered his fair copies of Lister's notes "the most valuable possession I have"[135] though in a letter to the Wellcome in 1928 he gave this honour to the certificate from his Lister class.[136] Cheyne transcribed the notes as though Lister himself were speaking through the page, and they retain a striking immediacy. The great similarity in content and style between surviving student notes from the lectures has been taken to mean that Lister expected his students to write down everything verbatim [137], and their ability to do so was aided by the fact that Lister spoke slowly and clearly in efforts to overcome a stammer.[138] Cheyne's notes even retain the first person and we can almost hear Lister's voice reaching down to us through the centuries. We have from Cheyne's notebooks a sense of the pioneer in Lister and the enthusiasm with which Cheyne absorbed his words.

Lister's work had brought hope that sepsis could be prevented by excluding microorganisms from wounds if he could find a suitable substance to do the job. After a series of tests on possible chemical antiseptics, Lister had settled on carbolic acid, then known as German creosote, which had been used successfully in mitigating the damaging public health effects of the Carlisle sewerage system.[139]

He had run his first main clinical trials on cases of compound fracture, arguing that they came directly to the surgeon, and putrefaction had not had time to set in. In a compound fracture, the bone has penetrated the skin, forming a deep wound, often caked with dirt and bacteria from the scene of the accident. His first case was not a success, and he characteristically blamed his own procedures.[140] He had to wait another five months, until August 1865, for the next one. He finally treated a compound fracture on the leg of a boy who had been run over by an empty cart. Lister dipped a piece of lint in undiluted carbolic acid, and laid it on the wound. The leg was then left alone in splints for four days. When he lifted the lint, there was no real suppuration, and with further dressings of diluted carbolic, then water, the wound healed.[141]

He had made his findings public at the 1867 meeting of the British Medical Association in Dublin. His address involved an important departure from the way most surgeons considered sepsis at the time, with significant implications for wound treatment. He noted that Pasteur's work showed that sepsis was caused by living microorganisms, and considered that they were partly introduced into the wound from the air. Surgeons argued about whether to exclude oxygen from wounds or whether it was actually beneficial to the healing process. Lister was saying it had nothing to do with oxygen, and that the *microorganisms* in the air were the problem.[142] How could they keep these germs out? He reasoned that, if he could keep wounds free from bacteria until they healed, he would have solved the problem of wound infection.

The use of chemical antiseptics on wound dressings was only part of what Lister called the Antiseptic Principle, and what his followers soon began to term 'the Listerian Principle'. This involved a series of careful steps to exclude bacteria from the operating environment. The first step was to destroy the germs on the patient's skin, on the surgeon's hands, on the instruments which were to be used and on everything surrounding the area of operation. The second step was to prevent living germs from entering the wound from the air or the surrounding objects *during* the performance of the operation, and the final step was to prevent germs from spreading into the wound *after* the operation.[143]

To exclude germs from wounds, Lister not only impregnated dressings with an antiseptic, he also relied on a carbolic spray. This pervaded the air during his operations, and was sprayed around the patient when wounds were dressed. Apart from being toxic and harmful to the patient, surgeon and dressers, there were, from an early stage in its use, questions about how effective it was in killing microorganisms. Lister himself was aware of these issues, but for many years continued to believe that germs were present in sufficient numbers in the air to warrant the use of the spray in some form. Carbolic acid also caused considerable irritation to the surrounding skin, and to the surgeon's hands when it was used for disinfection.

There were other important issues with Lister's methods. They were expensive and time-consuming,[144] so many hospitals were not particularly happy with them, especially as Lister himself insisted on keeping his ward patients in hospital until he considered the wound out of danger of further infection. This often took months. The spray was messy, and contributed to extra costs and operating time.[145] Moreover, success permitted not even the slightest error or omission on the part of the practitioner, so surgeons found themselves involved in an elaborate set of precautions which they were required to follow to the letter. Lister also shifted responsibility for failure unequivocally onto the surgeon. If anything went wrong, he blamed an omission by the operator, including himself, not the method or its underlying assumptions. Some surgeons, reluctant to become involved in theoretical debates, believed the issue of carbolic antiseptic was made more confusing when it was linked to theories. They found it less useful to connect its effectiveness with chemical experiments involving microorganisms, and preferred instead to judge it on clinical results - the number of wounds with carbolic dressings which did *not* turn septic.[146]

Lister relied heavily on his assistants to collate his statistics.[147] He himself continued to conduct experiments, and his wife Agnes documented them. In 1869, he had left Glasgow to take up the post of Professor of Surgery at Edinburgh, and by the time Cheyne enrolled as a student, Lister had developed a cult following in the lecture theatre. He was popular with students, though by no means all of them, and not all his colleagues accepted his theoretical stance. John Hughes Bennett, Professor of Medicine and Clinical Medicine, was a staunch upholder of spontaneous generation, and could not see how microorganisms could be the cause of disease or sepsis, once famously remarking, "Where are these little beasts? Show them to us, and we shall believe in them. Has anyone seen them yet?"[148]

Cheyne, however, had seen in Lister's ideas, and particularly in the implications of his experimental methodology, the enormous potential for change in medicine and surgery. Along with it, he had found a focus for his considerable intellectual energy. His studies remained general for the time being, but were gradually moving towards an interest in the *science* of medicine. His devotion to Lister's work was increasingly rooted in the theory and scientific method underlying the practices. In short, even at this early stage in his career, Cheyne was developing an interest in the embryonic science of bacteriology.

4

Edinburgh to Vienna

An introduction to microbes

Apart from his burgeoning interest in Lister's experimental work, Cheyne later highlighted how interested he and other students were in the broader range of operations Lister was now able to perform. Listerian antisepsis permitted more and more interventions which had hitherto been considered too dangerous. Cheyne felt his lecture notes had demonstrated "the advanced state of surgical treatment in [Lister's] wards at that time."[1] The index lists twenty-three conditions, and each one outlines the details of a case and the proposed treatment. For a case of caries (bone decay), affecting a child in both feet, Lister had noted that, where they would have had to remove both feet in an adult, "if the joint be kept at rest and general health be attended to"[2] the feet could be saved. General health measures included "cod liver oil and a good nourishing diet,"[3] and were to be observed alongside the application of an "apparatus composed of Plaster of Paris to keep the joint at rest".[4]

Significantly, alongside work on dangerous and life-threatening conditions, Lister was encouraged to attempt "the remedying of deformities"[5], work Cheyne attributed to Lister's broad interest in humanity. He performed osteotomies for "knock-knee, curved tibiae, malunited fractures, etc."[6] and "incisions were also made freely into joints for the removal of loose cartilages or in the case of tuberculous joints with the idea of relieving tension."[7] The treatment, as far as Cheyne was concerned, knew no bounds, and even stretched to "a case of piles treated by injection of carbolic acid."[8] He later considered that many of these interventions were "at that time looked upon by other surgeons as unjustifiable procedures – if indeed they had been thought of at all."[9]

Particularly in years to come, when Lister's students recalled their university days, they portrayed themselves in their memoirs very much as pioneers in Lister's cause. Stephenson considered himself "a pupil and a worshipper"[10] of Lister. Nevertheless, as we have seen, there were other charismatic lecturers in the medical faculty at Edinburgh in the 1870s, and becoming a Listerian was a matter of choice.[11] It was certainly not an easy one. Leeson noted that "Lister did not teach for examinations, they did not enter into his scheme, and he declined to be fettered by them ... he sought to instil principles."[12] Cheyne was a brilliant student, who could have passed his exams and even become a revered member of his profession without Lister, had he chosen to take a path with fewer risks. On the other hand, his growing fascination for Lister's ideas had given him the advantage of participating in something worthwhile, a final distraction from his resolve to go to sea. His enthusiasm was reflected in his exam performance, whether the lectures were Lister's or not.

He later noted that

> ... class examinations, which all students were expected to attend, were held from time to time (in winter generally twice during the session), and at the end of the session the list of those who got 75 per cent. of marks was announced in order of merit, and bronze medals were given to the first two or three men according to the number of students who competed. [13]

By the time he graduated, Cheyne had taken 12 of these medals, in almost every subject he studied. He took the medal in chemistry with 98%, the third prize in junior surgery with 96% and the first prize in clinical surgery with 93%.[14] He also took the first medal in practical physiology, with 93.5% in the 1872-3 session.[15] Not surprisingly, he came to the attention of Lister himself, but largely as a result of an incident which involved an even greater effort on Cheyne's part.

Lister generally presented a prize, a case of silver catheters, to the student with the highest marks in his senior clinical surgery class, on the basis of exams at the end of the session.[16] When the 1872-3 session came to a close, Cheyne badly wanted the Lister prize, but he was still only a junior, and the class was generally attended by fourth and fifth year students. He decided to sit the examination, however, to give him an idea of the questions so that he could begin his fourth year in a good position.[17] The only problem was that the examination took place the same day as his physiology exam,[18] which he needed to pass, and he wanted to beat his nearest rivals. Not to be deterred, he took physiology in the morning, tying with his chief competitor for the first medal, and rushed off to sit the Lister exam in the afternoon, taking the coveted prize.[19] He said that he could "recall vividly my notes as if I had the notebook at my side. I wrote and sent in my papers. To my great astonishment, and that of the other students, at the end of the session my name was read out first with 83 per cent. of marks."[20]

Cheyne's mental energies were astounding during this period. He was given to staying up all night to study the day before an exam[21], which makes all the more exceptional his stamina in taking the Lister prize. However, it was not particularly unusual for medical students to take on too much. Students sometimes took exams in subjects they had studied in previous years, often because of the timing of the exams, and professors recognised that this could leave them "completely worn out and overtaxed".[22] Although Lister could not have failed to notice Cheyne, he was interested in much more than his high exam marks and the fact that he had taken senior exams in his junior year. He was more specifically impressed by Cheyne's understanding of Listerian procedure and the theory behind it, and was quick to recognise him as both a potential assistant and a loyal advocate.

Nevertheless, Lister seems to have been largely unaware of Cheyne until he took the senior exam as a junior, and the incident was probably a catalyst. He later wrote,

> Mr Watson Cheyne first attracted my notice by the extraordinary excellence of his answers in the examination for honours in my class of Clinical Surgery in the University of Edinburgh. Among the very numerous students who competed, Mr Cheyne stood out from all the rest both by the extent & accuracy of his information & by his masterly grasp of the subject. On inquiry I learned that he had been equally distinguished in other classes; & his student career was closed by graduation with first Class Honours, implying very high attainment in all departments of the medical curriculum.[23]

Lister called him to an interview, in order to present his prize, and asked Cheyne why he had never applied for a dressership. Despite his success, Cheyne was still not particularly self-confident, and had clearly not realised that his dedication alone stood out to a man like Lister, who may have seen aspects of himself in the younger man. Standard procedure in obtaining a dressership was rarely left to chance. A student was expected to apply to Lister personally, to be placed on a waiting list which "was always filled some two years ahead".[24] Leeson recalled how he undertook the procedure with great trepidation, despite an introductory letter from his old professor at St. Thomas' in London. He took himself to Lister's house, braved the butler to gain entry and was relieved when Lister agreed to place him on a list, giving him a start date.[25] Cheyne was privileged not to have to go through these nerve-wracking preliminaries.

As he could hardly have failed to do, Lister selected him out of a class of 200-300 students [26] to become a dresser in the summer of 1873, and William Watson Cheyne found himself in the front line of Lister's daily work. We have the impression he was almost nonplussed at the speed at which things were moving, but he entered into the activities with gusto. As a dresser, he was expected to assist Lister in dressing patients, and also had to learn to dress wounds antiseptically himself. He held tins of lotions, lifted patients and acted as a general assistant to fetch and carry.[27] The immediate ambition of most dressers was to move on quickly to a clerk's duties [28], and in just over a year, Cheyne had progressed to a clerkship too. Clerks were a more senior form of dresser, chosen from amongst the junior dressers for a six month period, and there were usually three on Lister's team, compared to twelve dressers.[29] One operated the general carbolic spray, one the spray for the instruments and one administered chloroform, rotating between duties after each two-month period.[30] Leeson noted that "four dressers were assigned to each clerk, who had to break them in, teach them their minor duties, and initiate them into the mysteries of the antiseptic ritual."[31] Because Lister still believed at this stage that a significant number of bacteria could enter a wound from the surrounding air, his clerk sprayed carbolic in the area around the patient throughout the operation.

Cheyne's notes from Lister's lectures provide a vivid illustration of the use of the spray in changing dressings:

> ... Have the spray at hand and a piece of calico dipped in the watery solution of carbolic acid to put on the wound called a guard. Now as you raise the dressing it is of vital importance to have the spray playing in the angle between the dressing and the skin of the patient and not on the side ... where the dressing is raised. If you have a drainage tube in take it out and wash it. Then introduce the drainage tube if necessary and in putting on the dressing put the centre of the dressing over the wound spraying, till it is covered, then securely bandage in this case trusting essentially to figures of 8.[32]

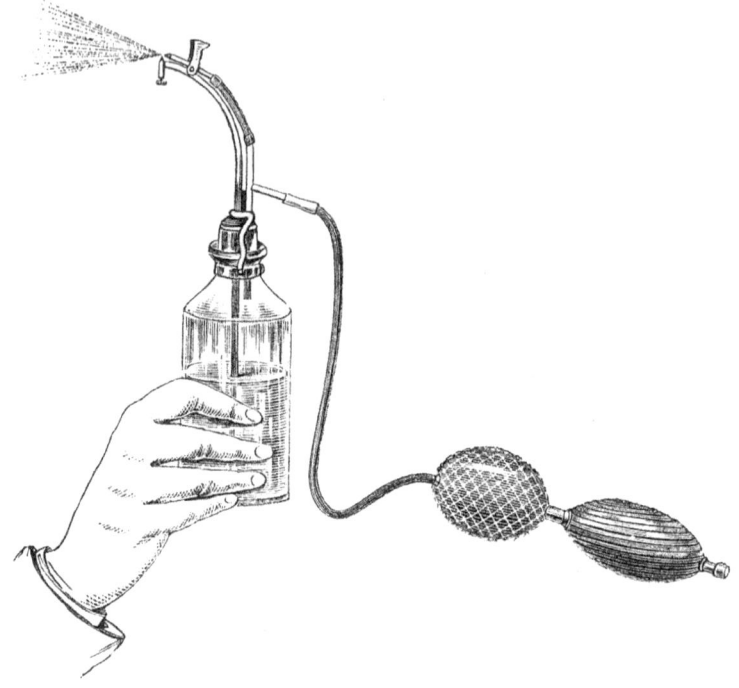

FIG. 15.—HAND SPRAY PRODUCER.

Lister's carbolic hand spray.

Ultimately, a steam spray was developed, but in the very early days, the dresser had to use a hand-operated one, which became more and more tiring the longer Lister took.

Lister was becoming a legend abroad, especially in Germany where there was interest in the theoretical base behind Listerian antisepsis as well as the practice, but he was still experiencing a level of scepticism about the theory from colleagues in Britain, even amongst practitioners who were happy to use his methods. He was often accompanied on his rounds by a variety of foreign visitors, particularly from Germany, who were anxious to experience his methods first-hand. To Cheyne, the conspicuous lack of British surgeons clamouring to see Lister at work remained a source of embarrassment [33], but the students seem to have taken a certain pride in the foreign visitors. In fact, Leeson noted that, although the foreigners could be an unwelcome distraction to the students if they took too much of Lister's time, they "were proud of them all the same, for other professors had no foreigners."[34]

On these occasions, Lister gave his visitors such a patient and careful demonstration of his antiseptic dressings that the dressers were left exhausted. Cheyne was clearly speaking from experience when he said that "the poor dresser, who was almost and, indeed, sometimes actually faint from the pumping of the spray, was for the time being completely forgotten. But, however exhausted he was and however much his wrist and arm ached, not one of his dressers would give in and let Lister down."[35] The visitors, on the other hand, issued gasps of surprise when they were handed dressings to smell, and discovered they were devoid of the usual rancid odour.[36] Then "a violent conversation would break out among them, accompanied by equally violent gesticulations, so that one became alarmed lest the peace of the nations was going to be endangered."[37] There were "Scandinavians, Danes, Germans, Russians, Poles, French, Italians, Americans and Japanese."[38] and Leeson noted that it involved a "great deal of bowing and card presenting, Lister most graciously receiving them all, smiling kindly, and inquiring, now in French, and now in German, about his foreign friends who had given them their introductions."[39]

As young men already familiar with the charisma of Lister's teaching, however, Cheyne and his fellow students jostled to be the ones to accompany Lister in his work for other reasons too. They wanted to witness first-hand the enormous variety of operative procedures which had hitherto been dangerous or inadvisable because of the likelihood the wound would turn septic, particularly in injured and badly-set joints. Cheyne, from the first moment he heard Lister speak, considered himself privileged to be part of what he knew at the time was a revolutionary era in surgery.

In the 1873-74 session, he came away with 96% in general pathology and the first medal.[40] He was by now also class prosector [41], which involved preparing dissected material to illustrate points of interest. From then until he completed his degree in 1875, he swept the board, taking the first prize in all his subjects: 96% in midwifery "and diseases of women and children"[42], 99% in senior surgery [43] and 99.5% in the practice of medicine.[44] Even at this early stage in his medical career, he demonstrated a leaning towards medical science and experimental work in medicine, and was gradually becoming aware of the theoretical debates and allegiances behind Lister's work.

In Edinburgh, Cheyne was lucky. Lister had relatively strong support from students, and reflecting Lister's views was not necessarily prejudicial to a student's chances of a good degree, provided he recognised that other lecturers expected a different approach. Lister set and marked his own exams, so students could at least mention Listerian principles in these. There was nevertheless opposition to Listerian theory from some of the Edinburgh professors [45], and there was clearly still a risk involved in students attaching themselves openly to it. It was, in effect, an informed choice and a leap of faith at the same time.

Cheyne quickly became involved in "the rivalry between Lister's students and those who continued the old system"[46], which reached its height around 1876-7. The enthusiasm of Lister's supporters, who were in the majority, bordered on contempt for the unenlightened, and Cheyne was receiving a grounding in what was to become a central part of his legacy, defending Lister against his opponents. He was not afraid to use wit and sarcasm, much of which he may well have developed during this period. He explained:

> The Listerians, who were considerably the more numerous, looked on the non-Listerians as lost souls, Tories of the most die-hard description, and not likely, when they came to practise their profession, to be able to give their patients the best chance possible. The non-Listerians, on the other hand, looked on the others as crazy believers in vain things like germs, rash to a degree, blinded with their enthusiasm, placing their patients in the greatest danger by the outrageous treatment that they proposed, and, as they said that their wounds did not suppurate while those of the other side did, liars of the first water.[47]

Cheyne considered the rivalry to have been "of the greatest value to the cause of asepsis".[48] It made the Listerian students exceptionally careful in their surgical practice; any errors on their part conceded a point to the non-Listerians. For Lister's followers, and indeed for Lister himself, this went far beyond point-scoring. They feared, with good reason, that any mistakes which resulted in suppuration of wounds would be blamed on the methods and not the practitioner. If a student failed to check a small detail in the Listerian procedures, and a wound suppurated despite receiving antiseptic treatment, the opposition would seize on the opportunity to call into question the methods in general.

The Listerian students assumed a collective responsibility for any errors, but each one of them also took it personally, especially if he had been involved in the mistake. The punishment was self-imposed, and not necessarily instigated by Lister himself. Lister had simply encouraged the students to take responsibility for errors in procedure and not to blame the principle. The students developed their own form of responsibility and corporate control, and this was to have a number of consequences for the future. It was the foundation of Lister's vanguard.

Cheyne later described vividly how the young idealists felt if they conceded a point to the opposition:

> ... we went about as if everything were lost, ashamed to raise our heads, and subjected to merciless chaff. Our only topic of conversation among ourselves on such occasions was how the infection had got in, what was the error in technique which had been committed and whose fault it was; most excellent training and never forgotten; nothing could make us more careful than to be made to feel that suppuration in an operation wound was due to some error in our technique and was a serious reflection on the surgeon or the dresser.[49]

In 1875 Cheyne completed his degree with 81% in Medical Jurisprudence and took the first medal.[50] He graduated M.B., C.M. and had won a total of 12 medals. He had to consider where to go from here. He had spent nearly five years focusing entirely on his studies, and suddenly they had come to an end, returning him to the crossroads he had experienced in Aberdeen. He badly wanted to apply for the post of Lister's House Surgeon, but was too shy to ask.[51] Despite his enormously successful student record, and the fact that Lister recognised his talents and loyalty, Cheyne was not confident of his ability to compete for the position. His previous indecision and difficulties prior to studying medicine had damaged any self-confidence he might have had. The result was that Lister had appointed someone else.

Lister, by this time held Cheyne in particularly high regard, and sought him out to ask him why he had not applied. When Cheyne responded that he had badly wanted the post, but had not had the confidence to put himself forward, Lister must have been relieved to learn that he had not lost one of his most promising students. He asked Cheyne if he would be prepared to wait a year, until the post came up again, and then it would be his.[52] This gave Cheyne an enormous personal boost. He now no longer needed to deliberate about his future. It clearly lay with Lister. Most importantly, however, he now knew the extent to which Lister valued his services, and he could feel more confident in making a full commitment to a life in medicine.

Patronage was important, especially for those without family connections or adequate financial means. A patron could be a significant help to a recent medical graduate in building a career in the 1870s and '80s. There were procedures for selecting house physicians and house surgeons, for example, which were not always particularly indicative of a graduate's skills, so if men like Lister wanted the best, they had to find a way of choosing them personally.[53] Family connections could catapult a man through the early years of his career [54], but Cheyne had no such connections, nor any means of cultivating them. Lister's patronage was probably the only thing standing between Cheyne and renewed thoughts of the sea.

Naval service would not have been an unusual choice for a newly-qualified practitioner, and many went to sea for a few years, to gain experience and amass some savings to launch their career on their return. The post of ship's surgeon did not carry a great deal of prestige [55], however, and incumbents generally moved on to a land-based career within a few years. The 1855 Passenger Act had made a "duly qualified"[56] medical practitioner a requirement for passenger ships with over 300 people on board for short voyages and over 50 on board for longer journeys. Merchant ships were required to carry a surgeon if there were more than 100 people on board.[57] In other words, there were plenty of opportunities on emigrant ships, vessels carrying people to work in outposts of the Empire, or liners carrying the growing numbers of Americans embarking on Grand Tours of Europe.[58]

It is clear that, by this time, it was no longer Cheyne's first choice of career, as he had "badly"[59] wanted to apply to become Lister's house surgeon. He had learned a lot more about life since he had been in Edinburgh, however much he had buried himself in his studies. He had learned that a ship's surgeon carried little weight in the profession, and he had proved himself capable of more, even if Lister had needed to confirm it for him. Interestingly, when he considered what to do during his year's wait, he could have chosen the sea. It was, after all, a temporary solution for many. He could have amassed some funds to take him through the early years of his career and done something he badly wanted to do at the same time. Instead, he chose another route, normally open only to those with sufficient means. He decided to try and raise the money to allow him to tour some of Europe's foremost medical institutions, and to learn more about recent developments and methodology.

He returned briefly to his aunt and uncle in Fetlar for the summer to see how he could "scrape together a small amount of money."[60] He had clearly realised the year in Europe would be more commensurate with where he was now heading in the profession, and it would have to be funded somehow. Others who graduated in his year in Edinburgh did the same, including Stephenson [61], and Lister probably recommended it. The work of French, German and Austrian scientists and surgeons was at the centre of developments in surgery and they were interested in Lister's methods, sometimes keeping a watchful eye on them.[62] Vienna was one of the most common destinations for Cheyne's contemporaries at Edinburgh [63], no doubt because of its association with Professor Billroth, whose own work involved antiseptics. In fact, there were so many British graduates in Vienna around this time that they met twice a day, building networks which would serve them well in their future work.[64]

At the end of June 1875, Lister himself had returned from a "triumphal march"[65] of some of Europe's major medical institutions. He had effectively been able to see his methods in action across the Continent, particularly in Germany, where he was given an enthusiastic welcome in Munich, Leipzig, Halle and Berlin.[66] It was by no means the first time Lister had toured these institutions. His brief honeymoon in the English Lake District had been followed by a three-month visit to European medical centres, including Vienna.[67] Cheyne's educational tour was a logical step if he were to become Lister's house surgeon on his return.

Cheyne's registration document from the University of Vienna.

Cheyne's class attendance certificate from the University of Vienna.

Funding it, however, required some thought. A sojourn in Vienna could cost around £40 a month, including the cost of travel, and attending short courses with European specialists cost from £2 to £5.[68] The cost of living in most European cities was lower than in Britain, but Crowther and Dupree note that even this was "beyond the means of the needier practitioners."[69] Cheyne had reached the age of 21 two years previously, and was now eligible for the land his grandfather had left him in trust. He owned a third share of twenty-four merks of land at Bouster, in the island of Yell; his two aunts, Christian and Grace, owned the other two-thirds. In 1875, to allow Cheyne to go to Austria to study, the Trust was broken so that Cheyne could borrow money, using his portion as security on the loan. They borrowed £150, which he bound himself to repay by Martinmas, 11th November 1875.[70] The land would later be sold in 1883, as his guardian, David Webster, did not leave a great deal of money or assets when he died in 1881.[71] Webster had a son of his own in the Ministry, and Cheyne's aunts almost certainly became dependent on their share of their father's Trust.

The fairly constant process of trying to find funds to launch him in a profession clearly had a lasting effect on Cheyne. As an older man with a long career behind him, he was on one occasion opening a parish sale of work in Lerwick Town Hall. He gave an impassioned speech on how those pursuing professional qualifications were not always better off than those doing manual work, and that, "he wondered ministers went into the profession at all, as they were condemning their families to poverty."[72] He commended the congregation for setting up a fund to ensure an adequate annual increase in their minister's stipend.

In Vienna, he attended lectures delivered by some of Europe's foremost medical teachers and practitioners. For physiology and higher anatomy he had Professor Ernst von Brücke [73], a Prussian physiologist who was keen to eliminate speculation from medical research by providing evidence from physics and chemistry towards an empirical base. Von Brücke is also remembered as the mentor of Sigmund Freud, who was, coincidentally, also studying in Vienna in 1875,[74] and probably attended the same lectures as Cheyne, though there is no indication they were aware of one another.

General Pathology in Vienna was delivered by Professor Stricker. Practice in microscopy was under Professor Exner, who became known for his work on the physiology of the brain, and physiological and pathological chemistry was taught by Professor Ludwig.[7]. Cheyne also mentioned von Hebra [76], the founder of modern dermatology, completing an illustrious list of famous names.

Cheyne's most significant experience, however, almost certainly came from Professor Theodor Billroth [77], generally regarded as the man who established modern abdominal surgery. While Billroth had been in Zürich prior to moving to Vienna, he had introduced the concept of audits, where he published all his surgical results, whether they were good or bad, and analysed them objectively. More interestingly still, he was also working on his own form of antisepsis, but did not accept the bacteriological foundation of Listerism, that microorganisms were the cause of suppuration. Billroth had a powerful influence in Vienna, and Godlee, Lister's nephew and biographer, suggested that his opposition to Listerian antisepsis may have slowed down its acceptance in the city.[78] By the late 1870s, however, Billroth did not consider himself completely opposed, though he continued to have doubts about its foundation and universality.[79] Overall agreement or not, there was considerable interest in Listerian methods in the European institutions Cheyne visited.

Cheyne's first-hand experience of Lister's work gave him a special status. He recalled: "I may say that I was very much impressed with the enthusiasm that there was for Lister wherever I went. (The fact that I had been one of his pupils and would be house surgeon on my return to

Scotland was an Open Sesame, and I was even asked in Strasbourg to instruct the Professor of Surgery there in Lister's methods!) I could not but contrast this with the state of matters in our own country."[80]

Rickman Godlee, in his biography of his uncle, suggested that the differences between the German and particularly the English medical qualification systems may have had a bearing on the difference in attitude. The German system had for some time placed more emphasis on a university education, so that by the 1870s most surgeons and medical teachers had studied in a university. By way of contrast, he notes that, "in 1867 there were 64 surgeons and assistant surgeons attached to the eleven great London hospitals with medical schools. Only 7 of them had medical degrees."[81] The implications were that German surgeons had fewer practical skills than English ones, but were more likely to understand and appreciate the relevance of the scientific theory behind the practice. It was therefore noteworthy that Billroth's objections to Lister were largely based on disagreements about theory.

However, within Britain, Edinburgh had developed along different lines to the London universities, and had been "the only one in the British Isles in which anything at all comparable to the German system existed."[82] It is therefore understandable that so many of Edinburgh's medical graduates chose periods of study in mainland Europe after graduation. Britain, on the other hand, was an acknowledged leader in sanitation and hospital reforms, and Lister's methods of treatment were of considerable interest on the Continent.[83] Godlee, like many Listerians, considered any failure to reproduce Lister's results in some of the continental hospitals or laboratories, a result of "slovenly"[84] habits and practice, or a failure to understand Lister's "guiding principle"[85] correctly.

Cheyne, too, was to subscribe to this explanation, though it did not always account for discrepancies. In the spring of 1876, he arrived in Strasbourg, where he was lucky enough to spend three months in the laboratory of the pathologist Friedrich Daniel von Recklinghausen [86], who had once been assistant to Virchow, generally regarded as the founder of modern pathology. It was here Cheyne had his first sustained experience of laboratory work in bacteriology, and he was to take back with him some perplexing questions. He was interrogated by professors and their assistants alike, who were anxious to learn about Lister's methods, but who could not necessarily replicate the results. Professor Lücke's assistant in Strasbourg claimed to see "all sorts of things swimming about"[87] in the discharge from the professor's cases treated with carbolic acid, when he examined them under a microscope. Cheyne, in true Listerian fashion, suspected this was because the procedures had been faulty in some way,[88] but the apparent anomaly that microorganisms were to be found in dressings which had been treated antiseptically was to trouble him, and would be a concern to Lister when the results were relayed to him on Cheyne's return.

Cheyne returned to Shetland briefly after his studies on the Continent, in order to bide his time until he could take up his post as Lister's House Surgeon. He made himself useful in a medical capacity while he was there, and earned his keep in a period where his studies had already cost the family more than anticipated. In the autumn of 1876 he was involved in vaccinating children in East Yell, on the island adjacent to his home in Fetlar. Vaccination against smallpox had become compulsory in Scotland in 1863 for infants (and earlier in England and Wales). Not all parents presented their infants to be vaccinated at the appropriate time, but the system of recording and reporting 'defaulters' meant that they risked legal action if they did not. Vaccination was not generally free, and all but paupers had to pay, though reports of smallpox in London caused the authorities in Shetland's capital, Lerwick, to offer free vaccination in the town in January 1877.[89]

In the rural district of Shetland where Cheyne was conducting vaccinations only a few months earlier, the registrar had a duty to record and report the infants not presented for vaccination. Cheyne was enlisted by Dr Robert Cowie to assist him in the North Isles of Shetland as he was temporarily resident there at the time. Dr Cowie, the vaccinator [90] for the whole of Shetland, had a Lerwick-based practice. Deborah Brunton has highlighted the differences in application of compulsory vaccination in Scotland compared to England and Wales. In Scotland, rather than being conducted through the public authorities, it remained in the hands of private practitioners, and was generally more successful.[91] Vaccinators were paid a fee per vaccination, and as soon as it was clear the vaccination had been successful, they were required to issue a certificate known as Schedule A. It was the duty of parents or legal guardians to ensure infants were presented [92], and they were given vaccination forms when they registered the birth.[93] Postponements of no longer than two months were granted if the child was ill during the required presentation period, in which case the vaccinator had to issue a certificate known as Schedule B.[94] Parochial Boards granted travelling expenses as they saw fit to vaccinators who had to visit rural districts. In Dr Cowie's case, the Board in Yell, for example, was expected to grant him travelling expenses on top of the standard fee he was paid per vaccination. However, he probably found it convenient to enlist a qualified practitioner already in the area. A previous incumbent of Dr Cowie's post, Dr Loeterbagh, had engaged in a battle with the same Parochial Board in 1865 over what he considered inadequate expenses for travelling from the town, and had almost resigned over the issue.[95]

Cheyne was probably glad of the temporary work. Furthermore, Dr Cowie was a graduate of Edinburgh University, knew the Royal Infirmary well where Lister was based, and had been a student of James Young Simpson[96] who had introduced anaesthesia by chloroform. Dr Cowie was therefore very likely to have been aware of Cheyne and his studies, and made use of his presence in Shetland. They had mutual acquaintances such as the Edmondstons in Unst, and it could have been through these contacts that Cheyne secured the appointment.

Logically in a place with no resident qualified practitioner, people in outlying districts made use of the vaccinator's presence to consult him about other matters. A.D. Mathewson, the registrar for this particular district of Yell where Cheyne was vaccinating, had mentioned in earlier correspondence that he was considering asking Dr Cowie to look at his daughter's shoulder when he came to vaccinate.[97] His daughter Margaret, who was to feature briefly in Cheyne's early career, ultimately consulted Cheyne himself[98]. His encounter with her is detailed in the next chapter.

Cheyne's real interests, however, were already beginning to lie beyond surgery or general medicine. In the continental laboratories he had been introduced to an environment which was leading him into science. It was not until he was back in Edinburgh, and had taken up his post as Lister's House Surgeon in October 1876, that Cheyne had an opportunity to take advantage of the time he had spent abroad to look into the matter. His general responsibilities were already time-consuming. House surgeons were always graduates, "and had charge of the wards in the Professor's absence."[99] They were busy men and "rarely off duty"[100]. On the other hand, both Lister and Cheyne were concerned about the anomalies in the laboratory in Strasbourg, and the Professor was delighted [101] when Cheyne asked if he could examine some of the wounds in the wards.

With Lister's blessing, Cheyne therefore initiated his own laboratory experimentation into "the vexed question of the relation of organisms to antiseptic dressings"[102]. In terms of raw material, he had a major advantage over his colleagues in Germany. Where they had the benefit of advances in laboratory methods, Cheyne could test dressings he knew Lister had controlled as closely as possible, and therefore he could be sure there were no errors or omissions. He was keen to discover whether the doubts he had heard in Strasbourg would manifest themselves in his own experiments, and if so, whether they were due to real issues with the antiseptic, or simply an imperfect understanding of Listerian procedure.

He set up a crude laboratory "in a little passage behind the operating theatre in the old Edinburgh Infirmary."[103] He did not have the benefit of advanced techniques for staining bacteria, which Robert Koch and others would later develop in Germany, and which would make the microorganisms easier to examine with the naked eye. In fact, it was a very basic research environment, with "no oil-immersion lenses, no solid cultivating media, no proper incubators."[104] He later said, by way of excusing the makeshift nature of it with the benefit of hindsight, that

> everything was in its infancy, and I had to carry on my observations by the aid of fluid cultivating media, though it was not long before I was able to go over all the work again with a proper microscope and other appliances.[105]

Cheyne had embarked on the course which would make him a pioneer of medical bacteriology in Britain. He had trained in Edinburgh in medicine and surgery, but Germany, Austria and Lister had made him into a bacteriologist. He entered into the role with characteristic enthusiasm, if not a little trepidation about what Professor Lücke's assistant's "all sorts of things"[106] might turn out to be in the supposedly antiseptically-treated discharge.

Cheyne's future now clearly lay with Lister, but as house surgeon, he needed clinical experience on the ward and in the operating theatre, a higher priority than his bacteriological research. By remaining with Lister, he was able to develop both side by side. In Cheyne, Lister had a valuable combination of skills, and importantly, could rely on him to interpret his findings within the terms of Listerism. It was to help Lister adapt his principles and practice through successive challenges.

5

Mr Cheyne enjoys a lark

A patient describes Lister's surgical ward

Some of Lister's students, dressers and family, such as John Rudd Leeson, Rickman Godlee and Cheyne himself, have provided us with retrospective notes and memoirs of their work on Lister's wards. However, a remarkable account by one of his ward patients leaves us with a unique impression of Cheyne himself in his role as Lister's house surgeon in Edinburgh, and the period immediately following it. The previous chapter alluded to an incident where Cheyne was consulted during a vaccination campaign on the island of Yell in Shetland by a young woman with a tuberculous shoulder. This chapter examines this woman's own account of her treatment in Lister's ward at Edinburgh Royal Infirmary when she made the decision to seek help there. During her eight-month stay at the Infirmary in 1877, Margaret Mathewson, whose father had been Registrar in East Yell during the vaccination campaign, maintained a constant correspondence with her family. She also wrote a detailed account of her experiences afterwards, which she called the Sketch [1], and which she intended for publication. Her case is particularly interesting for a number of reasons.

Firstly, she gives a laywoman's perspective on Lister's methods in action, and was sufficiently interested in them to pay attention to important details. Secondly, her account provides evidence of the more human aspects of life on the ward, where history has tended to focus on the scientific and medical advances. Perhaps most importantly, however, it illustrates some of the almost 21st century concepts of patient care Lister practised on his patients. Margaret Mathewson came from Yell in Shetland, the neighbouring island to Cheyne's home in Fetlar. They knew people in common and spoke the same dialect, which helped to reassure Margaret in the unfamiliar environment of the ward and make her feel a little closer to home. She not only came from Shetland, she came from the same *part* of Shetland as Cheyne. Her narrative and correspondence also shed light on the relationship between health and medical practice with which Cheyne had been familiar when he was growing up in Fetlar. Most significantly, the narrative provides a uniquely detailed account of the daily routine which had become Cheyne's life, and provides a valuable portrait of him as a young man.

Margaret Mathewson, Lord Lister's Shetland patient.

Microbes and the Fetlar Man

Cheyne was just 24, and had not long completed his studies, but had already spent time in some of Europe's most forward-looking medical institutions, and returned to find himself in a privileged position as Lister's house surgeon. He was at the forefront of the development of Listerian methods when they were still considered new and the theory behind them had not been universally accepted. He had come from a modest, but professional background, where his family's status in the community afforded him a position of respect. He was one of only two children in the household, and financially, had managed to escape the consequences of his father's adventures by the good graces of his grandfather, who had more or less provided for his education.

Margaret Mathewson, on the other hand, was the second youngest of twelve children of a schoolmaster in Yell, the second largest island in the Shetland archipelago. Her father, like Cheyne's, was something of a local legend, but for quite different reasons. He had been taught reading, writing and rudimentary arithmetic by an aunt, and was teaching others by the time he was six.[2] He went on teaching local children, at one point on Cheyne's neighbouring island of Fetlar [3] as well as in Yell, until he was 78.[4] Margaret was taught entirely by her father, and knowledge and reading took second place only to religious devotion. Her father belonged to the Church of Scotland, but she herself had converted to Wesleyan Methodism.[5] It has been argued that she used her account of Lister's ward partly as a vehicle for her religious zeal and to convert people to the idea that hospitals were not as bad as people thought [6], but this in no way does justice to the detailed record she left of everyday life on the wards. She was genuinely interested in medical matters, and was quick to grasp them.

Around 1873, she developed tuberculosis in the lungs, and she later experienced great pain in her shoulder. It prevented her from dressing herself easily, and she was barely able to knit. Knitting was an economic activity for Shetland women. It was an important supplement to a family's livelihood, generally bartered for other goods in the economic system known as Truck.[7] It ensured the women rarely rested, even when they sat down at night, and photographs recall women knitting as they walked down the road.[8] Margaret Mathewson's correspondence reveals that her family battled constantly with debt, and that her knitted "haps", which could bring in around seven shillings, made the difference which allowed them to acquire staples such as tea, sugar and bread.[9] In autumn 1876, things took a turn for the worse when Margaret injured her arm trying to enter the loft of the byre. She later told Lister, "the ladder slip't from my foot Sir,"[10] and that a board of wood had struck her shoulder.

A Shetland woman walking down a road knitting, probably early 1900s.

She visited the Reverend Mr Barclay, the local Church of Scotland minister. There was no resident doctor on the island, and, as she noted later: "Mr B[arcla]y is all the Practitioner that's in our Island."[11] The relatively small and scattered island population was hardly a very lucrative prospect for general practitioners. The Royal College of Physicians of Edinburgh had noted in the 1850s that the rural region of Scotland's Highlands and Islands was seriously short of doctors, and that, in some places, Shetland in particular, "the ministers and the landowners still gave medical help."[12] Cheyne's own grandfather, the Reverend William Watson, had "dispensed medicines"[13], and on one occasion noted in his diary that there was "no day without application for medicine, external or internal."[14] Even gifted local people sometimes offered a rudimentary medical and dental service, and on occasion, this produced some rather remarkable results. On the Shetland Mainland, a man called John Williamson, known locally as "Johnnie Notions", had successfully inoculated a whole community against smallpox in the 18th century, years before Jenner's safer vaccination techniques became widespread.

This form of inoculation was not unknown in Britain at the time, but it was risky, as the occasional patient could contract a stronger form of the disease and die, and for a short period after inoculation, the disease was contagious. Williamson's achievement seems to have been in his perfect success rate, presumably down to the care he took, and his method of preparing the serum and inoculating the patient. Edmondston, quoting the Rev. Mr Dishington, noted about Williamson's method:

> Several thousands have been inoculated by him, and he has not lost a single patient ... He is careful in providing the best matter, and keeps it a long time before he puts it to use – sometimes seven or eight years; and, in order to lessen its virulence, he first dries it in peat-smoke, and then puts in under ground [sic], covered with camphor. Though many physicians recommend fresh matter, this self-taught practitioner finds, from experience, that it always proves milder to the patient, when it has lost a considerable degree of its strength. He uses no lancet in performing the operation, but, by a small knife made by his own hands, he gently raises a very little of the outer skin of the arm, so that no blood follows; then puts in a very small quantity of the matter, which he immediately covers with the skin that has thus been raised. The only plaster that he uses for healing the wound is a bit of cabbage leaf.[15]

Williamson, who lived in Northmavine where Cheyne's father grew up, turned his hand to a number of things. Ironically, he had made a bust of Cheyne's great grandfather, from black oak driftwood, which the family acquired and used for many years as a wig stretcher.[16]

In Fetlar, as late as the early 1900s, the principle landowner, Sir Arthur Nicolson (by this time a different individual from the one who had conducted agricultural experiments and with whom Webster had corresponded), extracted teeth for his tenants and lanced tooth abscesses for his wife and children. On 25th January 1913, his wife wrote in her diary, "Arthur drew a tooth for Charlie [probably the postman] after tea, & he got rid of his toothache."[17] There was nothing unusual in this in a place where, even if a doctor were close enough to pay a visit (by this time there was a resident doctor in the neighbouring island of Yell, Dr Taylor [18]), the weather could easily prevent him from making the crossing to the island until the medical emergency had passed. The Nicolsons were careful about health, and took as few risks as possible, particularly as their youngest son had been born with brain damage and was susceptible to infection. On one occasion in the depths of winter, when the doctor called in on his way back from visiting a family, they learned he had been amongst people with influenza, and as soon as he left, they sprayed with sulphur. Lady Nicolson's diary gives a vivid description of the doctor's arrival and departure:

> Five men went over [to Yell by small boat] for the Dr for one of the Coutts children. He did not come in - rode over on his bike & called coming back but, as he was not sure if he had influenza as he had been amongst it in Yell, he did not sit down. Arthur saw him & gave him some refreshment, then fumigated with sulphur after he left. The men waited here for him, & had tea before he came back from Hubie [where the Coutts family lived, on the other side of Fetlar from Brough Lodge].[19]

In the period just before Cheyne qualified, Dr Cowie on the Shetland Mainland, in his dissertation for the University of Edinburgh published in 1874, provided strong statistical evidence of greater longevity and better health in the islands compared to the Scottish mainland, and noted that people were very aware that killer epidemics such as smallpox were brought in from elsewhere.[20] My husband, a native of Fetlar, informed me that, as late as the 1980s, the generation of people who grew up on the island in the early 1900s were given to emptying drops of 'Zoflora' (an antibacterial disinfectant) onto their handkerchiefs, particularly if they were speaking to any 'strangers' in the island, to counteract what they considered an increased risk of infection.[21] Though there has been a resident nurse in Fetlar in more recent times, there has never been a doctor. The nearest one is on the island of Yell, but when Cheyne was growing up, there was no official medical presence there either, and people were careful.

Barclay, on the other hand, who had treated Margaret Mathewson in Yell, had a level of formal medical knowledge, as his father had been a doctor on the neighbouring island of Unst.[22] He was able to operate on

Margaret's shoulder and provide her with ointment for the dressing. Lister later commented, after he had seen the shoulder, that the operation had saved her life, as it had diverted discharge and prevented the tuberculosis from returning to her lungs.[23] When the arm worsened, however, after the fall, Margaret decided to go to Edinburgh for a consultation at the Infirmary. Her family, aware of the reputation of hospitals in general, was violently opposed to the idea.[24]

In correspondence which has recently come to light between Margaret and her brother Arthur, dated October 22nd 1876, Cheyne himself examined Margaret's arm only a few months before she went to Edinburgh, while he was in the district vaccinating children against smallpox. The advice he gave her comes over as slightly restrained, and there is no mention of his suggesting to her that she travel south to consult Lister, so unless she omitted this from her correspondence, her decision to travel was the result of other advice or information.

She says, "[He] told me to take a tea spoon of co[d liver] oil 3 times a day & always increase the quantity to a tablespoon. I told him I could not take any as it came so much back on me."[25] She had apparently been persevering with it for seven months before she finally gave up. Cheyne admitted that it did not agree with him either, and when she told him she took a tablespoonful of cocaine every day instead, he suggested she take a spoonful of brandy and a cup of cream with it. He called this his "own medicine"[26], by which he may have been indicating that he not only suggested it to people, but took it himself, as he remained constantly on the edge of tuberculosis-related problems. He also suggested she put a 'blister' on the shoulder joint.

Tuberculosis in the bones and joints would later become Cheyne's speciality, but at this stage he had done none of the bacteriological research which would inform his views on treatment, nor may he have been aware of the very specific diagnosis Lister would later give for the pain in Margaret's shoulder. His advice, as a result, appears to be general and a little haphazard. Margaret, at this stage, preferred to place her trust in Mr Horrell, a Methodist minister who had suggested she take medicine which turned out to be cocaine. She had spent three weeks in Lerwick under his advice and treatment in April 1876.[27]

When Cheyne asked whether she had put any "rubbing stuff"[28] on the sore spot, she mentioned "Dr Bow's Linament"[29] and "paraphine oil"[30], but Cheyne had heard of neither of these remedies in this context. He muttered a characteristic "Dear O me"[31], but was nevertheless interested.

He asked her about the effects of the medicine, and what she had expected it to do. It was apparently supposed to bring her out in pimples, but had not done so, nor had it benefited her particularly. It illustrates the range of treatments people were trying, and Cheyne thanked her for having introduced him to them. He ascertained that she had not been given this advice by a doctor, but then left it at that, and does not appear to have offered any further advice from a professional perspective. Margaret noted to her brother that she intended to write to Horrell again, which indicates she may have expected more substantial advice from Cheyne than he offered, but at some point within the next few months, she made the decision to travel to Edinburgh Royal Infirmary for a consultation.

Cheyne's restraint in offering Margaret any substantial professional advice could be the result of a number of factors, as he was not known for his reticence to help those in need. He was newly-qualified, and by his own admission, shy and not particularly self-confident. He had not yet begun his real work with Lister, and had none of the self-assurance he would later acquire as a result of it. He was very much used to following orders and instructions to the letter, and was there to deliver vaccination which was conducted under strict documentation and bureaucratic procedures, so he may have been uneasy about going into much detail about anything else. It is also very possible that he was not only approached by Margaret on this occasion, but by a number of other people taking advantage of the presence of a doctor on the island. Moreover, at this stage, Margaret's shoulder may not have suggested to Cheyne the need for surgery, as the conversation was largely about pain relief. His not mentioning the possibility of consulting Lister is more likely to be due to one of these circumstances than a belief that Lister could not do anything for Margaret's situation. As we have seen, Cheyne was nothing short of evangelical as a student about the variety of new operations Listerian antiseptic procedures had enabled.

Margaret was not alone in being prepared to travel a good distance for medical treatment. Dr Cowie mentions that Shetland patients sometimes visited Edinburgh Royal Infirmary [32] (though not necessarily Lister in particular), and by no means all of Lister's patients lived in the vicinity of Edinburgh. The English poet William Henley, a friend of Robert Louis Stevenson, travelled by boat from London to consult Lister on his tubercular foot, on the strength of the accounts he had read in newspapers of the controversy over Listerian treatment.[33] Henley also left an account of his treatment in verse, but was a fee-paying patient, unlike Margaret, and his impressions were literary, rather than particularly objective.

Microbes and the Fetlar Man

Margaret's journey to Edinburgh had been one Cheyne himself knew well, as he continued to visit Fetlar most years of his life. On the other hand, he is unlikely ever to have travelled in steerage like Margaret. The sea passage was rough, though she did not feel sick on this occasion. On the other hand, when she was returning home to Shetland months later, she would note that she arrived very seasick.[34] Travelling from Margaret's home in Yell to a location on the Shetland Mainland (the largest landmass in the Shetland archipelago) where she could leave for the Scottish mainland was already a journey by sea, and from there she would probably have had to wait a day or more for a packet to take her onwards. She would not have known, when she set off, how many days it was going to take her. Even today, the sea journey takes 12-14 hours just to Aberdeen, and in extreme weather conditions, the ferry has been known to wait for long periods outside Aberdeen before it could dock. Margaret travelled to Leith, further south still.

In steerage, she was lucky enough to find a local acquaintance who took care of her luggage for her. Daniel Scollay was "going out to Australia"[35], and it was as well he was 'kind and attentive' to her, as the conditions were unimaginable, even for a healthy person. She notes that "we were crammed up with passengers [most of them men] in the steerage and we got 250 cows aboard at Kirkwall [Orkney], also lots of Sheep, Horses & Pigs."[36] When they landed, Scollay left her at Leith Walk Railway Station when he proceeded to Glasgow, and Margaret found her way to her "usual boarding when South"[37] at the house of a Mrs Barclay, who came originally from Unst, one of the three North Isles of Shetland (Yell, Unst and Fetlar). It is very likely she was related in some way to the Reverend Mr Barclay. Most Shetlanders had a network of 'exiles' to call on in the major cities, as they still do. After two days, Margaret called on her cousin Martha, and asked her to accompany her to the Infirmary. By the time she reached the hospital, she had been away from home for two weeks.

Admission to the Infirmary was at the discretion of the Professor of Surgery, and decisions were made solely on medical grounds, regardless of where the patient came from.[38] Margaret wisely felt that a letter of introduction would be useful, and she had written to Barclay to ask for one. When her cousin accompanied her on that first day, she expected no more than a consultation. It was common in Shetland families for the sons who could not be supported locally to go away to work at sea or to mainland Scotland, and until daughters married, some went into service in Scotland and England. Margaret had spent time in service in Edinburgh and Liverpool and had only returned when her shoulder began to trouble her.[39] She was by no means a naïve countrywoman coming to the city, and was aware of the growing reputation of surgery at the Infirmary, though her account indicates she was not aware of Lister by name. She heard of him only when she arrived at the Infirmary for the first time, hoping for a consultation, and asked the name of the best surgeon.[40] The gateman recommended Lister, for which Margaret was later to be grateful when she saw the ill effects in patients treated by other surgeons.

The porter directed her to the "waiting place."[41] They went up two flights of stairs, and on the second landing, the first person she encountered was Cheyne. Margaret notes that she "recognized him at once, but he did not recognize any of us."[42]. She appears to have recognised him before he spoke, but curiously he did not remember her. She does not appear to have let him know immediately that she knew him, however. She asked instead to see Professor Lister, and Cheyne said she would need to wait an hour or two, so he proceeded to examine her shoulder. At this point, she handed him Barclay's note of introduction. This is where Cheyne began to deduce who Margaret was.

> Dr Cheyne seemed to know the write in the address, he then looked at me, then read the note & again looked at me, and said, "Do you know me? Yes Sir. Who am I? Dr Cheyne of Fetlar Shetland Sir. Yes, the same (Martha was surprised we were any way acquainted)[43].

This is curious on two counts, as the letter she wrote home to her father just after the encounter notes in brackets that "Martha had told him who I was before"[44], and presumably this means that Martha had spoken to Cheyne while Margaret was undressing to be examined. In other words, he recognised her neither on seeing her nor on being told her name by Martha, yet he had examined her arm in Yell only a few months before. It is possible his initial failure to recognise her was simply because he was not expecting to see her there, especially as he must have seen quite a lot of people in the course of a day, both in Edinburgh and in Yell on the day when he was vaccinating. On the other hand, when he saw the address on the letter, and Barclay's name, he began to put two and two together. It would have been strange if Cheyne had not known Barclay, since he and Cheyne's guardian were Church of Scotland ministers in neighbouring parishes, and his grandfather's diary also indicates that he visited Barclay regularly.[45]

Microbes and the Fetlar Man

The way Margaret describes Cheyne throughout her narrative indicates that he did not stand on ceremony, and seems to have been as delighted to have someone from Shetland in the Infirmary as Margaret was to hear a familiar accent. Though he had clearly not recognised Margaret at first, by the time he introduced her to Lister, she had become "an acquaintance of his from Shetland".[46] Her original correspondence marks his name erroneously as Chiene,[47] which is particularly confusing as Professor John Chiene was also at the Edinburgh Infirmary at the time. However, her spelling simply reflects how both names are pronounced in Shetland ("Sheen").

He examined her shoulder and told her it was not out of joint, as she had thought, but that there was an abscess, which was causing her pain. As it turned out, she would have to wait a month for her operation on account of the open abscess. When Lister examined her shoulder for the first time, he was astounded to learn it had been operated on by a minister of the church, and asked Cheyne if this was correct. Cheyne, clearly the acknowledged resident expert on the unconventional behaviour of island communities, confirmed that it was.[48] If he had since learned that there was anything unusual about a state of affairs he had known all his life, he did not say so. Lister commended the operation, which, he explained to his gathered students, had saved her life. However, Barclay's operation had "drawn off a lot of discharge,"[49] and an abscess had developed.

Cheyne was particularly attentive to Margaret while she was in the Infirmary, and seemed to feel a level of responsibility towards her as someone from home. Her letters and her narrative give us an impression of Cheyne as a sympathetic doctor, but also one who had not in the least forgotten where he had come from. When Lister had seen her, Cheyne gave her "a wink"[50] as a signal to follow him, and accompanied her to the ward where she would be admitted. He invited her to "take a seat at the fire"[51], and make herself at home to wait for the nurse. She had not expected to be taken into the ward on the same day as her first consultation, but Lister had said he could not operate on her shoulder until the results of Barclay's operation were clear, so she would be placed under observation.[52]

Margaret mentioned Cheyne regularly in her letters home, and she relied on him to give her information she would never have been able to approach the other doctors about, as Lister preserved a degree of formality in his wards.[53] She was an intelligent woman with an inquiring mind and as good an education as her family had been able to give her. After her operation, she eventually tired of staying in bed, and wanted to sit up for a while. She was confident enough of Cheyne to ask, "If you please Sir, will it be long ere I can get out of bed?"⁵⁴. Cheyne reassured her it would not be long, but did not tell her when, which clearly frustrated her. It was nevertheless useful to have common origins. Neither her account nor her letters indicate that she was given to speaking very familiarly with any of the other doctors, with the exception of words spoken in pain and frustration to a dresser, as I explain later. She spoke more familiarly to some of the nurses, though she distinguished even between them, and she generally only spoke to the doctors when they addressed her.

Margaret settled as well as she could into the ward, though on the second night she awoke at 11.40 pm "and began thinking on what strange scenery."⁵⁵ About midnight, Cheyne came into the ward, went past the beds checking on each patient "and seemed to be looking what position they lay in."⁵⁶ When he came to Margaret, he was surprised to find her awake.

> … dear O me, how are you awake at this hour alone? I awoke a little ago Sir. Have you any pain in the arm? Not at present Sir. Do you feel much pain when about going asleep? Very much Sir - at times, and very often it wakens me out of sleep. Just so, and what like is it? It is as if the arm was starting off. Yes so it is. O well I hope if you stay with us you will get free of all your pain and good night. Good night Sir and thank you.⁵⁷

When patients were not in bed, they sat by the fire to keep warm. She was first sent to ward no. 3, which had nine beds, wooden flooring with flagstones down the middle, and a long table with lotions and dressings. There was even a bookcase full of books for the patients to read, and she described the room as neat and tidy. All the beds "had nice white covers on, clean pillow cases, and clean sheets and the room so tidy and neat."⁵⁸ The patients were even kept busy winding bandages.

> At 9.30 a.m. Miss Logan came on duty. At 10 she came with an armful of gauze 40 yds to the patients to tear into lengths of 20 yds each, some about 3 inches wide and others less for bandages.⁵⁹

This was a more homely contrast to the scene she had witnessed earlier, while she was waiting to be examined. She had seen one of Lister's students exit the operating theatre carrying a leg "rolled in silk paper"⁶⁰ dripping with blood, and the woman sitting next to her had noted, pragmatically and in passing, that this was her husband's limb. In a letter to her brother, Margaret also mentioned the endemic overcrowding in Lister's wards, with sometimes three to a bed.[61] Lister was keen to keep his patients until he could be sure their wounds were healed sufficiently to reduce the possibility of recurring infection. In this sense, his own expectations often outstripped the physical capacity of his wards, and mattresses were sometimes placed on the floor to accommodate more patients.[62] The general atmosphere in the wards was, according to John Rudd Leeson, one of Lister's dressers, a combination of carbolic, stale tobacco smoke and boiled beef. "I have no reason to think there were bathrooms as I never saw any".[63]

On the whole, however, Lister's wards were luxurious compared to the popular impression of mid-19th century hospitals. In 1872 a Training School for Nurses had been established at the Infirmary, and three Nightingale-trained nurses were sent up from St. Thomas' in London to ensure it was set up along reformed lines. They were largely young ladies of the middle classes with a better than average general education and an abiding purpose in life. William Henley, Lister's poet patient, described one of them, Mary Logan, in verse. He highlighted her obviously genteel upbringing and cultural education:

> She talks Beethoven, frowns disapprobation
> At Balzac's name, and sighs at Mme. Sand's,
> Knows that she has exceeding pretty hands.
> Speaks Latin with due accentuation ...[64]

Another of the nurses, Angelique Lucille Pringle, gave in her diary an account of her first impression of Lister's ward. She noted that there was no trace of the common ward smell of wounds, but considered the atmosphere too close, which made her "feel faint"[65] and commented that not enough air was circulated. Insufficient attention was paid to the patients' diet, and she found the existing nurses "slovenly"[66], unwilling to start work until they had eaten breakfast. Cheyne, like other house surgeons in the hospital, had his rounds in the morning, and Lister followed with his at noon [67], including on Sundays.[68] The Nightingale nurses introduced the same kind of efficient machinery into the management of the wards that Lister had introduced into the management of wounds, and by the time Margaret Mathewson arrived, they were running like clockwork.

Miss Pringle was surprised to discover the surgeons did all the dressings [69], but was also wrong in this observation. Margaret herself notes that medical students were given practical instruction in dressings in the ward, and "had to put in 6 months at dressing cases in "The Surgical", then 6 months in "The Medical" ere they could graduate."[70] However, Miss Pringle may have been making the distinction that the *nurses* did not do the dressings at this stage. After Lister had demonstrated Margaret's arm to his students as "a very interesting case for us all"[71], he asked Cheyne to dress her shoulder antiseptically, after which she said it "felt much easier".[72]

Ward efficiency gave way to the occasional treat for patients, sometimes without the knowledge of the matron. Margaret describes a tradition where any patients who were being sent home cured stood the others a cup of tea on their last day. On one occasion, this stretched to a bowl of mutton soup, with Nurse Logan's knowledge. When the matron came in half way through the feast, they all hid their bowls, but were discovered when a little boy could not hide his properly, and spilled some.[73]

It was ultimately in the operating theatre, however, where Margaret's courage almost gave out, and her account reveals a sharp contrast between the clinical procedure of the teaching hospital and the experience of a patient. Leeson's memoir of his work with Lister suggests that "no pains were spared to make the patient easy and comfortable"[74]. He quotes an anecdote where Lister admonished a dresser for bringing into the operating theatre a tray full of uncovered instruments. Lister was incensed:

> How can you have such cruel disregard for this poor woman's feelings? Is it not enough for her to be passing through this ordeal without adding unnecessarily to her sufferings by displaying this array of naked steel?[75]

These unprecedented exhibitions of humanity are corroborated in other accounts. Leeson, however, was writing retrospectively, largely in praise of Lister, and was perhaps a little too enthusiastic. He goes as far as to suggest that patients were so at ease that they positively relished being paraded in front of a contingent of medical students. He says they were "never resentful and rarely seemed nervous, and that they even seemed to appreciate the fact that so many 'doctors' were interested in … their case."[76]

Margaret's account makes it clear, however, that Lister's concern in the operating theatre was not always duplicated in aspects of procedure. Even today, a patient awaiting their first surgical intervention is generally apprehensive, and has little idea of what will happen, or how they will feel. Margaret's experience would raise a few eyebrows in the critical atmosphere surrounding hospital care in the 21st century. More significantly, however, it highlights the fact that, while Lister's men were encouraged to take every care to alleviate physical discomfort on the part of the patient, they could not always allay their fears, and some of the hospital arrangements clearly achieved quite the opposite. Margaret was escorted to a dark room, where the terror she was trying so hard to suppress had ample time to breed:

> Next day, I was called upstairs to be lectured on, and was put into a dark room where I found by their voices there was [sic] others before me. There were 2 doors in the room and a porter at the front to let out any who was called into the lecture room or "Theatre", (this dark room was called "The amphi-theatre" but oftener it was called "The dark hole.")[77]

This first time, she sat about two hours in the 'dark hole', undressed and ready to be exhibited to the students, and then Cheyne emerged to tell her she "was not wanted today yet as there were so many to be done."[78] The same procedure was repeated the following day, albeit this time with a blanket, and again Cheyne came in at the end of her ordeal to tell her she was not required. By this time, she was "shaking with fear,"[79] despite the blanket she had been given.

By the time she was finally taken into the theatre, to be "lectured on before about 40 gentlemen"[80], she was breaking out in cold sweats. Lister responded with concern, and suggested she turn her back on 'the gentlemen'. What she encountered when she did so, however, turned fear into panic.

> ... on turning round my feelings and fears were more aroused by looking first on the blackboard and then seeing the diagram of my arm chalked on it in its then swelled state, also the natural shape, then special marks where it had to be operated on; seeing this I almost dropped down and felt faintish as, untill [sic] then - I had always had a hope it would not be so serious an operation altho I had a fear it would, but now my fear was about to be a reality after all. Dr Cheyne came and took my arm and helped me downstairs and told the nurse to put me to bed and stay with me a little.[81]

Mr. Cheyne enjoys a lark

It was at moments such as this that Cheyne appears to have played the Good Samaritan, and kept an eye on Margaret's welfare. He was concerned enough about her to walk her all the way to the ward, and to ask the nurse to give her special attention. Lister's oversight here may have been for a good reason. He considered it important that his patients understood his methods, and why he was using them, however unpleasant it might appear. On the other hand, there is ample evidence of his kind attention. Cheyne was clearly one of those moved by Lister's humanity, and he later quoted a passage from Leeson's book in one of his talks, though privately he questioned general aspects of Leeson's account [82]:

> The incident, which is not only narrated but also pictured, represents Lister coming into his ward followed by a long train of distinguished surgeons, British and foreign, and stopping at the bedside of a little girl, who had damaged her doll. She held it out to Lister, and he, falling into the humour of the scene, examined it very carefully, located the injury, asked the nurse for a needle and thread, and proceeded to repair the damage, to the great delight of the little patient.[83]

Cheyne's presence, a reminder of home, clearly heartened Margaret. She thought it "homely"[84] to see his face. While she always addressed him respectfully, they spoke familiarly, in the way she conversed with some of the nurses but certainly not with Lister or the senior surgeons. On one occasion, when she had missed her dinner after another spell in the 'dark hole', she ran into Cheyne on her way back to the ward. She says he:

> … came laughing and said, "O I'm glad to see you stood your operation so well today." (I saw it was [a] lark).
> I said "Yes Sir, I stood it fine I think."
> "Were you cold in the dark hole all yon hours?"
> "Yes Sir. I was cold and hungry too, but I see I'm like the "Lawyer and the Barber". I have got to wait."
> "Well, isn't it a shame to the nurses?"
> "Yes it's a shame to you Sir, as it's by the Doctors orders."
> Dr Cheyne enjoyed the Lark very well (but next day I got my dinner coupled with a Lark).[85]

The joke may seem brutal to us today, but highlights the fact that both Margaret and Cheyne accepted the horrors of the 'dark hole' as an unpleasant fact, and merely joked about it to alleviate the distress. However, she did not have 'larks' with the other doctors, nor indeed with all the nurses, and it was as though she and Cheyne were speaking in a sort of personal code, implicit in their common Shetland origin. It is difficult to tell from the way she wrote whether they spoke to one another in dialect though. Margaret may have done so, but her Sketch represents Cheyne's speech in normal English on the whole. The fact that she notes how she used the dialect elsewhere indicates that either Cheyne spoke to her in a Shetland accent, but not necessarily dialect, or she chose, out of politeness, not to represent him this way in her Sketch, given his station.

In fact, Margaret was proud of her ability to speak a dialect no one else understood properly on the ward, and we have the impression that Cheyne found it a diversion to speak to someone from home - perhaps a break from the usual formality of the wards when doctors were present. For Lister, the formality represented a form of respect. The surgeons were very much in charge, and formal address was part of the secret of their efficiency. Lister approached his dressers as "Mr.", for example, a convention which was dropped after his day.[86] We have the impression from Margaret's narrative that, for Cheyne, the 'larks', and speaking to someone from home, were a rare break from this. On one occasion when Margaret had Shetland visitors, one of the nurses complained she could not understand her, and assumed she was speaking Gaelic (not an uncommon assumption by those whose Scottish cultural geography is less than perfect, even today). She also assumed Margaret came from the Highlands. Margaret's answer was almost indignant: "No nurse, I do not belong to the Highlands, but to Shetland, over 100 miles north past the Highlands, and that was the Shetland language and not Gaelic."[87]

It is clear from the above incident that Margaret did not normally speak to anyone on the ward in full dialect, except to her visitors. Apart from the fact that others may not have understood her readily, there was also considered to be an element of respect involved in not using dialect with non-Shetlanders.[88] Margaret may have been proud of her 'language', but nevertheless felt the need to modify it if she wished to be understood outside the islands. This mixture of dialect and 'grammatical English' can sometimes sound over-formal, and is known to Shetlanders as "knappin" (the "k" is pronounced), a blend of the more natural dialect and a form of English the speaker feels obliged to use for politeness or to be understood. It is often greeted with a tinge of humour by other Shetlanders, and is not always considered appropriate today. Though it is not specifically represented in the way she wrote the dialogue, she probably did not always 'knapp' to Cheyne, given her accounts of the jokes they shared. It is evidence that she felt comfortable with him at one level, something few of the other patients could have said about their relationship with any of the doctors.

On the whole, however, even within their own private world of 'larks', they preserved a degree of formality. The dialect, at least in Margaret's day, retained two separate words for 'you'. 'Du' was used when speaking to familiars, where 'you'[89] was not only the plural form, but was also used for respect, in speaking to older people, those you did not know well, and non-Shetlanders. However familiarly they joked together, Margaret and Cheyne still used the polite 'you' form, where her Shetland visitors addressed her as "du"[90]. In replying to them, she used the plural form, as she was speaking to more than one person, but she addressed Cheyne as 'you' and 'Sir' as a mark of respect.

Because Margaret's account is full of dialogue, her banter with Cheyne gives us some idea of his speech as a young man. He had been in Aberdeen and Edinburgh for several years, and no doubt 'knappit' most of the time when he was speaking to non-Shetlanders, but like many exiles from the islands even today, may have reverted to more natural speech when he was speaking to someone from home. His Royal Society obituary notes that he was "brusque in speech and throughout his life spoke with a strong Shetland accent."[91] There is nevertheless a big difference between accent and dialect, so when it was noted that "his many years in residence in London had not made him forget nor forego the quaint dialect of his youth,"[92] the writer was almost certainly confusing the two. Cheyne may have continued to speak with a strong Shetland accent, but is unlikely to have used dialect words at the hospital on a normal day, for the same reasons as Margaret. For him, there were probably professional reasons too, as dialect was unlikely to have been considered in keeping with either his professional status or his education. Shetland dialect is complex, with influences ranging from Norn (a form of Old Norse) and Scots to Dutch and Low German[93], and as Margaret noted, at least one of the nurses at the Royal Infirmary could not understand it.[94]

Margaret was called upstairs every day for the next two weeks, and by the time they were finally ready to receive her in the operating theatre, she acknowledged that her "fear had given place to confidence"[95], though she had again been refused dinner [96], and was losing weight. Cheyne continued to keep up her spirits by joking about the meals, and Cheyne it was who finally accompanied her into the operating theatre. If Margaret appreciated Cheyne's reassuring familiarity, she was grateful for Lister's kindness and, to the tune of silent prayer, was prepared to place her life in his hands. Her account confirms Leeson's in terms of Lister's quiet respect for his patient's feelings, and evokes a theatrical scene in more ways than one. "Prof. bowed and smiled. I returned the bow ... I then felt Professor's hand laid gently on my arm as if to let me know he was near me, which I could and did confide in and this did encourage me to hope for the best."[97]. She recalled that Cheyne then "laid a towel-saturated cloth with chloroform over my face"[98].

After her operation, Margaret was in considerable pain from the chloroform (it had taken an hour and ten minutes to bring her round [99]). Her brother Arthur came in to visit, and Nurse Logan asked him to leave, as his sister was not well enough to see anyone. Cheyne posted a notice on the door to the ward: "No visitors allowed - W.W. Cheyne"[100]. A baby had also undergone an operation, and the ward was closed except to staff. When Margaret came around, she loosened her bandages to see whether her arm was still there (amputation was common for conditions such as Margaret's). Lister, fearing infection, later questioned this on his rounds, and Cheyne jumped to Margaret's defence by explaining that she had not realised what she was doing because of the chloroform, "thus we must excuse her."[101] After the operation, she was unable to eat without vomiting for five days and was administered morphine for the pain. At first, she tried to refuse it, but was eventually persuaded to take it. When Cheyne came in to enquire about the patients, he asked the nurse whether Margaret had taken the morphine:

> Dr Cheyne came in about 12 p.m. and said well nurse how is the patients? Margt is very feverish & restless owing to a lot of pain [...] did Margt take the medicine? Yes Sir after some persuasion. How was that? Well, I believe she thought it was other medicine but as soon as she heard it was really for the pain she took it at once. Oh I thought she had a good reason for saying 'No,' but seeing she has taken this I can't give more at present. But give her plenty of 'Ice', mind. Yes Sir.[102]

In a later version of the narrative, Margaret mentions that Cheyne also had trouble persuading another patient to accept the morphine:

Mr. Cheyne enjoys a lark

She continued very weak during the evening and seemed to get worse as the evening wore on. Dr Cheyne (our house Dr) came & took her pulse every half hour, and a special nurse was set at her bedside. Dr Cheyne came with a medicine glass full of morphia etc. and offered to her. She was very against taking it. He pressed on her to take it, and it would make her better. She did so after some persuasion.[103]

Cheyne returned the following day to encourage Margaret to suck ice, saying "Now will you suck as much "Ice" as you can? … Well you must suck as much as ever you can, as your life depends on what Ice you eat or sucks, and you may yet get better."[104] He was clearly very concerned about her at this point.

As soon as he had known the operation was imminent, he had written to Barclay in Yell, without Margaret knowing. She had not realised when she went into hospital that the intervention would be serious, and had not prepared her family for it. Barclay relayed the news to her father, who did not disclose the gravity of it to the rest of the family. After the operation, Margaret indicated to her father that she thought it "very kind of Dr Cheyne to write Mr Barclay about me, also for Mr Barclay to tell you ere I got my operation, as thus you were prepared for the news ere it came."[105] She was relieved, however, that he had not told the rest of the family, in case they worried. She was ordered to lie on her back for a full month after the operation, so could not easily send a letter of her own, though she gradually began to recover. She had one of the nurses put a piece of paper in a basket and bring it close to her chin so that she could scrawl a letter to her father, and another to her brother Arthur, who had been visiting her regularly. The doctors, however, found out what she had done when she took a significant turn for the worse and had to admit to writing letters. Which doctor commented is unclear, but he indicated that she risked losing her eyesight and it would put her recovery back a full month.

When her father wrote directly to Cheyne, she was concerned, and in a letter dated May 10th 1877, she said she was:

> … very sorry to see by it that you have wrote Dr Cheyne. If you have referred to me in it, and I rather think you have, as he was in here last night & I thought there was something singular (in the wind as the Scots say) up with him with regard to me. Would you send me a <u>copy of it</u> if you have it? I meant to say ere now that if

you did write him or Professor Lister not to say anything about me, as we don't know how things might be taken by them.[106]

She gave no indication what she meant by this. In a later letter in June 1877 she mentioned that she did not mind her father writing to Lister, just as long as he did not mention her operation, "the nature of it or the severity of it."[107] He was only given permission to ask when she was likely to be allowed home. This, in fact, she urged her father to do, as local doctors who had recommended patients apparently wrote to Lister regularly enquiring about their progress [108] and she indicates friends of patients wrote directly to Lister too. Barclay clearly did not do this on Margaret's behalf. Perhaps she had initially been concerned that her father writing directly to Lister or Cheyne would be taken as unacceptable, as Cheyne had sent a letter through Barclay, but possibly, on inquiry, she had discovered that others did it, so gave her father permission to follow suit.

By this time, she saw little of Cheyne, as his period as house surgeon ended in April, and direct consultation with Lister was not easy, especially as he was away for periods of time. Margaret also hints that Cheyne was preparing to leave for London with Lister, though she may not, at this point, have realised it would ultimately be permanent. In April, she wrote to her father:

Dr Cheyne dressed me yesterday (Sunday) for the last time. Appearantly [sic] as he's going away (I hear to London) tomorrow (I think for his holidays) and thus I took the opportunity of asking him, If you please Sir, will it be long ere I can get out of bed? No, it won't be long now, but he never said when. He told the Dr that has to dress us when hes [sic] gone that I was a countrywoman of his own thus you must treat her kindly. I think it was very kind of him to say so, and it shews he has not looked on me as a stranger, being from Shetland, & I always thought it homely to see his face among so many strangers, thus I am very sorry he's going away at all ...[109]

Cheyne continued to call in to see her when he was in town, and "would always speak and ask how I was getting on."[110] By late May, however, her arm was still "sorer than it used to be owing to being very tight bound up & that makes it feel uncomfortable together with an elastic bandage rolled over which tightens more ..."[111] She was further disturbed by a patient in a nearby bed, possibly suffering from an inoperable brain tumour, who rambled, plagued with severe headaches, and at times failed to recognise her or her own family. Margaret lived for letters from her family and copies of the *Shetland Times*.

Lister's patients necessarily remained in hospital for some time afterwards, to allow their wounds to be treated antiseptically, and Margaret was to be there around eight months from the time she had first arrived in Edinburgh, some of this period in a convalescent home. During this time, the nurse's refectory burned down, mercifully sparing the wards, and she witnessed accident victims and other tuberculosis patients come and go. She watched young children die, and people she had befriended. At times, all the death around them had a dampening effect even on Margaret's confidence in the doctors' skills.

Cheyne had been an enormous comfort to Margaret during what was a difficult and at times traumatic experience. His familiarity and good humour had made her feel more at home, and in the painful days after her operation, he paid careful attention to her. His concern about the operation had even prompted him to write to Barclay at home, to warn them. If the careful treatment Margaret had received from Cheyne was ever in doubt, it was brought sharply into focus when his period as house surgeon came to an end. A Mr Hart was assigned to dress Margaret's wounds, a dresser close to qualifying fully as a doctor. He wrenched her arm and treated it so roughly that it bled, and she was in considerable and lasting pain. Her letters home indicate that Cheyne, before he completed his period of duty, had considered her – and a number of other patients – almost healed and ready to go home.[112] On a visit to the ward in late June or early July, he "was much surprised to hear they were so far back after 2 months more time on their wounds, and it was easy to see that Dr Cheyne thought it was done on purpose".[113]

This last statement is important, given the prominence Margaret later afforded the incident, and it indicates that it may have been Cheyne who confirmed her belief that the dresser was deliberately trying to cause her pain. Margaret wrote to her father on July 2nd 1877 that "I believe had Dr Cheyne (Fetlar) staid here all the time I've been I should have been thinking of getting home now.".[114] Hart was playing dangerous games with one of Lister's patients, which he made worse by goading her, saying she "should be thankful that [her] arm [was] healing up so well".[115] She confronted him directly, however, and learned that he considered hers "a rare case and that's my chance for lessons".[116] He meant it was an opportunity for him to produce original evidence which could advance his career. Margaret, however, was not to be put off.

Well Sir, if you presume to dress me any longer so cruel, I am determined to inform on you, as I have that privilege if I choose, thus I am reminding you of that so as to prepare you for your dismissal Sir. Do you really mean it Margaret? I really mean what I say Sir, as I have suffered too long for your pleasure [the following part is crossed out] rather than to cause any gentleman to lose so important a situation as you are preparing to fill.[117]

Margaret was well aware that, as Leeson put it, "only three clerks would be chosen out of the twelve dressers, [and] the competition was severe.[118]

The dresser apologised, but did not improve his treatment. He clearly did not really believe either that Margaret would report him or that Lister would take it seriously. The Professor, however, was not fooled for a moment, nor was he prepared to allow bullying on his ward, and when he saw the condition of the wound, immediately knew it was the fault of the dresser (perhaps because someone had suggested it to him, possibly Cheyne). He had "expected to see its great progression at this date."[119] He conducted an investigation at her bedside, in front of the whole team, including Mr Hart. Had he chosen to speak to Margaret about it in confidence, the outcome may well have been different. At this point, Dr Roxburgh, Cheyne's successor as house surgeon, told Lister who the dresser had been. He did not force Margaret to tell him.

The incident shows how well Lister understood the mind of the bully. He took pains to make it clear to Hart that Margaret had not reported him, and questioned her in depth about whether she had considered the action deliberate on his part. She answered that, although he had told her he was conducting an experiment on her, she did not believe it had been intended cruelly, and that he "did it so as I should not have a stiff joint afterwards."[120] This was not enough for Lister, and he asked her how she could be sure it was not deliberate. Even when she replied that Mr Hart had joked about it, saying she "would be able to pull him across the bay near our place in Shetland when he came there to spend his Holiday yet someday perhaps,"[121] Lister was not completely convinced, and clearly found the matter disturbing. He admonished Hart: "You see this young woman has not said a word against you to any person & surely you will now treat her more gentle."[122]

She notes that nothing changed, however. Cheyne's interest in the matter was more than a passing concern. Recovery times affected his ability to return evidence of the success of Lister's procedures and, as he himself reiterated many times, Lister's students, dressers and house surgeons felt personally responsible if they made mistakes.[123] He did not, at this stage, know why the wounds were taking so long to heal, but clearly suspected the dresser's mishandling, and may even have mentioned it to Lister. Lister was away in London at the time, and only when he returned did he examine Margaret's arm and find that Mr Hart had "reopened what was set together".[124] Hart knew exactly what he was risking, and was clearly taking advantage of Lister's absence to experiment on her, as she herself was aware.

Lister probably encouraged patients to report to him any ill treatment or rough handling by dressers [125], and Margaret implies that they ran the risk of dismissal if the accusations proved to be correct.[126] The details of Margaret's case are interesting. They present a graphic account of how Lister dealt with cases of bullying and cruelty by his staff, and this has been highlighted recently by Mary Wilson Carpenter as evidence of 'patient welfare' in 19th century hospitals, posing the question as to whether Lister was unique in this, or whether it was more widespread.[127]

However, the details of the incident are complicated by the fact that a number of versions of the Sketch exist. I have largely quoted from a version dated August 8th, 1879, but in a version dated 27th September 1879, her *Preface* says that "in complying with the request of my friends to publish it I have written several copies having left out some insignificant items and put in others more interesting."[128] One of the items she considered "insignificant" involved details of the incident with the dresser. The later version is only partly in her hand, the remainder having been written out by Laurence Williamson, a resident of Yell, and known locally as a "self-trained scholar".[129] Moreover, two more versions have recently been discovered, dated 15th December 1878 and a complete version dated July 26th 1879.[130] In the latest of the manuscripts, Margaret's version of the events with the dresser changes significantly. In the 15th December 1878 version, she notes that she did not confront Hart until she was almost ready to leave the Infirmary, that he responded that she herself would be a "cruel lassie"[131] if she were to report him, and that he then invited her to come back and train to be a nurse. In the July 26th 1879 version, she suggests that Cheyne also asked her to return to Edinburgh to undergo nursing training.[132]

The discrepancies and progressive refining of the Sketch have been the subject of some debate.[133] Why did Margaret feel she had to change the

Microbes and the Fetlar Man

spirit of her account and omit the detail of this incident? In the September 1879 version, she says that she was concerned about informing on him "as there were a great amount of events might come out of it."[134] Cheyne may have been involved in her decision.

A letter has recently come to light which Margaret wrote to Cheyne in August 1879 about her possibly publishing her *Sketch*, in which she asks his opinion about her references to him. The letter may never have been sent, as it was found in a box of material containing correspondence by the Mathewson family, not in Cheyne's effects. However, if it was sent and she made a copy, particularly given the date of the letter, a month before the September version was produced, it is not out of the question Cheyne advised her against mentioning the incident. Margaret wrote in her letter:

> Perhaps Sir you have heard ere now that I have written 'A Sketch of my Eight Months in the Royal Infirmary' as a cousin of mine got a loan of it and went off to Fetlar, & read it to too many there. Thus you probably have heard of it. It was written in Dec. '78 only not at all in 'The Infirmary' as I had nothing noted down there of the many strange things I saw & heard (only 'The Convalescent' Meals) all the rest of it is entirely from memory & to be as a help to memory in after years D.V.[,] but of late I have been prevailed on to enquire about having it printed but I have not heard the printers terms as yet only I believe it will be too high for me. It is not written with a view to censure any person in connection with 'The Infirmary' as all deserves my highest approbation (with one exception viz Mr —— my 'dresser' for 3 months after you left there Sir. But that was experiments he tried I believe and that was the cause of my long stay & which also retarded my progress.) I had occasion to use your name Sir, very much in my first 3 months there in writing 'The Sketch', please Sir have you any objections therefore to its printing if the terms is moderate? I might have come over to Fetlar, & I would have taken it to give you a look through (and a laugh) but I cannot get as one of my brothers is very bad at present from 'pulmonary Consn.' Hoping you are well & wishing you every success.
> I am Dear Sir, Your humble patient, Margaret C. Mathewson.[135]

Margaret's letter to Cheyne is respectful, but remains familiar: she thinks her sketch would provide him with 'a laugh' and no doubt remind him of their 'larks' in the Infirmary. Carpenter has suggested that Margaret's concerns about mentioning the dresser "might have been hindsight after receiving some cautionary advice from Cheyne, and that it was because of such cautionary advice that she proceeded to revise the manuscript over the next six weeks. Perhaps he did even 'take a look through (and a laugh)' at her manuscript and suggested some editing."[136] He may have felt it inappropriate or open to challenge, partly because the dresser would have progressed in his career by then, and perhaps partly because it may have been taken out of context and reflected badly on Lister. Interestingly, if he saw the *Sketch*, Cheyne seems to have had no objection to the rather light-hearted references to himself, which tells us he had the sense of humour not to worry about them.

Lister's motives for catching the bully were commendable, and in keeping with the general impression of Lister's humanity written by so many of his students. However, his concern went beyond justice for the patient. He knew his critics would seize on any opportunity to discredit him, and if there were any failures because his procedures had not been followed, it would reflect badly on his antiseptic methods as a whole. He needed to be absolutely sure he trusted his assistants to carry out his instructions to the letter, and surrounded himself only with loyal advocates. Cheyne already ranked very highly in Lister's entourage, and he recognised how the human principles acted alongside the medical and bacteriological ones.

Margaret's dialogue provides a fascinating account of daily life in Lister's wards. She knew some the nurses well by the time she left the hospital, and considered Mary Logan to have become a friend. She knew, and had experienced, the nuances of the hierarchy in Lister's team of dressers and assistants, and she had undergone the painful preliminaries to surgery in an era of change. Her experience freezing in the 'dark hole' was terrifying. She had no idea what was going to happen to her, and was paraded before a theatre full of students, but she also presents a unique insight into Lister's medical team as one of the first to be trained in 'patient care' as we know it, on the principle that the more comfortable and informed the patient is, the more likely they are to get well. In another respect, too, Margaret Mathewson's account is unique. She was interested in medicine and how Lister's procedures worked, and was even encouraged to return to the Infirmary and train as a nurse.[137] Her powers of observation were exceptional, evidenced by her description of Lister's carbolic spray, to which she even adds a rough drawing:

> The top is a fountain, in the centre of which is a burner and wick, the bottom is also a fountain, and a burner and wick. This weeks [these

wicks] is lighted to heat the spirits of wine in the bottom fountain, also the Carbolic Lotion in the top fountain, then there's a little vial attached to the outside (in front) which is filled with another sort of Lotion, then there's three gutta-percha tubes joining together forming a spout which carries off the respective ingredients.[138]

One of the more interesting factual errors in her account attributes the invention of Lister's spray to Cheyne.[139] When Cheyne was taking notes at Lister's lectures as a student, Lister had already been using the spray for several years, and gave detailed instructions for its proper use.[140] We can only assume that, if Cheyne had told her this, he had meant that he made modifications to the spray to produce the specific version Margaret saw in use. It is more likely, however, that she had recalled it wrongly.

Her account provides an example of how Lister was keen that his patients understood his procedures, and why they were necessary. Margaret clearly went away with a better understanding than most. Lister did not attempt to cure tuberculosis. Instead, he treated aspects of the disease. He drained abscesses and cut away damaged flesh and bone, treating them antiseptically to avoid post-operative infection. However, he was acutely aware that, when the patients returned home after their convalescence, their living conditions were likely to leave them open to renewed infection. From a very early stage in his use of antiseptic treatment, he was at pains to ensure that, when his patients were discharged, they were either prescribed, or took with them, a supply of carbolic lotion to allow them to dress their own wounds antiseptically until they healed completely. Lister wrote to a Mr J.A. Collinson in 1870, who was clearly a private patient, given the "handsome fee"[141] he left:

> I am sorry you left without a supply of the material for the dressing. I send you more now by post, and I hope that the telegram will have prevented you from using any other dressing before it arrives. You saw me dress it so often that you will be able to use it quite well yourself. You wash the wound with some of the lotion of which I send you the prescription, and also wash the skin around, but do not wash the wound with dirty rag: always use a clean fresh piece for the wound itself. Then put on the wound a small bit of the green plaster dipped with the lotion and then a good large piece of the brown plaster.[142]

Most importantly, Lister was anxious that his patients understood *why* carbolic was such a key element in their recovery, and ensured that his students and medical team instructed them in the basics of the germ theory of putrefaction. The rather revolutionary assumption was that the more the patients understood, the more likely they were to comply with his instructions. Margaret was quick to grasp this, and although she referred to bacteria as "insects"[143], had clearly understood the gist. Just before she was discharged, Dr Roxburgh took her through a test of her knowledge and the antiseptic principle (I have modified the punctuation to make the dialogue more readable):

"Now Margaret, I must hear what you understand about dressing your own case (& all in general form that, which may come under your notice)."

"Well Sir, first I would wash it in the "Carbolic", then put a small bit of protective so big over the wound, then the dressing and bandages, taking care to bandage from me, and when touching an open sore not to touch it with my fingers."

"Can you make a dressing?"

"Yes Sir."

"Why bandage from you?"

"Because it goes closer and smoother on & keeps longer in its position Sir …"

"Why not touch a sore with your fore fingers?"

"Because, Sir, there's poison in the fore fingers & thumbs, also the great toes."

"Yes, very good. Now would you ever put wadding next a sore? It's nice & soft, isn't it?"

"Yes Sir. It's nice & soft but it should <u>never</u> be put to a sore as there's insects in it (tho unperceivable to the naked eye) and those insects would cause death if they got into a wound."

"Yes you are quite right, and had I known sooner you understood those things you should have been away ere now."[144]

Members of Lister's staff were, of course, expected to understand the bacteriological reasons for the use of antiseptics in preventing infection, but it was unusual for the medical profession to concede inside information to the uninitiated. For Lister, on the other hand, it was essential if he were to ensure patients did not return home and undo all the antiseptic work which

Microbes and the Fetlar Man

had taken place in the hospital. It was a pleasant surprise to Margaret, who had even expected Lister's lectures to be delivered in Latin[145], such was the usual divide between men of his background and people of hers. The incident emphasises a point Cheyne would later clarify many times, that Lister's main concern was prevention. He preferred to stop bacteria entering wounds rather than to wrestle with the consequences once they had taken hold.[146]

As we have seen, daily life on Lister's ward at the Infirmary was not only advanced in terms of treatment, but also in terms of a rather modern attitude to patient welfare and a human understanding of the needs and fears of his patients. Margaret Mathewson was by no means the only person to document her experience of Lister's wards, but she was the only ward patient, and the only 'laywoman' to leave such a detailed account.

She became so interested in the medical procedures she witnessed that she recalls rather proudly how Nurse Skene at the Convalescent Home in Corstorphine suggested she undergo nursing training, and would have been happy to recommend her to Miss Pringle.[147] However, she considered that her arm would preclude her taking this up, "for years to come."[148] She was nevertheless proud of the fact that she had impressed Nurse Skene with her knowledge of dressing wounds, and what she would do if she were "in a country place & miles away from a Dr and a bad accident happen from a scythe or such like"[149]. As we have seen, Margaret suggests a number of other people encouraged her to take up nursing, including Cheyne.[150]

By all accounts, she was a shining example of Lister's doctrine that, the more a patient knew when they left the hospital, the less likely infection was to take root and the condition reoccur when they arrived home. Nevertheless, when she finally reached Shetland, her "arm gathered three times over"[151] and she wrote to Dr Chiene in Edinburgh, who sent her a piece of tubing to drain it. In July 1878, Cheyne went home to Fetlar to visit his family at the Manse, as he did most years. While he was there, Margaret visited him from her home in Yell to show him her shoulder. "He probed it to see if it was sound at the bone,"[152] which left it sore for some days. She told him she had written to Dr Chiene in Edinburgh after her arm had gathered, and Cheyne was amused to discover that she had fitted the tube herself, in front of a mirror. He pronounced the shoulder "sound at the bone"[153] and was delighted to see such a successful outcome. There was still no resident doctor in Yell or Fetlar, and it was not unusual for Cheyne to find himself approached to help when he was home, but Margaret's case was different to most, as he had become so well acquainted with it in Edinburgh.

Margaret corresponded with Cheyne at least once more, in 1879, to discuss the details of her 'Sketch'. To all intents and purposes, her arm had recovered well. Two years later, however, the infection returned, and she died in September 1880 at the age of 32. Her father also lost two of her brothers the same year, probably also from tuberculosis.[154] She was buried in Mid Yell in Shetland on October 2nd.[155] In the meantime, Cheyne's life had been turned upside down by events which were to change the course of history.

6

With Lister to London

Pioneering British bacteriology

On his return from the Continent, Cheyne had very soon found himself in an interesting position. He had his new duties as Lister's house surgeon, but had also brought with him from his European tour valuable experience of laboratory work. It had fired his interest in what Lister called the "vexed question"[1] of the relationship of microorganisms to putrefaction in surgical dressings. With Lister's blessing, Cheyne had launched into an investigation of the "all sorts of things"[2] found in antiseptically-treated dressings in Strasbourg. The results from Lister's own hospital cases, all treated by his antiseptic procedures, presented persuasive evidence of the success of antiseptics in preventing putrefaction in wounds, but these figures did not in themselves solve the prevailing theoretical issues. They did not prove beyond doubt whether bacteria were responsible for sepsis in wounds, nor that antiseptics worked *because* they kept bacteria out. Besides, Lister was reluctant to allow Cheyne to publish his hospital statistics until 1879. In other words, surgeons – whether they were happy with the results of antiseptic methods or not - were still free to interpret these results as they pleased, and be loyal to one theory over another.

There was still a reluctance in some quarters to accept bacteria as the cause of infection. Cheyne's language summed it up when he used the word 'convince'. He said that "it was very difficult to convince surgeons that tiny pieces of protoplasm about 1/20000 of an inch in diameter could be the cause of the septic diseases, the surgeons of that day were interested in keeping up their anatomy ... and minute germs and processes of fermentation seemed very far removed from practical work."[3]

On the other hand, even when Lister began to win the argument against spontaneous generation, he still faced growing claims that micrococci were appearing in antiseptically-treated dressings. In Germany, some were even claiming to have found "moderate-sized bacteria"[4] in them. While Lister was aware of the damage carbolic acid did to skin, and would devote time to seeking a less harmful antiseptic, he was far more concerned by the possibility that it may not be as effective as it appeared to be, and the implications this had for the relationship of germs to sepsis. So far, he only had his unpublished clinical results with which to counteract critics.[5] He needed unequivocal proof of two issues: experimental evidence that specific bacteria were the direct cause of sepsis, and that the antiseptic procedures he had developed were the key to preventing them from doing damage in wounds. In encouraging Cheyne's interest in bacteriology, Lister was entrusting him with producing evidence towards solutions which would allow Listerians to adjust and progress their methods internally, rather than find them adjusted by critics who could choose to ignore the theoretical base underpinning the method. At this stage, Lister was also actively conducting experiments towards this end, and had a clear idea of the type of evidence that would be required, but was unable to devote all the time needed for such an extensive task.

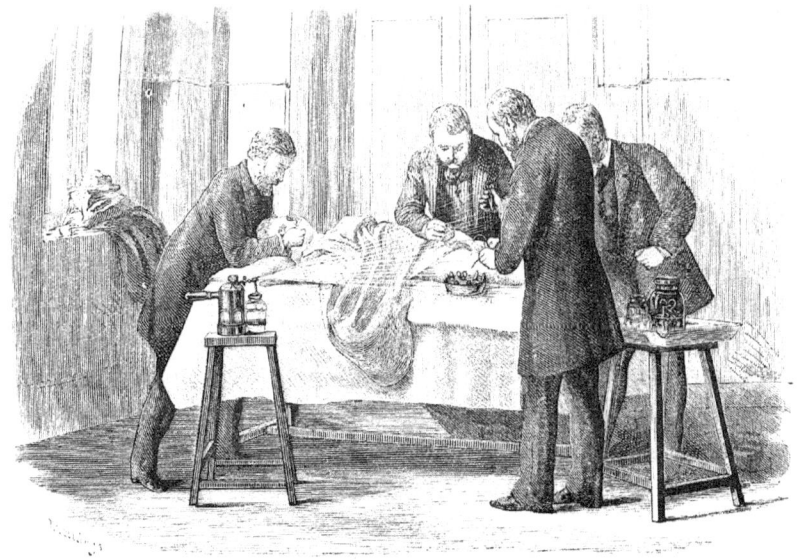

FIG. 23.

This figure represents the general arrangement of surgeon, assistants, towels, spray, &c., in an operation performed with complete aseptic precautions. The distance of the spray from the wound, the arrangement of the wet towels, the position of the trough containing the instruments, the position of the small dish with the lotion, the position of the house surgeon and dresser, so that the former always has his hands in the cloud of the spray, and the latter hands the instruments into the spray and various other points, are shown.

Illustration from Cheyne's Antiseptic Surgery, *showing position of the surgeon and his assistants during a Listerian aseptic operation, including the spray.*

Microbes and the Fetlar Man

The situation was further complicated by the fact that questions were arising about aspects of the theoretical base itself. There were doubts, for example, whether healthy tissues inside the body were always germ free, indicating that infection could develop inside the body, as well as being introduced by bacteria from without. If this were the case, Lister's elaborate procedures of sprays and dressings were only addressing a small part of the problem. If surgeons had found the procedures cumbersome in the first place, it must have seemed positively hopeless to try and keep germs from the air out of wounds if infection could arise in healthy tissues inside the body. Cheyne was in a good position to contribute to the experimental evidence Lister knew he needed to address all these doubts, as he could rule out any possible lapses in procedure by working only on dressings Lister had overseen:

> I saw that I had an opportunity which those with whom I had conversed abroad did not possess, namely, that my observations would be made on cases treated by Lister himself, and therefore would represent the actual bacteriological facts, and we could not plead, if the results proved unfavourable, that the treatment had not been thoroughly carried out. Lister was delighted and gave me leave to do anything I wished, and discussed plans with me and gave me much useful advice. It was, in fact, impossible for Lister to do this work himself, and therefore he was very glad to have it done under his eyes.[6]

In his makeshift laboratory in Edinburgh, Cheyne devised methods for obtaining the material for his tests, taking samples from the skin under the dressing. He introduced the samples into flasks containing a variety of liquids which acted as a growth medium, and noted whether microorganisms developed. If they did, he noted how long it took them to appear.[7] The results of his observations and laboratory tests perturbed him. He discovered that microorganisms were indeed capable of invading dressings which had followed an antiseptic course of treatment, even though the treatment had apparently been successful and there was no sepsis. Given his unequivocal loyalty to Listerism, this presented him with a dilemma. The revelation was not simply puzzling. It had the potential to jeopardise the credibility of Listerian theory.[8] He ultimately deduced that the microorganisms were micrococci, which he assumed were harmless, as opposed to bacilli, which he took to be noxious. He concluded that only the former were found in wounds treated antiseptically, as he could find no bacilli.[9] The wounds in which he discovered what he took to be micrococci did not seem to look quite as healthy as they should have, but neither were they septic. He was to continue his investigation for around three years before he published his results, and at some point took the experimentation a step further, reporting in a quite matter-of-fact way how he used his own body:

> I injected the cultivations in considerable quantity into rabbits and guinea-pigs, but they were apparently none the worse for it. To test it still more completely I injected some of them into my own arm and patiently awaited the result. On the first occasion I injected one minim of the cultivation, which must have contained myriads of these organisms. Next day my arm was painful at the seat of injection and a little swollen. A day later I went, from a feeling of duty rather than of pleasure, to look on at the Hospital sports, but I must confess I should have liked to have had my arm in a sling. But that was all: the whole thing passed off quickly without any local abscess or fatal result. I don't remember the details of the second injection, but I think that I did not put in such a large dose and that no special symptoms resulted. I was very young at the time, and I would not advise a repetition of the experiment, for the more virulent staphylococcus aureus or albus might be present also, a fact which I did not know at the time when I made this experiment.[10]

Before he could complete his investigations in Edinburgh, however, a series of unexpected developments were to open new doors for him. His leanings were clearly towards laboratory medicine by this time, but there were few, if any, professional opportunities for this kind of work in Britain during this period. The field was in its infancy. Cheyne himself said that "bacteriology had scarcely begun in this country; indeed, as active workers on bacteria at the time, I can only recall the names of Klein and Burdon Sanderson"[11], both of whom were working more on public health issues. Worboys has noted that "in the early 1880s, it seemed likely that Cheyne and Ogston would become the country's first bacteriologists …"[12] Alexander Ogston was another confirmed Listerian[13], based in Aberdeen.

Lister's encouragement was theoretically allowing Cheyne to pursue his bacteriological interests within the general confines of a career in surgery. Cheyne was still a young man, and a fiercely loyal one. On the one hand, he was genuinely fascinated by the scientific potential of laboratory experimentation in surgery. On the other, he was one of Lister's close circle, believed unequivocally in Listerism, and was at times a little apprehensive about whether his bacteriological experimentation would present him with a conflict of interest. He asked himself, "Could it be that the whole theory on which we were working was wrong, and that ... these minute bodies ... were only accidentally present in the wound ...?"[14] His first defensive reaction was to question the operator, thinking there had been an omission in Listerian procedure, but the issue with the micrococci had even excluded this possibility.

In 1877, barely two years after he had completed his degree, Cheyne wrote up his laboratory work on Lister's 'vexed question' into a thesis, which he called, "Researches as to the relation of microbes to wounds treated by Mr Lister."[15] "At this time," as he later told E.S. Reid Tait, "a sum of money had been collected in memory of Mr Syme, the interest of which (£100) ... was to be awarded to the best essay on a surgical subject sent in about the time of graduation."[16] Cheyne was awarded the Syme Scholarship [17], which brought him much-needed funds for two years.[18] It was not easy to find funding for research in Britain in the 1870s [19] and the scholarship was an important step forward in financing his work. Not long afterwards, William Turner, Professor of Anatomy, also appointed him one of his Demonstrators.[20]

Since there were no schools of bacteriology in Britain at the time, his living clearly had to come from surgery, but here he had an opportunity to remain at the forefront of the embryonic field of bacteriology too. In terms of ethics, Lister was well known as an anomaly in the medical profession, which was fiercely competitive. He was prepared to admit to his own errors of technique, as Cheyne and others would remark again and again.[21] In fact, Leeson found this aspect of Lister sufficiently noteworthy to comment: "fancy a Regius Professor at the largest medical school in the kingdom ... lecturing upon his mistakes! What a contrast to the general rule!"[22] His students were expected to follow suit and question their own work if anomalies appeared, but in making the statement about micrococci, Cheyne had ventured into the start of a long learning process.

He had barely embarked on his laboratory experimentation when his life changed dramatically. As we have seen, Lister had become something of a father figure to him. The professor's inner circle enjoyed a slight relaxation of the usual formalities, and it has been suggested that "the childless Lister took pleasure in their company, though they remained in awe of him."[23] Apart from his nephew, Rickman Godlee, however, Lister appears to have taken Cheyne more under his wing than anyone, and the professor had almost become the "surrogate father"[24] figure he had never really known. The result was that Cheyne would have done almost anything for him, and in the spring of 1877, he was given the opportunity to prove it.

One Sunday morning, he was asleep in the house surgeons' bedroom at Edinburgh Royal Infirmary when he awoke to find Lister shaking him. Lister told him he had received an invitation to become Professor of Clinical Surgery at King's College, London, jointly with John Wood. It was a difficult decision for Lister, and he could not make up his mind whether to accept the appointment. In the first place, he had to take into consideration the violent opposition to the invitation from his followers in Edinburgh. As soon as his students became aware of the possibility they might lose him, they held meetings and sent him pleas not to accept.[25] In the second place, Lister was more than aware of the general indifference to his methods in London. Initially, he declined the invitation.

In Edinburgh, he had a relatively comfortable existence, and loyal support was important to him for more reasons than one. The antiseptic principle had attracted a good deal of interest on the Continent, but not in England. Edinburgh was one of the strongholds in Britain, largely because it was Lister's base. He was uneasy about leaving his support base to venture into what he knew would not be an easy situation. Moreover, despite Cheyne's laboratory work and Lister's own experimentation, they had still to present any real solution to 'the vexed question'. What, in fact, caused sepsis in wounds?

In a personal capacity, Lister was persistent, but had suffered setbacks in his youth, in particular a breakdown following personal illness and the death of his brother.[26] Moreover, his solid academic training at home and at a Quaker school had combined to make him remarkably good at seeing the points of view of others. This was a rare trait in a rapidly advancing scientific world where people had strong opinions and remained fiercely loyal to their own views, but it could also be stressful to have to struggle against a body of opposition which must, at times, have seemed intransigent.

Cheyne insisted that Lister "spoke ... clearly and plainly"[27] in his lectures and talks, using "well-reasoned arguments, supported by facts observed in the course of experience and confirmed by experiment."[28] There is nevertheless evidence that Lister had difficulty expressing his written arguments clearly and, according to Leeson, he was well aware of his limitations. He apparently once said, "Well, gentlemen, it is no new thing for me to be misunderstood, somehow or other I do not seem to have the power of expressing what I mean in language that is capable of conveying my meaning."[29] If Lister were to accept the invitation to march into the lion's den in London, he needed to make sure he was surrounded by an armed guard of trusted assistants who knew and understood his work, and who could help him put the arguments over convincingly to the English medical establishment. He was not the first to recognise that being right, or even producing the evidence to prove he was right, was not enough. He had to understand battle tactics, even if he did not believe in war.

Lister's motive had little to do with self-image, and much more to do with ensuring his methods were not misinterpreted by well-meaning, but untrained practitioners, who would make mistakes and play straight into the hands of the opposition. This was why it was important to him that the link between bacteria and sepsis was understood, and that practitioners were conversant with the theory behind the practice, though he was also aware that bringing in theory could confuse the issue for many surgeons. Whether he mentioned theory or not, he still needed convincing experimental evidence to support it. Cheyne had become an important ally and assistant in this process, and Lister realised he was much too valuable to be left in Edinburgh. Apart from the young man's growing laboratory experience, Lister had been impressed by his determination and comprehensive understanding of the theory and practice of antisepsis, and recognised in him the spirit of the "crusaders"[30] who would help him tackle the "unbelievers"[31] in London with bacteriological evidence as well as statistics. When Lister shook Cheyne awake that Sunday in 1877, it was to ask him if he would accompany him to London should he decide to go, and to continue as his house surgeon for another six-month term.[32]

Cheyne's response was unequivocal, and has been quoted many times: "Go with Lister to London! I could not believe my ears! Of course I would go with him to London or anywhere else."[33] This burst of enthusiasm was to characterise Cheyne's future role as Lister's advocate, defender and friend. He and the handful of other colleagues Lister took with him were to provide the moral support Lister needed, as well as proven expertise in the new methods. Lister himself had been making frequent trips to London during this period, and despite his indecision still considered London his ultimate home. His nephew Godlee said, "If he occasionally hankered after his native city, it was because he was a Londoner at heart, and wished to extend his influence there."[34] He was away in London for several weeks in April, and only returned at the beginning of May after he had been approached in an unofficial way by members of King's College in February 1877.[35] He had responded positively, on condition he be allowed to change the teaching conditions there, but thought it unlikely they would agree. Meanwhile, a controversy had been unleashed in the English press when a reporter had picked up some remarks made at one of Lister's lectures about the superiority of medical teaching in Edinburgh compared to London. The London professors were outraged and Lister was forced to send a meek apology.[36]

Lister also stated publicly that he had no intention of accepting the offer, affected in part by a petition of some 700 signatures from students imploring him not to leave.[37] Another surgeon was appointed in London, but negotiations were renewed shortly afterwards and he ultimately accepted the new Chair of Clinical Surgery, which had been created especially for him. Apart from Cheyne, Lister asked Dr John Stewart, for whom he had enormous esteem, and two dressers, W.H. Dobie and James Altham, to accompany him. Stewart ultimately took Lister's methods to Canada, making a difficult decision to go into provincial practice and consultancy in Nova Scotia when his father became ill.[38]

Lister's final decision to accept the London offer was based on the reluctant belief that he could better serve humanity by converting the sceptics than if he were to remain in Edinburgh and preach to the converted.[39] In Edinburgh, he had hoped to produce a large number of trained advocates for his methods, who could then go out into the world and take the skills with them, but he ultimately believed he could speed up the process by going to London and tackling an influential establishment. In the meantime, he left trusted men in Edinburgh to ensure the transition did not jeopardise the use of his methods there.[40]

The consequences of the move were nevertheless felt almost immediately in the Scottish capital. Margaret Mathewson, Lister's Shetland patient who was still in Edinburgh at this time, was lucky enough to be one of only two patients who escaped one of the more ruthless effects of Lister's move to London. At Edinburgh Infirmary, a decision was taken to remove all Lister's long-stay patients, even those whose wounds had not yet fully healed. Lister's immediate successor, Dr Annandale, was reluctant to take responsibility for them, probably in case any died, which he did not want on his record.[41] It was handled at the Infirmary with insensitivity and

Microbes and the Fetlar Man

virtually without warning, but Lister himself went to enormous lengths to limit the damage. Margaret Mathewson described the announcement in graphic terms. "He [Professor Annandale] came downstairs & took all our names & said, "I want 8 empty beds tonight". 4 were turned out of bed & sent home as they were. Some got Lotions & dressings with them, others got prescriptions for to get it from the Druggist (some of them were crying at being sent away unhealed)."[42] Lister had six male patients moved to an Edinburgh nursing home and a female patient taken down to London at his own expense, but any remaining cases were discharged.

Margaret Mathewson described in her Sketch the extraordinary case of a parlour maid from Torquay named Lizzie Thomas, on whom Lister had operated in Edinburgh, and whose wounds had not yet healed. When the removals began, Dr Roxburgh wrote to Lister in London, and Lister "telegraphed back for her to be sent as soon as convenient but to travel in bed or in a lying position."[43] One of the nurses, Miss Logan, accompanied her, along with Dr Stewart, and Lizzie was bundled into "the operation basket"[44] so that six students could carry her to the station. The students created a rumpus at the station, and the whole incident caused "quite a commotion"[45] at the Infirmary. When they arrived at King's College, the sister in charge would not admit them as the proper admission papers were missing, so Dr Stewart appealed to a porter, who helped them in. Remarkably, Lizzie survived her ordeal and a second operation by Lister, and was able to greet him when he made a visit to Torquay in 1897.[46]

A few years later, Professor Spence would criticise Lister, suggesting that he left an estimated seventeen cases of uncured chronic abscess when he departed from Edinburgh.[47] Cheyne leapt to Lister's defence, and only succeeded in finding eight cases in Lister's notes, noting that they were initially placed under the care of Dr Bishop.[48] Lister, however, had second thoughts, presumably on hearing of the commotion in his former ward, and had asked Cheyne to "take charge of such cases as the house surgeon thought it desirable to place under"[49] his care. Cheyne took six of them, excluding a female patient who was probably Margaret's Lizzie, as he notes she was taken down to King's.[50] He remarked in his Boylston Prize essay submission:

> I had the satisfaction of seeing them <u>all cured</u> with the exception of the little boy, whose abscesses had become putrid before he left the Infirmary, probably from slipping of the dressings owing to the extreme deformity of his body. His father removed him in August

1878 to the West of Scotland. He was then considerably improved, having youth on his side to resist the septic influences; &, when he was last heard of, he was running about.[51]

Shortly afterwards, Cheyne too arrived in London, an event he later described as a moment he approached with trepidation, considering himself to have been a "raw and shy northerner with much to learn, and much to unlearn,"[52] whose "education was at once taken in hand most energetically by the then residents at the hospital …"[53] He said, "I was taught many things which were not included in the comprehensive curriculum at Edinburgh University, some of which, however, have been the greatest service to me since."[54] All those in Lister's party found themselves at the start of a process of adjustment, from the academic reception to personal matters, and even differences in local fashion. Stewart mentioned that he and Cheyne felt slightly out of place as they were both bearded, where all their London colleagues were "sprucely shaven".[55]

When the little group arrived in the lion's den, whatever fears they had about the reception of Lister's ideas were initially confirmed. Cheyne contrasted it sharply with his first experiences of Lister's lectures in Edinburgh saying, "There was no crowd running to get a good seat. Apart from Lister's dressers only a few students would stroll into the theatre quite casually, apparently taking little interest in the lecture and seldom taking any notes."[56] Cheyne, in fact, along with the dressers, often attended the lectures to make up numbers [57], as the only other regular attendees were "abler senior men"[58] or "those who had already obtained their degree".[59] The latter "had remained at the hospital with the view of acquainting themselves with Lister's work and if lucky of obtaining the resident appointment."[60]

If it was "disappointing and depressing"[61] for Cheyne after the exciting atmosphere of Edinburgh, Lister himself must have wondered if he had made the right decision, particularly given the alarming eviction of patients from his Edinburgh ward. They very soon discovered one of the main reasons for the general apathy in London. Students were not necessarily disinterested in the new methods Lister had brought with him, but knew they were unlikely to pass their exams if they openly subscribed to them. They knew that if they were to represent these views to examiners who were still resistant to them, they would almost certainly fail to obtain the qualifications they needed.[62] Cheyne implies here that this "backward state of matters"[63] was new to him in London. Leeson also explained that, "it was generally the best men who applied for the [Lister's] dresserships: those who had done well in the classes and were interested in their work, as they knew they would get little examinational help from him [Lister] and would be led into paths that were outside the curriculum."[64]

The Edinburgh examination system had in some ways allowed students to advocate Listerian antiseptic methods and still pass their exams. Firstly, professors, albeit with help from other examiners, set and marked the exams for their own subject, so Lister himself was in charge of clinical surgery. Moreover, the written questions in other subjects did not require the students to address the question of antiseptics.[65] Secondly, Lister and Spence, the Professor of Surgery, were on good terms, even if they differed on theory, or details of practice. Spence used carbolic to clean wounds,[66] so his students could get away with a loyalty to antiseptics themselves, and still pass their exams if they did not broach the professorial disagreements over other matters, except, of course, in Lister's clinical surgery exams. In fact, when Cheyne left Edinburgh for London, Professor Spence furnished him with a glowing reference, which described him as "one of our most diligent and distinguished students"[67] On the back of the letter was written, "With Prof. Spence's best wishes."[68] Relations between opponents had not, of course, always been this cordial, even in Edinburgh. The acrimonious dispute between Syme and Simpson [69], not always founded on medical arguments, had been a glaring example a generation earlier, but the system of examination did not, at least, tend to disadvantage the students in quite the same way as in London in terms of Lister's methods.

Even in London, as we have seen, Cheyne recalled that many of the students who were unwilling to be seen attending Lister's lectures at King's had actually entertained a private fascination for his methods.[70] Nevertheless, Cheyne's decision to follow Lister was a characteristically brave one, particularly given the fact that he could hardly risk any more changes of direction in life, if only for financial reasons.

Cheyne considered that Lister's inaugural address at King's had left his new professional colleagues confused and stunned. It was to prove, nevertheless, an important occasion in establishing his mark, and it has recently been shown that, far from showing an ingenuous disregard of his position, Lister planned the content and timing perfectly.[71] What may have been lost on some of his new colleagues at the time was, in fact, a pivotal moment in medical history: Lister presented experimental proof that a specific bacterium was solely responsible for fermentation in milk, and inferred from it concrete evidence that infectious disease was caused by a specific bacterium. To do so, "he devised a procedure to obtain a pure clonal population of B[acterium] lactis, a result that had not previously been achieved for any microorganism."[72]

He had conducted the experiments in the summer and autumn of 1877 [73], but as I have noted in a previous chapter, Cheyne's student notes show he was already laying the foundation for these experiments with milk at least as early as 1873, and was lecturing on them to his students. He ultimately published his results of 1877 in the Transactions of the Pathological Society of London in 1878.[74] He devised a procedure which became known as the "limited dilution method"[75] to isolate the bacteria. Jakob Henle had proposed in the 1840s that diseases were caused by living organisms, but had established that the mere presence of bacteria did not indicate their link to disease or fermentation, because they may have been there accidentally. To prove the link, a pure sample of a living microorganism would have to be introduced into an appropriate host, and if disease appeared, it could be taken as a definitive link. Until Lister's experiment, no one had succeeded in obtaining a pure sample of a living bacterium.[76] Robert Koch, a pupil of Henle's, also took his professor's theory to its logical conclusion around this time, and developed methods using solid media to prepare pure cultures.

However, despite the dramatic timing of Lister's address, it left some members of his audience puzzled. Cheyne later noted that, "Perhaps not unnaturally they expected to be told about the revolution in surgery which Lister had inaugurated, and … the majority of the surgeons present could not understand what the lactic fermentation of milk had to do with surgery."[77] Nevertheless, Lister considered it safe to infer from his results a definitive causal link between specific bacteria and infectious disease, and this was key to understanding the theory behind his practice.[78]

Cheyne's appointment as Lister's house surgeon was for six months, after which he was appointed Extra-Sambrooke Surgical Registrar.[79] However, he faced other important challenges in sustaining his position at King's. Surgeons were generally expected to make the bulk of their living from private practice, and he was not sufficiently well known to establish one. Theoretically, Cheyne's association with Lister should have been a guarantee of his reputation, but for many years he was viewed by both his colleagues and the public as Lister's bacteriological assistant.[80]

With Cheyne's general shyness and overall fixation with his bacteriological work, it is also possible he did not make it particularly clear that he was seeking private patients until he realised he was without a viable income. As no immediate way of making a living presented itself, he even at one point gave some thought to entering the Indian Medical Service [81], which offered a secure income and would not have been inconsistent with his bacteriological interests, given the Service's interest in public health. Cheyne had by now abandoned the idea of using his medical degree to take him to sea, and he was no longer contemplating naval service. Either the ever-present threat of tuberculosis precluded it, or he had finally decided that his future lay elsewhere. It is likely that he realised working as a ship's surgeon was rarely considered more than a temporary experience for recent graduates [82] and that, in fact, he was capable of more. Even with this decision, however, he needed to find a way to establish a living in medicine without having to abandon his laboratory work.

Cheyne ultimately approached Lister with the problem. Lister was reluctant to lose him, and provided him with a retaining fee of £200 a year to administer anaesthetics for him [83] (around £11,000 today).[84] He also made him his private assistant, a position he was to share with Rickman J. Godlee, Lister's nephew. Godlee was to remain a lifelong supporter of Lister and his methods, and wrote the first definitive biography of his uncle. Both he and Cheyne were ultimately to become Presidents of the Royal College of Surgeons, and Cheyne was to build a successful practice in Harley Street. This took time, however, and illustrates the importance of reputation in a period of medicine where ideas and methods were developing at an alarming rate. It was significant on two counts: his acceptability to the general public, and the way he was perceived within the profession. Many surgeons were struggling to keep pace and weigh up the variety of confusing arguments for and against practices. It can only have been more confusing still for the general public, and those with means relied on recommendations and reputation in making their choice of doctor.

In terms of general acceptance, he first had to make his reputation as a successful and reputable surgeon. Where his reputation with his colleagues was concerned, Cheyne probably saw *himself* as Lister's bacteriological assistant, and was quite comfortable in the role until he realised the consequences. Most significantly, however, it may have been the fact that he was seen as *Lister's* bacteriologist that introduced the real dilemma. Lister was interested in bacteriological evidence from a very specific perspective which not everyone considered relevant to surgery. Cheyne's role as his bacteriological assistant implied his loyalty to Listerism, and his publications in the late 1870s and early '80s more than confirmed his views in this respect.

In 1878, Cheyne finally presented an account of the findings of his bacteriological experiments to the May 6th meeting of the London Pathological Society [85,] which was published in 1879 as "On the Relation of Organisms to Antiseptic Dressings".[86] In Strasbourg, he had been confident that the bacteria found in antiseptically-treated dressings must have been the result of inadequate procedures, such as the incorrect use of carbolic, but it was not quite so easy to conclude this when he discovered micrococci in dressings which had been overseen by Lister himself. Since the presence of micrococci, which could clearly survive carbolic treatment, did not seem to mean the wound was septic, he concluded that they were most likely to be harmless. The statement was to have important consequences for Listerians.

It was ultimately, however, in his role as a surgeon that he was able to sustain the living which allowed him to continue his bacteriological work, and was realistically the only professional path open to him. In pursuing this course, Cheyne still faced an important professional hurdle. He had to sit the exams for Fellowship of the Royal College of Surgeons of England. Throughout the 19th century, progressive attempts had been made to regulate the medical profession to help exclude unqualified practitioners, who were all too prevalent. By the end of the century in England, the Royal College of Surgeons had a well-developed system of membership, and anyone wishing to progress in surgery at the levels to which Cheyne aspired needed to take the exams for admission as a fellow. In 1879, he first studied for the initial Primary Examination in physiology and anatomy, and then several months later, the Final Examination, as the two exams had been separated by 1876.[87]

Although Cheyne was responsible for aftercare in Lister's wards, they were half empty for some time, until Lister's reputation had time to spread.[88] It gave him time to study, but contributed little to his living costs or the cost of taking the exams. He needed to find a way to consolidate his position and fund his research as best he could. His laboratory work gave him the opportunity to submit essays for awards, but it seems there were more opportunities for these in America and on the Continent than there were in Britain.[89] Not surprisingly, therefore, Cheyne began in 1880 with a submission for the American Boylston Prize, awarded by Harvard University, and won it outright with his essay on "Antiseptic Treatment, What are its essential details? How are they best carried out in Practical Form?"[90] He signed his submission, "Truth, our Guide"[91], as it was a requirement of prize submissions that the manuscript itself should not contain the candidate's real name, so applicants had to use a pseudonym.

Any clue given to the judges on the identity of the candidate would lead to disqualification.[92]

> Each dissertation must be accompanied by a sealed packet on which shall be written some device or sentence, and within which shall be inclosed the author's name and residence. The same device or sentence is to be written on the dissertation to which the packet is attached.[93]

Cheyne's choice of pseudonym seems to have been either a reflection of his own version of Listerian objectivity or a tacit gauntlet to Lister's opponents. The *British Medical Journal* believed Cheyne was the first British person to win the American prize [94], which brought with it a very welcome $150.[95] The judges were exacting, and had, in fact, failed to award a prize the previous year as they did not feel they had received any essays of a sufficiently high standard. In Cheyne's case, however, the adjudicators had considered the work "of such great excellence"[96] that it "led the Committee to add to the award the gold medal of the Boylston Prize."[97]

*

Cheyne, writing retrospectively, painted a bleak picture of the situation he found in London. He was not impressed with the surgical results he read in the documentation:

> When I was Lister's house surgeon in London there was a recess in my bedroom with shelves in it on which were kept a number of notebooks of the surgical wards. If I were sleepless or awoke too early I used to take down these volumes and read them. The notes of the cases commenced with a careful history of the illness and a description of the disease or injury. In cases where operations were performed an elaborate account of the procedure was given. And then in an extraordinarily large number of the cases this was followed by a statement that the patients had died of one or other form of septic disease; healing by first intention seemed to have been quite a rare occurrence. To one trained in Lister's methods and in his wards these books gave an appalling picture of the surgical results of the period to which they referred, then only a short time before Lister went to London. This state of matters had lasted for centuries and had come to be looked upon as the natural and inevitable course of events after injury or operation.[98]

This highlights a striking contrast in the approach of different hospitals at the time. Leeson, moving from St. Thomas' in London to Lister's ward in Edinburgh, had experienced matters from a different point of view. St. Thomas', which then operated within a context of hospitalism, had recently been rebuilt on the pavilion system to which James Simpson and Florence Nightingale subscribed. After the emphasis on free movement of air at St. Thomas', Leeson was not initially impressed by "the gloom, the crowded wards and a bed in most of the corners" at Edinburgh Royal Infirmary.[99] He was surprised to discover that the patients were nevertheless "happy and comfortable, and were evidently doing very well"[100] in Edinburgh, even though he could see no "Nightingales"[101] (though by the time of Cheyne's house surgeoncy in Edinburgh, Nightingale nurses had been sent up to train a new generation [102]). The philosophy and training of the nurses had become a distinguishing feature in comparing ward procedures, even for medical students, depending on where they had been trained.

At King's, in fact, the fiercest opposition to Lister's methods came initially from the nursing staff. They were members of an Anglican religious community called the Sisters of St. John, who had been given sole responsibility for nursing at King's College hospital.[103] They had been formed in 1848 as a response to poor standards of hospital nursing, and their work was underpinned by moral and religious convictions as well as practical nursing procedures. Robert Bentley Todd, a founding father of King's College Hospital and advocate of improved standards in nursing, arranged for the foundation of a religious-based nursing community which would take on the nursing work at the hospital.[104] The community went through teething troubles, but finally found its footing when Florence Nightingale accepted its offer to help staff the wards of wounded during the Crimean war. They subsequently ran the wards at King's College hospital with nurses trained in the Nightingale tradition, and Lister and Cheyne fell out with them almost immediately.[105]

Cheyne later made light of it, rather derisively, by saying that "the hospital was nursed at that time by a sisterhood who looked upon the wards as their private rooms, into which no man should come without their permission, nor should he interfere with any of their arrangements, such as ventilation, etc. They could always produce a rule of their sisterhood why we should not do this or the other thing which we considered essential for the success of our work."[106] He suggested that the surgeons and their staff were forced to take it with a pinch of salt, as the nursing staff was sometimes given to imposing "childish"[107] and unnecessary restrictions in order to establish authority. He noted, with a hint of sarcasm, that he and his colleagues only got through with their "keen sense of humour (being more or less of Scottish descent)".[108]

To be fair to the nurses, who were operating under their own form of strict discipline, they were coming from a very different tradition to Lister's surgical staff. They had been in sole charge of dressings before Lister insisted on introducing his own (male) dressers [109], and they resented the intrusion on an area their training had taught them was sacrosanct. Trained in the Nightingale tradition, they concentrated heavily on well ventilated, bright wards and cleanliness, but Nightingale remained uncomfortable with germ theories of disease. The result was that the nurses trained in her methods did not understand the concept of asepsis in the same way as Lister's surgeons, and fiercely guarded their role in keeping the ward outwardly clean and aired.

Cheyne was used to his own discipline, and had learned the hard way as a student trying to uphold Lister's methods against the opposition. He was as unforgiving with the nurses as he was with himself. When he arrived at King's, he clearly took some pleasure in his superior understanding of aseptic arrangements, and from his accounts of it, the young surgeons half enjoyed their battles with the nursing staff, secure in the knowledge that Lister had taken over. They seem to have considered themselves to have been engaged in a permanent battle against ignorance as they understood it, and can have done little to dispel any popular notion that Listerians were an exclusive sect.

Lister's experiment on *Bacterium lactis* in the period leading up to the summer of 1877 had effectively provided evidence of the link between a specific bacterium and fermentation, which he extrapolated to be equivalent to infectious disease, including sepsis'.[110] Nevertheless, as we have seen, Lister did not always have much success in getting across the bacteriological case for the effectiveness of his methods. Instead, in 1879, he finally allowed Cheyne to publish his clinical statistics to show the rates of success in his hospital cases.[111] The paper caused an immediate controversy and a violent reaction in some quarters. Professor Spence in Edinburgh, one of Cheyne's old professors, claimed that his own statistics were comparable to Lister's, using antiseptics but without the elaborate procedures for disinfection, particularly the spray.[112] I shall discuss this episode in detail in a later chapter, looking at how Cheyne handled Spence's public criticism, but the argument illustrated how difficult it was to prove success definitively on statistical results alone, without the bacteriological evidence. Others in Britain had contributed substantially to the debate, such as John Burdon Sanderson and William Roberts, a physician at Manchester Royal Infirmary, who had conducted important experimental work in the 1870s to produce evidence against spontaneous generation.[113] However, breakthroughs in Germany attracted much more attention.

Germany in particular attracted an enormous number of scientific and medical visitors from Britain, including Lister and Cheyne. As noted in a previous chapter, Rickman Godlee suggested that the generation of German doctors working in the 1860s and 70s - the ones who were producing the breakthroughs - had largely undergone a university education, with a grounding in the scientific theory behind medical concepts. In Britain, on the other hand, Edinburgh was the only university to have "anything at all comparable to the German system,"[114] and in London, it "did not exist at all."[115] He was not suggesting English doctors were inferior, only that they had not had such a grounding in theory or laboratory work as the Germans. In trying to convey his antiseptic principle, Lister oscillated between manifestly linking it to theory and leaving it out of the explanation altogether [116] in case it led to confusion, particularly as there was little consensus on theory anyway.

Nor was bacteriological research very well supported financially in Britain, and German laboratories were less affected by legislation limiting experimentation on animals. This meant that they had become something of a mecca for British students and teachers of scientific medicine alike. It is then little surprise to learn that, where British researchers such as Lister and Cheyne had long understood the theory, and had produced experimental evidence of it, and where Lister had even anticipated some of the pending methods for verifying the evidence [117], they had nevertheless had difficulty convincing many of their colleagues. The Germans, on the other hand, had not only successfully claimed most of the advances in methodology, they were also to publicise ways of formalising the criteria for verifying the evidence. This latter development in particular was welcomed by Lister and Cheyne. If Lister's evidence was not acceptable to sceptics in Britain who took issue with Listerian theory, perhaps they would now be convinced of the theory by means of German methodology. Either way, it suited Lister and Cheyne's purposes when it came to illustrating the significance of the relation of bacteria to sepsis.

In 1876 in Germany, Robert Koch had made a breakthrough when he showed experimentally that a specific microorganism, *Bacillus anthracis*, was the cause of anthrax. Anthrax is a deadly zoonotic disease which generally affects livestock but can be transmitted to humans. It is still a significant problem in developing countries today. Koch's tests showed that, when putrefying substances such as infusions of meat or blood were injected into rabbits and mice, the animals developed infections. He also showed that anthrax bacilli could survive and cause the disease, even if they had no contact with animals. He first demonstrated his work in the laboratory of the botanist Ferdinand Julius Cohn at the University of Breslau.[118]

Unlike Pasteur, Koch was a doctor (a district medical officer in a rural district of Germany) before he was a scientist. Like Cheyne, he had cobbled together his first laboratory, in Koch's case in his four-room apartment.[119] Significantly, he had trained at the University of Göttingen under Jakob Henle who, as early as 1840 when the concept of disease caused by miasma predominated, had published work suggesting that infectious diseases were caused by living, parasitic microorganisms.[120] With the beginnings of bacteriological thinking in his medical background, Koch shared with Pasteur, Lister and Cheyne a belief in the importance of basing an understanding of disease on experimental evidence. He went further, however, and formalised a methodology for obtaining and verifying it which had the advantage of being particularly visual, so the behaviour of bacteria could be readily observed.

Firstly, he devised ways of cultivating bacteria on solid materials, such as potato, then gelatin and agar, rather than the liquid media used by Lister and Pasteur. The solid media more easily identified the specific bacteria responsible for the disease. He then developed methods for staining bacteria to make them more visible under the microscope, and pioneered the use of photographic images of microorganisms in his publications, as illustrative proof of his results. By publishing photographs of bacteria, everyone could actually see for themselves the structure of microorganisms and their relationship to specific diseases. Finally, he developed a set of standardised experimental steps, which became known as Koch's Postulates, for verifying that a particular organism was the definitive cause of a disease. The postulates provided a set of universal standards for the embryonic field of bacteriology, and this is where they were particularly useful to Cheyne.[121]

These steps required that a specific microorganism must be shown to be constantly present in diseased tissue, as opposed to being seen only occasionally, or by some scientists and not others. The microorganism had to be isolated from a diseased organism and grown as a pure culture, and this culture had to be capable of producing the disease when it was inoculated into a healthy animal."[122] Koch later added a fourth condition, which required that "the pathogenic microorganism had to be recovered from the infected animal and produced once again in pure culture."[123]

Koch's master work, *Investigations into the Etiology of Traumatic Infective Diseases*, was published in German in Leipzig in 1878, and was not immediately available in English. As we have seen, Lister's own experiments on Bacterium lactis pre-empted the logic of Koch's postulates, as did the work of Koch's assistant Loeffler, but Koch did not publish his definitive steps until 1884. Lister and Cheyne nevertheless already recognised in the 1878 work the potential of Koch's staining and photographic methods for illustrating the relation of bacteria to infective disease. Cheyne in particular was to be instrumental in bringing the methods to the attention of British practitioners, and he was as interested in Koch's laboratory and photographic methods as he was in the implications of the results. In 1880, at the request of the New Sydenham Society [124], he produced a translation of Koch's work, suggesting in his preface that the reader could not "fail to admit the beauty and importance which it records."[125] He particularly noted Koch's methods of illustrating his results with plates of bacteria photographed under a microscope. Koch had "forwarded a considerable number of the photographs to Professor Lister."[126] Cheyne remarked that they had "unmistakably been taken from sections of tissue"[127] and "when examined by a pocket lens or projected on a screen, show plainly that the drawings are faithful representations of what have been seen."[128] Cheyne had also "received from Dr Koch several of his stained sections,"[129] in which bacteria were highlighted in microscopic samples.

In the meantime, revelations by another major advocate of Koch's techniques in Britain led to a discovery which presented challenges for Listerian theory. Alexander Ogston, Surgeon at Aberdeen Royal Infirmary and a confirmed Listerian,[130] had visited Koch in Germany, and had learned staining techniques from him.[131] He supported the idea that specific germs could be linked to specific diseases. Where Koch had only been able to demonstrate the links between specific bacteria and specific diseases using samples from animals, convincing proof was needed that it also applied to human samples, and Ogston used Koch's techniques to fill this gap in an experimental study of abscesses.[132] He announced his results at a meeting of the German Congress of Surgeons in Berlin on April 9th, 1880, less than a month before Pasteur announced similar results using liquid cultures.[133] One aspect of his results, however, posed a challenge to Listerians, and Cheyne in particular. They showed that micrococci, which Cheyne had found in Listerian dressings but considered harmless, were in fact anything but harmless, and were the *sole* cause of abscesses deep within the body.[134] Significantly, Ogston's findings were initially published in Germany in 1880, and not until 1881 in Britain. Lister had been careful until this point to try and unite all advocates of the germ theory of putrefaction [135], and Ogston was no doubt aware of this.

Ogston's work showed that micrococci, as well as bacteria, could produce disease, and that Listerians had been wrong in thinking healthy, living tissues could not be attacked by microorganisms. His findings were taken by Lister's opponents to mean that Listerian procedures were not always effective, and that they were based on faulty theory.[136] This presented Lister with a difficult situation where he was forced to make a public choice between Cheyne's findings and Ogston's.

He sided with Cheyne, and attacked Ogston's views at the August 1881 International Medical Congress.[137] The results were nevertheless difficult to ignore. The implications were much more serious than Cheyne's assumption that inadequate antiseptic procedure had caused the introduction of micrococci, or that the latter were non-pathogenic. Cheyne and Ogston were among the very few working in experimental bacteriology at the time in Britain, and were both established Listerians, so the split could have caused damage. As it was, it had an unexpected side effect - it forced Listerians to adapt.

In 1880, Cheyne had submitted a dissertation entitled "The Principles, Practice, History and Results of Antiseptic Surgery", which was awarded the Jacksonian Prize [138], around £10-12 offered annually since 1800 to Fellows of the Royal College of Surgeons of England.[139] It was apparently paid through an endowment of £200 from a Colonel Jackson, who had once aspired to become a surgeon. The *British Medical Journal* commented that, had Colonel Jackson known of the illustrious pioneers in medicine who would compete for the prize, and the landmark nature of the dissertations it would fund, he may have contributed a larger sum.[140]

In March 1880, when Gerald Yeo resigned, Cheyne had been appointed Assistant Surgeon to King's College Hospital in his place.[141] His clinical reputation was not in doubt, nor was his reputation in bacteriological research. How, then, was he to deal with the discrepancy between Ogston's findings and his own? His ultimate response was to amalgamate and develop some of his prizewinning essays into a monumental work, *Antiseptic Surgery, its Principles, Practice, History and Results*, which he published in 1882.[142] Years later, he gave his reasons for publishing the work by saying that the time had now come for all the recent work on the treatment of wounds to be brought together and documented in book form.[143] He included updated material from his work on microorganisms in wounds and on temperature after operations, which had given him the Syme Fellowship in Edinburgh in 1877.[144] He added the material from his Boylston Prize essay for Harvard University on "the various methods of antiseptic surgery and the best modes of applying them to practice".[145] Finally, he included his Jacksonian prize-winning essay on the "history, principles, practice, and results of Antiseptic Surgery".[146] The amalgamated work was an exhaustive treatment, over 600 pages long, of the evolution of antiseptic methods in surgery to this point in time. It was translated into German by F. Kammerer and published in Leipzig in 1883 as *Die antiseptische Chirurgie; ihre Grundsätze, Ausübung, Geschichte und Resultate.*[147] The "salient points"[148] were also apparently translated into Chinese.

To address the implication that microorganisms could attack healthy tissue, he made a distinction between a germ theory of putrefaction and a germ theory of infection. The former involved the action of microorganisms on *dead* tissue, which he noted had always been the basis of Listerian theory, and the latter involved "specific, parasitic organisms causing disease in *living* tissues."[149] He attributed the difficulties in accepting Listerian methods to a confusion between the two theories. The distinction was not Cheyne's own. It had been suggested by Roberts in 1877 [150], but had not been accepted by Listerians at the time as it implied that the patient had to be protected both externally and internally from bacteria. It was as well that reviewers did not really pick up on the distinction, as Cheyne was to revise his position again a couple of years later.[151] Instead, the journals focused on his exhaustive retrospective on antiseptic development, which confirmed him as a leading chronicler of the subject.

In fact, in writing *Antiseptic Surgery*, Cheyne was in a good position to detail the ways Listerian treatment could be adapted, at the same time anticipating some of the objections of opponents. He incorporated a chapter on modifications to standard Listerian treatment, which already indicated the extent to which it had been adapted since Cheyne had taken notes in Lister's lectures as a student. He concentrated his advice on instructions for doctors in country practice, where the spray in particular might not be readily available, and emergency wound treatment in the fields of war, a subject he would develop more fully at a later date with his own war experiences in South Africa and the Dardanelles. First of all,

Microbes and the Fetlar Man

he took into account the objections provincial doctors had put forward to Listerism:

> The difficulties urged are that the spray is too heavy to carry: that it is not always easy to return a long distance to see the patient on the day after the operation, and that the dressings are too expensive for the lower classes.[152]

He addressed all three, including reducing the number of dressings, and gave instructions for dispensing with the spray altogether during after-treatment. He concluded that:

> By the means described, the difficulties in the way of the adaptation of this system in country practice may be overcome, and instead of causing additional expense to a poor patient, it saves expense in many ways. The dressings required are so few that the price of the materials employed is not greater than that which would be necessary even if water dressing were used; and expense is saved in many other ways, as I shall mention in the latter part of this work, notably in the rapid healing, which is of course of the greatest consequence to the bread winner.[153]

His section on adapting Listerian antisepsis to war conditions discussed a principle he was to assess and develop during his own war experiences, namely the value of early antiseptic treatment of a wound as soon as possible after injury. He detailed Lister's and Esmarch's methods. Esmarch had recognised that modern bullets tended to pass quickly through the clothes and therefore ran less risk of taking septic material with them into the wound. He suggested a soldier carry a basic antiseptic in his kit, and apply it immediately if he were wounded, but that he should not touch the wound with his finger, which could introduce bacteria. Thereafter, he could be carried to a field hospital, where he could receive more formal treatment, but in the meantime bacteria had been prevented from making the wound septic.[154]

Throughout *Antiseptic Surgery*, Cheyne paid due reverence to the methods of 'Mr Lister', but showed how much they had been adapted to suit the conditions of use in practice and advances in theory. A key factor in their adaptation had been the rapidly developing laboratory methodology. Worboys has noted that *Antiseptic Surgery*, from its moment of publication, established Cheyne in a role as "gatekeeper"[155] for the introduction throughout Britain of bacteriological research from the continental laboratories. He was to be particularly heavily involved in diffusing the methods developed by Robert Koch in Germany, and was to be consulted throughout the 1880s and '90s as the leading British authority on Koch's work. His evidence and reporting in cases of dispute were to be considered measured and responsible, and he was to make a major contribution to some of the central - in some cases - acrimonious debates of European bacteriology until the end of the century. Importantly, he saw most of the advances in continental bacteriological methodology as complementary to Listerian theory and practice, and a major factor in their continued adaptation.

7

Quite singular beauty
Illustrating microorganisms and tuberculosis

In 1882, before he had even reached the age of 30, Cheyne found himself a leading member of a small handful of pioneers in the new science of bacteriology in Britain. Through his work in the field, he was chosen to represent British medicine on one of the most important fact-finding missions in the early days of the formal development of bacteriological methodology.

In 1876, Robert Koch had risen to international prominence when he demonstrated a definitive causative link between the anthrax bacillus and anthrax. By 1880, he had moved to the Reichsgesundheitsamt (Imperial Ministry of Health) in Berlin, where he had the benefit of improved laboratory conditions, and it was here that he isolated the tubercle bacillus as the cause of tuberculosis in 1882. Tuberculosis was the biggest killer in industrial countries at the time.[1] Cheyne had long had an interest in the pathology of the disease, not least because some of his family had suffered and died from tubercular conditions. These particular developments were of interest to him for a number of scientific reasons, however. He was especially interested in the implications for tubercular bones and joints, but he was also on his way to becoming the flag-bearer for Koch's laboratory methods in Britain.[2] It was to be a period where he had the opportunity to investigate the methods further, in contrast to later years where we are left witnessing his sense of frustration in never quite being able to follow up to the degree he would have liked. Time-constraints, his duties as a surgeon and a general lack of resources would so often constrain him.

Within three weeks of Koch publishing his results, Cheyne had arranged a demonstration of the tubercle bacillus at King's. Koch's private assistant, Dr Goltdammer, had brought from Germany specimens prepared by Koch himself, which illustrated the bacillus in a human lung, tissue from an infected cow and a guinea pig "which had been inoculated with tubercle."[3] Goltdammer and E. M. Nelson assisted Cheyne in staging the demonstration. It was a well-attended event, testimony to the level of interest generated by Koch's discovery, and most of the best-known British authorities on germ-theory and antiseptic surgery were there.[4]

Cheyne was interested in demonstrating the bacillus and its characteristics to the British audience, but it was the methods of staining and observing bacteria which most arrested his attention. He could see their potential as a tool for rapid development in the field. He illustrated the bacillus chiefly by showing microorganisms for other diseases, including the ones that caused leprosy, anthrax and septicaemia, and how they differed from the tubercle bacillus when they were viewed as a stained section under a microscope.[5] There had been an enormous advance in laboratory methodology since Cheyne had begun to experiment in a corridor of Edinburgh Royal Infirmary as a recent graduate. The staining techniques, solid cultivating media and the slides used to display the sections containing the bacteria under the microscope, had all been developed in the last few years, largely in Germany.

In a demonstration Cheyne and Nelson gave at a meeting of the Royal Medical and Chirurgical Society a couple of weeks later, the staining methods were considered the highlight of the event. Professor Tyndall had received specimens of the tubercle bacillus from Professor Ehrlich in Germany. They were stained by a new method which illustrated the microorganisms with more "distinctness and beauty"[6] than in Koch's specimens. The report from the Society in the *British Medical Journal* considered the stained samples to be "of quite singular beauty and distinctness,"[7] and illustrated that "proper illumination was even more important than a high magnifying power."[8] It is interesting to sense the aesthetic appreciation in the medical world when the new methods were paraded before the profession. Journals presented a spectacle where practitioners betrayed an unmistakable sense of wonder as they watched the coloured microorganisms. They were captivated.

Demonstrations followed in other parts of the country, finally giving provincial practitioners the opportunity to observe for themselves the link between a specific bacterium and a specific disease. In June 1882, Cheyne displayed the bacillus at the Norfolk and Norwich Hospital before the Medico-Chirurgical Society [9], and would continue the systematic demonstrations into the following year.

In the meantime, however, an opportunity arose for him to investigate the tubercle discovery for himself, as a representative of British scientific medicine. Contemporary suggestions that British medical laboratory work was subject to greater constraints than in Germany had led to the establishment of the Association for the Advancement of Medicine by Research in March 1882. It was chaired by Sir William Jenner, and Charles Darwin was initially nominated as a committee member. Darwin died the day before the inaugural meeting, but he had taken "the warmest interest in the new Association from the first and was a magnificent subscriber to its funds."[10] His death meant that Cheyne narrowly missed coming before him to present his work. The Association's aim was "to promote those exact researches in Physiology, Pathology & Therapeutics which are essential to sound progress in the Art of Healing, and to remove any hindrances which obstruct these researches."[11] This last part of the aim was an implicit challenge to antivivisectionists in all but name, and the Association also criticised the lack of funding for medical research and researchers.

The 1876 Cruelty to Animals Act had recently restricted vivisection practices further in Britain by introducing a licensing system for laboratory researchers using animals, and was seen by many in the medical science field, including Lister and Cheyne, as a hindrance to British bacteriological researchers in making the kind of advances streaming out of European laboratories. Laboratory animals included dogs, and this in particular attracted strong protest. In Britain, the Brown Animal Sanitary Institute, founded in 1870, had been the centre of public health laboratory research using animals [12], and its work drew public criticism through a strong antivivisection movement [13], which ultimately led to the Act.

The Act stipulated that experimentation on animals had to be justified in terms of advances in saving or prolonging human life. Cheyne took out his first licence in 1880 or 1881 and later recalled that he held licenses for the next 12-15 years.[14] If any animal were used, it must be anaesthetised so that it should not feel pain.[15] This was not a new practice. Cheyne's notes from Lister's lectures in the 1870s contain evidence that Lister had been anaesthetising animals, illustrating how an experimental procedure on a donkey's heart had killed the animal simply by administering incorrect proportions of chloroform to air.[16] With the new system of licences, experimentation was subject to inspection, and the inspector who made the reports submitted them to the government, where they were kept on record. There was cause to resurrect the records from Cheyne's 1887-8 experimentation in 1891. A Mr Smith asked the Home Secretary in Parliament whether Cheyne had obtained a certificate to use anaesthetics when he conducted experiments which "involved boring holes in the knees of living rabbits".[17] The Home Secretary replied that these experiments were done under Certificate A, "which exempted him from anaesthetics"[18], but that the animals were either inoculated or sedated with chloroform. Lister and Cheyne were vociferous advocates of the use of animals in experimentation, and Cheyne later spoke in Parliament on the subject, attracting vehement criticism from antivivisectionists.[19]

The issue remained sufficiently controversial with the British public for Lister's supporters to feel the need to explain why a man who otherwise clearly felt moved by the sufferings of humanity should behave with such apparent disregard for the welfare of animals. Leeson suggested that it was Lister's "very pity"[20] for suffering humanity which led him to "inflict the smaller pain upon the lower creatures in order to avert a greater pain upon the higher"[21], and illustrated his love for animals by detailing his distaste for hunting, and describing the tins of water he put out in summer for the sparrows.[22] Cheyne, too, justified his support for the use of animals in the laboratory by separating the greater good from what he saw as a lesser cause.

In the wake of Cheyne's high profile demonstrations, the newly-formed Association for the Advancement of Medicine by Research saw an opportunity to put its aims into action. Koch was not the only European scientist to claim evidence for the cause of tuberculosis, and not everyone was immediately convinced of his arguments. National rivalries in Europe, particularly between France and Germany, also tended to mask the clarity of evidence. In France in 1881, Professor Toussaint had published a paper which claimed he had succeeded in cultivating from the blood of tuberculous animals and from tubercles an organism he regarded as a micrococcus. He claimed it produced tuberculosis when injected into animals [23], and he regarded it as essential to the disease.[24] Other researchers brought arguments against Koch, and some were not convinced that he had shown the tubercle bacillus to be the sole cause. An objective evaluation was clearly required, and the first major activity of the Association was to commission its own investigation into the matter. A sub-committee presented a recommendation to the Executive Committee on 12th July 1882 (only the fifth meeting after the establishment of the Association) that Cheyne be sent out to investigate the various claims in France and Germany as to the cause of tuberculosis.[25] They could hardly have chosen a higher profile subject.

We should be clear, however, that this was not simply an attempt to verify continental research. It was seen as a first step in a much broader investigation into the causes and characteristics of tuberculosis, laying the foundations for investigating not only bacteriological origins of disease but other factors too. The sub-committee "thought it best for the present to limit themselves to reporting on the best method of organizing & carrying out pathological enquiries"[26] on tuberculosis. However, the ultimate investigation "should be as wide as possible, that is to say, ... it should not be confined to verification of the results obtained by any particular method of enquiry, but ... attention should be given to each of the great aetiological questions referred to, namely those of heredity, previous disease and infection."[27] They were approaching the matter from the point of view of assessing the field initially, and gathering the tools and information for moving forward with their own research. It was partly because they were anxious not to pre-empt the relative importance of the bacillus or the conditions of the body which might make it more susceptible to the disease. The Koch school in Germany in particular placed enormous emphasis on the role of bacteria, and the Committee was aware of the various opinions in Britain, where "the non-specific view of tuberculosis" was widespread.[28] This was based on earlier experiments conducted by Burdon Sanderson and Wilson Fox, and by Cohnheim and Fraenkel in Germany (Cohnheim had later become convinced of a specific microorganism theory, however).[29]

It had important implications for policy. If a single bacillus were solely responsible, research into prevention could simply develop ways of stopping it from attacking the body, such as vaccination. If, on the other hand, more complex forces were at work, and bodies weakened by inadequate living conditions and bad nutrition were shown to be an equally important factor in the development of illness, measures would need to be broader and involve social improvements. The Association was anxious to recognise the broader possibilities, but focused specifically on a bacteriological assessment of the competing claims in Europe as it could not, at the time, identify anyone with all the requirements needed for a more extensive investigation.[30] When it came to documenting and assessing the current work of the continental laboratories, however, Cheyne was their obvious choice. He was interested in the methods used, had a good record in documenting the developments in the embryonic science of bacteriology, and had translated Koch's key work on the cause of diseases.

Cheyne was despatched to Toulouse to see Toussaint, and then to Berlin to investigate Koch's work, calling to see others on the way. He was granted £300 for the investigation, and was required to submit his report by January 1883.[31] During this period, they envisaged he would need to be out of the country for "two working months of the year"[32], and that, during the rest of the time, he would be "completing and arranging his work for publication".[33] He was fortunate to be released from his other duties in order to undertake the investigation, and Lister was almost certainly instrumental in recommending the move. It was announced to the medical profession in the *British Medical Journal* as follows:

> The Executive Committee recommended that, as a first step in the direct promotion of research, Mr Watson Cheyne should be requested to undertake the verification of the results lately obtained by Koch on the subject of tuberculosis, and the comparison of these with the results obtained by Toussaint and other observers[34]

It was a further opportunity for Cheyne to study continental laboratory methods first hand. It was funded, and he had been given leave from his surgical duties to carry it out. He arrived in Toulouse on July 21st 1882, where he was "kindly received by Professor Toussaint,"[35] who explained his methods in detail and showed him the results of his experiments on cultivating tuberculosis in laboratory animals. He used liquid cultivating media. When Cheyne left, Toussaint gave him specimens of the micrococci he believed to be the cause of the disease, "and also portions of various organs of tuberculous animals for examination."[36] Cheyne took them back to Britain to test for himself.

From Toulouse, Cheyne moved on to Berlin, where he first visited Dr Max Schüller, who considered Koch to have shown beyond doubt that the tubercle bacillus caused the disease, but that it was probably not the only cause. He had also described micrococci, which he considered to play a role.[37] Cheyne then moved on to visit Koch himself. He spent ten days in his laboratory, and said he could not "sufficiently acknowledge Dr Koch's kindness, nor the readiness with which he placed the results of his experiments and his methods at [his] disposal."[38] He "was able to appreciate his extreme care and accuracy in experimenting and his entire want of bias in drawing his conclusions."[39] Once again, he was given thorough instruction in the methods and equipment. He presented a detailed account of these, and the results of the experiments he witnessed, in his subsequent paper, including a particularly thorough comparison of Koch's staining techniques with other methods. He was also able to conduct experiments of his own using the methodology, and was again given samples to take away with him to undertake further experiments on his return. He said, "I made sections of a variety of cases of tuberculosis, and Dr Koch gave me several pieces of tuberculous tissue for further examination."[40] Though he was unable to include Professor Klebs in his visit, his report incorporated a detailed letter from him as part of the evidence. Klebs was also unconvinced that the specific microorganism isolated by Koch was the sole cause of tuberculosis, and pointed to "finely granular masses"[41] in Koch's cultivations which had the character of micrococci.

Drawing of Bacillus alvei, found inside one of Cheyne's books at Leagarth House

When Cheyne returned to London, he used the samples and equipment he had brought from Koch and Toussaint, but made it clear that he himself had considered it essential to conduct his own tests, not only to assess the results of Toussaint and Koch, but because he felt it important to understand why their results had differed.[42] He apologised at one point to the Association for delays in presenting his findings, but the results of his own cultures were not yet ready, and he had needed to send tuberculous material from eyes to another researcher, a Mr W. Jennings Milles, Curator of the Museum at Moorfields Ophthalmic Hospital, as he had not had enough time to investigate it himself.[43] Moreover, some of the animals on which he had conducted experiments had been stolen from the laboratory, presumably by antivivisection activists, but he added that "the result of the experiments had already been ascertained pretty fully in most of these."[44]

Cheyne presented his full report "On the Relations of Micro-Organisms to Tuberculosis" to the Association on February 1st, 1883. It showed that he had been unable to reproduce Toussaint's results, and his definite and positive results using Koch's samples and methods led him to the conclusion that Koch had shown a specific link between the bacillus and tubercular disease. It was an exhaustive "consideration of all the facts"[45] which demonstrated that "tuberculous processes in the lungs are due to the tubercle bacilli, and, so far as I know, to them only."[46]

It seems, however, that he had been too thorough. At their meeting on 22nd February, the Committee agreed that a number of amendments should be made, and noted that Dr Payne and Dr Burdon Sanderson had agreed them with Cheyne himself. They asked him to abbreviate the historical introduction which, if previous work such as *Antiseptic Surgery* was anything to go by, was probably too exhaustive. They also asked him to omit "references to opinions stated in conversation"[47] with Toussaint, as well as some of the "practical suggestions"[48] he had made at the end of the report. Cheyne had also admitted he was a poor scientific artist[49], so could not furnish drawings with the work, and the Association approved expenditure so that "the Report should be illustrated by microscopical drawings to be executed by Mr Murston, at a cost not exceeding £5 and that these drawings be engraved on stone."[50]

Importantly, Cheyne went one step further than he had originally been asked, and offered explanations for why Koch, Toussaint and the other investigators had produced different results. He had observed that the tubercle bacillus tended to develop in the epithelium of the alveoli of the lungs, and could lead to different conditions depending on the number of bacilli and how quickly they grew. It also helped explain why it produced different effects in rodents and in humans. His conclusions were welcomed with enthusiasm by supporters and with respect by sceptics, whose doubts, if they had any, were qualified by Cheyne's reputation. Henry Green, for example, wrote "let me add - and surely this must enhance its claim on our acceptance – Koch's investigations have been verified by one of our own countrymen, one who has a special reputation for this kind of work, Mr Watson Cheyne."[51] Spencer Wells in his Hunterian Oration for 1883 praised Cheyne for going further in his experimental observations, and offering explanations for the different results.[52] The debate nevertheless continued, and as late as December 1884, Dr Percy Kidd expressed doubts at a meeting of the Royal Medical and Chirurgical Society that "the connection between tuberculosis and the bacilli, to the exclusion of all other organisms, had been shown".[53] A lengthy and detailed response appeared from Cheyne in the *British Medical Journal* on January 24th 1885.[54]

Cheyne's European visit had established him as the leading British expert in Koch's methodology and one of the most respected of Britain's early bacteriologists. He had been entrusted with assessing one of the most important bacteriological discoveries of the decade, and in Britain his investigations were a major contributory factor to the general acceptance of Koch's claims. His further investigations, however, had shown that, with the appropriate tools, he and other British scientists were capable of making further inroads into the nature of disease. He continued to give talks and demonstrations through the spring of 1883, taking the methodology outside the hospital and university to medical practitioners and the interested public. At a meeting of "the Northern District of the Metropolitan Counties Branch"[55] of the British Medical Association on March 9th, 1883, held at the house of Dr Henty, Cheyne gave a paper on "Tubercle: Its Etiology and Modern History"[56], and illustrated it "with a remarkable and highly instructive series of preparations."[57] In May, he gave similar demonstrations at the new Town Hall in Hackney, accompanied by a paper on "Tubercle-Bacilli in relation to Tubercular Disease".[58]

His work led to an important research scholarship which allowed him to consolidate the achievements. In October 1883, the research committee of the British Medical Association

> ... unanimously decided to recommend to the Council of the Association to appoint to one of the research scholarships Mr Watson Cheyne, F.R.C.S., of London, whose reputation and work are of a high order, and whose recent researches on the bacillus of tubercle have excited much interest, and have been highly approved for the accuracy, industry, and skills which they display. The subject to which Mr Watson Cheyne will devote himself is that of the "Relationship of Micro-organisms to Disease".[59]

The following year, he was presented with a major opportunity to showcase to the general public not only his own skills, but also contemporary developments in bacteriology. On May 17th 1884, the announcement in the *British Medical Journal* of a major International Health Exhibition in South Kensington, with the Prince of Wales as patron, noted that it would have "special provisions for scientific instruction".[60] Cheyne was asked to erect one of two working laboratories to display the latest developments in public health and medicine, the other falling to Professor Corfield, who represented the environmental side of public health. The announcement noted that Cheyne would

> ... illustrate in a practical manner the kind of investigations with which it is important that medical officers of health should henceforth be familiar in the investigation of the germs of disease which multiply in air, soil, and water; the methods, the materials, the instruments, and the results of Pasteur, Koch, Klein, and others will be shown.[61]

The laboratory was situated on the top floor of the City and Guilds Technical Institute adjoining the exhibition. It was "somewhat difficult of access,"[62] but this was considered an advantage in attracting genuine visitors, where, according to the *British Medical Journal*, "the merely casual and ignorant visitor would find insufficient interest."[63] It was nevertheless considered to be "full of the most profound significance to medical and sanitary visitors of the ordinary preliminary knowledge."[64] For the first time, ordinary practitioners could see the results of some of the pivotal developments which had taken place in the last few years.

The laboratory seems to have been "largely fitted up by the aid of Dr Koch, and of Dr Koch's laboratory in Berlin,"[65] and Cheyne made two trips to Berlin to collect equipment and photographs. In explaining the background to the laboratory to Ernest Hart, Chairman of the National Health Society who was to give a lecture on the exhibition to the Society of Arts, Cheyne chose to emphasise the government support Koch received for his work in Germany, probably in an attempt to highlight a contrast with his own difficulties in financing bacteriological work. He noted that "Dr Koch's laboratory is subsidised by the government. It consists of director, library, biological department under Koch and several assistants, and a chemical department. All expenses of investigation are paid."[66] He nevertheless noted that "Koch's salary is only £300."[67] The laboratory Cheyne set up at the International Health Exhibition was effectively showcasing the kind of research environment available to researchers like Koch.

When it was up and running, the laboratory included a biological library which afforded "a kind of information which is not to be obtained anywhere else"[68] in Britain. It allowed visitors to study current bacteriological methods, and specimens of organisms grown in gelatine, including the tubercle bacillus, were on display. The "well lighted room"[69] also demonstrated antiseptic technology, with "more than one kind of incubator and disinfecting-chamber, as well as all the ordinary appliances of a microscopical laboratory."[70] Cheyne had transformed it "into one of the most interesting portions"[71] of the exhibition. He gave demonstrations at intervals, showing "cultivations illustrating nearly all the known pathogenic microphytes and bacilli connected with the infective organisms now so much recognised as causative elements in disease."[72]. The glowing report on the laboratory in the *British Medical Journal* noted that "Mr Cheyne's demonstrations were eagerly followed by health-students from all parts of the kingdom. A certain number of tables were set apart for study and research, and these were fully occupied from the first to the last days that the Exhibition was open"[73] Other speakers used the laboratory too. On July 26th, Mr C.B. Plowright gave a talk on "Potato Disease: its Propagation, and its Prevention".[74] In short, the project was a huge success.

Ernest Hart, in his lecture on the exhibition to the Society of Arts on November 26th 1884, advocated the permanent establishment of laboratories like this "as the best possible sequel to this exhibition."[75] He hoped for an Institute of Public Health, which could bring together the "numerous voluntary organisations"[76] working in public health with resources such as "libraries, class-rooms, and meeting-rooms."[77] He noted that, although Britain had led developments in sanitation, Germany and France had led in "sanitary research"[78] in the last decade, and he hoped that the laboratories at the exhibition had paved the way for "wiping away"[79] this criticism.

Another interesting consequence of the exhibition was what turned out to be some rather important work on disease in bees. Cheyne later explained in a letter to E.S. Reid Tait how he came to be involved. "One day … in connection with this laboratory … a gentleman, Mr Frank R. Cheshire, F.R.M.S. brought me a specimen of what he said was foul brood of bees. On examination a definite bacterium was found and we claimed that it was the cause of the disease. This has been disputed in America, but the matter does not seem to be fully settled yet." [Cheyne wrote this down in the late 1920s] [80]

They had effectively isolated a microorganism (Bacillus alvei), which they believed to be the cause of the disease,[81] and which led Cheshire to develop a treatment using a mixture of carbolic acid and syrup in the bees' food.[82] They published their results jointly in the *Microscopical Journal* in August

1885. On the basis of tissues brought to his laboratory in the autumn of 1884, Cheyne was also able to publish the results of investigations into a condition known as idiopathic purpura haemorrhagica [83], which produces red or purple blotches on the skin.

The exhibition produced one of the earliest manuals on the practical application of bacteriology, *Public Health Laboratory Work*, which Cheyne wrote with W.H. Corfield and C.E. Cassal. It was printed and published in London in 1884 for the Executive Council of the International Health Exhibition, and for the Council of the Society of Arts. Later editions of the book, written by Henry R. Kenwood, became the standard text used by the British Ministries of Health throughout the 1880s and 90s.[84]

People continued to send Cheyne samples to examine long after the exhibition, including an interesting case in 1886 which involved the suspected poisoning of guests at a wedding party in Carlisle. Dr Walker of Newcastle, the public analyst for the city, suspected the ham which had been eaten at the wedding dinner, and isolated a small bacillus which he then cultivated in different media. He had sent portions of the ham to Cheyne and to Dr Greenfield in London, and his discussions with them had been instrumental in his conclusions.[85] It is an interesting illustration of the extent to which bacteriology was now playing a role in police investigations.

When his report to the Scientific Grants Committee of the British Medical Association was published in the *British Medical Journal* in October 1884, Cheyne was finally forced to address his error in the interpretation of micrococci in antiseptic dressings, in the wake of the methods he had learned from Koch. He admitted that he had "missed the discovery ... that micrococci are present in all acute abscesses"[86] and that they were the cause. He suggested that, had he had the benefit of the new staining techniques, he may have reached a different conclusion. Nevertheless, he continued to doubt Ogston's view that micrococci were the *sole* cause of inflammations. When he recalled this paper in his correspondence with E. S. Reid Tait in the late 1920s, he chose not to mention this point, and focused instead on its "several points of great importance"[87], including the fact that "microbes in the blood are generally got rid of through the kidneys and this may explain the frequency of Pyelitis in connection with various septic processes."[88]

He followed the report with a *Manual of the Antiseptic Treatment of Wounds for Students and Practitioners*, published by Smith and Elder early in 1885. In the preface, he pointed out that it was complementary to his

[Reprinted from the BRITISH MEDICAL JOURNAL, September 20th, 27th, and October 4th, 1884.]

REPORTS to the SCIENTIFIC GRANTS COMMITTEE

of the BRITISH MEDICAL ASSOCIATION.

REPORT ON MICROCOCCI IN RELATION TO WOUNDS, ABSCESSES, AND SEPTIC PROCESSES.

By W. WATSON CHEYNE, M.B., F.R.C.S.,

Assistant Surgeon to King's College Hospital, Surgeon to the Paddington Green Children's Hospital, etc.

THE constant presence of micrococci in wounds not treated antiseptically, their constant presence in acute abscesses, their frequent presence in wounds treated aseptically and following an aseptic course, their presence in pyæmia in the form of emboli in the blood-vessels, and in the pus of pyæmic abscesses, and their association with other septic and inflammatory processes, are facts so striking and important as to merit the most careful consideration and investigation. The most remarkable of these facts is the constant presence of these organisms, often in large numbers, in the pus of acute abscesses before they have been opened.

The fact that the pus of acute abscesses may contain micrococci was pointed out several years ago by Billroth[1]; and, in 1879, I published somewhat similar observations in a paper on the presence of micro-organisms in wounds treated aseptically.[2] My observations were made on thirty-two cases by means of cultivation-experiments; flasks containing cucumber and meat infusions being inoculated with the pus with various precautions. In only seven of these cases did micro-organisms develop, and they were always micrococci. From these experiments I concluded that micro-organisms were not present in all acute abscesses; and, judging from the fact that they did not, as a rule, seem to interfere with the aseptic course of wounds treated aseptically, and from other facts mentioned in my paper, I held that they were accidentally present in the abscesses, and were not the cause. I did not, however, at that time, stain the pus by the methods of late brought into notice by Dr. Koch, and I thus missed the discovery made by the use of those means of demonstration some time later by Dr. Alexander Ogston, namely, that micrococci are present in all acute abscesses.

In his original paper in the BRITISH MEDICAL JOURNAL, and in a later paper on micrococcus poisoning in the *Journal of Anatomy and Physiology*, Dr. Ogston comes to very important conclusions, which may be shortly stated as follows.

1. Acute inflammation, so acute that suppuration is present or imminent, is always due to micrococci, save in the exceptional cases where a burn or blister, or some similar cause, has been at work.

2. The common micrococci that exist around us and in our intestines, are one and the same with the virulent cocci that cause inflammation.

[1] Coccobacteria septica.
[2] See *Pathological Transactions*, 1879, and also my book on *Antiseptic Surgery*. Smith, Elder, and Co. 1882.

Page from Cheyne's own collection of his Reports to the Scientific Grants Committee, which he cut and pasted into the volume by hand.

earlier *Antiseptic Surgery*, and he later told E. S. Reid Tait, "… the publishers asked me to write a smaller volume up to date."[89] However, the preface also noted that it dealt with "matters which were not ripe for discussion"[90] when he had written the earlier work. He abandoned the two germ theories he had noted in *Antiseptic Surgery* [91], and now believed infection in wounds was caused by "the growth of minute vegetable organisms in the discharges from the wound or in the tissues of the body, these entering the wound from without."[92] Traditional Listerism believed pathogenic microorganisms could not survive in living tissues. Cheyne had effectively adapted his own error of judgement into an updating of Listerian theory, which he termed "enlarging"[93].

He was to 'enlarge' it even further in investigations from 1885-6 on the broader conditions determining whether a microorganism caused infection, such as the number involved and how receptive particular tissues were to infection. He said that he had been puzzled by Lister's revelations in 1881 that he rarely found putrefaction in blood clots if there were only a few septic bacteria in it, and that this had led Lister to speculate that the number of bacteria helped determine whether or not a wound turned septic. The phenomenon continued to puzzle Cheyne, and he "decided to look into the question of whether this would apply to disease as well as blood clot."[94] He was surprised to discover in his investigations that "the number of bacteria which are introduced into a wound at the same time is a very important factor in the severity and rapidity of the disease."[95] He published the findings in the *British Medical Journal* on July 31st 1886 under the title, "Report on a Study of Certain of the Conditions of Infection."[96] Some of the more unsettling questions his research had engendered had led Cheyne to branch out from a focus on wounds, dressings and the causal relationship of bacteria to disease, into much broader studies on how this relationship worked, and how it affected questions of public health. This was happening at a time when Lister was largely confining himself to experiments to find an antiseptic which would be effective without the poisonous side-effects of carbolic or sublimate.[97]

Cheyne's translations of pivotal European work, and his visits to the Continent, particularly Germany, had also made him a recognised published authority on contemporary developments in bacteriology as a whole. In 1885, the New Sydenham Society had "decided that the time had come to select the leading papers, dealing with the relation of micro-organisms to disease, which had been published abroad."[98] The Society, known for publishing translations of European works, considered itself "fortunate enough to secure the services of Dr Watson Cheyne as Editor"[99], and the work was published as *Recent Essays by Various Authors on Bacteria in Relation to Disease*.

Taken as a whole, the essays concentrated on the "causal relation"[100] of specific bacteria to disease. Cheyne himself translated the paper on "Micro-organisms in Human Traumatic Infective Diseases", by Dr Rosenbach and "On the Aetiology of Acute Purulent Inflammations" by Dr Garré. The work also included essays on recent methods of disinfection and vaccination, but this was secondary to the important issue of the causal link. Cheyne noted in the preface that Pasteur's work on the rabies vaccine was included largely because of its importance, but that it did not, in itself, provide absolute proof of the link between bacteria and disease in the sense that they could not always ascertain whether the people vaccinated had been bitten by a dog that was rabid.[101] As we have seen in Lister's experiments with *Bacterium lactis* in the 1870s, the demonstration of a causal link between bacteria and disease in Britain was at least contemporary with the German evidence, yet the emphasis of the New Sydenham Society on foreign translations may have given the impression that investigators in Britain were lagging behind the Continent.[102] In fact, the emphasis of their work was different, and within both Britain and Germany there were internal differences in terms of the relative importance researchers attached to the way bacteria influenced disease. Lister himself is likely to have encouraged Cheyne to translate important work from the Continent. Cheyne went on to translate Flügge's 1886 edition of *Micro-organisms with special reference to the Etiology of the Infective Diseases*, published in 1890, once again by the New Sydenham Society.[103]

An important illustration of the differences in approach involved the search for the cause of cholera, and it led Cheyne into an acrimonious debate with Koch's opponents in Britain. The debate had implications for public health policy, and was sufficiently high profile to provoke an enquiry. Successive outbreaks of cholera in the 19th century had brought about public health reforms in Britain which were dictated not so much by any accepted understanding of what caused the disease, but the need to contain outbreaks.[104] In the earlier part of the century, measures had largely involved improvements to the environment, informed by a belief that cholera was a product of unsanitary conditions such as overcrowded housing and decaying waste which produced poisonous gases, essentially a miasma theory rather than a germ theory.[105] By mid-century, John Snow, a Medical Officer of Health in London, had recognised the role of contaminated water supplies in cholera epidemics, and had even succeeded in having a pump in London removed when he pinpointed it as the source of an outbreak, but the public health authorities as a whole only recognised water supplies as one possible source. Snow was sceptical of miasma theories, and suspected a specific agent, which was introduced accidentally into the body and multiplied somehow in the gut,[106] but he nevertheless still held to a chemical reaction, rather than bacterial origin for cholera. As early as 1854 in Florence, however, the same year as the London cholera epidemic, Filippo Pacini identified a comma-shaped organism and linked it to the disease, but his discovery was not generally taken up.

When cholera arose in Egypt in 1883, the British authorities, newly installed in the country, adopted the prevailing British public health approach to containing the disease.[107] They collected epidemiological evidence and spoke to local doctors, concluding that cholera was endemic in Egypt and that the solution was to improve sanitation, establish hospitals and control ports. The assumption was that the disease was arising from localised sources in the atmosphere and was not contagious. The German and French authorities, in contrast, sent bacteriologists to Egypt to investigate.

The Germans and French viewed it as an opportunity to show that a specific bacterium was the sole cause of cholera, and Robert Koch, for whom bacteria were all-powerful in defining a disease, was sent to find it. He identified a small rod-shaped bacterium but was unable to convince the French that this was the bacillus they were looking for, and he could not prove it definitively as he could not reproduce the disease in animals, fulfilling his own 'third postulate' (that an isolated, pure culture must be able to produce the disease when inoculated into a healthy animal). When the epidemic in Egypt subsided, he went in search of the bacterium in Calcutta, and this time identified a comma-shaped bacillus, but he had no more success in reproducing the disease in animals than he had achieved in Egypt, so once again, he could not definitively prove it according to his own standards. He nevertheless continued to claim success on the basis of the fact that he found this bacillus in Asiatic cholera, that it was always present in patients who contracted the disease, and that he had found it nowhere else.[108] The British authorities were not only unconvinced of Koch's evidence in its own right, but they also recognised that it would imply a change in their policy. The British government therefore sent out a commission of public health researchers to India to investigate Koch's claims. A principle member of the commission was Edward Klein.

Klein was a Croatian-born British microbiologist who had trained in Germany and had been working alongside Burdon Sanderson at the Brown Institute. Quoted in the *Gazette of India*, the Commission's preliminary findings were published in the *British Medical Journal* on January 3rd, 1885, even before the Commission had returned from India. They had concluded that there was no causal link between Koch's comma bacillus and cholera. The Commission had noted that "'comma-bacilli' occur also in other diseases of the intestines"[109], that they could not find them in any number or with any frequency "in acute typical cases of cholera"[110], nor did they "behave in any way differently from other putrefactive organisms"[111] when cultivated. Klein considered that bacteria did not cause cholera directly inside the body, but instead grew in unhealthy external environments, producing a chemical ferment which entered the body as a virus.[112] He differentiated this from the "true infectious diseases"[113] like tuberculosis, which he agreed originated with a specific organism in the body.

"The friends of Dr Koch"[114] reacted quickly. Two weeks later, Dr Heron brought, fresh from Koch's laboratory, samples showing evidence of another comma bacillus, discovered by Finkler and Prior in Bonn, which was not immediately distinguishable from Koch's bacillus by microscopic observation alone, but clearly behaved differently in cultivating material.[115] The debate intensified over the following months. Cheyne stood up publicly in favour of Koch in what has been called "a battle with Klein for the leadership of nascent British bacteriology."[116] In fact, it was a characteristically brave battle on Cheyne's part, and born out of pure conviction about the strength of the evidence. Where Klein was supported by the bulk of key opinion leaders in British public health, including Burdon Sanderson, Cheyne was backed by his colleague Crookshank and by Lister[117], a formidable team, if few in number. He was also supported, however, by Dr Heron's evidence of Finkler's bacillus.

On March 24th, 1885, Cheyne took part in a discussion of the cholera question at a meeting of the Royal Medical and Chirurgical Society.[118] He had loaned specimens and cultivations, as had Klein, and the all-important specimens from Dr Heron were also on display. Cheyne answered Klein's critique of Koch's comma bacillus, particularly defending the fact that Koch's methodology had aimed to distinguish it from very similar bacilli, which they could now demonstrate. Cheyne noted that the only reason he had been able to distinguish the organism was that he had used Koch's cultivations and advanced staining techniques, as opposed to microscopic examination alone. The discussion, as had been expected [119], was adjourned to a special meeting on March 31st, and the matter was given due prominence by an earlier start.[120]

On April 4th in the *British Medical Journal*, Klein criticised Cheyne's reading and interpretation of what Koch had actually said, suggesting that Koch had no idea about Finkler's bacilli when he made his claims for the comma bacillus, and therefore did not indicate that they could be distinguished only in cultures. He interpreted some of Koch's words to mean that he had only used microscopic methods. Klein reiterated that his own evidence did not bear out Koch's claims.[121] Cheyne responded a week later by quoting Koch in detail, and pointing to evidence that Koch had indeed identified similar bacilli, which nevertheless looked and behaved slightly differently, and had recognised the need to distinguish these by cultivation methods as well as microscopic observation. It was a battle of words, and a wonderful example of the importance not only of the new methodologies, but of scientific translation in guiding the arguments on both sides of the issue.[122] Klein was a native German speaker [123], and Cheyne a fluent one.

A few weeks later, in his capacity as Research Scholar to the British Medical Association, Cheyne published the results of a special investigation he had himself made of the cholera bacillus. The report was published in four exhaustive parts in the *British Medical Journal* in April-May 1885.[124] He had travelled to Paris in November 1884, timed, as he thought, to catch the outbreak at its height, but it had subsided more quickly than he expected, and he had experienced difficulty obtaining samples. This was compounded by problems accessing hospital laboratories, but he was ultimately offered the facilities at the Quinze-vingt Ophthalmic Hospital by Paul Haensell. His analysis of the few samples he could obtain showed that the only bacilli present which did not seem to be those always present in the intestine were Koch's comma bacilli. On his return, he sent samples to Koch to confirm that he was indeed working with the same organism.[125]

In his report, Cheyne argued a number of points in Koch's favour. Firstly, he noted the difficulty of pinpointing the specific agent of cholera because, unlike other diseases which had been traced to a particular microorganism in the blood, cholera arose in the intestine, which was full of other bacteria, making it difficult to pick out the one actually causing the disease. As animals did not appear to be susceptible to it, proving a causal link, by trying to reproduce it in animals, was difficult. This made it hard for Koch or anyone else to prove definitively that the bacillus caused cholera on the basis of Koch's own postulates for defining the cause of a disease.[126]

Cheyne also reiterated that, because a bacterium had to be distinguished by a whole range of factors, not just its apparent form, and because a number of very similar bacilli had been identified, observation under the microscope was generally not sufficient to distinguish between them, and cultures had to be employed. To make his point, he entered into a clear and exhaustive description of the various methods of cultivating bacteria to distinguish between them. He considered the explanation necessary as their details were "not sufficiently appreciated"[127] in Britain. The whole report said quite clearly that Cheyne considered the work of the British Commission to have been based on incomplete evidence, relying too heavily on microscopic observation. On closer examination, Cheyne and others had shown that the 'other' bacilli the commission found were not identical to Koch's one.

He was saying, in effect, that incomplete or inappropriate methodology had brought their evidence into question. By implication, he was challenging British public health methodology and saying that "the precision of Dr Koch's methods" and "the care which he has devoted to the subject"[128] were not emulated in Britain. Cheyne was not denigrating British scientists. He was simply defending the importance of comprehensive laboratory methodology and testing, and making full use of recently-developed tools.

The New Sydenham Society produced a prolific set of translations of German work during this period which nevertheless gave the impression that all bacteriological advances were emanating from Germany. Cheyne was one of their most valued translators. Most of his contemporaries had spent at least some time in German or Austrian laboratories, either during their studies or afterwards [129], and German had been an important part of their early education, so there was no shortage of experts in the field of bacteriology when it came to translation. Koch's work and new methods had been a gift to those in Britain seeking definitive evidence of specific bacteria as the cause of a disease, and had also allowed more detailed investigation of the nature of this relationship. Whereas in Germany, the national pride it generated filtered into government policy, and laboratories and researchers like Koch were well funded, the debate in Britain was more multifaceted. There may have been more than an element of 'challenging the Germans' in some of the reactions of the British scientific community.

When Cheyne pitted himself against Klein, all these nuances were involved at one level or another. There were those who were pleased to see British bacteriologists standing up against their powerful German counterparts [130], and may have deflected attention away from the very detailed arguments Cheyne set forth. Others simply held to the received chemical explanations for cholera.[131] Finally, there were those who highlighted the pointlessness of debate without proper facilities to test the hypotheses, and criticised the lack of investment in sufficient and appropriate laboratory facilities: "Is it not almost a national disgrace that this country, whose responsibility with regard to cholera is so heavy, affords to investigators of this new aspect of the question so few opportunities of studying it."[132] The debate reached proportions which led those interested in bacteriological questions to worry about the damage the ongoing public controversy could do.[133] It also had implications for domestic and colonial policy, as Ruth Richardson has shown in the case of the Indian Medical Service. Despite demonstrations of Koch's comma-bacillus in Bombay in May 1884 by Henry Carter, the administration stuck to its policy that cholera was a 'local' disease and needed to be contained by environmental measures.[134] Interestingly, Carter considered the organism a vibrio, rather than a bacillus, a fact which had been recognised by Pacini and which is now understood to be correct.[135]

The Medical Officer of Health in Bradford, Thomas Hime, wrote to the *British Medical Journal* on May 23rd 1885 bemoaning the fact that indecision on the matter could fuel misdiagnosis, raising alarms about Asiatic cholera when it may have been other, less virulent forms, with consequent effects on trade. He called for Klein to express his views clearly on whether Koch's bacillus, even if it could not be proved to be the sole cause of Asiatic cholera, could at least be a diagnostic indicator of it, as it was always found in cases of the disease, even if, as they seemed to be saying, other bacteria were present.[136] Hime considered Cheyne one of Britain's "ablest bacteriologists"[137] and suggested that the only way to arbitrate in the dispute between Cheyne and Klein was to repeat their experiments "before a competent commission"[138]. He went on to say that "Neither of the gentlemen named need consider such a reference to arbitration as in the least derogatory, and a decision of the disputed points would be of the utmost importance."[139] He suggested that Tyndall, Lister and Burdon Sanderson were eminently qualified to adjudicate the experiments.

Cheyne responded by accepting the challenge and throwing down the gauntlet, declaring emphatically, "Let us repeat the experiment side by side before the Commission."[140] He was confident of his ability to prove his point publicly:

SIR,-I have read Dr Hime's letter, in last week's JOURNAL, with great pleasure. I fully agree with him, and think that the only way in which this matter can be settled is by repeating the experiments in dispute before a competent commission. The subject is of such immense importance, that it certainly should not be allowed to rest in its present unsatisfactory state.[141]

He said that the experiments

> ... should be carried out in a laboratory in which cultivations of cholera-bacilli have not previously been made; that the apparatus used - more especially the syringes, needles, etc.- should be new; and that every step we take should be in the presence, and subject to the criticism, of each other and the members of the commission.[142]

Klein responded simply two weeks later, on June 13th,

SIR,-In answer to a letter by Mr Watson Cheyne, printed in your issue of May 30th, I beg to say that I shall be very happy to assist in any inquiry that has for its object the investigation of the relation of Koch's comma-bacilli to Asiatic cholera, provided Dr Koch takes part in the inquiry.
I am, your obedient servant, E. Klein.[143]

It was a clever way of declining the invitation, as it was clear Koch was unlikely to be released from his duties to agree to this public test of strength. It was generally viewed in this way by others. A letter appeared from the original correspondent, Hime, to say how disappointed he was in Klein's response, calling it a covert refusal to submit his experiments to the test of a commission of experts, and calling his condition of Koch's attendance "manifestly impossible (as well as unnecessary)"[144]

Koch himself, at the Second Conference on Cholera in Berlin, from May 4th-8th 1885, noted the various experiments he felt had repeated his results, including those of Cheyne, to whom he attributed Klein's eventual u-turn. Although it was not published in Britain until January 1886, Koch noted that:

> Klein's report has also in England been subjected to a very thorough and able criticism by Dr Watson Cheyne. Klein was compelled, in

consequence of the unanswerable objections made by Dr Watson Cheyne, to withdraw most of his assertions, or almost all which are of importance, and thus to record, in a drastic manner, the untrustworthiness of his former statements; more especially, he had to admit that the cholera-bacilli differed from those occurring in phthisis, in dysentery, and in the mouth; and he has further admitted that he has found true cholera-bacilli in all cases of cholera. Thus he finally comes, under compulsion however, exactly to the same result as I did - namely, that the cholera-bacteria are a specific variety, and seen exclusively in cholera. Klein will not be able to escape from all the conclusions which follow from these facts, unless he again involve himself in contradictions.[145]

Cheyne was initially vindicated by Klein's refusal to take up the gauntlet, but this did not mean he had won the war in Britain. The Royal Society contributed to the arbitration in quite a different way, and set up a Commission of Inquiry under C. Sherrington, C.S. Roy and Thomas Brown. The Commission was sent to Spain, where cholera was still in evidence, to identify Koch's comma bacillus. They did not find it, and suggested instead that the disease was caused by a fungus.[146] Cheyne must have felt cheated of a sound victory, given the direction of the debate, and all the detailed work he had put in to setting out the evidence. He was not a man to dwell on setbacks, however, and "bore no ill feeling on defeat".[147] He was too busy, and besides, it was not in his nature.

The debate in the journals returned to bacteriology in general, and the lack of support for its development. Many questioned the value of so much emphasis on discrediting Koch and not enough research into the real causes of diseases. *The Lancet* targeted the vivisection issue, and questioned why the government allowed the culling of rodents simply to remove them, but did not consider their experimental use in laboratories acceptable, even if the purpose were to save human lives.[148] There was also an economic dimension to disease control which affected public health policy in Britain. Quarantine, generally favoured in international ports, was a thorny issue for the British authorities, who recognised its damaging effects on trade when ships were detained in port for long periods. The 1872 Public Health Act had gone some way to improving administration with the creation of Port Sanitary Authorities. Where "non-quarantinable infectious diseases" [149] were concerned, the sick were detained in an isolation hospital, while the healthy were simply "monitored after disembarkation"[150]. The ship on which they had arrived was able to move on after disinfection. However, as Krista Maglen has noted, quarantine continued to operate alongside the new system for over 20 years after the Act, and "the ambiguous definition of 'quarantinable' and 'non-quarantinable' disease" [151] meant that cholera often fell between the jurisdiction of the old and new authorities. The result was that ships suspected of cholera often underwent inspections by both the old and new authorities, causing delays to shipping and trade.

The bacteriological debate on cholera was to rage on into the 1890s, when a vaccine was developed by Waldemar Haffkine at the Pasteur Institute in 1893 [152]. A significant part of Cheyne's bacteriological legacy, on the other hand, was his work in embedding laboratory testing and bacteriology into the clinical records and teaching at King's College Hospital. As Demonstrator of Surgical Pathology, he "secured the methodological microscopic examination of all tissues removed at operations, and he incorporated the results of these investigations into the histories of the cases."[153] To provide the facilities for this, he "inaugurated a department of his own at King's College Hospital,"[154] which he paid for largely with his own money. It was, in effect, an Institute of Clinical Pathology.[155] In 1886, he worked with Edgar M. Crookshank to set up an extra-mural course in bacteriology at King's College in London. It was the first time the subject had been taught formally in an English medical school, and was years ahead of its time. It would be many years before bacteriological teaching became routine in the medical curriculum [156], though Scotland had been ahead in this respect.[157] In fact Cheyne's obituary in *The Lancet* later recognised this when it noted that he was "responsible for the great advance in clinical teaching at the hospital."[158] The course was significant in theoretical bacteriological debates in the mid-1880s between supporters of Koch and Pasteur in Britain, and by 1888, Cheyne had also helped Crookshank deliver the first lectures in bacteriology at the Royal Veterinary College.[159]

Nevertheless, the basis of Cheyne's career remained a clinical one. By the mid-1880s, he was Assistant Surgeon to King's College Hospital, and Senior Surgeon at Paddington Green Children's Hospital, and it is remarkable how he found time for the bacteriological work he was still undertaking. His clinical commitments nevertheless reminded him daily of the consequences of the lack of resolution over the understanding and treatment of diseases like cholera and diphtheria, and no doubt inspired him through the long hours. The Paddington Green Children's hospital had moved to a new building in August 1883,[160] and Cheyne was there from at least 1884.[161] His work there was to be the basis for notes he published on the treatment of diphtheria, then a disease particularly affecting children, and Cheyne was involved in a high profile case of the infection, which affected him deeply.

Once again, Germans had isolated the bacillus they considered to cause diphtheria, Klebs in 1883 and Loeffler in 1884, and again, there was widespread scepticism in Britain. This time, Cheyne too was reluctant to stand up and say that the matter had been resolved definitively. He did not include it in bacteriological demonstrations he gave on March 27th, 1884 at the Parkes Museum [162], but by March 1887, he was prepared to say that Loeffler's research had shown the disease was due to a microorganism, most probably a bacillus.[163] He qualified it by saying that "the exact pathology of diphtheritic paralysis still remains a complete mystery."[164] However, even these guarded statements were not sufficient for everyone.[165].

There were a number of complications in verifying the cause of the disease, including the fact that, as with samples taken from the gut in the case of cholera, samples from the mouth were full of a variety of microorganisms, making it difficult to distinguish, and the public health establishment preferred once again to dedicate its resources to environmental measures. It could not be shown in all cases of the disease, and it sometimes appeared when there was no disease. Emil von Behring is credited with the discovery of a successful antitoxin in 1890, but it would not be until 1894 that laboratory diagnostics were successfully introduced in Britain to help reduce the costs of misdiagnosis and hospital admissions, and render use of the antitoxin more effective.[166]

Prior to these developments, a standard treatment for arresting the development of the disease was tracheotomy - opening a passage in the trachea to allow a patient to breathe, when it was clear the larynx was obstructed. Cheyne, however, advocated tracheotomy earlier, before the false membrane reached the larynx, in order to arrest the progress of the disease. His logic rested on judicious use of antiseptics. Tracheotomies were considered dangerous, particularly in children, and the success rate was not particularly good. They were generally performed when a 'false membrane' formed at the back of the mouth and the throat, and threatened to asphyxiate the patient.[167] Cheyne considered the lack of success in tracheotomies almost entirely down to infection, which he showed could be overcome by sensible use of antiseptics, and that the operation could save a patient's life. He demonstrated how this had worked in the case of a two year-old boy at the Paddington Green Children's

Hospital in September 1886, which he indicated had helped him test his theory successfully.[168] His real motive for pursuing the solution so carefully, however, came from an earlier case of which he had charge. He says he "was led to reconsider"[169] the question specifically because of what happened. Unusually for Cheyne, his professional writing on the subject leaks the impression that this sad case affected him deeply.

It concerned a woman called Frances Helen Prideaux. She was one of a new generation of women who had qualified for medicine and were proving their worth in a range of different hospital disciplines. In fact, Cheyne's cohort of students in Edinburgh had been witness to the fight to incorporate women medical undergraduates in the 1870s. The first women matriculated at Edinburgh in 1869, and attended mostly segregated lectures. Many professors felt mixed classes would restrict their ability to indulge in the "bawdy medical jokes"[170] which were "a traditional initiation into a masculine profession".[171] Lister was one of the strongest opponents to the inclusion of women in the study of medicine in Edinburgh, the first British university where it was permitted. The first women matriculated just as Lister took up the Chair of Clinical Surgery, and he had little influence, but he later campaigned actively to persuade the British Medical Association to prevent female membership and to expel its two existing female members. He even threatened to resign from the Association if they did not.[172] Lister was not a misogynist, and valued the input of his own wife, Agnes, in his work to the point where he felt he could not go on after her death. His opinions were rooted in his understanding of the role of women, and he could not accept the "unseemliness and impropriety of having medical topics discussed without restriction in a mixed company of men and women."[173].

Lister's views never really changed, and Godlee notes that he remained in opposition to women in the medical profession all his life.[174] His pupils, however, invariably mellowed. Stephenson, Cheyne's contemporary at Edinburgh, though he worshipped Lister in all other respects, was positively glowing on the subject of women in medicine by the time he wrote his memoirs in 1918.[175] Later in his career, Cheyne was very much in favour of women practitioners too. When he spoke at King's College Hospital Medical School in October 1918, he noted the presence for the first time of 16 women students, saying that "women had already shown that they did right to enter the medical profession. During the war women doctors had been of the greatest service to the country. Not only had they released men for medical service at the front, but had carried on base hospitals as efficiently as men."[176] If indeed Cheyne had initially been influenced by Lister's opinion, the case of Helen Prideaux, alongside whom he worked at the Paddington Green Children's Hospital, was probably instrumental in changing his mind.

She had graduated from the London School of Medicine for Women, which had been founded by Sophia Jex-Blake, a member of Cheyne's cohort at Edinburgh. The Committee of the New Hospital for Women in Marylebone Road was anxious to appoint Prideaux an assistant physician as soon as she graduated, but to gain experience, she applied for the post of house surgeon at Paddington Green Children's Hospital. Her first application failed, but she was elected a few months later and began work there in November 1885.[177]

Barely three weeks later, she contracted diphtheria. An abscess probably broke, and Cheyne, as Senior Surgeon at the hospital, suspected diphtheria and sent her home. The following day, however, when the disease was confirmed, Cheyne applied antiseptics (corrosive sublimate) with some success, but two days later the problems worsened and by the following day Cheyne discussed the possibility of tracheotomy, though a decision was taken not to go ahead. Miss Prideaux "failed rapidly",[178] which meant the tracheotomy had to be performed the following day anyway, by which time Cheyne implied it was too late, and she died shortly afterwards. Her obituary in the *British Medical Journal*, without saying so directly, implies that the decision on whether or not to perform the operation was pivotal in whether Miss Prideaux survived. It notes that "towards the end Mr R. Parker brought his special knowledge of tracheotomy to bear on the discussion of its advisability."[179] Cheyne is described as "unremitting in his care, never quitting the house, night or day, after the first spasm of the glottis had occurred, without leaving his colleague, Mr Stanley Boyd, in charge."[180] For some reason or other, perhaps because Miss Prideaux was such a promising member of staff, Cheyne took a special interest in the case and it affected him deeply.

Important developments had taken place through the 1880s in the way Listerians understood the relationship between bacteria and disease, and Cheyne's experimental work had contributed to the evidence which clarified these changes. He visited Berlin in 1886 at the invitation of Robert Koch, with the specific aim of conducting experiments to consider broader factors in the causes of infection and disease. Lister believed strongly in the power of the body to resist bacteria. He had already raised this issue in response to why ovariotomies could often be performed without antiseptic treatment and no sepsis ensued. At the Cambridge meeting of the British Medical Association in 1880, he had discussed the results of experiments which showed that the rate at which bacteria multiplied was not constant, and that different cultivating media, or 'soil' gave the microorganisms either more or less of a chance of growing and causing putrefaction.[181] His conclusion was that the blood of animals, including humans, had a greater or lesser resistance to particular bacteria, and that "when too few in number, the bacteria never grew at all".[182]

Cheyne was not present when Lister delivered his paper, as he had been assigned to accompany Koch, who was also present at the conference, "and take him about".[183] It was some time before he read the paper, but when he finally did so, he could see the logical implications. It meant that bacteria were not necessarily all powerful in themselves, and their ability to infect could depend on a range of factors, such as the resisting powers of the blood and tissues, and the number and form of the bacteria. In Berlin in 1886, Cheyne was able to test this using Koch's methods, showing that the severity of the disease "depended partly on the number of bacteria introduced in the first instance, and partly on the resisting power of the species of animal used for the test."[184] He presented his results as part of a report on his general experimentation on the subject to the Scientific Grants Committee of the British Medical Association in 1886.[185]

He consolidated his role as reviewer and commentator on bacteriological matters in a series of lectures at the Royal College of Surgeons in London in February 1888, entitled "Suppuration and Septic Diseases".[186] He summarised the current theories of sepsis in terms of their relative emphasis on either bacteria themselves or other factors, such as the resistance of tissues in the human body. He concluded that neither one stood alone, and that germs were not, after all, the all-powerful enemy, to be kept out at all costs, whatever the consequences, but instead only part of the problem, and were in fact relatively easy to deal with.[187]

The concepts of 'seed' (bacteria) and 'soil' (the levels of resistance of the human body) were central to bacteriological arguments in Britain in the 1880s, where the German schools were more inclined to see germs as all-powerful.[188] In Britain, the acknowledgement that both could act together was by no means new, but it was the relative importance advocates attached to each which had often divided them, and Cheyne's laboratory research had gone some way to bringing together the broad variety of factors involved in sepsis. For all his emphasis on the ubiquity and potency of germs, Lister had never doubted that the resisting powers of the human body were of significance and, by consequence, that rest was an important issue in the healing of wounds. Margaret Mathewson, his Shetland patient in Edinburgh, had been severely chided when she tried to write letters, putting her arm under strain during the healing process. She had been told it would put her recovery back a full month.[189] As early as 1879-1880, Cheyne had given this prominence in the introduction to his Boylston prize-winning essay:

> I do not mean to imply by this essay that attention to the antiseptic details is the only principle to guide us in the treatment of wounds. On the contrary the great principle is Rest. Rest from mechanical effects such as tension, movement &c. and Rest from chemical disturbances such as the products of putrefaction or other chemical substances used in the treatment of the cases. Of these no doubt the antiseptic details are the most important for by strict attention to them septicæmia & pyæmia may be avoided but without attention to the other means of obtaining physiological rest the best results cannot be obtained.[190]

In his subsequent *Antiseptic Surgery*, into which the Boylston Prize essay was incorporated, he devoted the concluding words of his exhaustive discussion of the subject to the value of rest: "The whole principle of wound treatment may be summed up in the one word - REST."[191] In his 1888 lectures, however, Cheyne was echoing a theoretical middle ground which helped keep pace with the advances in experimental evidence. Perhaps the summary nature of his 1888 lectures was an indication that he was contemplating turning his attention fully back to his clinical career. He would have had good reason, as, at the beginning of 1887, he had decided to marry.

Cheyne's family life had not begun auspiciously. He had barely known his parents, and had grown up not even knowing his full name or the identity of his father. He had probably grown up in a relatively strict household where we can only assume he was not an average member of the community in any social capacity. The only other child in the household was much older than him, and they probably had little in common. He had experienced periods of indecision and depression as a teenager, particularly when the bizarre circumstances of his father's death were revealed, and when he left for London he had no family members around him at all.

By the early 1880s, when he was barely 30, he was almost alone in terms of immediate family. On 13th May 1881, his uncle, the Reverend David

Webster, died leaving his two aunts, Christian and Grace, to find a home outside the Manse. They may have lived for a while on the neighbouring island of Unst, but Christian followed her husband to the grave two and a half years later on 15th December 1883. Cheyne would continue to visit his remaining aunt, Grace, until she died in 1906. By this time, she was living in Fetlar in a small dwelling registered as Hillside Cottage in the 1901 census,[192] almost certainly the house known locally as "The Chapel".[193] It had been a Methodist Chapel but had fallen into disrepair and was sold.

In other words, Cheyne had few immediate family members left by the mid-1880s, and the extent to which he maintained any close contact with his father's family is unclear. His friends and colleagues were marrying. In July 1884, he had been best man to a friend, James Sanderson of Galashiels [194], and it may have turned his own thoughts towards marriage. In many ways, by the time he had finished his studies and moved to London, Lister and his medical team had become Cheyne's family. As Lister's house surgeon, he had regularly slept in the hospital, though he had lodgings at 6, Old Cavendish Street. Leeson remembered with great fondness how Lister, like his colleagues, was given to having his house surgeons and dressers to dinner at his house. His language suggests that Lister's wife Agnes was a source of admiration on these occasions, if not also a surrogate mother figure for the students far from home.[195] Stephenson mentions meeting Cheyne again at a dinner, accompanied by Lister, many years after their student days, at the house of Sir Lauder Brunton.[196] Lister remained by all accounts a surrogate father to Cheyne, and along with Lister's nephew and assistant Rickman Godlee, the younger man may even have become a surrogate son to the childless Listers.

Cheyne's life, in fact, revolved entirely around medical circles. He can have had little time for anything else. By the mid-1880s he was Assistant Surgeon at King's College hospital, later Senior Surgeon, and was also Surgeon at Paddington Green Children's hospital, not to mention his experimental bacteriological work and publications, his regular lectures around the country and his work in private practice. It is therefore perhaps not surprising that, when he chose to marry, he chose a nurse. She was more likely to understand the pressures of his work and his regular absences from home.

Her name was Mary Emma Servanté, the daughter of the Vicar of Plumstead. She was four or five years younger than her husband, and had been born in London around 1857.[197] She had trained as a nurse at St. Bartholomew's, and was a resident probationer there in 1881,[198] so had been involved in nursing for five or six years when she married. She was one of the relatively new breed of middle class, educated gentlewomen who entered nursing in the wake of the Nightingale reforms, and part of the even broader movement towards a professional role for women in medicine. There is some indication she may have been estranged from her family. One of Cheyne's grandsons, who unfortunately never met his grandmother, recalled hearing that her father had disapproved of the marriage. He apparently had strong opinions, and disapproved of Cheyne's religious affiliations [199], rather an ironic coincidence given that Lister was forced to take similar difficult decisions when he left the Quakers to marry Agnes.

The Cheyne wedding took place in midwinter, on 29th January, 1887, in the district of Lambeth in London [200], notably not in Mary's father's parish, which may be further evidence of his disapproval. Mary was around 30 and Cheyne had just turned 34. There is every indication that she helped Cheyne in his work, and echoes of Agnes Lister and her role in her husband's work may have been part of what attracted Cheyne to Mary in the first place. She would later write to Lister on Cheyne's behalf when he was ill in bed "with his old trouble"[201], sending the results of experiments her husband had been conducting for Lister on anthrax.

Cheyne had been helping Lister test the effectiveness of specific antiseptic substances, and was working closely with him on experiments on iodide gauze. Anthrax contains spores, and carbolic in particular had been shown to be ineffective in killing them. Lister wrote instructions on the methodology for some of his experimentation "at Mr Cheyne's dictation"[202], but Cheyne also undertook experiments of his own and sent Lister the results. He reported two experiments on 6th September 1887, but did not consider either satisfactory.[203] He clearly intended to report the results of further investigation when he fell ill, and Mary responded for him on 16th September.[204] Her notes are clear and exact, and probably at her husband's dictation. Perhaps he really had found another Agnes Lister.

Their first child was born exactly a year after they were married, on 12th January, 1888,[205] and was appropriately named after Lister: Joseph Lister Watson Cheyne. He would be known as Lister, later Sir Lister, and the name would appear in subsequent generations. Their second son was born just over a year later, on 26th June, 1889,[206] and was named after another medical giant. He was called William Hunter Watson Cheyne, but in the family, was always known as Hunter, or sometimes Ha. In many ways, Cheyne seems to have seen his new family as an extension of his medical world, and for a little while, it appeared to be thriving.

8

Where tubercle bacilli most do congregate
The tuberculin affair

Professionally, Cheyne was now a recognised British expert on the bacteriology associated with tubercular bones and joints. As a recent graduate, he had already had experience in the field, and had followed the progress of Margaret Mathewson, Lister's Shetland patient, when she underwent an excision of the shoulder joint. It had been an unusual case, as the tuberculosis had moved from her lungs into the joint, offering an unexpected opportunity to prolong her life.[1] Cheyne continued to specialise in the field, and it would lead to an opportunity to demonstrate his integrity in one of the highest profile international medical scandals of the late 19th century.

In 1888, he submitted an essay on "Tuberculous Diseases of Bones and Joints", for which he was awarded the Astley Cooper Prize the following year. The prize had been established by the will of Astley Cooper, an early 19th century English surgeon, for the best original essay or treatise in anatomy, physiology or surgery, and a committee at Guy's Hospital in London set the subject.[2] It was only awarded every three years, and carried with it a sum of £300.[3] In this particular round, the theme was one where Cheyne had comprehensive and detailed experience. He could hardly have failed to apply, and even a quick glance at his essay shows why he could not have failed to be awarded the prize. He concentrated on recent developments in illustrating the bacterial relation to the disease, and the consequences for treatment.[4] The prize money may well have been an incentive, now that he had a wife and child to support, but the meticulous detail and comprehensive coverage were more typical of Cheyne's commitment to the subject.

His introduction to the essay is revealing, and he devotes considerable space to recent innovations in bacteriological methodology. For Cheyne, this methodology had provided the kind of graphic, visual evidence required to demonstrate to practising surgeons and physicians the relation of microorganisms to disease. Even better, he could now demonstrate *how* bacteria caused disease, and the possibilities must have seemed endless for further research. New staining techniques highlighted the form of bacteria under the microscope, and allowed researchers to document their behaviour in more detail. Most importantly of all, the methods were helping to show the link between *specific* bacteria and a *specific* disease, and to light the way towards a cure or appropriate preventative measure.

Cheyne's essay details the laborious process by which he sawed off pieces of tuberculous bone to give him the samples for investigation, decalcifying the bones and making thin sections of them. He noted that "only those who have attempted it can have any conception of the labour involved"[5]. He advocated a complete process of investigation, beginning with the naked eye, following up with a hand magnifying glass and then, particularly if he suspected anything unusual, examination under a microscope using advanced staining techniques. He considered this the only thorough method for observing tubercular material in bone, saying, "I have more than once thought that there was nothing of any special interest in the bones when I have sawn them, and yet on making thin but complete sections and staining them most important appearances have been brought out; and also it will be seen that my description of the appearances and the process of the disease differs considerably from that ordinarily given because I have been able to see and study the whole of the affected part from one side to the other."[6] Hence the difficult process of obtaining the samples, as he needed ones which, as he would later say, were "suitable both for naked eye and for microscopical examination."[7] Cheyne had learned what he considered the most effective staining techniques from Paul Ehrlich in Germany [8], who had published his method of staining the tubercle bacillus in 1882. Cheyne noted:

> The specimens are first stained with Ehrlich's haematoxylin, the formula for which is now published, and after it has become blue it is then stained in a mixture of acid rubine and orange. The result is at least a treble stain, and as different tissues take on these stains in various degrees of intensity and combination, we have often a variety of tints in the same section.[9]

He went on to show how this must then be viewed under an appropriate microscope in a specific way in order to see the stained bacteria to best advantage. He sent original specimens with his essay, and even sent his "Maltwood's finder",[10] with detailed instructions for using it, to ensure the judges located the item he was citing in the same way as he did, so they could verify his data. A "finder slide" is a reference grid for locating objects on a microscopic slide. A Maltwood's finder consisted of "a series of squares, one-fiftieth of an inch across, each containing two reference numbers enabling the square to be located by giving its 'latitude' and 'longitude'."[11]

Most interesting of all is Cheyne's incorporation of photography, and he submitted original images with the essay. Robert Koch had made extensive use of photography in illustrating the tubercle bacillus.[12] In Britain, Edgar M. Crookshank, Cheyne's colleague at King's, had become a major pioneer of the method, and had written extensively on the subject.[13] Cheyne was also enthusiastic about incorporating it into investigative procedure. People could doubt drawings, but here were photographs which, in those days, were still generally considered not to lie, though even photography was not without its sceptics. Cheyne himself was fascinated by photographic processes, but for his essay he engaged "a distinguished photographer who has lately taken up photo-micrography, having learned how to use the microscope in the best way, and I think it will be admitted that these photographs are far superior to any drawings and show a future for photography which must lead to its almost total superseding drawings as a means of representation of matters where absolute accuracy is required and where all individual though unconscious bias must be excluded."[14] He later revealed his collaborator as Andrew Pringle [15], a well-known Scottish photographer who had recently experimented with micro-photographic methods.

Cheyne noted that reproducing the photographs on opal glass would have improved their transparency, but he and the photographer had decided that, as the images would frequently be handled in the course of examining the essay, the slides would have had to be mounted in frames, a method suitable for exhibition and museums but inconvenient for his current purposes. He also had to consider the expense. He therefore "had them printed on albumen paper and mounted on separate sheets."[16] He even sent duplicates of some of the prints, on which he had "marked the various points".[17] He was afraid the committee would find the photographs "more confusing than a diagrammatic drawing"[18] at first, until they adjusted to the method and appreciated its advantages.

In his work with Lister, Cheyne had learned to anticipate a barrage of doubts and scepticism from a number of quarters. His meticulous instructions to the prize committee were an opportunity to showcase his use of the most accurate and up-to-date methodology, and to anticipate as many questions as he could think of. He would probably not have the opportunity to explain in any other way.

He was awarded the prize in 1889 [19], and proceeded to demonstrate around the country the illustrative power of the methods he had used. At the annual meeting of the British Medical Association in Birmingham on August 9th, 1890, he demonstrated his work using lantern slides made from the photographs. The *British Medical Journal* reported that "Mr Watson Cheyne showed what a degree of perfection microphotographs may reach, considering the difficulties which have hitherto been met with by those who have attempted to use microphotography as a recording method in microscopical work ..."[20]

He gave another lecture at the Harveian Society on October 16th 1890 [21], using specimens and lantern slides to illustrate the pathology, general conditions and treatment of the disease. He was praised for recommending expectant treatment in defined circumstances, particularly for children, with careful observation in preference to an automatic decision to operate at an early stage. Mr Edmund Owen felt that "British surgery up till now had been too much under the influence of German pathological laboratories,"[22] and he was under the impression "Mr Cheyne's practical essay might do much to place the question on a more satisfactory footing."[23] This is an interesting statement in the light of Cheyne's own prolific citation of German authors though, as we shall see, he too was selective in his use of German methods.

Cheyne had been appointed Hunterian Professor of Comparative Anatomy and Physiology at the Royal College of Surgeons of England in 1888 and again in 1890 and 1891.[24] He returned as Hunterian Professor of Surgery and Pathology in 1892 [25], and gave a number of lectures during these periods. He progressively refined the work he had presented for the Astley Cooper Prize essay, and gave a series of lectures to the Royal College of Surgeons, which were extracted in the *British Medical Journal* from November to December 1890, throughout April 1891 and finally from June to July 1892.[26] He illustrated the lectures using "photomicrographs made by Mr Andrew Pringle,"[27] and Pringle accompanied him to the lecture to show the slides on screen.

One problem with the acceptance of tubercle bacilli as the cause of tubercular disease was that, on occasion, relatively few were found in samples of bone. Cheyne addressed this in the first set of lectures by suggesting a range of characteristics he considered sufficient to diagnose the disease without the necessary conditions of the tubercle bacillus itself. The second set of lectures, in 1891, focused on the characteristics of different forms of tuberculosis, presenting evidence from experiments on animals, and factors contributing to the disease. He included an impressive range of biological, environmental and social conditions, down to the fine detail of the effects of age and geographical trends in diet. His final set of three lectures, in 1892, were an extensive analysis of types of treatment, where he placed as much emphasis on rest and responsible exercise as on surgical intervention. As we have seen, this concept was by no means new to Listerism or even to Cheyne's own work. He was in favour of expectant treatment where possible, detailing the circumstances where it was worth waiting and observing, with an emphasis on rest, in case the condition receded on its own. He went so far as to assess the relative merits of British east and south coast resorts.[28] The detail in the set of lectures is staggering in itself, but his careful consideration of the interplay of the range of factors - by no means all medical - which affected tuberculosis and its treatment in the 1890s, was more than noteworthy, and was in keeping with his growing understanding that bacteria alone were not the only consideration in diseases and their treatment.

> Many patients undoubtedly do best at the seaside; but there are others, and these are by no means few, for whom somewhat high inland places are the most suitable. In these cases I believe the best thing to do is to send them to a cottage or farm house in some suitable inland place, and let them be as much in the open air as possible. Where a seaside place is chosen, I think that in the early stage, or where the patient is weakly or cannot walk, some of the South Coast health resorts are preferable, as a rule, to those on the East Coast, although after the disease has improved or been removed by operation, the East Coast stations often brace up the patient extremely well, and, so to speak, give the finishing touches to the treatment.[29]

Cheyne continued to develop his methodology and understanding of the subject. In 1895, he finally published his book on *Tuberculous Disease of Bones and Joints: its pathology, symptoms and treatment*, and explained in the preface why the publication had taken so long after the success of his Astley Cooper essay. "The delay in the appearance of this work ... has been due to my desire to assure myself, from sufficient experience, that the treatment I was recommending was founded on a sound pathological and practical basis."[30] This type of caution did not go amiss, particularly in the light of subsequent incidents in the treatment of the disease. He had also hoped to be able to reproduce Pringle's photomicrographs in the volume, but the cost proved prohibitive.[31]

Cheyne was still lecturing and adapting his methods when he delivered a series of Harveian Lectures in December 1899 on surgical aspects of the disease: tuberculous gland disease, tuberculous peritonitis and genito-urinary tuberculosis.[32] He continued to give talks around the country to give provincial doctors a taste of the bacteriology behind the conditions they were treating. He sent a paper and slides on tuberculosis along to a meeting of the Chelsea Clinical Society on April 29th 1899, though he himself could not be there.[33] He gave another talk on the Natural History of Tubercle on June 29th 1899 at the 44th Annual Meeting of the South Midland branch of the British Medical Association in Kettering. As was now quite usual, his talk was "profusely illustrated by microphotographs exhibited on the screen, showing synovial membrane, cartilage, and bone, and other tissues affected with tubercle, the speaker pointing out on the slides the life-progress of the disease through the different tissues".[34]

Despite his crippling schedule, Cheyne apparently also found time to be interested in the new work of young doctors. Many years later, Harold Stiles, known for his work on tuberculosis and cancer, recalled how Cheyne had encouraged him, particularly during the time he was working on the Astley Cooper prize essay. He noted that "his house in Harley Street was always open to me, and I shall never cease to be grateful to him for his help and encouragement during my struggling days. I had the same admiration and reverence for him that Lister's pupils had for their great master."[35]

When Cheyne's translation of Carl Flügge's work *Micro Organisms, with Special Reference to the Etiology of the Infective Diseases* was finally published in 1890 [36], it unleashed another round of the debate on the position of bacteriological research in Britain in relation to other countries. The translation was considered accurate, though the New Sydenham Society was criticised for the engravings, which were unfavourably compared with German, French and American ones.[37] The ubiquitous criticism resurfaced of the lack of financial support for laboratory experimentation in Britain. Cheyne was fortunate enough to have been involved with Crookshank in setting up the first official laboratory in a English hospital at King's, but had needed to use a good deal of his own money to provide the facilities he required,[38] and this expense extended to anyone subsequently using them. He noted in a letter in June 1891 to Albert Carless, who was to use the laboratory while Cheyne was in Edinburgh, that, "The College provides

... gas & fire: you must provide anything else you want for yourself."[39] In general, "before the 1890s most bacteriological work was performed in the corners of domestic rooms, doctors' surgeries or in "microscope rooms" in hospitals."[40] Doctors were expected to use them as a tool, but they were not staffed by specialists.[41]

As we have seen, however, Cheyne was interested in the potential of the methodology emanating from the continental laboratories to advance theory and practice. As I have noted, from at least 1882, he had been helping Lister experiment on the effectiveness of iodide gauze and other chemical antiseptics. Lister was anxious to find a substance that killed bacteria effectively, but did not do the same harm to skin as carbolic acid, and Cheyne's knowledge of the advanced staining techniques were invaluable in testing them. Cheyne conducted a number of experiments of his own, but also dictated detailed instructions to Lister on preparing and staining bacteria. He explained in one letter: "Expose the glass for about an hour to a temper[ature] of about 110°... Ehrlich does this by means of a metal plate with a Bunsen's burner under one end ... Then put the cover glass (after heating) face down in staining solution, where it is left floating. Staining solution may be gentian violet methyl violet, fuchsine or methylene blue ..."[42] Cheyne had gone on to note how Koch did it, and Lister added to his notes, "Cheyne prefers the following ... [which] gives a better staining."[43] This illustrates the extent to which Lister made use of Cheyne's expertise on staining methodology, and gives us an idea of its significance in Lister's work.

His ongoing interest in these processes, involving improved observation of bacteria and laboratory experimentation, led Cheyne beyond the direct question of a causal link between specific bacteria and diseases, to an interest in *how* bacteria influenced disease. This was to prove important, as it was now clear there were no simple answers, and a variety of conditions could come into play, such as the number and strength of the bacteria and the resisting power of the tissues in the body. Cheyne's previous publication, two years before, *Public Health Laboratory Work*[44], had focused on methodology, and Flügge's preface, in the work translated by Cheyne, noted the prominent role of the rapidly-developing methodology in understanding bacteria and disease more comprehensively. Flügge noted:

> By the discovery of numerous disease germs, and of the methods of their pure cultivation, it has become possible to study experimentally the conditions of life of the infective organisms, their mode of life,

their relation to their surrounding, their transportability and the mode of their entrance into man, and thus to obtain information as to the causes of the peculiar mode of spread of epidemic diseases, in a manner incomparably quicker and more trustworthy than by empirical and statistical methods.[45]

By the mid-1880s, although Cheyne was moving towards a broader understanding of the relationship between bacteria and disease, he nevertheless continued to look to German and French sources for advances in methodology and staining techniques. It is also important to note that, when he selected authors for his 1886 collection of essays, it was the publisher who restricted the selection to continental works. The New Sydenham Society's rules did not allow him to include 'English' works. The prefatory note makes it quite clear that, while Cheyne was allowed to select the papers and illustrations himself, he had been "restricted by the Society's law from taking any by English authors."[46] It might even lead us to speculate whether this note was added at Cheyne's request, if he were sensitive to the fact that nationalistic elements in debates could distort the scientific arguments.

This should be seen in contrast to others, who appeared to quote exclusively German sources because they genuinely believed they were superior or unique. When Dr George F. Crooke mentioned that "it must be acknowledged that German pathologists were not only the first in the field, but have ever since maintained their precedence,"[47] the *British Medical Journal* editorial criticised him for not doing "justice to the labours of his own countrymen in elucidating the connection of the tubercle bacillus with the pathology of phthisis, and would cite as an example the omission of all reference to the papers of Mr Watson Cheyne and the work of the Brompton Hospital as recorded by Dr Percy Kidd, Mr H. H. Taylor, and others."[48] The following year, the journal similarly criticised an American publication by Professor N. Senn for mentioning *only* Cheyne among the British contributions to the treatment of tubercular bones and joints, when others merited a mention.[49]

These national sensitivities were never far from the surface. Despite his familiarity with the methodology and theoretical leanings of German schools, Cheyne was more than prepared to rise above national rivalries, and was even prepared to override personal loyalties if he felt the evidence demanded it. A graphic illustration of this was to present itself in the case of Robert Koch's tuberculin, which highlighted the damage done by untimely publicity and national expectations.

Robert Koch was a hero in his country, as was Pasteur in his, and rivalry in the scientific domain was promoted in the interests of national pride. When Koch announced in 1890 that he had discovered a cure for tuberculosis, which became known as tuberculin, it caused an immediate national and international sensation, given the scale of the disease. Koch's basis for the vaccine was that it killed infected tissues, preventing the growth of the tubercle bacillus by starving it. He apparently did not claim that it killed the bacillus itself. However, from the start, he was unwilling to release information on the composition of the treatment, and said himself that he would not completely understand how it worked until trials were complete. He noted in the *Deutsche Medizinische Wochenschrift*, translated for the *British Medical Journal*:

> It was originally my intention to complete the research, and especially to gain sufficient experience regarding the application of the remedy in practice and its production on a large scale before publishing anything on the subject. But, in spite of all precautions, too many accounts have reached the public, and that in an exaggerated and distorted form ...[50]

The treatment produced strong and uncomfortable reactions in patients before signs of recovery were seen. These included fever, rashes and even collapse.[51] Some apparently died. For the general public, none of this seemed to matter. People flocked to Berlin to receive the treatment, and British hospitals sent delegations out to see for themselves or convened special meetings to consider how they should proceed.[52] From Dublin, the "flight of medical men to Berlin"[53] left readers of the *British Medical Journal* with the sensationalist impression of a city left virtually without medical expertise. The general impression from some of the reports of activity was that anyone who did not at least consider the issue would be judged by history, but it unleashed policy ranging from enthusiasm to caution in British medical establishments. The *BMJ* even included a list of hospitals which had *not* taken any steps, cautious or otherwise.[54] Lister himself took his niece for treatment, and was very favourably impressed.[55] Conan Doyle, author of the Sherlock Holmes series and himself a doctor, rushed to Berlin.[56] There was also a flurry of letters to the British medical press, some calling for caution [57], some expressing doubts about Koch's reticence to reveal the chemical components of the treatment,[58] and some calling for a more systematic approach by the medical profession, to assess it more objectively.[59]

Among the many reactions in November 1890, a notice appeared in the *British Medical Journal* from Cheyne and G.A. Heron, to say that Koch had approached them both "to demonstrate in London his method of treatment of tuberculosis, and has supplied us with a small quantity of fluid for this purpose."[60] The emphasis on the 'small quantity' was no doubt an explicit attempt to ward off any rush of people expecting treatment. Koch's request could theoretically have been seen as putting Cheyne and Heron in a difficult position. While the British medical establishment was still considering how to react to the conflicting evidence emerging from all quarters, Koch had asked them personally - on his own behalf - to demonstrate the method in Britain when he himself had no definitive evidence that it worked.

Cheyne, however, did not consider this to be in any way a difficult position, and chose instead to demonstrate his close relationship with Koch. He emphasised the integrity of Koch's motives, suggesting that he wished to ensure the treatment was reproduced faithfully, so that the medical profession in England could judge for itself. No conditions had been placed on the demonstrations except that they be arranged soon, and that as many medical men should be present as possible.[61] In his public response to the invitation, Cheyne said, "he had been for many years an intimate friend of Dr Koch's, and it was only natural that in looking about for persons to whom to entrust the duty of bringing the nature and properties of his remedy before the British medical public, Dr Koch should have chosen two old pupils who were well known to him."[62]

Koch had been dissuaded from publishing the characteristics of the fluid beyond the general manner in which he had described it. When he had discussed the mode of publication of the method with the Prussian Education Minister von Gossler earlier that month, on November 7th, the latter had "prevailed upon Koch to reserve publication of the nature of the remedy and took the entire responsibility of the step upon himself."[63] Koch tried to control production, and to ensure quality tests by making sure all supplies went through Dr Libbertz, tested by Dr Pfuhl [64], but it was all slow and cumbersome, and demand far outstripped supply. Moves were made to ensure production of the lymph was entirely under Prussian state control to prevent unsupervised replication.[65] This was heavily criticised, and Koch's ethics were later brought into question when he failed to recognise the work of British and other researchers on the treatment, particularly Abraham and Crookshank, and the joint work done by William Hunter and Cheyne.[66]

By November 29th, 1890, Koch was struggling to control matters. The *British Medical Journal* noted the general alarm in the more discerning scientific community when it said that "[a]t present there is a somewhat uncomfortable feeling that the precious lymph is not being utilised fully for the benefit of medical science", and that "[t]oo many of the cases are being treated merely as out-patients",[67] with little systematic observation. An "occasional correspondent"[68] to the *Journal* reported the general exodus to Berlin as though it were an act of homage:

> Berlin at the present moment must be the place in all Europe where tubercle bacilli most do congregate. Hotels, lodging houses, and hospitals, public and private, are full of patients in every stage of phthisis, and suffering from every form of tuberculous affection, who have come to the German capital as to a Pool of Bethesda. The mystic elixir is injected into all who present themselves - in many cases apparently without much attempt at discrimination between suitable and unsuitable subjects for treatment. It is whispered also that the tiny rivulet of lymph at present available is being transmuted into a broad Pactolus of gold by some of the happy few who were privileged to obtain an early supply. No shadow of such a suspicion rests on Koch himself, and he is said to know nothing of the despicable use to which his magnificent discovery is being put by certain persons who have abused his confidence. He no longer inoculates patients himself; indeed, he seems - for the present, at least - to have withdrawn from the clinical field into his laboratory, where he remains somewhat like Achilles in his tent ... He is thoroughly disgusted with the appearance of a vulgar 'boom' which the whole thing has assumed, and he is particularly annoyed at having been driven by circumstances, very much against his will, to bring his results, unfinished as they are, prematurely before the scientific world. He has now firmly made up his mind to remain silent till such time as his investigations are completed.[69]

Seen in this light, Koch may have been enlisting Cheyne and Heron as part of a final attempt to cut through a situation which had spiralled out of control on the Continent, and to conduct tests for him in Britain, whatever the outcome might be. He knew of Cheyne's reputation for objectivity in Britain. A handful of prominent German physicians, including von Bergmann [70], were beginning to report the results of clinical tests, systematic observation of patients who had been given tuberculin. Initial conclusions were largely positive, depending on the severity and type of tuberculosis treated, but the secrecy behind the composition of tuberculin did not help bacteriologists test it in laboratories, particularly as it was in such short supply. The hype had already escalated to the point where scepticism abroad was leading to demands for visible evidence, despite the fact that the Prussian Emperor had recognised Koch's achievement with the Grand Cross of the Red Eagle [71], and Pasteur had officially congratulated him.[72]

In choosing Cheyne, Koch was recognising his reputation for objectivity but also trusted his comprehensive understanding of Koch's own methodology. He must nevertheless have also known that Cheyne was unlikely to be satisfied with clinical trials alone and would take the tests into the laboratory. Cheyne had, after all, visited him in Berlin on behalf of the British scientific establishment only a few years earlier in order to assess his claims on the tubercle bacillus, and had looked much further into the behaviour of the bacillus in his report to the Association for the Advancement of Medicine by Research. Not only did Cheyne understand the need for transparency, Koch must have known that simple loyalty was unlikely to cloud his scientific judgement if the results were to come out against tuberculin.

As Cheyne and Heron travelled to Berlin, the outcry in the British press escalated over the lack of real provision in the German capital and the secrecy behind the composition of the treatment. The *Pall Mall Gazette* complained that "the *Vossische Zeitung* cries out that money for providing beds, barracks, and hospitals for the poor would be preferable to eulogies on Koch's genius. Temporary hospitals have been established in empty lofts, devoid for the most part of any sanitary arrangements, even of pure air, and there is no one to offer money or help."[73]

Cheyne and Heron spent a short time in Berlin with Dr Koch, collected a supply of tuberculin and left again for London on the night of November 20th 1890.[74] A reporter for a Manchester newspaper [75] sent back word that the supply of tuberculin was almost exhausted in Berlin, and treatment at the Charité Hospital had been stopped until more supplies could be got. On the other hand, a number of private "inoculators"[76] still had supplies and were charging a fortune for them. Cheyne and Heron were back in Britain to begin their demonstrations by early December. King's College and Victoria Park Hospital for Diseases of the Chest, where Dr Heron was based, took what the *British Medical Journal* considered "a wise and broad view of their duties"[77] in allowing their staff to demonstrate under their roof, given the public débâcle in Berlin. Dr Heron gave his demonstration in the first week of December 1890.[78] Cheyne gave morning demonstrations at King's College Hospital at 9.30 am each day of the first week of December, then again on the Monday and Wednesday of the following week, then each following Wednesday, in order to observe the progress in the first cases.[79] He also gave demonstrations at Paddington Green Children's Hospital [80], where he continued as surgeon. Places at the demonstrations were limited, and by ticket only.[81]

The initial demonstrations were followed by clinical observation over a longer period. Cheyne was testing the fact that, if the treatment killed affected tissue, and starved the bacillus, the results would not be instant, but would need to be observed over time under the effect of repeated injections. There was no shortage of volunteers among the patients, who rushed to "avail themselves of the benefits"[82] in the same way "as their richer fellow countrymen and fellow countrywomen who during the last few weeks have hurried in such numbers to Berlin."[83] Importantly, the conclusions he ultimately made public about tuberculin were based not only on clinical results from his trials on patients, but also on laboratory experimentation.[84] Cheyne was once again taking matters a step further than was, strictly speaking, required of him, but the extra work was to prove important.

In the middle of this frenetic activity, he left for St. Petersburg. Lister, along with Robert Koch and Louis Pasteur, had been invited to the opening ceremony of the new Bacteriological Institute there, which had "been founded mainly through the exertion of Prince Alexander of Oldensburg."[85] Lister was "unable to accept the invitation"[86] and Cheyne went in his place. However, given the central role Cheyne was now playing in assessing tuberculin, it is not out of the question that Lister asked him to go in his place regardless of whether he himself could attend. One newspaper reported that, "Dr Watson Cheyne ... who has been in charge of Dr Koch's patients at King's College Hospital, started for St. Petersburg tonight to give the Russian physicians the benefit of his experience in London. It has been ascertained that Dr Koch's treatment is applicable to leprosy, and leper patients have been treated with marked success."[87] In sending Cheyne in his place, Lister was not only recognising his leading position as a bacteriologist, he understood the current relevance of his work for the new institute.

On his return, Cheyne continued his assessment of tuberculin in London. He and Heron were by no means the only ones to give demonstrations in

Britain. In fact, they were taking place all over the country, from Bristol to Edinburgh. C.T. Williams, a London physician, had even taken himself off to Berlin to make a systematic report on around 100 cases, noting, "It is a curious spectacle, and one by no means flattering to the votaries of science, to see the medical world of Berlin given up wholly to the worship of 'Koch injection' and 'reaction process.'"[88] He could find no more than eight cases where tuberculin had brought about genuine improvement, but urged further investigation in deference to Koch. There was no shortage of doctors and scientists willing to test the treatment. The significance of the testing by Cheyne and Heron was that Koch had requested it himself, whereas others had gone to Berlin to request a supply for testing or to observe hospital cases. Amidst reports that the treatment was not only ineffective, but could even make the condition worse, Cheyne and Heron were taking on a test of their integrity, as well as their scientific objectivity. After all, the world was waiting with bated breath. People *wanted* it to be effective.

Cheyne was approached for private supplies of tuberculin as soon as it was known he had some, but he only had enough for his immediate work. The day before he left for St. Petersburg he replied to an unknown correspondent that he could not supply him with any of his own quota of tuberculin, but that the enquirer could try writing to Dr Libbertz at 28 Lüneburger Strasse, Berlin, through whom all official supplies were being processed. Cheyne suggested he mention his name if it helped him to obtain a supply, but his own was clearly expressly and only for the purpose of the trials.[89]

In an address to the Royal Medical and Chirurgical Society on April 28th, 1891, he presented his extensive results on use of the treatment in a wide variety of tuberculous conditions. He noted the value of administering it continuously in the hope that it would eventually remove the "soil" in which the tubercle bacillus could grow. He also took into account existing treatments for the disease. He tested a variety of conditions, including the form of tuberculosis, the stage of the disease and the size and frequency of the dose.[90] At a special meeting of the Society, held a week later on May 7th to discuss his paper, he was widely congratulated for his cautious conclusions and his objective approach.[91] When asked specifically about the value of the treatment in general, he replied that it could not attack tubercle directly or single-handedly, and that, as he would not advise using it in the early stages of the disease, it was unlikely to be an effective direct treatment in later stages.[92] In lupus, he suggested a preliminary course of "tuberculin was of use in bringing into sight all the foci before more radical measures were undertaken."[93]

In terms of treating tuberculous bones and joints, he continued to advocate expectant treatment, observing the patient through periods of rest and changes in diet. This took time, however, and he had been hopeful the tuberculin treatment would have been valuable in reducing the period, though he now felt it was unlikely. He conceded that, in cases where operations were chosen as treatment, it might help to reduce the extent of them.[94] He was careful to conclude from his results that alarmist reports about adverse effects were "greatly exaggerated",[95] particularly those focusing on whether or not tuberculin actually *caused* the disease.[96] In a later paper, in August 1891, he modified this to suggest that the "fluid not only fails to produce immunity, but actually in many cases apparently predisposes the body to the tuberculous infection, and … it was apt to favour the spread of sepsis."[97] This view may have been precipitated by evidence from others who had administered the treatment. W.G. Spencer commented many years later, after Cheyne's death that, "[h]e gave at King's College Hospital a flamboyant account of Koch's tuberculin to those invited, including C. MacNamara, F.R.C.S. and myself. There were two children in MacNamara's ward at Westminster Hospital with advanced hip-joint disease. On repeating Watson Cheyne's prescription and injecting tuberculin, both had acute suppuration and quickly died: no further use was made of the remedy."[98]

On the other hand, when Cheyne published "Tuberculous Disease of Bones and Joints" in 1895, though he no longer cited tuberculin in terms of treatment for these conditions, he suggested that general dismissal of the treatment had been too hasty.[99] We have the impression that, on the whole, the mood of hysteria acted as a form of pressure to produce definitive statements on the treatment, and Cheyne's caution, along with what must have seemed at times like contradictory comments, could easily have been dismissed. His initial conclusions were, in fact, received with enthusiasm, and he was applauded for his conspicuous lack of bias:

> As a model of how a scientific investigation into the use of a drug should be conducted, and of how the results should be made public, Mr Cheyne's paper fully deserved the enthusiasm with which it was received by the audience to which it was addressed. Carefully accurate in facts, cautious in inference, and eminently judicial in tone - the unbiased report of a man anxious only to discover the

truth and to utter it without fear or favour - it is the most important contribution that has yet been made, in this country at least, towards the solution of the momentous question with which it deals.[100]

Cheyne had once again taken the matter further than was strictly required of him, and had reported his results without bias, despite his evidently close friendship with Koch. It is not only evidence of Cheyne's integrity, it also indicates the levels of caution in the British scientific establishment, no matter how sensationally the media had reacted, and no matter how excited researchers and the general public may have been about the prospects of the new treatment. Cheyne had provided credible evidence which cut through the nationalism and hype behind the scandal. Other medical meetings throughout Britain discussed clinical results and came to the same conclusion.[101] Nor was this initial use of tuberculin any more successful in Germany, and gradually its use was restricted to diagnosing tuberculosis, though its efficiency in this respect was also questioned.[102] Cheyne did not, however, totally despair of it, and had hopes that any harmful effects could be isolated and removed. William Hunter did some work on this, producing modified versions, and Cheyne was involved in testing the resulting products, but he was not hopeful about them in the conclusions to his August 1891 paper.[103] His general conclusions were nevertheless to resurface in future debates, centring on the fact that a more suitable way of using tuberculin could emerge from further investigation.

Cheyne's professional relationship with Koch does not appear to have suffered, but despite his clearly-established reputation in experimental bacteriology, his active involvement in it was gradually coming to an end, a situation he would later blame partly on the lack of professional opportunities for laboratory research in Britain. In 1888, he resigned his research scholarship at the British Medical Association [104], and in 1889 became Surgeon and Teacher of Operative Surgery at King's, along with his duties at Paddington Green.[105] On the other hand, he had clearly seen how quickly bacteriology was advancing, and how his split workload would make it difficult for him to keep pace with developments. He later commented to E.S. Reid Tait:

> In 1891 I was getting overwhelmed with work which was increased to a very large extent when I became, in 1893, Professor of Surgery at King's College. I was already Examiner in various Universities and

> private work was beginning to increase after years of patient waiting. So far, my published work had been essentially of a bacteriological nature and that science was making such rapid advances, that I was forced to make a decision between surgical work and the further study of Bacteriology. I decided without hesitation to follow up a surgical career. Bacteriological work now involved an increasing knowledge of chemistry, while surgery had still many problems which had not yet been attacked.[106]

Cheyne was writing this in the late 1920s, with the benefit of hindsight, but at the time he would demonstrate on more than one occasion a hint of regret at not having been able to continue in bacteriological research. He remained at the forefront of attempts to fund it, and often implied that lack of funding opportunities in the field held research in Britain back. He was nevertheless justified in assessing his workload during this period and making a choice one way or the other. A quick assessment of Cheyne's posts and honorary duties gives an idea of the sheer extent of it, and shows why he was feeling so overwhelmed. His reputation had grown enormously, so that he was in constant demand, and he continued to work long hours to fit it all in, somehow finding the time to keep up his crippling rate of publication. On 19th August 1890, he was informed that he had been formally and unanimously elected Honorary Consulting Surgeon at Bexley Cottage Hospital in London, and thanked for the services which he had "already unofficially rendered."[107] The Cottage Hospital had been open since 1881, and Cheyne had clearly been consulting there informally for some time. By at least the early 1890s [108], he also had a growing private practice at 75 Harley Street, where he was to remain until he retired.

We cannot help but wonder whether Cheyne had a private life at all, yet by the beginning of the 1890s, he had a wife and two young sons who were tolerant of his busy timetable. He had entered one of the most stable periods he had ever known when he married, but his professional and public engagements were beginning to require more time than he had at his disposal. He had been forced to limit himself in terms of a bacteriological career, but could have been forgiven for thinking there were few limits to his remaining activities if he organised them adequately. With Mary's background in nursing, it is also possible she was able to assist in the practice in some way, just as Agnes Lister had done for her husband. On 10th July 1891, however, Mary gave birth to a daughter, whom they named Mary Frances.[109]

His period of domestic stability was to come to an abrupt end when tragedy struck. On 4th May 1892, Mary Frances died.[110] She had lived less than a year. With Cheyne's confidence as a surgeon, how devastating it must have been for him to know he could not save his own child. He conveyed the sad news to Lister straight away, and Lister sent a letter of condolence to Mary the following day:

> My dear Mrs Cheyne,
> I was grieved to hear from Mr Cheyne the sad news that you had lost your little daughter. I cannot but write a few words to express my sincere sympathy in [sic] Lady Lister will I am sure write.
> Believe me
> Yours very truly,
> Joseph Lister.[111]

Only weeks later, it was announced that Cheyne would be leading a discussion in the last week of July on the surgical treatment of spinal abscess at the Annual Meeting of the British Medical Association in Nottingham.[112] Mary never fully recovered from the death of their daughter, but the details of her health are unclear. On 6th February 1894,[113] she followed her daughter to the grave (she was in Caxton in Cambridgeshire when she died), leaving Cheyne with two young sons to bring up in the midst of his busy timetable. He had suffered two losses in less than two years, and was faced with a household to organise alongside his surgeon's and consultant's duties, and the private practice which sustained the family's income. On 2nd December 1893, on top of becoming Professor of Surgery at King's, he had taken on yet another consultancy when he was unanimously elected Honorary Consulting Surgeon at the North London Hospital for Consumption in Hampstead.[114]

It all came less than a year after Lister had lost his own wife on a holiday in Italy [115], which saw him withdraw gradually from hospital duties and public life. This left Cheyne and other flag bearers to take the lead in ensuring Listerism remained on the surgical agenda as Lister himself had intended it to be understood, particularly in the wake of developments discussed in the next chapter. It is difficult to know how Cheyne coped with this rapid succession of events. Only days before Mary died, he had delivered a Harveian Lecture on the treatment of cancer of the breast [116], but her death appeared to slow him down, at least for a few months. He does not seem to have resumed his usual gruelling schedule until around September, though he was clearly trying to complete work on his next publication throughout this period, as well as contributions to other works. Perhaps this helped him to take his mind off the situation, but it cannot have been easy. Besides, life went on around him, beyond his control. Only a few months after Mary died, Cheyne heard on 7th June 1894 that he had been elected a member of the Royal Society.[117] It was evidence of his remarkable catalogue of achievements in a relatively short space of time. He was, after all, only 42. His decision to leave behind an active involvement in bacteriological research was clearly not enough. If he were to continue to fulfil his professional and public engagements in the surgical sphere too, it was clear he would have to do something to keep his household running in his absence.

In 1891, while Mary was still alive, the family had benefited from three household staff: Rose Bannister from Chorley Wood in Hertfordshire, Emily Redding, perhaps a relative of Rose, as she came from the same place, and Florence Ansell, from Boxstead in Essex.[118] They were all described as domestic servants in the census,[119] and may not have been trained to manage an entire household in the same way as a housekeeper, nor to have been left in sole charge of his two young sons. Joseph Lister Watson Cheyne, the eldest, had just turned six when his mother died, and Hunter was only four. Clearly Cheyne would need the services of a housekeeper, and probably also a nursemaid or governess. Family tradition suggests that he employed a woman called Margaret Smith as housekeeper, though it is unclear when she became involved with Cheyne's household. It is reasonable to assume, however, that it was around this time, to alleviate the management problems.[120]

Margaret was a Shetlander, born in Lerwick, the capital of the islands, around 1859.[121] Her father, George Smith, was a solicitor, Sheriff Clerk for the "County of Zetland"[122], and Margaret grew up in a middle class household at Braeside House. When she was two, she had a nursery maid [123], Charlotte Irvine from Whalsay, a neighbouring island. The household also had a tablemaid, Margaret Leslie, from Quarff [124], not far from Lerwick, and a cook, Margaret Irven, from Delting [125] in the north of the Shetland Mainland. Margaret grew up in a relatively large family, with an older sister Christina, a younger sister Georgina, and a twin brother, James.[126] She also seems to have had a much older brother, John, who had been born in Edinburgh, before her parents had moved to Shetland.[127] The family suffered a blow, however, when her parents died in quick succession, leaving her an orphan in her early teens. Her sisters,

Christina and Georgina went to live with their brother, John Scott Smith, who had taken over as Sheriff Clerk of the County [128], and the two women were there in 1881. Margaret, however, probably moved 'south' when her parents died, and may have worked in other households before she found her way to Cheyne. No doubt her Shetland origin was instrumental, and she may even have been recommended to Cheyne, through his Shetland connections in London, when he obviously needed help. She may even have already been a personal acquaintance.

As soon as Cheyne had someone to help run the household, assuming he had indeed added Margaret to his staff, he buried himself once again in his work, with ever more publications and public duties. In February 1894, around the time of Mary's death, it had been announced he would be contributing to Treves' volume, *A System of Surgery*, a manual "intended primarily for students"[129], which was eventually published in June 1895. This meant that he probably did considerable work on it as soon as Margaret took over the running of the household. He contributed an essay on inflammation, five essays on suppuration and three on ulceration. For Volume II, which appeared in 1896, he contributed a further chapter on "Diseases of the Breast"[130] in keeping with his recent work on operations for breast cancer. In the first year after Mary died, he also worked on a publication of his own, a practical book for his students entitled *Treatment of Wounds, Ulcers and Abscesses*, which was published in September 1894.[131]

His public engagements rocketed too, and it may be further evidence of the fact that he was burying himself in his work to lift his spirits, or in order to consolidate his position in surgery now that he had made a firm choice. His work was still the place where he felt most comfortable. On Monday October 1st, 1894 at Limmer's Hotel in London [132], he chaired the annual dinner of old students from King's College Hospital at the beginning of the winter session, focusing on Lister's recent retirement from active hospital work.[133] Lister's retirement had tremendous implications for Cheyne at a time when even the "bulldog"[134] was feeling overwhelmed. Agnes Lister's death had devastated her husband, and his decision to retire followed fewer and fewer public appearances.[135]

A few weeks later, on Friday, Oct. 26th, Cheyne travelled to Taunton to deliver a speech at the West Somerset branch of the British Medical Association, on the Development of Modern Methods of wound treatment, "upon which an interesting discussion" was expected.[136] Meanwhile, things in the household appeared to be running smoothly, and Cheyne, still at a vulnerable time in his life, made another important decision. Less than a year after Mary's death, at the end of 1894, he married Margaret Smith. The wedding took place at Burnt Ash Hill in South London, and the vicar of the parish, the Reverend F.W. Helder, assisted the Reverend T.H. Henderson, vicar of Farley and Pitton, Salisbury, who performed the marriage ceremony [137].

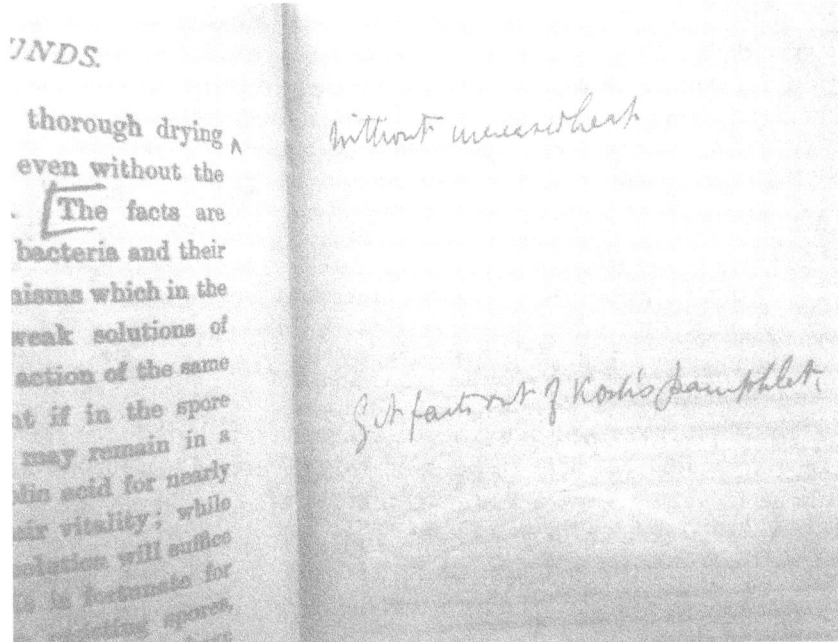

Pages from Cheyne's own amendments to The Treatment of Wounds, Ulcers and Abscesses, *probably for the second edition, which was published in 1897 (first edition 1894). The lower note on the right-hand photograph reads: "Get facts out of Koch's pamphlet".*

Margaret was six years younger than her husband, but was still almost 36 when they married. By all accounts, their marriage was not a particularly happy union as they grew older, and we cannot help but speculate that Cheyne saw the act of formalising his relationship with Margaret as a practical solution to his domestic pressures. He was no stranger to family tragedy, and perhaps since childhood had learned to approach it with a practicality that helped overcome grief. Mary's death must nevertheless have brought home to him how alone he was in the world, and Margaret's Shetland origins probably reminded him of a home he had largely only known on holiday since he had left at the age of 11. Margaret remained with him until her death in 1922, but there are indications they later spent long periods apart, perhaps because Cheyne's work consumed all his energy, and perhaps because he used it as a retreat from domestic life.

On the other hand, he could not have kept up his gruelling schedule if Margaret had not been managing the household and the children, including Mary's two sons. She may not have been aware of just how much her husband would be away from home, or at least locked in his study with his writing. In March 1895, Cheyne helped edit the first Annual Report of King's College Hospital.[138] In July-August the same year, he gave a paper on "Operations for Malignant Disease of the Pharynx and Naso-pharynx with Cases" at the 63rd Annual meeting of the British Medical Association, in the Surgery section.[139] On September 30th, 1895, he was approached by Major General A.L. Playfair to act as one of several Consulting Surgeons to the Siddons House Private Hospital, a facility for "members of the theatrical profession".[140] Others acting as honorary consultants there included Lenthal Cheatle, who had become Assistant Surgeon at King's in 1893.[141]

Shortly afterwards, on 30th October, Margaret gave birth to their first daughter, and named her Grace Ella Margaret Watson Cheyne. She was always known as Meta, a shortened form of Margaret, to distinguish her from her mother. They were ultimately to have three children, and the naming convention differed noticeably from Cheyne's first marriage, where his offspring had been named to honour the illustrious history of British medicine.

The same year, he was invited by the Medical Society of London to deliver the three Lettsomian Lectures "on some surgical subject … not to exceed one hour each."[142] In February and March of the following year, he delivered the three lectures on "The Objects and Limits of Operations for Cancer", not long after Meta was born, an irony which would become clear in later years. The lectures were well received. One reviewer referred to the tables of data as "monuments of industrious research"[143], noting that "their conspicuous merit" lay in "their strong personal note,"[144] telling "the story of an accomplished surgeon honestly grappling with exceptional difficulties."[145] Even so, it is likely the author had little idea of the extent of them, and was probably not including the rapid changes in Cheyne's domestic life.

On 27th June 1896, had been appointed Examiner in Surgery at Cambridge, for examinations on 8th December 1896 and 27th April 1897.[146] He was generally considered "a model examiner, though more lenient than his reserved manner suggested."[147] Apparently, in the examinations he would oversee in his time at the Universities of Edinburgh, Oxford, Cambridge and London, "no candidate had reason to feel that he had been deprived of chances of making good."[148]

He did not apply for a place on the Council of the Royal College of Surgeons again in 1896,[149] and perhaps he was reassessing his ability to fit in so many public appointments. If so, he had clearly re-thought the situation by the following year, when he decided to stand, and came in 4th [150], with 290 votes [151], giving him a place to represent King's College Hospital on the Council. There had been 5 places and 11 candidates.

His travelling resumed in earnest in 1897, when he went with Lister and a number of other prominent medical men to Canada. He had been invited to act as President of the Section of Pathology and Bacteriology at the British Medical Association's Annual Meeting in Montreal from August 31st to September 3rd 1897.[152] It was the first time the annual meeting had been held outside the United Kingdom [153], and was the second invitation sent by the city of Montreal to the Association. The first had been an attempt to secure the 1896 meeting, but the honour had already gone to Carlisle. The second invitation, for 1897, was accepted [154], and prompted an exodus of Britain's medical élite to Montreal at the end of the summer. Cheyne left in the *Numidian* from Liverpool on 12th August 1897, bound for Quebec, and his wife Margaret accompanied him.[155] On their arrival, Lister was presented with an honorary degree from McGill University, and Cheyne gave the opening address in the Pathology section, where he seized his opportunity to bemoan the lack of funding for pathological research at home. It contained a strong personal note:

> ... while pathologists are thus working out problems which affect the general well-being of mankind, and the solution of which can be of no personal gain to themselves, is it too much to ask mankind to furnish the means for such research? The English are looked on as a thoroughly practical people, and yet it is a very remarkable thing that England is almost the only country which does not realise the importance of scientific research, and the result is that in England, with very few exceptions, men who might otherwise have thrown much light on these matters are compelled to turn their attention to practice in order to make a living. Unless work of this kind is done, how can we hope to advance with any rapidity in the treatment and cure of disease? The surgeon or physician must wait till the information of which he is urgently in need has been acquired for him by the pathologist. Such apathy can surely only be the result of ignorance. A rich man affected with an obscure or incurable malady cannot understand how it is that he fails to obtain the definite opinion or the relief which he so earnestly desires, and for

which he is prepared to pay any price. Surely if he understood the meaning and importance of pathological research, and that the practising physician can only apply and carry out what is taught by the pathologist, he would bestir himself to aid research in order to gather information which might be of much use to him and to others.[156]

Cheyne's return to surgery in the 1890s may have been partly a general reconsideration of his capabilities, but part of him clearly regretted the need to have to leave experimental bacteriology behind, and he made no secret of the fact that he deplored the under-funding of the field in England. The fact remained that, apart from the many prizes he applied for (and usually won), he had always had to struggle to keep this side of his work funded. He clearly felt cheated, and perhaps also weary of having to fight so hard to be able to pursue what he considered essential work. His surgical career, however, could hardly have been better. On 1st July 1897, he was notified that he been elected a member of the Council of the Royal College of Surgeons of England.[157]

His workload continued to grow. On 27th November 1897 he was informed he had been recommended and accepted as an honorary surgeon to the Scottish Hospital in London, and asked to attend a banquet on St. Andrew's Day to raise funds. Tickets were one guinea each.[158] Finally, on 10th April 1899, he was unanimously elected President of the Pathological Society of London [159], an office he was to hold until 1902, when John Burdon Sanderson took over. This overview of his frenetic activities is a graphic illustration of why he had been forced to take his domestic arrangements in hand after Mary's death if he wished to continue as one of Britain's leading authorities. Years later, however, Sir Harold Stiles would remark: "For years, he was one of the busiest of men; and while he shouldered a heavy burden of responsibility and got through a prodigious amount of work, he never seemed to be in a hurry."[160]

Many of his surgical publications and activities in the 1890s had turned to the treatment of cancer, and in February and March 1896, he once again delivered three Lettsomian Lectures on "The Objects and Limits of Operations for Cancer", at the Royal Medical Society of London.[161] He examined in detail surgery for cancer of the breast, throat and intestinal tract, and his main consideration in whether and how to operate was determined by the comfort of the patient and whether the case was considered curable. It has recently been shown how Lister was ahead of his time in considering the patient's comfort high on the list of prerequisites for a practising surgeon [162], and his pupils seem to have emulated this in their own practice. Cheyne's lectures on cancer are an interesting example of how his humanity as well as surgical good sense played a part in his work. He was concerned about the mental effects of operations on both men and women. In his Harveian Lecture on tuberculosis, for example, he did not advocate double castration of testes because it had a tendency to depress male patients, who, he reasoned, could easily become less acute men of business, lazy, irritable, and morose, and in some cases may actually become demented, or maniacal."[163] Some, he thought, became suicidal. The following extract from his work on breast cancer illustrates his views on consideration of the patient's wishes:

> There is at the present time a tendency with some surgeons to careful selection of cancer cases for operation, that is to say, only to operate on quite simple cases. This is not, I think, a proper point of view. No doubt it is the way to get good statistics, but what of the poor patients who are not operated on because the prospect of success is not good? I do not think that patients should be refused operation unless the disease cannot be removed, unless early recurrence is very highly probable, or unless operation means almost certain death or yields a hopeless functional result. Of course, if one has something better to substitute for the radical operation, such as colotomy in extensive rectal cancer, the matter is quite different, but where this is not the case the patient should be told all the circumstances and allowed to take his choice.[164]

Cheyne judged the evidence on its own merit as he always did. His criteria, again, were the comfort of the patient and the chances of curing the condition. When it came to treatment for intestinal cancer, he noted "I quite agree with the tendency of the English school of surgery as opposed to the German, and would exclude from the radical operation cases of rapid growth, cases where the disease forms large masses in the intestine or is deeply ulcerated, cases where it has passed through the wall of the rectum and invaded surrounding parts, as indicated by fixidity, and rapidly-growing tumours high up, even though not yet fixed. In none of these is there any real prospect of cure or of marked prolongation of life."[165]

He had gathered data for an impressive 172 cases which, as I have noted, the *BMJ* regarded as "monuments of industrious research".[166] In

Microbes and the Fetlar Man

cases of cancer of the mamma, he had adopted Volkmann's three year limit, taking 'cured' to mean any cases where the condition had not returned after three years. Using this parameter, he was able to "show a larger percentage of cures than is recorded in any previous report, while the mortality from the operation is minimal."[167] The lectures were well received, with a few queries about details which Cheyne was able to set straight. Where results were not as straightforward or impressive as those for breast cancer, critics conceded that "Mr Cheyne greatly aids the reader by a careful, impartial, and complete account of his own cases."[168]

In May 1898 he reported two cases which tried to replicate successes in removing ovaries as a way of relieving inoperable breast cancer, but his results indicated that there was no particular advantage. This disappointed him, and he found it "discouraging".[169] Some surgeons, as he had stated, were interested in performing operations only where they could be reasonably sure the results would not harm their statistics and personal reputation. Cheyne's more patient-focused criteria do not mean he was a reckless operator, however, and he was, in fact, regarded as a "good and safe surgeon",[170] no doubt encouraged initially by the wider range of operations made possible by antiseptic surgery but also, perhaps, from a degree of humanity in attempting to ease suffering. His Lettsomian Lectures were published as a volume in 1896.[171]

It is interesting that later writers dwelt heavily on the safety of his operations, but did not consider him particularly innovative. One noted, "He was not endowed by nature with a great degree of originality ..."[172] This can be taken to mean he was not reckless in his choice of operation, but does not quite fit with some of the rather interesting procedures he adopted. The development of antiseptics had allowed Lister to introduce innovations in plastic surgery, and Cheyne appears to have followed suit. In January 1898, he repaired the damaged nose of a boy who "...eight years ago, met with an accident through which he lost a large portion of his nasal bones."[173] He "raised a flap of skin from the right side of the nose, exposing the periosteum. He then removed the femur from a rabbit, split it into five or six pieces, and simply laid these on the periosteum and replaced the flap. The wound healed without suppuration, though a small amount of serum was let out a few days after the operation, and the result, as shown to the Society, was excellent."[174] He also invented a number of surgical instruments, as did many surgeons, and in 1890, had developed a wire frame for incompletely descended testicles, as a way of improving treatment for certain instances of the condition.[175]

The advances in medicine and surgery had, in fact, given rise to a thriving industry of instruments and equipment. Cheyne had a probe made, which thereafter became known as the 'Watson Cheyne probe' "intended for fine dissections such as isolating the veins in operations for varicose veins, and in exposing small veins."[176] It was displayed in 1894 at the Annual Museum, which showcased the medical product industry. His 'cage for undescended testicle' was exhibited at the stand of Messrs. Down Bros. of St. Thomas' Street. Borough, S.E., who manufactured aseptic surgical instruments [177], along with his dissector and probe again, and hernia needles which had been designed by Cheyne and Macewan. The stand also displayed "Mr Watson Cheyne's Retractor for use in Intestinal Operations".[178]

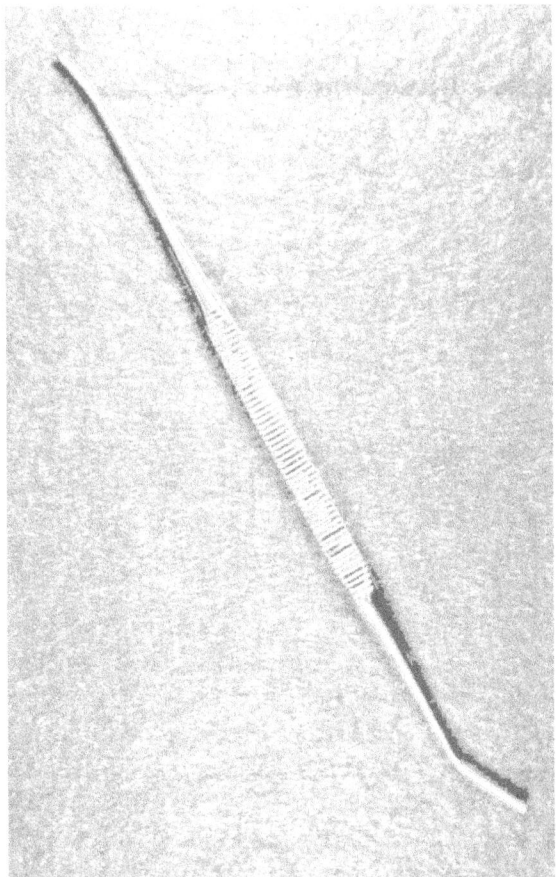

A 'Watson Cheyne' probe

A result of one of Lister's negotiations with manufacturers, however, led Cheyne to an important surgical innovation at the end of the 1890s. Lister had experimented with different forms of suture for stitching wounds together. Silk sutures were common, but he ultimately preferred catgut (which actually had nothing to do with cats, but was derived from the small intestines of sheep, goats and cattle). Antiseptics had also revolutionised the use of sutures, as they could now be soaked in solutions prior to use and harboured less danger of infection. Cheyne suggested to Lister that these sutures could also be used as a drain [179], and in 1894, Lister asked J.F. MacFarlan and Co. for information on how they produced sutures of catgut which had been soaked in chromic acid. The company, however, could not make the product themselves, and war conditions had disrupted supplies of imports. A former employee of the company, George Merson, became aware of Lister's requirements, and began to produce catgut sutures at his home to meet this demand. In the early 1900s, Merson sold his company to Johnson and Johnson; this was the beginning of their major production of sterile medical supplies.[180]

In 1898, Cheyne and G.A. Sutherland, a physician and colleague of Cheyne's at Paddington Green, used the catgut sutures as a drain to test a new intervention for hydrocephalus, a condition which affected children in particular.[181] Also known as *water on the brain*, it is caused when an unusual level of cerebro-spinal fluid accumulates in the cavities of the brain, making the head appear larger. In 1896, a physiologist, Leonard Hill, had identified a possible way of approaching congenital hydrocephalus by re-routing the accumulated liquid so that it could be reabsorbed by the veins.[182] Cheyne and Sutherland confirmed this observation on a six-month-old child by using a catgut drain to open a channel of communication for the fluid to drain out internally, rather than previous techniques, which had drained the liquid outside the body, with a risk of infection.[183]

> Before the operation a catgut drain was made as follows: a bundle of finest catgut, containing 16 strands and about 2 inches long, was prepared, one end of the bundle being tied together and the other being free. As soon as the dura mater was incised, the tied end of this bundle was seized with a pair of sinus forceps and pushed downwards and slightly backwards between the brain and the dura mater for about an inch. The other end of the drain, which projected through the slit in the dura mater, was then grasped with sinus forceps and pushed through the substance of the brain into the

expanded lateral ventricle. The brain was very thin at this point, and clear fluid escaped immediately. Having thus arranged one end of the drain in the subdural space and the other in the ventricle, three fine catgut stitches were employed in completely closing the opening in the dura mater, and the skin was stitched up with a continuous silk suture … When the dressings were removed on the fifth day after the operation the wound was healed. The head was distinctly smaller in all its dimensions, and there was a complete absence of the tension which had been present before operation.[184]

On August 6th 1898, Cheyne and Sutherland spoke on hydrocephalus at the 66th annual meeting of the British Medical Association, probably with lantern slides.[185] and they reported this and two other cases in the *British Medical Journal* in October 1898.[186] The child died of complications, with symptoms of basilar meningitis, three months after the operation, however.[187] The mortality rate remained high for the intervention, and it would be the 1950s before significant inroads were made. The operation remains a difficult one. The technique used by Cheyne and Sutherland was nevertheless a turning point [188], and aspects of their work are retained in the treatment used today for the condition. Antiseptics had considerably broadened the range of operations possible, and Cheyne was nothing if not inventive in developing them.

His most prolific output, however, remained his publications. Towards the end of the decade, he contributed to the *Encyclopaedia Medica*, published in a number of volumes in 1899,[189] and the same year he helped edit volume three of King's College Hospital reports.[190] He also began work with F.F. Burghard, a colleague at King's and at Paddington Green Children's Hospital, on an extensive *Manual of Surgical Treatment*, which would be published in a number of parts, the first in 1899. Cheyne recalled how he had become "much dissatisfied with the text books"[191], as he could not find in them all the information he required. He had been asked to remove a malignant larynx and "wanted to know what to do with the wound when I had finished the operation, how I could arrange the dressings, how to deal with the oozing of blood, how to provide for his breathing, and many other matters which had evidently been looked on as trivial by former writer[s]. Curiously enough, most books seemed to have been copied from one another."[192] He "determined to write a complete volume on the treatment of the various surgical diseases and operations, starting from the point that the reader knew nothing to begin with and that I had to treat him as though he were facing the operation for the first time."[193]

The "volume" was to turn into a gargantuan six-volume work, over the course of several years, and was to prove immensely popular, particularly with students. In the 1950s, G. Grey Turner remembered how he had awaited with "eager anticipation"[194] the successive publications of each volume. We should bear in mind that, by this time, Cheyne was in his upper forties, and the tale of how he managed to write the work on top of his surgical duties and the many talks he gave is testimony to his stamina. Turner had met him at the Lister Centenary celebrations in London in 1927, and Cheyne had told him how it had all come about.

> At that time Cheyne was a busy London surgeon on the full staff at King's and at Paddington Green Children's Hospital, on the court of examiners at the College of Surgeons and on the council there, and with many other duties. To find time for the work on the book he rose at six o'clock each morning, when he spoke into a dictaphone and corrected what had been typed out from his dictation of the previous day. This must surely have been a very early use of the dictaphone; but even with its assistance it was a great undertaking, for the work totalled nearly 3,000 pages, and with the exception of a chapter on anaesthesia it represented the personal effort of the two authors. There were many illustrations and each volume had its own index.[195]

It was very much intended as a practical guide, drawing on the personal experience of both surgeons, rather than a discussion of the merits of one method over another, in order "not to confuse"[196] the practitioners and students at whom it was aimed. The full set of published volumes would not be complete until 1903. Volume II had barely been issued when Cheyne set off for several months' war service in South Africa, discussed in a later chapter, but Volume III came out on July 23rd 1900, only a few days after his return, so he had clearly completed at least his own part in it before he left.

The extensive experience of the two authors did not prevent them from acknowledging the work of others [197], though there was some criticism they were illustrating only their preferred options. When Volume III was issued, the *BMJ* said that, "the chapters dealing with amputations suffer to some extent through the authors' self-imposed conditions of describing only such operative methods as they believe to be best,"[198] but Cheyne stuck to his guns throughout the series, as he wished it to be very much a manual with clear directions. He responded to the criticisms in the preface to Volume IV by noting once again that:

> ... the object we originally set before ourselves was to describe as fully as possible only those methods of treatment. To this plan we have adhered; and therein also will be found our justification for the omission of certain newly-introduced methods of treatment which may prove of great value, but which are as yet too recent to enable a definite verdict to be pronounced upon them. To describe all methods of treatment would be to convert a manual into an encyclopaedia.[199]

They nevertheless took account of the criticism, and by the time Volume V was issued in 1902, it had become clear that they would not complete it adequately in five volumes, so added a sixth. The *BMJ* reviewer considered this to have made it into the "most important and instructive"[200] of the series. This, however, was still not sufficient, and they were forced to divide the last part into two volumes.[201] In his preface to Volume VI, Cheyne apologised to his readership for the fact that "this volume has been somewhat longer in making its appearance than its predecessors. This is partly due to its size and the larger amount of material requiring to be worked up, and partly to the fact that, owing to the absence of one of us in South Africa, the printers have been able to overtake us."[202] If this was not tongue-in-cheek, Cheyne must have genuinely believed that producing so many volumes of highly technical material in three years, taking an excursion to the war in South Africa and maintaining a gruelling schedule of professional and personal appointments was a perfectly normal achievement. The later comments from Turner about its popularity with students is borne out by the fact that he felt the need to apologise for it taking a little longer than he had expected and that, by 1901, Volume I had already been reprinted twice.[203] Even when the work had only been two-thirds completed in 1901, the *BMJ* had praised the two authors for their "energy and industry" in the "prompt publication"[204] of the various volumes.

Basil Cheyne, Sir Watson's youngest son.

Margaret had clearly taken charge of the household by this time. On 12th October 1898, she had given birth to a son, who was named George Basil after her father.[205] She had also been busy organising the household along different lines, and at some point, brought down new staff from Shetland; by 1901 there were at least two Shetland staff, Andrina and Catherine Slater.[206] The former had been a servant in Margaret's brother John's house in Lerwick in 1881.[207] By 1911 the staff at the Harley Street residence would all be from Shetland except Rose Chapman; along with Andrina and Kate Slater, Christian Anderson had been brought down from Fetlar.[208] Cheyne had bought land in Fetlar by this time, and returned regularly to the house he had built on it. It is possible Christian's family had asked him if a position were available. She was later to play an important role in preventing an unpleasant incident in the household.

It is significant to note the extent of Cheyne's workload in the late 1890s, as it was to have implications for his health as he entered the new century. He was still juggling an extensive schedule of publications, public appearances and surgical work. Though he had largely given up his active role in experimental bacteriology by the 1890s, he continued to comment and write prolifically on subjects where it had a bearing, and it remained the foundation for his surgical activity. In fact, when he gave the address at the Annual Meeting of the British Medical Association in 1897 in Montreal, it was as President of the section on Pathology and Bacteriology [209], and he was still revered as one of the leading men of the field in Britain. Moreover, one persistent bacteriological issue was to dog him in one way or another for the remainder of his career. It involved a return to where he had started: the relationship between bacteria and the treatment of wounds.

9

Surgical towels in the chamber pot

Antisepsis or asepsis?

Research and discourse on the relationship of microorganisms to wound treatment spanned - in one way or another - the best part of Cheyne's career, and it is the subject for which he is chiefly remembered. His role in the development, adaptation and documentation of Listerian asepsis was one of the most significant in the field. Objections to the use of chemical antiseptics had involved a range of issues, at first a prolonged belief in spontaneous generation and then a recognition that many of the effective antiseptic substances damaged the skin of both patient and surgeon. Through the various adaptations in Listerian theory and practice, Cheyne defended each turn of events by returning again and again to a simple premise: where a number of factors came into play in terms of the healing of wounds, measures taken to disinfect the surgical environment must take account of bacteriological issues. In this chapter, I shall examine how Cheyne approached his arguments in defence of Listerian antisepsis, which ranged from anecdotes to witty condemnation of his opponents.

After his Harveian Lecture in London in October 1890 on the "Treatment of Tuberculous Joint Disease", it was noted that Cheyne "might almost be the biographer in England of micro-organisms".[1] His reputation as one of Britain's first bacteriologists was rarely in question, but his publications, particularly *Antiseptic Surgery*, had also extensively chronicled developments in the field, many of which he had witnessed personally. In other words, he had an influence in a number of ways: firstly through his original experimentation and testing the work of others, secondly through his documentation of developments in general, and finally through his dissemination of laboratory techniques in medicine. In 1886, his contribution to Gant's *The Science and Practice of Surgery* was considered superior in content to the other chapters. The *BMJ* noted, "Mr Watson Cheyne has given a chapter on Bacteria in Surgery, which will be read with interest, but we do not feel that the other parts of the work at all equally advanced when treating of subjects in which micro-organisms have to be considered."[2] In the same way, he was considered well-suited as 'the biographer' of antiseptic surgery. A review of his contribution to the *International Encyclopaedia of Surgery*, edited by John Ashhurst, Professor of Clinical Surgery at the University of Pennsylvania, read:

> Dr Ashhurst has wisely secured as author of "The Antiseptic Method of Treating Wounds", the chief assistant of the founder of the system, Mr Watson Cheyne, whose competency to summarise Mr Lister's method, with all its networks of theory, fact, science, art, and detail, in the form of a treatise, has already been shown in his Jacksonian Essay on that subject. We congratulate Mr Cheyne on the conciseness and brevity of his contribution, and the entire absence of needless and tedious repetitions about traumatic inflammation and other questions relating to wounds, and to subjects fully discussed in other papers.[3]

In the late 1870s, not long after he had graduated, he was already developing clear arguments against opponents in Britain such as Charlton Bastian, who had carried the flag for spontaneous generation to the end of the decade. When Cheyne wrote his Boylston Prize submission, parts of which later became *Antiseptic Surgery*, the copy he kept for himself contained notes. When it came to a discussion of Bastian's experiments in favour of spontaneous generation, assuming the notes are his, he allowed himself one of his many moments of wit, which he seems to have tempered at least a little in published sources:

> We have learned from Dr Bastian & other experiments that the introduction of cheese into vessels is almost essential for the occurrence of spontaneous generation. We need not therefore fear the spontaneous origin of organisms in wounds nor the spontaneous occurrence of ferment, so long as we do not introduce cheese into them. Cheese à la Bastian therefore is a bad dressing ... I for one have no hesitation in accepting the germ theory of putrefaction & in looking on the true principle of antiseptic surgery – surgery directed against the <u>causes</u> of putrefaction – as a battle against the entrance into & growth of organisms in the discharges of wounds ... So long as we do not introduce cheese à la Bastian.[4]

Once he and other Listerians had dispensed with Bastian and spontaneous generation, other, more persistent, problems arose. Some opponents, like Sampson Gangee, contended that the constant changing of Listerian dressings was a hindrance to the natural processes of healing, which required rest, first and foremost.[5] Others based their opposition on the fact that bacteria had to be introduced into wounds to be harmful and could not simply enter from the air. Burdon Sanderson, for example, argued "that germs did not just fall into wounds from the air, being far too few and fragile; rather, they mainly came from other diseased sources and had to be "introduced" into the body."[6] The logical conclusions of what became known as the 'Cleanliness School' were that, provided everything touching the wound were kept externally clean, putrefaction could be avoided without the necessary use of chemical antiseptics, particularly Lister's spray. Listerism was to adapt considerably to take account of some of the evidence from these arguments.

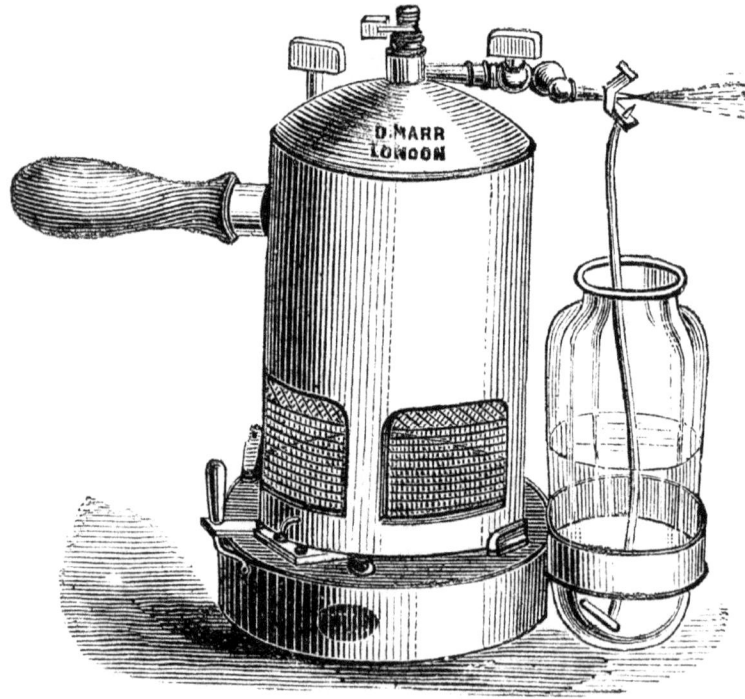

FIG. 16.—THE ORDINARY STEAM SPRAY PRODUCER.

Lister's carbolic steam spray.

Even in the early days, there had been those who were happy to use antiseptics, but claimed equally good results without the full procedures Lister advocated, and based their claims of success on their clinical data alone. Lister distrusted statistical evidence on its own, as it did not, in itself, prove the Listerian principle.[7] He was afraid others might attempt to replicate the methods with an imperfect understanding of the reasons for them, which could put patients' lives in danger. He had been reluctant to allow Cheyne to publish the statistics of his antiseptically-treated cases until he was finally forced to respond to critics who had taken his silence to mean he could not produce evidence of the success of the antiseptic method.[8] When he finally decided the time had come to publish, he "unexpectedly found the necessary materials in Mr Cheyne's laborious analysis of"[9] his hospital case-books, and allowed Cheyne to submit them for publication in November 1879.[10] It prompted an acrimonious argument in the *British Medical Journal*, where Cheyne, still young and relatively inexperienced, acquitted himself with dignity against one of his old professors. Professor Spence in Edinburgh launched a detailed attack by suggesting his own statistics for success were comparable to Lister's, using antiseptics without the elaborate Listerian procedures, particularly the spray.[11] It came in response to a "challenge"[12] launched by the *BMJ* to anyone who could show similar success using "simpler"[13] forms of dressing, implying that any evidence of this would constitute a definitive test of Listerism.

Cheyne and Lister planned a coordinated counter-attack the following week. Lister responded to Spence, outlining the reasons why he considered his argument invalid and objecting to Spence's slight on his abilities as a surgeon.[14] Cheyne, in the same issue of the *BMJ*, tackled Spence head-on, saying, "I have ... preferred to meet Mr Spence on his own ground - on his numerical results ; and even though such statistics have not as a rule the power of proof possessed by cases related in detail, yet I think there can be no doubt as to their force in this instance."[15]

In other words, he wisely did not enter into arguments about bacteria at this stage, as he knew it would confuse the issue, particularly as so many surgeons were not interested in the germ theory of putrefaction. Instead, he based his argument on the dates of Spence's statistics, suggesting he had chosen a specific period which was "the only bright spot in a long and melancholy record of disastrous amputations".[16] Cheyne trawled publications for more recent results by Professor Spence in the *Medical Times and Gazette*, and adjusted his figures more closely to the period of Lister's results, from 1874-1876. His adjusted figures clearly showed statistically that, in fact, Lister procedures had yielded the most successful overall result. It was a masterful and carefully detailed response. He had obviously devoted any spare moment to a week of intensive study of the figures, no doubt with Lister's guidance in the background, and presented his results with a degree of confidence, but no hint of the self-satisfaction in Spence's original. It was a particularly mature response.

The *BMJ* considered Cheyne to have shown conclusively that: "Mr Spence has entirely missed the points at issue. We do not require statements as to general results. What we do want is evidence to enable us to judge whether the same constant results, which are so notable a feature of strict antiseptic treatment under any circumstances, can be obtained by 'simple' methods of dressing under unfavourable circumstances, such as bad hygienic conditions, or after certain operations, such as incisions into joints, psoas abscess, etc. If this cannot be done, then, in spite of any number of favourable results in favourable cases, we must recognise that the surgeon who treats his patients on the principle of total exclusion of living septic particles is acting on a new principle, and is able to obtain results better than those which can be got otherwise ... the result of the discussion has only been to bring out in more striking contrast the imperfections of 'simple' methods of dressing in the hands of a most distinguished surgeon, when compared with the brilliant results of treatment conducted on strict Listerian principles."[17]

Spence was one of the most eminent surgeons of the day, one of Cheyne's ex-professors in Edinburgh and the object of his admiration, particularly as he had given him a very favourable reference to take with him to London.[18] In his response to Spence, Cheyne had been careful to note: "It is with considerable diffidence that I have entered into the controversy against one for whom I entertain great respect and to whom I owe so much."[19] Despite the security of Lister's backing, and despite the fact that he had little choice but to answer the criticisms of his own paper on Lister's statistics, it was a brave move on Cheyne's part, and he displayed nerves of steel in the manner in which he responded to his ex-professor, implying no hint of disrespect. Spence, on the other hand, objected strongly to being taken to task by one of his ex-students, and published an angry response:

> The editor could hardly have been unaware that, at the beginning of this controversy, I stated in the JOURNAL that I would take no

notice of such articles, as I considered Professor Lister was better able to speak for himself. I certainly do not think I am called on to notice every young friend of Mr Lister's who may choose to thrust himself into notice in a controversy between Mr Lister and myself. I do not consider that the correspondent referred to possesses either the independent experience in practical surgery, nor the amount of experience in different methods of treatment, to qualify him to deal with such statistics.[20]

The editor took Cheyne's side and considered the debate to have taken an unacceptable turn:

> We cannot avoid thinking that Mr Spence's remarks with regard to Mr Cheyne's contribution to the statistical question are scarcely worthy of so eminent a surgeon. Such remarks can in no way detract from the significance of the facts recorded. Mr Spence's great fear seems to be that attention to local measures may lead the surgeon to disregard constitutional conditions. Supposing this idea to be well-founded, it appears to us that if Mr Lister, while paying but little attention to the latter, can obtain under unfavourable conditions better results than can be got under simple methods of dressing with close attention to the constitutional state, additional evidence is afforded in favour of the method which he advocates. It is, we think, time that this controversy - into which the personal element has been introduced to too great an extent - should for the present cease.[21]

Part of the misunderstanding was perhaps due to the fact that they were using statistical results without any reference to what Cheyne would later refer to as "the bacteriological requirements of the case."[22] It was not uncommon for those who misunderstood Lister's ideas to assume that his main preoccupation was the use of a particular antiseptic or a specific piece of equipment. Spence, like many other critics, at one point suggested that the most important basis for Lister's methods was germs in the air, and that consequently, the spray was the most essential piece of equipment:

> I asked Mr Lister how he explained the escape of these my successful cases in an operation in which the wound is so long exposed whilst arresting bleeding, etc., if the germs floating in the atmosphere were

Microbes and the Fetlar Man

the cause of all the evils attending operations? To this, the most important part of my question, as it involved the principle on which his theory is based, Mr Lister vouchsafed no reply.[23]

This was a misinterpretation of Lister's principle, and Cheyne went to great lengths to set the record straight. Even worse, however, was the fact that Spence's assumption was a blatant disregard of what Cheyne had stated clearly in his original paper on Lister's statistics:

It will thus be apparent that the cases here recorded are examples of the results of the practical carrying out of a principle, and not of the application of any particular medicament in any special form to the tissues. The use of the spray is not more important as an antiseptic means than the purification of instruments, etc., in the antiseptic lotion. The spray merely purifies the atmosphere; and, as there are undoubtedly fewer organisms floating about in the air than are present on any instrument which has got dust or water on it, if one were compelled from any cause to give up one or other antiseptic means, one would at once choose the spray as the least important and that which could be the most easily dispensed with. In fact, the antiseptic treatment was carried on for years by Mr Lister without any spray; the wounds being washed out with carbolic lotion and carbolic putty, lac plaster, or other dressing acting on similar principles immediately applied … Hence, to say that, because a surgeon uses the spray, or, indeed, all the materials for antiseptic work, he therefore employs the antiseptic treatment, is incorrect, unless the various means are used with the definite aim of excluding the putrefactive ferments. On the other hand, to say that, because a surgeon does not use a spray or even carbolic acid, he therefore does not employ the antiseptic method, would be equally incorrect. So long as the result of his treatment is to exclude or render inert the causes of putrefaction, he has been practising antiseptic treatment.[24]

Lister himself never suggested the spray was anything but a part of an overall process. In fact, Cheyne later maintained that Lister had always seen the spray as a temporary element in Listerian procedure, until the process could be perfected sufficiently, and he noted that "from quite an early period, Lister was contemplating its abolition"[25] He saw it largely as a precautionary measure, given that, in the 1880s, he believed surgeons "had not yet become very expert with their antiseptic methods"[26] and did not all accept – or understand the significance of - the germ theory of putrefaction behind it. He also gave more prominence to bacteria in the air than would be the case later on. In a more specific sense, Listerian operating procedures took account of the dangers from a surgeon's breath, should he speak or even open his mouth during an operation. Given the visitors and students present at operations, it was almost impossible for them not to speak, and Lister considered the spray a safeguard here too.[27]

The reality was that the viability of carbolic acid, Lister's principle antiseptic agent, had long been called into question, not least by Lister himself. He was not only aware of the toxic qualities of carbolic and its ability to damage the skin, he was also aware of evidence that it was not always effective. As early as 1874, Ranke had questioned the ability of carbolic to kill all bacteria,[28] and in 1880 and 1881, Alexander Ogston in Aberdeen showed that it was ineffective in killing micrococci.[29] In 1881, Robert Koch also provided experimental evidence to show that carbolic was ineffective in killing anthrax spores.[30] In 1883, John Duncan reported the results of bacteriological studies where the spray had proved comparatively ineffective in excluding microorganisms.[31] Mikulics, too, had issued similar findings.[32]

In response to a paper by Professor Bruns from Tübingen, Germany, at the International Medical Congress in 1881, Lister had noted that he would like nothing better than to do away with the spray.[33] However, as his arguments at the time rested on his belief that bacteria could enter wounds from the air, he felt he could not guarantee all other parts of his procedure would be followed sufficiently to exclude bacteria, and he continued to advocate the spray as a safeguard. Even amongst those who considered it important to kill germs in the air there were concerns about the toxic effects of carbolic poisoning. In September 1882, Mayo Robson, in discussing the merits of eucalyptus spray, noted that he himself had seen people die of carbolic poisoning during abdominal sections on more than one occasion.[34] He had also come to the conclusion that, although air which had passed through carbolic may be aseptic or sterile, it was not antiseptic, and was not killing germs when it reached the wound.

Cheyne's obituary by the Royal Society was later to credit him with providing evidence which ultimately led Lister to abandon the spray.[35] He detailed his experiments in his speech at Lister's centenary, many years later, showing how he had opened sterile flasks to the spray at intervals to observe the reaction. He noted that he had lost the notebook by this time, as the experiments were never conducted with a view to publishing, only to further their own knowledge of the usefulness of the spray.[36] It illustrates the extent to which Lister was serious about acquiring objective data on the spray simply to inform his own decisions about it. By 1894, Cheyne was not even using carbolic dressings on children, as he considered them dangerous, and only used them on children for "short purification of the skin".[37]

According to Cheyne, Lister eventually abandoned the spray in 1887,[38] recognising that bacteria in the air were not the main problem, and counteracting the ones that were there probably did more harm than good. Cheyne nevertheless persisted his whole life in warning against complacency in thinking airborne microorganisms were of no consequence at all.[39] Lister, on the other hand, made a sensational announcement at the International Medical Congress in 1890 that he was ashamed to have recommended using the spray.[40] Once again, this kind of admission was unusual from so eminent a surgeon in the 19th century; it was not easy for them to admit their mistakes so publicly. In one thing he remained absolutely clear however. Perhaps not the spray, and perhaps not carbolic, but some form of chemical antiseptic was advisable as a safeguard if a fully aseptic environment were to be created and maintained for surgery and subsequent dressings.

Lister continued his search for more suitable antiseptics. When Koch suggested that corrosive sublimate was the only compound which definitively killed resistant spores and micrococci, Lister reacted positively, and welcomed the conclusions as similar to his own. He continued to believe in it for the rest of his career.[41] Its irritant and noxious qualities were, however, as bad if not worse than carbolic, and it was too toxic to use in the spray. The combined issues with chemical antiseptics ultimately encouraged the development of methods to avoid sepsis without the use of these substances altogether.

Non-chemical procedures began to gain rapid ground, and developed into what became known as aseptic surgical methods. Aseptic methods centred largely on disinfection by heat and steam, and were once again spearheaded in Germany. Pennington, in assessing the uptake of these methods against the background of Listerian procedures, notes how Cheyne used the terms "aseptic" and "antiseptic" very differently,[42] and quite specifically. In *Antiseptic Surgery*, in 1882, he used the word "aseptic" to refer to the results of Lister's antiseptic procedures. In other words, it was an end rather than a means; if surgeons did as Lister specified, their wounds would be aseptic. By 1889, when he contributed the section on Antiseptic Surgery in the *Dictionary of Practical Surgery*, he defined aseptic surgery more in preventative terms: where the object was "to prevent the entrance of living organisms into the wound"[43]. He used "antiseptic" methods, on the other hand, for measures to kill or "interfere with the growth"[44] of bacteria.

In the 1880s and '90s, antiseptic surgery was generally taken to mean the use of *chemicals* "to kill or prevent the growth of bacteria"[45] and aseptic surgery as methods which "use heat transmitted via water, steam or air, or physical methods"[46] for the same ends. Cheyne's own distinction was important to traditional Listerians, as it encapsulated their fundamental objections to steam methods, which they considered inadequate to cover the full Listerian precautions.

Pennington has shown with the example of three hospitals in Britain that by the early 1890s a majority had moved over to the "aseptic" method of disinfecting the surgical environment. He indicates, nevertheless, that the transition was by no means immediate or uniform, and a combination of methods continued to be used in many cases. "St Bartholomew's went on using complex antiseptic dressings into the twentieth century."[47] As the new 'aseptic' methods gradually came into more widespread use, many surgeons employed a combination of chemical antiseptics and steam. Moreover, in the early stages, there were a number of practical issues to be overcome by manufacturers of steam boilers. They had to be designed in such as way that they would leave dressings dry as well as disinfecting them, and the possible rusting and blunting of instruments had to be taken into account. Nor did early boilers easily cater for bulky items or answer the question of how to disinfect the surgeon's hands.[48] For Lister and Cheyne, the steam methods posed a number of more fundamental problems. They failed to see how advocates of the system proposed to keep the disinfected dressings aseptic over the course of healing. Where he had always been pragmatic about the need to find a suitable alternative to carbolic, Lister reacted decisively when it came to steam:

> The operation being concluded, an external dressing such as shall effectually prevent the access of septic mischief till healing is accomplished is, of course, a matter of essential importance. For this purpose some surgeons have of late years employed materials merely aseptic, such as cotton wadding sterilized by heat. But such a dressing having nothing in it to counteract any accidental

defilement, must demand an almost impossible degree of care in its manipulation in order to ensure that it is truly aseptic as left upon the patient. The mere aseptic dressing has also the fatal defect that it is liable to be occasionally soaked to the surface with discharge, in which septic development will then be free to spread inwards to the wound. I believe, therefore, that a dressing, in order to be trustworthy, must be charged with some chemical antiseptic substance.[49]

Despite the general uptake of steam methods, Cheyne continued to be sceptical about them, which gave him a growing reputation of being simply a diehard [50], though he was, in fact, anything but a reactionary and genuinely believed the new methods were inadequate on their own. His main objections to the "aseptic surgeons" are outlined succinctly in the following passage:

In the first place, this so-called aseptic method ... does not in any way meet the bacteriological requirements of the case ... To lay instruments on a dry sterilized towel, and to work with them for perhaps half an hour, or an hour or more, without re-sterilizing either them or the hands of the surgeon or his assistants, is utterly opposed to any real attempt at keeping wounds aseptic...[51]

Apart from the bacteriological arguments, Cheyne - as a true student of Listerism - believed the "aseptic method" laid the surgeon open to unintentional error, and that chemical antiseptics were an important element in safeguarding against human mistakes. He noted in 1895 that in the aseptic methods "used more especially by certain German surgeons, in which the use of antiseptics are, as far as possible, avoided ... the risks of error are so great - and from the absence of antiseptic solutions there is no possibility of correcting them - that an ordinary surgeon, who has not had a prolonged bacteriological training, will find it a matter of the greatest difficulty, indeed, almost impossibility, to obtain results which are at all comparable with those which he would obtain were he to use the Listerian method ."[52]

This position was reiterated by others, such as Cheyne's colleague Burghard when he had to take Cheyne's place at short notice and open a discussion on the subject at the British Medical Association in 1904.[53] Any failure of antiseptics to exclude infection was, according to Listerians, the fault of some omission on the part of the surgeon, not the fault of the antiseptic method itself. Cheyne would go on making the point again and again that "… failures in avoiding suppuration are apt to be attributed to the use of this or that material, or of this or that method of sterilization rather than to the true reason – namely, failure on the part of the surgeon to meet the bacteriological requirements of the case."[54]

Lister, and his own mentor Syme, had been notable advocates of this ethic of personal responsibility in an era where medical men of standing were reluctant to accept their own mistakes, largely because their positions may have been called into question. If this happened often enough, they could be ruined, particularly in their private practice which gave them their living. Syme, however, had spoken publicly on "the duty which devolves on every member of the profession who is so unfortunate as to commit a serious error"[55] to rectify it "by faithfully stating the circumstances which misled him, so that they be rendered less likely to cause similar mistakes in future."[56]

The reasons for Cheyne's entrenched position on Listerian asepsis were regularly misunderstood. In 1893, he visited the Edinburgh Hospital for Sick Children to see aseptic procedures carried out without chemical antiseptics by Harold Stiles, who had recently worked under the German bacteriologist, Kocher. German schools of bacteriology continued to pioneer dry aseptic methods such as steam, rather than soaking instruments in an antiseptic solution. Cheyne was apparently alarmed at seeing his colleague "place instruments on a dry, sterilised sheet instead of into a 1:20 solution of carbolic acid."[57] Stiles put this down to the fact that he was "evidently still afraid of the germs in the atmosphere."[58] However, Cheyne was concerned instead with the overall process of achieving and maintaining wound asepsis, and was particularly concerned that surgeons might have forgotten the bacteriological reasons why all these procedures were necessary. Where his scientific arguments were sometimes misunderstood, however, his anecdotes rarely failed to get the point across:

> I am always very suspicious of so-called sterilized towels and swabs, especially in work in private. They are often placed in a sort of potato steamer for a few minutes, an apparatus which would take hours to cook a potato, and therefore if I do not know the nurses I always order the towels to be boiled in a pan for half an hour and not touched before I arrive. Well, in one case, nurses were sent out

from a large aseptic hospital who were said to be fully trained and most excellent. They received my instructions about the towels, and on my arrival I found a very small, poky room with very few dishes in it (not the palatial apartments of which some of my more fortunate *confrères* are accustomed to operate). I looked about for my towels, but could see no trace of them. I therefore called the nurse and asked her for them. She lifted the lid of the commode and there were my aseptic towels reposing at the bottom of the nightstool – another source of *Bacillus coli* infection of wounds which is worth remembering.[59]

He would go on with his anecdotes on the subject well into old age. As late as 1927, he recalled a cruise he had taken where passengers gave him a whole range of reasons they had been given for septic wounds after an operation, including influenza. "A smaller number," he noted with some amusement, "had been attacked by gout which had settled in the wound, and in these cases the patients felt that they had been very unjustly treated by Nature because they had never drunk enough port to give them that disease!"[60]

He had stated his position clearly in *The Treatment of Wounds, Ulcers and Abscesses*, as early as 1894.[61] The preface stipulated unequivocally that if suppuration occurred in wounds made by surgeons in unbroken skin, and they were using antiseptic methods, the surgeon must have made an error in his procedures, and "it is only by honestly acknowledging this to one's self, and by searching for the fault, that such an occurrence can be avoided on another occasion."[62] When the second edition came out in 1897 he engaged in a debate with Charles Barrett Lockwood, a prominent advocate of aseptic methods in Britain [63], on the disinfection of a surgeon's hands with chemical antiseptics. He disputed the accuracy of Lockwood's experiments, which had concluded that the skin could rarely be disinfected adequately. Cheyne furnished counter claims that, with particular care, it was perfectly achievable.[64]

Nor was he in favour of relying on chemical sterilisation *without* meticulous attention to Listerian procedures, as omissions would equally invalidate the results. It was what he considered the rigorous nature of Listerian antisepsis he was fighting to keep on the surgical agenda, not antiseptics *per se*. With this in mind, for many years to come, Cheyne attacked the emphasis of 'aseptic surgeons' on materials and modern operating equipment at the expense of attention to detail in excluding microorganisms from the wound:

They follow the fashion in having a well-finished operating theatre, in dressing themselves up in the most approved and elaborate fashion, and in having instruments and dressings sterilized; but they end there. They forget the ubiquity of bacteria, and that their instruments, gloves, &c., may lose their sterility during the course of the operation; they deride antiseptics.[65]

Such was Cheyne's prominence in the debate that an observer from the *British Medical Journal* was sent in June 1893 to witness him in action performing a laparotomy. The observer was particularly interested in highlighting "the precautions which are undertaken by those who follow strict antiseptic methods."[66] It was an illustrative example of the need for the surgeon's hands to be aseptic, as it involved, at a certain point, the need to insert a finger into the abdominal cavity. Lister's spray was no longer part of the procedure for operations, and the Listerian operator focused instead on ensuring the instruments and the surgeon's hands remained aseptic throughout the operation. It is a clear example of typical Listerian procedures of the 1890s.

> The patient was under chloroform when he was wheeled in. The abdomen and lower part of the chest were exposed, the clothflies covered by blankets, and these again by mackintosh, the edges of which were tucked in. The abdomen was well washed with soap and lotion (1:20 carbolic lotion containing 1/500 part of corrosive sublimate), and scrubbed with a nailbrush, then it was washed with turpentine, and after that with the strong lotion again. The mackintosh was then covered with towels wrung out of carbolic lotion. Mr Watson Cheyne and his assistants had previously scraped their nails, and used soap and water, nailbrushes, turpentine and strong lotion to their hands and arms and to the elbow. An incision 2½ inches long was made over the region of the gall bladder, parallel with the linea alba; a couple of strokes reached the peritoneum, which was then picked up and opened by a side stroke …"[67] The observer noted that the patient "suffered no ill effects from the operation.[68]

Realistically, it was rarely a question of all or nothing. As we have seen, hospitals adopted the new equipment piecemeal, and continued to use aspects of Lister's procedures, such as washing their hands in an antiseptic substance, for instance, even if instruments were disinfected in steam apparatus. In *The Treatment of Wounds, Ulcers and Abscesses*, Cheyne maintained that the new aseptic methods were little more than an unnecessarily complex development of Listerism.[69] He believed that a thorough acquaintance with the science would mean that "the manipulations necessary to keep bacteria out of ... wounds" would become "automatic,"[70] and that, with the knowledge now available about the natural resisting powers of the body, Listerian procedures were sufficient to ensure safe surgery.[71]

Signed photograph of Sir William Watson Cheyne.

Recent commentators have tended to agree that the new aseptic methods grew out of Listerian principles more than the arguments of the Cleanliness School [72], as they continued to be based on bacteriological arguments and insisted on their own "rigorous procedures for ensuring sterility."[73] They all essentially agreed on the need to remove bacteria from the surgical environment, but simply differed on how it should best be done. Where Cheyne continued to advocate antiseptics, it was where he considered them a simple safeguard against the loopholes he feared the new methods would introduce. At the turn of the 20th century, surgical manuals recognised that the two methods were not mutually exclusive.[74]

It was against this background that Cheyne earned himself an undeserved reputation among his contemporaries for upholding Listerian antisepsis for its own sake. Many commentators implied that he, as well as antiseptics, had become an anachronism. Some accounts of his achievements are pervaded by an air of good-natured tolerance, as though the author is humouring a respected old man, slightly stuck in his ways. Sir Harold Stiles called him "the last of the diehards to abandon antiseptics for a strictly aseptic technique,"[75] but was forced to admit that "Cheyne got better results by his antiseptic methods than did many surgeons who professed to carry out an aseptic technique."[76] He added sardonically that Cheyne's ultimate 'conversion' to aseptic methods happened to coincide with the removal of the beard which had characterised him throughout his career.[77] This may have been a rather optimistic interpretation of the aspects Cheyne was prepared to accept, however. It is probably more accurate to see his so-called 'conversion' as his own judicious and cautious contribution to a process of adapting Listerism, though only as far as he felt the evidence permitted.

After Cheyne's death in 1932, Arthur Edmunds, who had worked closely with Cheyne, responded to Harold Stiles' assessment in *The Lancet* with a short, succinct rebuttal which, in retrospect, is a much more accurate appraisal of Cheyne's attitude towards aseptic surgical methods:

"Sir Watson was never a diehard. It is true that he was not carried away by every breath of theory, but he did most sincerely endeavour to incorporate in his work anything that was of value in newly introduced methods, and his adherence to the older methods of technique was not due to obstinacy, but to a definite, considered judgment that it was better ... He and I often discussed this question, and he always deplored the teaching that chemical antiseptics do such an infinity of harm. He quite realised the excellent result that experts, working under good conditions, could obtain by the more modern technique; but he felt that, in less skilled hands, under less perfect conditions, the results would not be so satisfactory."[78]

He went on to comment on Stiles' suggestions that Cheyne had been 'converted', by noting that, "controversy dies sometimes from mere weariness, not because the protagonists of a theory have proved their point. Perhaps some day "local chemotherapy" will be reinvented."[79] In short, the emphasis on Cheyne as an unremitting champion of Listerism belies his extraordinary adaptability, even as he grew older. He was anything but a reactionary, having pioneered the use of microphotographs and lantern slides in his lectures. He had been one of the first in Britain to recognise the importance of new methods for staining bacteria in the 1880s, and the house he eventually built for his retirement in Fetlar was a paragon of modernity. He kept abreast of the arguments, and adapted his methodology where he felt it necessary and prudent, in his publications as well as his surgical practice. In 1910, Lister, by then an old man who was quite withdrawn from the public stage, admonished Cheyne for mentioning surgical gloves in an article in *The Lancet*, "because it may convey to some minds the idea that you distrust carbolic lotion for the disinfection of the hands."[80] Lister nevertheless trusted no one but Cheyne to treat a cut on his groom's hand, and had sent him a letter asking him to see to it, making it quite clear he was concerned about the methods a GP might use: "My groom has cut his hand badly this morning, a local medical man has seen it, but I should be greatly obliged if you would look at it."[81]

In his publications, Cheyne was cautious, as ever, until he felt he could support his arguments with appropriate evidence. As late as 1909, although he had been keeping records on Listerian antisepsis for over three decades, he did not feel he had sufficient data on success rates with purely 'aseptic' methods to make a convincing comparison with Listerian results. He said he had "no actual statistics at hand as regards the frequency of suppuration in wounds treated by the recent methods, and consequently I have always felt that, though I was right, my argument was not altogether convincing, seeing that I could only speak generally".[82]

He made what he considered a breakthrough, however, when figures from St. George's Hospital were published in *The Lancet* in November 1909 by H.S. Pendlebury and Ivor Back.[83] The figures showed a 6.4% drop in septic cases in 1908 with 'aseptic' methods compared to 1906 when the practitioners had been using 'antiseptic' methods. Cheyne acknowledged the improvement in his response in 1910 [84], but presented evidence that his own figures showed lower failure rates overall. He had been using

Listerian antiseptic procedures, and suggested that, since the authors did not state whether or not they were following *Listerian* antisepsis in the earlier figures, with no omissions in the process, he could only deduce their antiseptic procedures had been faulty in the first place, and that this had affected their baseline data. He referred to "the extraordinary craze at the present time to banish the use of antiseptics,"[85] and attributed the results at St. George's to "faulty manipulations on the part of the surgeon".[86] It was an old argument, one to which Listerians had resorted since Cheyne's student days, but he insisted that this, and this alone, accounted for his own more successful figures.

In reality, Cheyne had taken on the formidable task of becoming Lister's spokesman long before the Professor retired. Cheyne was a shy man, and his own humility meant that he rarely took credit for anything if he saw it as anyone else's (particularly Lister's) achievement. Where Cheyne came into his own was as a chronicler and orator who was capable of presenting a clear, illustrative account. He was able to deliver concise explanations of Lister's work in plain language, make it sound exciting, and patiently disentangle any misinterpretation until even the most persistent opponents could hardly fail to be at least aware of the main points. His writing was described as "clear and practical"[87] and his teaching was always based on experience. He was not afraid to face the critics head on, and, as we have seen, often used graphic wit and humour in his counter-attacks.

We should also consider the extent to which Cheyne carried the flag for antiseptic surgery because he truly believed it, and not necessarily only on Lister's behalf. In 1910, he considered the disagreements between 'aseptic' and 'antiseptic' surgery so confused, and so detrimental to surgery that he "advocated the formation of a small committee of surgeons from different hospitals – say half-a-dozen, with the addition of one or two bacteriologists … to study how each works in practice. They might thus be able to arrive at an authoritative statement as to what was efficient and what was inefficient, and what was practical and what was not. In this way a sort of standard of wound treatment might be laid down which would be of the greatest value to the profession."[88] He could not resist adding that "his only desire in the matter is that the foundations which Lord Lister has so successfully and laboriously laid down are built upon with due care and simplicity and in a permanent manner."[89]

Contemporary accounts give Cheyne's own performance as a surgeon a mixed press. His entry in the Lives of the Fellows of the Royal College of Surgeons notes that he was a "good and safe" surgeon but "not a brilliant operator".[90] Others considered him a model surgeon and mentor. Sir Harold Stiles suggested that, "as an operator [he] was superior to his great master [Lister]"[91], and noted how, "in my earlier days I took him as my pattern of an ideal surgeon. I was fascinated not so much by his dexterity and rapidity, but more particularly by his calmness when in a tight corner, and above all by the thoroughness which was seen at its best in his operations for tuberculosis or malignant conditions. He had the confidence, born of real knowledge, but he was as modest about his work and capabilities as he was quiet and unassuming in his manner."[92]

He also left his mark as a teacher at King's where he was considered to be "in his element"[93], and "lenient"[94] as an examiner. His colleagues considered him "a true man of science"[95] as well as a great clinician, and it is arguably as a bacteriologist that he enjoyed his finest hour. His surgical achievements all involved a bacteriological foundation. As we have seen, when he was appointed Demonstrator of Surgical Pathology at the hospital, he arranged for all tissues removed during operations to be examined under the microscope, and included the results of the examinations in the case histories. To provide an adequate environment for the procedures, he set up a department in the hospital, at his own expense.[96]

The question of aseptic surgery was where Cheyne's career culminated. It brought together his bacteriological work and his clinical training. It was, moreover, where he had started, and it was fitting that he should spend the rest of his life defending Listerian processes as he knew Lister had intended them to be understood. In the fight against proponents of spontaneous generation, Cheyne had found himself on the winning side. When he took on the 'aseptic' surgeons, on the other hand, he was embarking on a campaign he would wage for the rest of his career, rarely with the results he would have liked, through the Boer War and World War I, and well past Lister's death. He never tired of reminding his audiences of the enormous advances bacteriology had brought to medicine and, with perhaps a hint of allowable self-satisfaction, put it into perspective as early as 1890 when he recalled his first 17 years at King's:

> Seventeen years ago the so-called practical surgeon laughed at these little organisms, men doubted their existence, or looked on them as curiosities and of no practical importance, and regarded those who devoted time to their study as unpractical men with a decided mental twist; and yet we have now, as a result of this study, a new science which almost overshadows all the other medical

sciences, the knowledge of which is essential to the practitioner, and the study of which promises therapeutic results in the case of the infective diseases of surpassing importance.[97]

As the century drew to a close, Cheyne was in his late forties. He had come a long way since his first lectures in Edinburgh with Lister, both in terms of bacteriology and surgery. He noted in the same speech at King's that "… it is just seventeen years ago today that I came to King's, a raw and shy northerner with much to learn, and much to unlearn."[98] His workload continued to dominate his life, and his second wife, Margaret, now had the responsibility for his household. He had four children, though the two from his first marriage were away at boarding school by the late 1890s. Cheyne's oldest son Lister, who was 13 by 1901, boarded in East Grinstead.[99] Hunter was 11, a boarder at a school in Berkshire.[100] He had one remaining aunt in Shetland, Grace, who had been living on her own means in the former Methodist chapel in Fetlar [101] since her sister and brother–in–law had died, and it is likely Cheyne helped support her. Until this point, Cheyne had been relatively fortunate in his health, given the way he drove himself, but circumstances in the first two decades of the 20th century were to be very different, and would gradually wear him down. The first of these was his brief, but exhausting involvement in the second Boer War in South Africa in 1900.

10

Microbes at the Cape

The Boer War

At the very end of the 19th century, a second British conflict with the Boers (Afrikaans-speaking farmers, the descendants of Dutch settlers) in South Africa filled the front pages of newspapers. Troops were pouring out of Britain, and their progress was followed closely in the press. The conflict was the result of years of tension between Afrikaaners, British and Africans in southern Africa. By 1854, Britain had recognised two independent Afrikaans-speaking republics, the South African Republic (Transvaal) and Orange Free State, populated largely by Boer farmers. In 1877, Britain annexed the Transvaal in an attempt to consolidate the southern African states under British rule, but was forced to reverse the decision in 1881, following a Boer rebellion known as the first Anglo-Boer War.

Diamond mining in southern Africa had, by this time, made the fortunes of English businessmen, particularly Cecil Rhodes, who had ultimately become Prime Minister of the British-ruled Cape Colony. With fellow multi-millionaire Alfred Beit, he had founded a new British territory north of the Transvaal. The discovery of gold in the Transvaal in 1886 precipitated an influx of Uitlanders (English-speaking prospectors) who enjoyed no political rights in the Boer-run republic, where a 14-year residency rule prevented them from voting. The Uitlanders were restless, and far outnumbered the Boers by this time. Cecil Rhodes saw his opportunity to claim the Transvaal for Britain and consolidate his financial interests by attempting to incite the Uitlanders to rebel, sending in Dr L.S. Jameson with a force to support them. The Uitlanders failed to rise to the occasion, and the raid came to nothing. Four years later, in October 1899, Britain declared a second war on the Boers. Jan Smuts, who later became a Prime Minister of South Africa, traced the beginning of the conflict to the Jameson Raid, however, which he called "the real declaration of war in the Great Anglo-Boer conflict."[1] Early in the war, the Boers besieged the British-held towns of Ladysmith, Mafeking and Kimberley, and resisted successive attempts to relieve the sieges. On 18th

December 1899, Lord Roberts was appointed to go out to South Africa [2] to take over Redvers Buller's command, amid criticism of the handling of operations. He proceeded to Modder River, where he planned to march the troops towards Kimberley to relieve the siege.

Medical and surgical conditions were to play a central role not only in the conflict itself, but also in the highly publicised controversies over the authorities' handling of operations. Sir Winston Churchill made his name as a war correspondent during the conflict, and a number of medical men, including Sir Arthur Conan Doyle, later published their experiences.[3] The influence of the press made this a different kind of war, characterised in part by criticism of the British authorities. The War Office came under particularly heavy fire over outbreaks of enteric fever (typhoid), which caused more suffering and death than battle injuries in some parts of the campaign.

The press had almost wholeheartedly backed the government's intervention in South Africa, and contributed to anti-Boer feeling among the British public. *The Telegraph* had commented, "Kruger's asked for war and war he must have."[4] The directors of *The Times* were strong supporters of Cecil Rhodes, whose economic interests had precipitated the war, and the newspaper's correspondent Flora Shaw had already played an important role in drumming up support for his cause.[5] Notable exceptions to pro-conflict sentiments were issued in the *Manchester Guardian* and the *Daily Chronicle*[6], but they were in a minority, and the mood of the country was generally in favour of the conflict. Once operations were underway, telegrams meant that editors could receive reports from journalists on the front line for the following day's issue, and newspapers competed for sensational aspects of the action. On more than one occasion, the agenda was a medical one.

In the early part of the war, newspaper editors were frustrated by British military censors, particularly those based in Cape Town,[7] who were careful to ensure any of the more unpalatable details were omitted from telegrams and letters sent back to Britain. Lord Roberts, however, who arrived in the Cape in January 1900, was generally considered to be more media friendly,[8] even if his attempts to control the images were simply a little more subtle. He was happy to interact personally with the press, but often made sure his own carefully-worded reports reached Britain before those of the reporters on the ground.[9] Depending on their editorial politics, newspapers were in a position to deviate from the initial stereotypical images of an efficient military machine defending British economic interests, and take a more critical approach to the authorities' handling of the war. This was to be particularly relevant when it came to hospital conditions and the treatment of the wounded.

The British government had initially underestimated the scale of the medical requirements for the military campaign. The regular army medical service had been recognised as the Royal Army Medical Corps (RAMC) relatively recently in 1898, and was overwhelmed by the scale of the operations.[10] Low pay and poor conditions had made it difficult to attract the best medical teams in the first place, though any criticism of their ability to cope with the complex conditions in South Africa was sensitive. Late in 1899, the government recognised the need to supplement the ranks with civilian surgeons.[11] An announcement was made in October that the RAMC would be "supplemented by the employment of 56 civilian surgeons and 11 nurses of the Army Medical Nursing Reserve."[12] They would receive one pound a day, a horse and a Captain's allowance, as well as two months' pay on completion of service. The nurses were to receive £40 a year, with allowances, and £20 on completion of their tour of duty.[13] When the announcement was made in the House of Commons that the President of the Royal College of Surgeons, Sir William MacCormac, had volunteered to go out to South Africa, it was greeted with "loud cheers … and … caused the greatest satisfaction throughout the country."[14] He was no ordinary civil surgeon, and this prompted more grandiose plans. The Director-General of the Army Medical Department thought it "desirable to have at least one consulting surgeon with each force"[15].

In early November, it was announced that George Makins of St. Thomas' Hospital and Frederick Treves had been appointed to join Sir William MacCormac, but that each surgeon would act independently, as a consultant, and have the services of a military secretary "selected from the surgeons on the spot".[16] It was announced at the same time that these three surgeons would receive remuneration equivalent to £5,000 a year,[17] presumably to compensate for losses from their private practice while they were away.

Nearly two months later, when the army was experiencing difficulties, and Lord Roberts was sent out to relieve Sir Redvers Buller, it was announced that four additional surgeons had been appointed: Sir William Stokes and Mr Kendall Franks, both practising in Dublin, and the Senior and Assistant Surgeons at King's College Hospital in London, William Watson Cheyne and Lenthal Cheatle.[18] Cheyne was to receive the same remuneration as Treves, £5,000 a year, and apparently Cheatle too, despite the fact that he was a junior, as a government clerk had made an error.[19]

Just as he turned 47 years old, no longer a young man, Cheyne prepared for his first direct experience of war. The decision came relatively suddenly. He was forced to cancel an invitation he had accepted to the opening of a new sanatorium in Palermo which was to take place in February 1900 [20] and, considering the intensity of his engagements during this period, probably a host of other appointments. He clearly considered himself physically up to the task, perhaps thinking his ability to maintain a gruelling clinical workload at home equally equipped him for the rigours of war. Despite his ill-health as a young man, there seems to have been no indication to the contrary.

According to one newspaper,[21] Cheyne's appointment came about through one of his patients. Sir George White, who later became General White, had apparently suffered a leg injury when his horse fell on him in India, before he went out to the war in South Africa. Cheyne had operated on the leg on 27th May 1898,[22] and the procedure had been "so successful that the General and the surgeon became friends."[23] The writer suggested that this was partly why Cheyne was asked to go to South Africa. The appointment of Cheyne and Cheatle was considered an honour for King's, as well as for the men themselves, and the subsequent celebrations were testimony to the general popularity of the decision.

On January 4th 1900, a complimentary dinner was given for both men in the Whitehall Rooms of the Hôtel Métropole in London. It was attended by members of the council and hospital committee of King's College, and a troupe of students and alumni.[24] Cheyne was popular with his students, and for all his apparent shyness, always seems to have been prepared to rise to their fun. On this particular evening, he was clearly in fine fettle. In what the *British Medical Journal* referred to as "an admirable little speech"[25], he alluded to initial fears that the surgeons of the relatively new Royal Army Medical Corps would resent the presence of civilian colleagues, but that, on the contrary, "they had welcomed the consulting surgeons most cordially."[26] Cheyne was appropriately complimentary towards the Corps and its work, though he cannot yet have been fully aware of the extreme conditions they faced, and the details of their collaboration with their civilian colleagues were to prove more complex. He and Cheatle were then "chaired around the room by the students"[27], who called on Cheyne to sing. His choral abilities were well known by this time, and he responded by giving them a rendition of "Annie Laurie" [28], a Scottish ballad he had probably sung many times before.

They either could not envisage the conditions they would meet in the field, or chose not to, perhaps as a question of morale, and the gathering

Microbes and the Fetlar Man

of King's staff and students preferred light-hearted banter to talk of battle wounds. It was, after all, a farewell dinner, and the general mood of patriotic fervour in the press could hardly have encouraged an atmosphere of caution. Whatever his exalted position, Cheyne was to be working in a difficult environment. He nevertheless seems to have treated it with his customary taste for new experience, and a rare opportunity to test his methods under novel conditions, perhaps even slightly inquisitive about how others were using them.

The students focused instead on familiar ground, and delivered a poem they had written to commemorate the departure of the two "surgeons of the antiseptic kind."[29] In a light-hearted glorification of their impending service, reminiscent of the music hall patriotism the war had engendered [30], they sang that they would "ride on real live horses, as Consultants to the Forces."[31] They would be "taking gauze of cyanide of a bright cerulean hue" and "a vat of pure carbolic, perchloride a ton or two, to make the usual lotions …"[32]

In the same way as Cheyne portrayed Lister as a knight riding out to meet an army of bacteria,[33] the students, more than a little tongue-in-cheek, saw their professors in the luxury of civilian operating conditions, and they sang on:

> They are foes to microparasites of every form & shape
> No "Bacillus Pyogenicus" their notice will escape
> & they'll put the fear of death in all the microbes at the Cape
> Well, you know the drastic treatment that I mean.
>
> They're so careful don't you know
> With their antiseptic towels and their gauze.
> Their incisions I might mention
> Will all heal by first intention
> In all cases that they'll deal with at the Wars.[34]

Aseptic operating conditions could hardly have been further from the situation they were to encounter in the field, even if the hospitals were better equipped than they had been in most previous conflicts. The students' implication that Cheyne and Cheatle were taking large quantities of antiseptics out to the war with them to fight the great bacterial foe was also a far cry from the rather more complex arrangements they would be faced with in the field hospitals. The majority of British newspapers joined in the lighthearted camaraderie of this unprecedented move to send the country's top surgeons off into battle, with only the slightest suggestion that they would be leaving civilian hospitals with reduced expert capacity. One paper, showing a bearded Cheyne, went so far as to say that he would leave "a gap in the medical life of London that nobody else" could "hope to fill"[35], rhetoric that largely missed the point.

Two days after the dinner at the Hotel Métropole, Cheyne and Cheatle set off for Waterloo station in two carriages to catch the 11.40 Castle Line Express to Southampton [36]. As they neared the station from York Street, they were met by a large contingent of King's College medical students who had come down to the station "in char-a-bancs, armed with horns, rattles, whistles &c., which were freely used on the way"[37]. The students removed the horses from the carriages and proceeded to pull Cheyne and Cheatle into the station themselves. They then hoisted the two surgeons onto their shoulders and carried them as far as the train, to the accompaniment of "toy trumpets and rattles"[38] and waving of flags. While they waited the half hour for the train to arrive, the students held forth once again with "God Save the Queen", "Rule Britannia", "For He's a Jolly Good Fellow" and "Auld Lang Syne"[39]. A number of Cheyne's colleagues from King's were also assembled on the platform to wish him "God speed"[40]. He was loaded onto the train with his luggage, and delivered a note of thanks to the assembled crowd, leaving him only a few moments to say some last words to his family.[41] The reports fail to indicate whether Margaret, and possibly the children, had been alarmed by the antics, though perhaps they were not bounced along to the station with him in the hijacked carriage.

Gratified as Cheyne was by this show of camaraderie and support, it is hard to know what he expected to find in South Africa. He cannot have been wholly unconscious of the fact that he was going to war. His experience until now had been entirely in civilian hospitals and laboratories, and his chief preoccupation whether or not chemical antiseptics were essential for preserving an aseptic surgical environment. Here he was, embarking on the steamer *Carisbrooke Castle* [42] for a major conflict. Whatever he was thinking as he left, he cannot have had a particularly accurate picture of the conditions he would meet, though he was to do his best to treat them with his customary pragmatism. He arrived in South Africa only a few weeks after Lord Roberts, whose contingent he was to join, and the first military objectives were to relieve the sieges.

Lord Roberts landed at Cape Town on January 10th, where he remained until early February before proceeding to Modder River.[43] On Friday,

February 9th, Colonel Stevenson of the Royal Army Medical Corps, stationed with Lord Roberts, sent a telegram to Cheyne, who was at Orange River, asking him to join the headquarters' staff. Cheyne set off immediately, and reached Modder River on Sunday, February 11th.[44] At Modder River, Cheyne noted reasonable conditions, close to a railway line with "admirably-fitted hospital trains … to carry off the wounded with comparatively little disturbance, and lodge them in well-appointed stationary hospitals within a short time".[45] He found the field hospitals in the area "more resembled stationary hospitals",[46] and "were able to accommodate the more seriously wounded till such time as it was safe to move them."[47] Significantly, he also commented favourably on transport conditions at this stage, noting that "there was ample transport from the field of battle, and only a short distance for the wounded to travel; hence, they were rapidly collected and quickly lodged in tents, where they could remain, till it was safe to send them to the base."[48]

Other British surgeons stationed elsewhere commented on the generally well-equipped and efficient hospitals. George Makins, one of the civilian consultant surgeons sent out to South Africa before Cheyne, considered the field hospitals to be marked by "a very high degree of excellence"[49]. The medical teams were reasonably well equipped to deal with a normal rush of wounded, who were taken from the field by stretcher to a battalion aid post, where they received attention from a medical officer and a medical NCO. They were then taken, again by stretcher, to a collecting station, and generally by ambulance to a field hospital. The field hospitals provided space on ground sheets on the floor for around 100 wounded. From the field hospital, they were sent to a stationary hospital somewhere along the main lines of communication, which could accommodate around 100 men on stretchers before they were moved on to a general hospital. These latter establishments were reasonably well provided for and could take 250-500 patients.[50] Makins noted that "the equipment, with small exceptions, proved equal to the demands made upon it"[51] and "the mobility of the camps was proved again and again."[52] However, they were occasionally overcrowded during an encounter, with "beyond more than three times the regulation limit."[53]. Under these circumstances, "the additional patients were … accommodated in marquees and bell tents, according to the nature of their diseases."[54]

Makins also noted that food supplies were generally good. Vegetables were scarce, but fruit, jam and lime juice helped prevent scurvy. Makins said he "never saw a case"[55] of the condition at all during his entire time in South Africa. In the early part of the campaign, meat had been brought in from Australia and New Zealand, and was considered good quality, and plentiful supplies of ice helped preserve it in the camps. Since it arrived ready to use, there was no "butcher's offal"[56] to provide a threat of contamination. Makins added that the bread and biscuit were also good quality.[57]

Probably on the march to Bloemfontein.

Cheyne was to see quite different conditions as soon as the convoy left Modder River. He set off the following day, Monday 12th February, first by train to Enslin, and then about 12 miles on horseback to Ramdam.[58] Lieutenant-General John French's division proceeded northwards to relieve Kimberley, in the wake of pressure from Cecil Rhodes. Kimberley was the centre of the de Beers diamond mining operations, and when it was laid under siege, Rhodes, who had a major stake in the operations, had barricaded himself in the town and then bombarded the government and British press with pleas for its relief. He sent "characteristically reckless"[59] word to Lord Roberts to "make the relief of this town your first priority or I shall surrender it to the Boers."[60] He had clashed seriously with the military command under Kekewich, who felt he was placing operations in jeopardy, and who was at one point given permission by Lord Roberts to "take Rhodes and clap him in irons if he defied him".[61] The press, however, took Rhodes' side, especially *The Times*, whose reporter Flora Shaw had been implicated in the Jameson Raid.[62] *The Times* and other papers continued to exert public pressure on the War Office, and had been instrumental in ensuring the force diverted to Kimberley, which "greatly complicated"[63] the invasion of the Free State.

French was to take Kimberley on February 15th. Lord Roberts, on the other hand, intending to fool Boer General Cronje, who expected him to follow the railway line, proceeded instead across the veld. Conditions on the march were to be nothing like the comparatively ordered situation Cheyne had described at Modder River. Ramdam, their starting point, consisted of "one square, whitewashed, tin-roofed farmhouse astride a low ridge in the almost featureless veld."[64] It was the height of summer, and on the first and second days, considerable numbers of men collapsed from heat stroke so that "some 60 had to be sent back."[65] Cheyne noted that "many who had campaigned in India said that the thirst here was much worse."[66] The vegetation was sparse, and the passing troops had "stirred up the dust, so that dust storms were frequent."[67] They arrived at De Kiel's Drift late on Tuesday, February 13th, and many of the men, "unaccustomed to this sort of campaigning,"[68] had failed to take provisions with them in their saddlebag, and had to go without supper. Because of the lack of tents, they then had to sleep in the open, "without covering of any kind"[69] if the baggage wagons did not arrive, and temperatures at night could drop very low. There was little to do in a surgical capacity, though they encountered four wounded left by French following an engagement the previous evening.[70] Most of the cases were not serious, except for "a little bugler boy who was wounded in the foot - a perforating wound through the front part of the foot - while running up the river bank, as he said, to get a shot at the Boers."[71]

The following day, in equally hot temperatures, they arrived at Waterval Drift. They left there around 5am the following morning, Thursday 15th, but had difficulty crossing the Drift under appalling conditions. The "steep, broken river banks and the soft, sticky mud"[72] was so gruelling that 200 ox wagons carrying the convoy's supplies were left on the north bank of the River Riet for the oxen to graze and gather their strength.[73] Cheyne later noted that "after the army had gone on, a considerable force of Boers attacked the rear part of the convoy, which had not yet moved on, and captured a considerable number of wagons."[74] This was an understatement. Meanwhile, Cheyne reached Wegdraai, and the medical teams had sixteen wounded to treat from the Boer attack early on Friday morning, along with some others brought in from Waterval Drift. As the wounds were slight, they proceeded to Jacobsdal on Friday afternoon, taking the wounded with them, and from there, sent them on to Modder River.[75]

The ambush at Waterval Drift was hardly as passing an incident as Cheyne's account suggests. The Boer raiding party was led by no less a figure than De Wet, who was to prove himself "the outstanding guerrilla leader of the war".[76] He had stampeded most of the contingent of 3,000 oxen, leaving the 200 wagons behind. Nearly a third of the transport available to the convoy was stranded with its African drivers beside the Riet River. They were carrying food supplies, medicines and bandages, which were now "at the mercy of De Wet"[77], and the episode threatened the entire progress of the march on Bloemfontein. Lord Roberts almost turned back towards the railway, which would have meant giving up his opportunity to capture Cronje. Had he not heard that French had sped on towards Kimberley, and had they not been able to take charge of some of French's mule carts, the whole course of the war may have changed.[78]

Lord Roberts' transport problems were generally considered to have been the most debilitating issue of the march. Pakenham notes that "it was not Cronje that proved the main obstacle to rapid progress, nor the Riet, nor the Modder. The obstacle was the transport arrangements for the infantry - or their absence."[79] The lack of suitable transport can equally be said to have been one of the greatest obstacles for the medical teams, and Cheyne was later to be one of a number of prominent medical men who publicly criticised the lack of dedicated transport. They were dependent on the army for ambulance wagons, and the army was in a position to requisition them for military purposes as it saw fit. "As a matter of military necessity," Cheyne commented in the *British Medical Journal*, "it was considered essential to cut down the number of ambulances with each division from the normal number of ten to two. The possibility that very serious difficulty and delay in collecting the wounded from the battlefields might result from this decision was repeatedly and emphatically pointed out to the military authorities by the medical department, but without avail."[80]

Cheyne indicated, by way of an example of good practice, that the New South Wales Ambulance Corps [81] had their own dedicated transport and experienced none of the shortages that plagued the RAMC. For the British convoy, medical tents and equipment were reduced, as they could not be carried, and this was to mean severely restricted operating areas. The transport difficulties had implications for health far beyond the transport of wounded, and the consequences were to make themselves apparent when the medical teams on the march were faced with more serious battle casualties and disease.

In the meantime, at Jacobsdal, Cheyne and his colleagues took a few moments out to pay a visit to the German Ambulance, which was engaged on the Boer side. They spoke with Drs. Kettner and Hildebrand, who, at

Microbes and the Fetlar Man

home, were assistants to Dr Bruns and Dr Esmarch.[82] Cheyne was very familiar with the work of Esmarch, a German expert on war wounds, and he had discussed his work in *Antiseptic Surgery* and other publications.[83] The German doctors showed the British contingent around their premises, explaining their cases to them. It was clearly one of those moments that highlighted the contradictions inherent in war, as the medical teams on opposing sides exchanged information on the frequency of the different types of wounds they were encountering. It seemed the German doctors were dealing largely with shrapnel wounds, in contrast to the British, who were experiencing mainly bullet wounds. The shrapnel wounds were generally larger skin wounds, and often arrived in a state of suppuration. The German doctors were not having a particularly successful time with abdominal wounds in particular, and were tending to leave them alone, to avoid infection.[84]

Cheyne left Jacobsdal on Monday morning, 19th February, at 4 am, and rode with the contingent the 27 miles or so to Paardeberg. They managed to have a break of two hours at Klip Kraal Drift, and arrived in Paardeberg about 1 o'clock in the afternoon to hear artillery fire. They discovered that over 800 had been wounded in the previous day's action.[85] Cheyne established his quarters close to the field hospitals on the banks of the Modder River, about three miles below Cronje's laager, where the Boers had retired after the fighting of the previous day. The field hospitals of the Sixth Division, close to where Cheyne camped, were run by Majors Pike and Ford, and close by was the Australian field hospital under Major Fiaschi. The Ninth Division field hospitals had been split up in the wake of the previous day's fighting. One, under Major Murray, remained close to where Cheyne camped, and the other, under Major Sawyer, had been forced to move to a drift about a mile downriver when it was shelled by Boer troops, though thankfully no one had been hurt.[86] The hospital tents were pitched above the high slopes of the river bank, where trees provided shade for the patients who could not be accommodated in tents. Cheyne found the location a striking contradiction to the fighting in the distance, commenting that "[w]e bivouacked halfway down the bank, and might have been picnicking on the banks of the Thames but for the occasional booming of big guns and the rattle of rifles."[87]

This was, nevertheless, where the full consequences of the lack of transport began to unfold. There were insufficient tents to accommodate the wounded from the fighting at Paardeberg, and they had to lie under trees for shade where they were available, but as Cheyne pointed out, there were very few trees on the veld except directly along the banks of the river or around farmhouses. Patients who had no tent and no tree had to lie "under temporary shelters made with blankets."[88] They were to remain in this camp for around ten days, until March 1st.[89]

What appears to be a destroyed railway bridge.

The consequences of the lack of transport did not end there. Because they had not been able to carry enough vessels for boiling water, the majority were used for cooking, leaving insufficient receptacles for boiling water either for drinking or for surgery. The water from the Modder River was muddy and unusable for any purpose without filtering and boiling, but the filters they had brought with them were not up to the job. They "very soon became choked up"[90] and the water ran through them too slowly. Even if the water could be filtered, however, there were rarely vessels for boiling it, and besides, there was next to no wood on the veld to use as fuel. Cheyne was later to comment that the predominant source of fuel was cow dung, which was not always available in sufficient supply, even if it were deemed suitable. He later said:

> The absence of trees renders one dependent on stray pieces of wood that one can pick up, such as the posts of wire railings, enclosures round the houses, and so forth, and the staple fuel used in that country is cow dung, and was not always obtainable.

It is thus evident that during the progress of an engagement, if lotions and sterilised water had not been made ready beforehand, it was really quite impossible to obtain proper lotions or even for the surgeon or the attendants to keep their hands decently clean. In many cases no lotions were available at all, or if they were they consisted of muddy water into which a little carbolic acid or sublimate was introduced - lotions which, I believe, were quite incapable of destroying even the most delicately constituted bacterium.[91]

Attending to bodies during the Boer War

Drinking water had to be brought in water carts from a farm about five miles away, as anyone foolish enough to attempt to fill their water bottle from the river found that "it went off with a pop"[92] when the bottle was opened, and let out gas with a "distinctly disagreeable odour".[93] This was due to more than the general muddiness of the water. Dead animals and effluent from Cronje's laager were floating downriver and contaminating it. All the water brought in on the carts was required for drinking, so surgeons had to make do with water from the Modder, and cope the best they could.

If the lack of clean water did not pose enough of a challenge for surgeons, antiseptics ran out in one of the field hospitals at Paardeberg, and "owing to the generosity of Lord Lister",[94] Cheyne was able to help out with a temporary supply of double cyanide powder"[95] with which they "were able to impregnate sufficient material."[96] For some days, there were so many wounded that the operating tent in one of the field hospitals had to be used to house them, and operations had to be conducted in the open.[97]

British surgeons serving in other parts of the war noted that surgical conditions were bad enough when tents *were* available. George Makins noted that

> ... in the tents the draught carrying the dust from the camp was one of the commonest troubles. The exclusion of dust was impossible, and it not only found its way into open wounds, but permeated bandages with ease. Often when a bandage was removed, an even layer of dust moistened by perspiration covered the whole area ... with a coating of mud. Again, in dust storms a similar layer of mud sometimes covered the whole of the exposed parts of the bodies of patients lying on the ground in the tents.[98]

He also noted the destructive power of the wind, and "once saw a whole hospital, fortunately unoccupied, levelled to the ground in the course of some twenty minutes."[99] If the wind was accompanied by rain, the damage was greater still. On one occasion, he witnessed flooding in a field hospital where "the patients were practically washed out of the tents."[100] He generally applauded the use of iron huts to counteract some of the less predictable environmental factors, but these were clearly not an option in Cheyne's case, on the march to Bloemfontein.

Microbes and the Fetlar Man

Erecting a temporary bridge.

Microbes at the Cape

The men in the lower part of the photograph are clearly in some distress, and may be suffering from fever.

Microbes and the Fetlar Man

And then there were the flies, which were a constant nuisance both inside and outside the operating areas:

> In a fresh camp they were sometimes not abundant, but after two or three days they multiplied enormously. Not only hospital tents, but living and mess tents, swarmed with them, the canvas appearing positively black at night. Even when dressing a wound, without unceasing passage of the hand across the part, it was impossible to keep them from settling, and during operations the nuisance was much greater.[101]

The extreme temperatures themselves made matters worse. Makins said that

> ... at times the temperature was sufficiently high to make either dressing or operating a most exhausting process to the surgeon. The heat of the day was not on the whole so disadvantageous from the point of view of the operator, as the cold of the nights during the winter in Orange River Colony. On one or two occasions serious operations had to be left undone, as it was only possible to consider them in camp, where, as we arrived at night only, the temperature was too low to justify the necessary exposure.[102]

Building probably used as a temporary hospital, with ambulance alongside

Cheyne also mentioned the problems of night surgery. By the time the wounded were brought in to the field hospitals, it was often late and most of the operating and dressing had to be done after dark. Arrangements for lighting were inadequate, and consisted mostly of candles

> ... which either had to be held by an orderly, or, owing to the small staff, had to be stuck in a dirty lantern or on a bit of wood. The only two abdominal sections which I did were both done late in the evening with the aid of a candle stuck in a piece of wood, and placed as near the wound as possible. In several of the foreign ambulances the acetylene light was provided, and in the cart which I organised at Bloemfontein I was fortunately able to procure and carry with me a small acetylene generator, which gave the most brilliant illumination. I would call attention to this point because I do not see that there can be any great difficulty in providing illumination of this kind to the various field hospitals, and it is a matter of the very greatest importance.[103]

Under these conditions of stress, it is therefore perhaps not surprising that things could, on occasion, go spectacularly wrong. Even in the more tranquil operating conditions of the hospitals in Bloemfontein, there were incidents. In his autobiography, Sir Arthur Conan Doyle, the creator of Sherlock Holmes, who had graduated from Edinburgh Medical School just after Cheyne and was working in a private hospital during the Boer War, mentions a rather unusual operation he witnessed by Cheyne on the wounded Dutch military attaché to the Boers, who had been paralysed "after some engagement"[104]

> A shrapnel bullet had broken one of his cervical vertebrae, the bone pressed on the nerves, and they had ceased to function. Watson Cheyne of London was the operator. He had cut down on the bone with a free incision and was endeavouring with a strong forceps to raise the broken arch of bone, when an amazing thing happened. Out of the great crimson cleft there rose a column of clear water 2 feet high, feathering at the top like a little palm tree, which gradually dwindled until it was only a few inches long, and finally disappeared. I had, I confess, no idea what it was, and I think many of the assembled surgeons were as taken aback as I was. The mystery was explained by Charles Gibbs, my mentor in such

Microbes and the Fetlar Man

Difficulties medical teams faced with transport.

matters, who said that the cerebro-spinal fluid, which is usually a mere moistening around the cord, had been greatly stimulated and increased by the pressure of the broken bone. It had finally distended the whole sheath. The forceps had punctured a small hole in the sheath and then the fluid had been pressed through and shot into the air as I had seen it. Perhaps the release was too sudden, for the patient died shortly after he was removed from the table.[105]

Even earlier in the war, surgeons had struggled on occasion to keep pace with the pressure of wounded. Frederick Treves had arrived in South Africa just before Cheyne. He followed the Ladysmith Relief Column, and operated during the battles of Colenso and Spion Kop, one of the worst encounters of the war, which brought around 800 wounded men into the surgical tents.[106] He was one of eight civilian surgeons supporting the Royal Army Medical Corps in the area. Two of these surgeons died, and a number of the others were injured during the three months Treves was there.[107] His graphic illustration of an operation in a field tent at Colenso illustrates some of the difficulties:

> It is needless to say that the operation tent is very unlike an operating-theatre in a London hospital, but then the open veldt is very unlike the Metropolis. The floor of the tent is much trodden grass, and, indeed, much-stained grass, for what drips upon it cannot be wiped up. There are not bright brass water-taps, but there is a brave display of buckets and tin basins … There is little room in the tent for others than the surgeon, his assistant, the anaesthetist, and a couple of orderlies. The surgeon is in his shirt-sleeves, and his dress is probably completed by riding breeches and a helmet."[108] Instead of nurses, the surgeon was assisted by male orderlies, with a variety of knives and tin openers hanging from their belts. Treves considered they were at least "certainly not hampered by the lack of a precise professional garb.[109]

Patients were brought through the tent one by one, and examinations were carried out under anaesthetic. Those awaiting their turn outside often had to witness it in graphic detail, as the tent flaps remained open in the heat. There was "limited capacity"[110] in each tent, and on one occasion, a man with a head wound lay underneath the operating table, mistaken for dead, while the surgeon amputated the limb of a soldier on the table. The leg fell into the face of the man on the floor, who promptly revived, rather shaken, though he survived.[111]

Makins noted that, at times, "the number of patients was extended beyond more than three times the regulation limit."[112]. He also hinted that, where the RAMC surgeons were trained precisely for these conditions, "the civilian surgeons were placed at a great disadvantage"[113], a subject Cheyne was to discuss later. Cheyne was, on the whole, careful to commend the skills of the army surgeons where he felt praise was due, but Treves noted that, in general, those in charge of RAMC field units were the scapegoats for any failure, and were rarely honoured when things went well.[114]

If transport had created problems for supplies, it was an even more serious issue when it came to transporting the wounded. By the time Lord Roberts' convoy had reached Paardeberg, they were a considerable distance from the railway line, and were moving further away from it the closer they moved to Bloemfontein. Yet, as Cheyne pointed out, when the convoy was on the move, the wounded could not be left on the veld, and somehow had to be sent back to a stationary hospital. He later wrote:

> It is a very different thing to transport wounded a short distance to a comfortable ambulance train, well supplied with every convenience, from putting them into buck or ox waggons, and sending them for two, three, or (in the case of Poplar Grove) four days' journey over the veld, with no roads, in springless carts, without suitable protection from the sun or rain, and without the comfort which a recently-wounded man requires. The agony which must be endured under such circumstances by many cases such as compound fractures, and the harm which may result to the patient, must be very great. In more than one case the result of the movement was to precipitate a fatal issue. Under the existing circumstances, however, no other course seemed possible. Military considerations prevented our carrying the wounded on with us (which, indeed, would have been almost as bad) and were equally against leaving them behind when we moved on. The ideal plan would have been to have sent back all those whom it was safe to move, and leave the serious cases behind at a field hospital at each stopping place with the necessary medical and other attendants. It was not, however, considered safe to do so at Paardeberg; there would have been difficulty in providing food and attendance, there was no house, and tents could not be spared.[115]

Despite the general difficulties, Cheyne was largely complimentary about the RAMC organisation and operations. On February 23rd, he and Major Bond of the RAMC rode up from the main camp at Paardeberg to see the field hospital which had been forced to move upriver from Cronje's laager, and discovered that the river water there was not in such poor condition. Cheyne commended the operation of the field hospital under Major Franklin, saying the wounded were "excellently looked after ... as was the case in all the field hospitals."[116]

Lord Roberts and Boer General Cronje, at the moment of Cronje's surrender.

While the medical teams coped as best they could during the wait at Paardeberg, military decisions had to be taken. Lord Roberts had arrived at Paardeberg only hours before Cheyne, and had seen the extent of the British wounded. He was prepared for an attack on Cronje's laager, but Cronje requested an armistice. Roberts refused, as the Boer general had also asked for British doctors to remain with him to attend his wounded since he had no medical provision of his own.[117] Only Kitchener was in favour of a renewed attack, and Lord Roberts was considering retreating to Klip Kraal, which would have allowed Cronje to escape. Pakenham speculates that "perhaps he was shaken by the sight of the British wounded. It was they who had had to pay the price of Roberts's and Kitchener's new transport system."[118] Once again, Lord Roberts was saved by a twist of fate: De Wet himself beat a retreat from his ridge, and the following

Microbes and the Fetlar Man

Tuesday, 27th February, General Cronje surrendered, and came into Lord Roberts' camp with 4,069 Transvaalers and Free Staters (including 150 wounded and 50 women).[119] Pakenham has noted that "Cronje's blunders had outmatched Kitchener's and Roberts's after all. It was the first great British victory of the war."[120]

In talks after General Cronje's surrender. Seated, left to right: Cronje's secretary, General Cronje, Lord Roberts.

Many years later, when Cheyne had retired, and for as long as anyone can remember after that, seven photographs inset into two long frames hung on the wall of the main stairway of his house in Fetlar. Guests climbing the stairs stopped to look at them. One of the photographs

Microbes at the Cape

Lord Roberts (very faint, to the right of the photograph) leaving Cronje and his secretary (seated).

shows Cronje shaking hands with Lord Roberts and three of them show him sitting in conversation with British officers. The remainder show the devastation he left behind in the laager.[121] After Cronje had surrendered, Lord Roberts sent Cheyne to the abandoned Boer laager with Majors Stevenson and Sylvester of the RAMC. They were to assess the condition of the wounded, and plan what was to be done to help them. They encountered "an appalling picture of devastation, and the stench was frightful"[122].

Microbes and the Fetlar Man

Cronje's secretary (far left) speaking to Lord Roberts.

Dead animals in all stages of decomposition were lying about everywhere; a house close to the drift, in which a large number of horses had been stabled, had had its gable blown in, and was full of dead horses. The greater number of the waggons were in ruins, some overturned, some completely burnt, while others were represented by a mass of broken wood and iron. The ground around was pitted and yellow from the lyddite shells, and strewn with fragments of shell, shrapnel, broken carts, and all sorts of *débris*.[123]

Microbes at the Cape

Scenes of devastation in the Boer laager when Cheyne and the medical staff went in to investigate after Cronje's surrender.

They discovered the wounded holed up in trenches, more like burrows, with various advanced conditions of wound sepsis. They discovered, moreover, the full story of why Cronje had asked for British medical assistance. His doctors and ambulances had left the camp and gone to Jacobsdal following the retreat after the battle of Magersfontein

Microbes and the Fetlar Man

on 11th December.[124] By this reckoning, Cronje's wounded had been without medical support for two and a half months. Cheyne noted that

> the wounded had been lying in the foul laager without medical attendance or dressings of any kind, and as a consequence the wounds were very foul and many of the patients were in a deplorable condition from sepsis.[125]

According to Cheyne, Lord Roberts had offered to take over the Boer wounded several days before Cronje's surrender, but the Boer general had refused both this and a subsequent offer to send in medical teams to attend to the wounded *in situ*.[126] Cronje was apparently only prepared to accept this on condition the British medical men remained with him, which Roberts would not allow. Cheyne unequivocally blamed Cronje, and applauded what he considered Lord Roberts' humanity,[127] clearly unaware of the near military disaster when he had contemplated turning back,[128] and focusing instead on Roberts' minimal offensives during the wait at Paardeberg.

Cheyne's party retrieved 159 wounded, and discovered a few more later.[129] A number of RAMC medical men who had accompanied them into the laager applied dressings, and the wounded "were carried out across the drift on stretchers, for the water was too deep for ambulance waggons to cross …"[130] They were left "on the riverbank, in a nice shady spot about a quarter of a mile higher up the river."[131] At this point, a number of the Boer doctors who had gone to Jacobsdal arrived to take charge of the wounded, and "they were provided with dressings, food and such covering as could be spared."[132] When Cheyne and the RAMC men handed the wounded over to the Boer doctors, he feared "there was not very much done for them in the surgical way, for several cases - such as gangrene - requiring immediate operation were left alone."[133] The patients were transported a few days later to the German hospital Cheyne had visited at Jacobsdal, though the majority were sent on to Kimberley, even further away.

Amongst the photographs hanging in Cheyne's house are three showing the burnt out wagons and devastation in Cronje's laager. In one, the débris is so tangled it can hardly be made out.[134]

Devastation, probably in Cronje's laager after his surrender.

The British convoy left Paardeberg on March 1st, and travelled about five miles to a farmhouse at Ossfontein, where they stayed a few days to build up supplies. Farmhouses tended to be sources of fresh water. Transport had gradually became available and the sick and wounded had been sent back to Modder River and Kimberley in detachments, so they had not needed to take any with them to Ossfontein.[135] They set off again a week later, on March 8th, and resumed the march to Bloemfontein, having missed, a day earlier, a major opportunity to capture the Boer leaders De Wet and Kruger at Poplar Grove.[136] Supplies once again had important implications. There are conflicting views on how the army's rations were holding out at this stage. Makins commented that there was only a few days' food shortage during the march from Modder River to Bloemfontein.[137] Pakenham, however, notes complaints that the 6th Division had been "starving"[138] ever since De Wet's raid on Lord Roberts' food convoy at Waterval Drift, and Cheyne indicated that, "the fresh meat was often so tough as to be quite uneatable".[139]

The lack of food for the horses of the cavalry, however, had also had consequences at Poplar Grove, where a planned charge failed to take place. By this time, not only the medical teams, but also Roberts' and Kitchener's own commanders were in conflict with them over transport arrangements, which they considered had hampered their ability to consolidate the action at Poplar Grove. Roberts' Chief-of Staff, Douglas Haig, said he had "never seen horses so beat as ours ... They have been having only 8 lbs of oats a day and practically starving since February 11th."[140] This was compounded by Lord Roberts' miscalculations about the condition and morale of the Boer force,[141] which helped prolong the war for another year. Cheyne noted that there were few wounded to deal with at Poplar Grove, as much of the intended action had not taken place, but that the cases they did have involved twenty cavalry men,[142] who had particular reason to be critical of the transport arrangements, as the situation had deprived their horses of sufficient food. There were six serious cases: a bad head injury, two compound fractures and three abdominal cases. The latter cases remained with the convoy for a day's observation, though they were on the move by then, and the remainder were sent back to Kimberley on March 9th.[143]

On March 10th, they moved on to Driefontein, where they inflicted a major defeat on the Boers, leaving the way clear to Bloemfontein. For the medical teams, however, the problems only began with the aftermath of the battle. With about 400 wounded to collect from the battlefield, and only four ambulances to carry them to the field hospital, the challenges were compounded by the fact that the fighting had not ended until nightfall.[144] Each ambulance could only take "4 lying-down cases and 2 sitting up,"[145] so it took a good part of the night before they could get the bulk of the wounded to shelter. It was also impossible to see all the cases in the darkness, so by the morning, some were still left to be picked up.

The situation had an unexpected consequence, which may have helped ease the situation. When Lord Roberts set off to his next stop the following day, he had to ride out across the battlefield, and came across a group of wounded who had not yet been collected. He "ordered up"[146] the rest of the ambulances, but still insisted his transport arrangements during the march had been necessary. However, he was effectively absolving the RAMC from blame, and ordered that if similar incidents were to occur in future, the RAMC should be given help in removing the wounded from the field of battle.[147]

There was no transport to take the patients on with them to Bloemfontein, and the transport lines to Kimberley were now closed, so the patients had to be left behind at the field hospital, which was moved to a farmhouse and left with supplies and medical staff.[148] Cheyne insisted that, given the condition of the transport in the wagons, this "was a very fortunate thing for the patients"[149] who were spared an uncomfortable, three-day journey to Bloemfontein until several days after the injury, allowing them time to recover as much as possible. He believed that "[w]herever possible, the severely wounded should be left in the field hospitals for some days and not hurried back to the base."[150] When wagons were finally sent to fetch them from Bloemfontein on 16th, Colonel Stevenson sent word to Major Pike at the field hospital, asking him to use his discretion about who was sent on, and he advised him to keep the serious cases back, even then. Pike agreed, and the serious cases were not sent on for another week.[151] The bulk of the army finally entered Bloemfontein on March 13th, over a month after they had left Modder River.

Exhausted as the men and medical teams were, the worst was yet to come. The full consequences of the inadequate transport arrangements set up by Kitchener and Roberts were about to unfold. The lack of suitable water, and utensils for boiling it, had led to an outbreak of fever, and most cases were, at the time, considered to be typhoid. Cheyne noted that it was endemic in South African towns,[152] particularly at certain times of year, but he [153] - and others, including Conan Doyle,[154] traced the epidemic to the conditions at Paardeberg, though Cheyne was later to question how many of the cases were actually typhoid.[155]

Throughout the war, not just on the march, one of the greatest problems faced by surgeons and military personnel alike was clean water. Makins noted that their supplies of drinking water were the only potable fluid available, though in the early days, beer had been available in the canteens.[156] Treves commented that at Colenso, water in general had to be shipped in quantity from Frere, over 20 km away by road.[157] Makins noted that ground water was used for surgery at Orange River, and that it was muddy but "wholesome"[158] and had to go "through the complicated processes of precipitation by alum, boiling, and filtration."[159] A stock of prepared water was kept for emergencies in a small room in the house of one of the railway servants "and fitted as a rough operating room by the Royal Engineers. The necessary utensils were provided by the Red Cross Society."[160] These conditions were obviously not available as soon as the troops left the proximity of the railway line, as in Cheyne's case on the march to Bloemfontein. As more and more troops were sent out, drinking supplies became an increasing problem, especially in the wake of the growing transport difficulties.

Makins, however, chose to see it as a question of discipline in the ranks and that, if the men had rationed their supplies more carefully, large quantities of water were not really necessary.¹⁶¹ When he was first at Orange River, he drank water with lime juice all day long, but broke himself of the habit when water was scarce and "by a very slight amount of determination",¹⁶² drank only at meal-times. He did not believe the men in the ranks tried hard enough to exercise this level of self control. Especially on marches, they "emptied their water-bottles during the first hour of the march, and the rest of the day endured agony, seizing the first opportunity of drinking any filthy water they met with"¹⁶³. He considered the answer to be constant supervision, but although officers took care to ensure horses and pack animals drank only in certain parts of clean watering holes, the men were in such desperation for water that they could not wait to walk the few extra yards to the clean water, and filled their water bottles "from amongst the animals' feet."¹⁶⁴

Showing the difficulties medical teams faced with transport, the image also shows the extent of the contaminated water.

Cheyne recalled that, even in the first few days of the march, the lack of drinking water and the tremendous heat meant that "all good resolutions about boiling or filtering water were thrown to the winds; water, no matter how muddy, had to be got as quickly as possible."¹⁶⁵ He believed this had led to problems such as diarrhoea, the particular variety of which had been renamed "Modders" or "Riets"¹⁶⁶ after the river water. "Modders" was considered worse than "Riets". Cheyne considered most of these cases to be manageable, but a few were so bad they required hospitalisation. He was careful to distinguish this complaint from the outbreak of enteric fever, which he blamed on the period entrenched at Paardeberg, adding that the dust and flies had helped it to spread.[167] In general, it was to be the source of greater losses among the troops than the fighting, and Cheyne commented that, given the state of the Modder at Paardeberg, it was "surprising that the result was not much more serious."[168]

By way of illustration, Cheyne recalled two nights when it had rained at Paardeberg:

> On Saturday, February 24th, and during the night, it rained heavily, and the river rose considerably, with the result that during the whole of Sunday there was a constant procession down the stream of dead and putrid horses and other animals from the Boer camp. Several hundreds were counted, and as a rule as many as fifteen or twenty were to be seen at one time. The stench from these animals was very bad, especially if they happened to be temporarily arrested by branches of trees, and so on. Hence the greatest care had to be exercised in the sanitation of the camp, and it says a great deal for the arrangements adopted that no illness, except the ever-present 'Modders', developed. It was, however, impossible to prevent the men drinking the river water: and that is, I believe, the source of the considerable number of cases of enteric fever which have developed since our arrival at Bloemfontein.[169]

By the time the army reached Bloemfontein, medical staff and soldiers alike were down with fever. The field hospitals moving with the army were supplied with groundsheets and blankets, and stationary hospitals along the route had stretchers, but the medical corps, already struggling with reduced resources on the march, was unable to cope with the unusually large numbers of sick. This onslaught had been unforeseen. All available public buildings in the town were requisitioned as typhoid hospitals, including churches, schools and even the Raadzaal, or parliament building.[170] One source notes that, throughout the war, of 556,653 British soldiers, 57,684 contracted typhoid and 8,225 of them died. In comparison, 7,582 were killed in action.[171] *The Daily Telegraph*, quoting published figures, claimed at the time that, between March 13th and June 21st alone, the period following the arrival in Bloemfontein after the march, 6,369 cases were classed as typhoid and 1,370 of them died.[172] The figures were nevertheless confusing. The *BMJ* noted that "the exact figures are not available",[173] and that, in the government statistics issued, the "number of deaths attributed to enteric fever [was] not distinguished."[174] Nevertheless, if Lord Roberts had needed confirmation of the implications of inadequate medical and food transport, this was testimony enough.

Bloemfontein was an educational centre for the state and had a number of robust school and college buildings at its disposal. Colonel Stevenson of the RAMC immediately requisitioned them for hospitals. Cheyne commented, with a note of admiration, how "[i]n one week Colonel Stevenson and his Secretary (Major Sylvester) took over, fitted up, and fully equipped with medical and nursing staffs, buildings holding, in all, 510 beds."[175] They included 100 beds in a large girls' boarding school, which had been run by Anglican sisters, and the latter stayed on to help nurse the sick, along with the RAMC orderlies. Beds were also fitted in a number of other schools, the state college, the old state barracks, a convent and the Parliament building.[176] It was impressive organisation, given that people were dropping down like flies with fever and the battle wounded continued to flow in from the encounters on the march.

There were also private hospitals in Bloemfontein, manned by volunteers, and supported by donations from Britain: the Langman, the Irish and the Portland. Private donations to help victims of the war in Britain and the colonies were estimated at around £6 million,[177] and included help for widows and families, disabled officers and men, and convalescent homes, as well as equipping private hospitals in South Africa itself. Some £750,000 were donated towards the private hospitals: £381,050 for equipment and miscellaneous items, and £219,385 for "extra comforts".[178] The private hospitals were considered to be well-equipped, and even luxurious, but the army hospitals tended to come in for criticism whenever anything went wrong. This was to be the case in one of the most high-profile scandals of the war involving the outbreak of fever at Bloemfontein. It was to rage on for several months, and many of the medical staff fell victim as well as the soldiers. Conan Doyle, attached to the Langman, one of the private hospitals, reported in the *BMJ* on June 7th 1900 (written on June 5th), while the outbreak was still raging:

> We have had 6 nurses, 5 dressers, 1 wardmaster, 1 washerman, and 18 orderlies, or 32*[sic]* in all, who actually came in contact with the sick. Out of the 6 nurses, 1 has died and 3 others have had enteric. Of the 5 dressers, 2 have had severe enteric. The wardmaster has spent a fortnight in bed with veld sores. The washerman has enteric. Of the 18 orderlies, 1 is dead, and 8 others are down with enteric.

So that out of a total of 34*[sic]* we have 17 severe casualties - 50 per cent. - in nine weeks. Two are dead, and the rest incapacitated for the campaign, since a man whose heart has been cooked by a temperature over 103 [degrees] is not likely to do hard work for another three months. If the war lasts nine more weeks, it will be interesting to see how many are left of the original personnel.[179]

The war was not to last nine more weeks. It was to last nearly another two years, though Cheyne was only to accompany Lord Roberts as far as Pretoria. He and other surgeons kept notes as well as working to cope with the situation. In fact, Cheyne, as though these circumstances were simply a continuation of his normal activities, kept detailed notes of all aspects of the march and stay in Bloemfontein, medical or otherwise. George Makins was later to be grateful for Cheyne's case notes when he came to write his account of the types of bullet wound encountered during the war.[180] Cheyne sent an account of the march home to Britain, which was published in the *British Medical Journal* on May 5th,[181] and he followed it up with another on May 12th,[182] specifically on the treatment of wounds. He was to take home with him tremendous lessons on wound treatment in war conditions, as well as the organisation of medical teams. In case his graphic descriptions of the march were hard to imagine, he also took a remarkable set of photographs which he later used to illustrate his talks on the war. They show the difficulties faced by the horse-drawn transport in scaling muddy inclines, the problems associated with the fateful decisions on the transport and communications infrastructure, and the general suffering of the men and medical teams.[183]

On March 30th, Cheyne received a telegram from Colonel Gormley of the Seventh Division (RAMC), informing him of a fight at Karee Siding, about twenty-one miles north of Bloemfontein, on the previous afternoon.[184] He was asked to go out there to attend to a number of serious cases, so he left by rail to Glen and then took a horse to Karee. He arrived there around 1 pm. There were around 154 wounded, and in 15 cases the abdominal cavity had been penetrated; five had already died.[185] He operated on one case, and left the other nine cases alone, as there was considerable debate about the advisability of operating on them in the difficult conditions, given the dangers of sepsis. He left Karee the following day, on April 1st, and discovered on April 2nd that his patient had died. Four of the cases he had not operated on had also died, leaving only five still alive.[186] The case illustrates the difficult decisions faced by even the most experienced surgeons in cases like this. There was to be a good deal of discussion after the war about the advisability of operating on abdominal wounds, because the operating conditions could be more dangerous than leaving the bullets in situ.

Lord Roberts left Bloemfontein on 3rd May, heading for Johannesburg and Pretoria, but an account of Cheyne's testimony to the Hospitals Commission on his return suggests Cheyne himself had left there by April 21st or 22nd, and had "gone away [for] a few weeks"[187], presumably called out to hospitals, such as the case at Karee Siding. When he left, there had been 300 deaths in a 50-bed hospital.[188] He then returned to Bloemfontein before leaving again for Pretoria with Lord Roberts' convoy, and he later noted that he had discovered 400 cases of fever when he arrived back.[189] He probably left again on 3rd May with the convoy, and this time, it appears Lord Roberts had attempted to address the transport issue, at least where Cheyne was concerned.

> After we left Bloemfontein I was provided with an ambulance cart of my own in which I carried an operating tent and all the requisites for performing operations, and I made a point of having at hand a number of rum jars which were kept filled with strong lotions, which could be diluted as required. I also had a jar full of swabs soaking in lotion and a large eight gallon tin which I always kept replenished at every opportunity with filtered and boiled water. It really took very little time to keep these vessels replenished, and one was always ready for any reasonable emergency.[190]

They were forced to halt for ten days at Kroonstad, and once again, this was precipitated by transport and supply difficulties. The Boers had blown up all the railway bridges from Bloemfontein to the Vaal, and Lord Roberts stopped in Kroonstad to wait for them to be repaired, to allow food supplies to be restored. The ultimate effect on health at Kroonstadt was even worse than at Paardeberg; typhoid developed again and "caused more British casualties that month than all the battles of Black Week"[191] (the succession of defeats which had prompted the War Office to send Lord Roberts to relieve Buller). Lord Roberts made changes to the RAMC command.[192]

Lord Roberts' convoy eventually marched into Johannesburg on May 31st, and took Pretoria on 5th June, marking the end of Cheyne's personal involvement in the war. From here, he returned to Cape Town, where he boarded the Royal Mail steamer *Moor*[193] on June 27th 1900. He had been away nearly six months. On board, coincidentally, was a veteran war artist of *The London Illustrated News*, Melton Prior, who almost lost the sight in his left eye when he was injured by a cricket ball on deck during the journey. Cheyne operated on the eye, and saved it, providing the press with a pleasant anecdotal interlude from the typhoid debate.[194]

Meanwhile, the medical teams at Bloemfontein continued to struggle with the epidemic. If this were not enough, a Unionist MP, Mr Burdett-Coutts, had taken it upon himself to visit the troops at Bloemfontein during the outbreak of fever, and sparked a bonanza for the press.

11

Hospitals or barracks?
Lessons from the War

William Burdett-Coutts was Member of Parliament for Westminster, and his fact-finding mission to assess medical conditions among the British forces in South Africa could not have happened at a worse moment for Lord Roberts. The visit was supported by his wife, the influential philanthropist Baroness Burdett-Coutts, whose surname he had taken when he married. She was heiress to the fortune of the banker Thomas Coutts, her grandfather, and was at one point the wealthiest woman in England. She was a friend of Dickens, and supported unfashionable charitable causes. Her marriage to her much younger secretary provoked a society scandal, but her philanthropic work was respected, and her approval lent weight to her husband's investigation.

There had been such coverage and debate on the war in the press that Burdett-Coutts had decided the British public should be independently informed about medical conditions. He considered the role of the press was to cover the progress of military operations and, he implied, to maintain an image approved by the war authorities. He suggested that the average war correspondent had neither time nor, necessarily, the inclination, to "follow a convoy back from the front, to spend the day in crowded bell-tents, or to walk the wards of a hospital."[1] He also suggested that the press was more likely to censor reports according to its stance on the war, and that "disagreeable truths"[2] about hospital arrangements were unlikely to find favour with editors.

On 18th January 1900, he informed his Westminster constituents that he would be setting sail for South Africa within a few days. He believed his personal experience of the Russo-Turkish war qualified him to inspect the medical arrangements in South Africa, and at the same time, with no attachment to the armed forces, he considered himself "an independent and reliable source."[3] He had the support of Florence Nightingale, not least because his wife was one of her benefactors. Not surprisingly, therefore, he launched an attack on the lack of female nurses at the front, noting later that, "the three General Hospitals were as good as a system which practically excludes female nursing, and is strangled by red tape, could make of them."[4]

The military authorities were not particularly impressed, and unfortunately for Lord Roberts, and for both RAMC and civilian medical practitioners, Burdett-Coutts reached the troops at Bloemfontein at a most inauspicious moment, at the height of the typhoid outbreak and the end of the march from Modder River. He had arrived in Cape Town on 6th February, but was initially refused permission by the Press Censor to proceed to the Front. In the version of events he published himself (as opposed to the letters he submitted to the editor of the Times), he suggested that he later realised the higher authorities may not have been to blame for this, nor for the lack of response he received to his letter to Lord Roberts, as he suspected the Censor had taken matters into his own hands. "I have always remained in doubt as to whether and how far they were the outcome of a personal or a delegated authority."[5] He also later admitted that Lord Roberts had more pressing matters to attend to at the time.[6] He nevertheless took the refusal as fortuitous, as he might otherwise never have found himself in Bloemfontein at the moment of the outbreak of fever, and this is where he considered himself to have real experience. He had apparently been a victim of typhoid himself in Turkey and had almost died.[7] When he finally arrived in Bloemfontein, he was appalled by what he saw in one of the hospitals, and sent a series of letters back to Britain, which were published in *The Times*.

Royal Army Medical Corps group in South Africa during the Boer War.

The government was forced to react. The full statistical horrors of the typhoid outbreak were being paraded before the public through the British press, showing that, far from being a passing medical emergency, the outbreak had killed more of the British force than wounds in battle. It caught the public imagination, and testimonies began to pour into the press from the families of those at the Front, which apparently corroborated Burdett-Coutts' account.[8]

He used emotional language and imagery, but was not always accurate, and made unfortunate statements which were taken to mean an attack on the authorities, or even allegations of cruelty towards patients. He later had to retract or explain them, noting that the details which appeared in the Times "broke off at a hospital at Bloemfontein"[9], by his own admission the worst affected of the eight hospitals he eventually saw, and a final chapter he had intended to publish on the other seven hospitals in the town had never transpired. It was nevertheless taken by Lord Roberts and the medical teams alike as an unwarranted attack on the authorities and the Royal Army Medical Corps, and provoked what became known as the hospital scandal.

On June 29th 1900, two days after Cheyne had set off for home from Cape Town, Burdett-Coutts spoke in Parliament.[10] He stuck by most of his allegations, but took the opportunity to rectify the omissions in *The Times* correspondence, largely to dispel any suggestion that he was being less than objective. Parliament appointed a Commission on July 24th to look into the allegations, and the Commissioners sailed for South Africa on 4th August.[11] To make absolutely sure omissions in his statements, or the interpretation of them by others, did not detract from his allegations, Burdett-Coutts rushed through a print run of his full account,[12] so that the public would have a chance to see it before the publication of the parliamentary enquiry. This gave him an opportunity to refute some of the allegations against his report in advance. His main criticisms focused on overcrowding, staff and materials shortages, and lack of beds, suggesting that the conditions in the field hospitals were not the result of a long march but bad organisation. It was also to suggest that it was responsible for the *spread* of enteric fever, which was endemic in Bloemfontein anyway, and had been the reason why a crisis had become "a disaster".[13]

> On that night (Saturday, the 28th of April) hundreds of men to my knowledge were lying in the worst stages of typhoid, with only a blanket and a thin waterproof sheet (not even the latter for many

of them) between their aching bodies and the hard ground, with no milk and hardly any medicines, without beds, stretchers, or mattresses, without pillows, without linen of any kind, without a single nurse amongst them, with only a few ordinary private soldiers to act as 'orderlies,' rough and utterly untrained to nursing, and with only three doctors to attend on 350 patients. There were none of the conditions of a forced march about this. It was a mile from Bloemfontein ... with a line of railway to two seaports, along which thousands of troops and countless trainloads of stores and equipment of all kinds, and for every one except the sick, had been moving up during the whole of that leisurely halting time.[14]

He continued with emotional language:

> ... the coarse rug grated against the sensitive skin burning with fever. The heat of these tents in the midday sun was overpowering, their odours sickening. Men lay with their faces covered with flies in black clusters, too weak to raise a hand to brush them off, trying in vain to dislodge them by painful twitching of the features. There was no one to do it for them. At night there were not enough to prevent those in the delirious stage from getting up and wandering about the camp half naked in the bitter cold ... Men had not only to see, but often to feel, others die...[15]

> "... 20 of the worst cases were removed to a more permanent hospital a mile and a half off. How were they taken? They were lifted out of their tents and put into rough ox-waggons—all typhoids and many of them dangerously ill - and then jolted across the veldt, which in this place is much broken by spruits and gullies. One case was in a state of 'hemorrhage' when moved.[16]

To add to the problems for the military authorities and the government, a *Daily Mail* correspondent, Julian Ralph, not only appeared to corroborate Burdett-Coutts' report with his own eye-witness account, but added another charge, regarding the lack of peace-time training the government afforded the medical officers of the RAMC, and saying he had been informed that, as they were army officers, working to army discipline, they were "tyros"[17]. This was also to suggest that there were rifts between the RAMC officers and the civilian surgeons, a claim vigorously denied by the latter, though Cheyne later suggested ways around any difficulties, indicating he did not believe the mixed arrangements were practical.[18] Ralph did not say who his informant was, except that he was "one of the greatest English surgeons."[19] It is unlikely to have been Cheyne, as he was perfectly aware of his position. There were any number of 'great English surgeons' in the vicinity at the time, and there was clearly *some* tension arising from the restrictions on them by army arrangements, so opinions on the subject were likely to have been fairly well known and well discussed among the medical teams.

Criticism in the press included the new typhoid outbreak at Kroonstad, where Lord Roberts had delayed for ten days to allow repairs to the railway line to let supply trains through. He no doubt found the criticism of his transport arrangements particularly ill-timed, since he had finally begun to address them. The main outrage on the part of the surgeons and military authorities was not that the facts were necessarily always wrong about what the journalists were reporting, but that they had been misrepresented and over-generalised on the one hand, and that they were implying unfounded allegations against the medical corps which was operating under unprecedented conditions. Lord Roberts responded in Parliament through his main political advocate, the Secretary of State for War, Lord Lansdowne.

> I ordered that the requirements of the sick were to be first taken in hand, as soon as the railroad had been repaired. The principal medical officer proceeded with the first train to Kroonstad, with surgeons and nurses. No. 3 General and Scotch Hospital had been held in readiness at Bloemfontein to be sent to Kroonstad directly the line was open. This was done, and they received 180 patients within twenty-four hours of arrival.
> I repeatedly visited the hospitals during the time I was at Kroonstad, and impressed upon the principal medical officer to do all that was possible to remedy matters. A few days afterwards I received a report from the medical officer that the medical arrangements were good, and Lord Methuen has since informed me that the medical arrangements were perfectly satisfactory.
> I was deeply distressed at being unable to make more perfect arrangements on first arrival at Kroonstad. But it was inevitable that in the rapid advance of our great army when the railway had been destroyed the suffering would have been enormously increased had

it not been for the prompt manner in which the medical authorities made use of the scanty accommodation available at a place little larger than an ordinary English village.[20]

Lord Roberts also recommended, crucially, that the word of surgeons who had been with him on the march to Bloemfontein be taken into account by the enquiry, particularly Cheyne.[21] Cheyne had seen how the situation had developed on the march and in the period the troops were at Paardeberg, where the experience of other informants had largely been closer to railway lines. Frederick Treves had responded to a request for clarification of the situation from the *British Medical Journal*, saying he was "shocked and surprised"[22] by the allegations, but Treves had not actually been present on the march to Bloemfontein, so was only in a position to present second-hand testimony. Other surgeons had written to describe the conditions, however, including Anthony Bowlby, attached to the Portland Hospital. He largely corroborated Cheyne's doubts as to how many of the cases of fever were typhoid, and how many were sand diarrhoea caused by contaminated drinking water.[23]

When Cheyne arrived back in London, he lost little time in delivering his own verdict on the conditions in South Africa. For some reason, he saw fit to pre-empt the evidence he was to give to the South African Hospitals Commission by delivering a detailed letter to *The Times*.[24] Unfortunately for Cheyne, the editor took some of his remarks out of context, and presented them as evidence that his statements to the Commission were likely to support Burdett-Coutts' claims in general. Cheyne generally took a more measured and considered view than Burdett-Coutts, and was also mindful of his responsibility to Lord Roberts in weeding out the facts from the sensationalism. Nevertheless, on one issue *The Times*, Burdett-Coutts and Cheyne broadly agreed, namely the consequences of the transport arrangements. In his letter, Cheyne had compared the British transport arrangements unfavourably with those of the New South Wales Ambulance service, which had its own, dedicated medical transport. The Australian press, understandably, had a field day. The *South Australian Register* noted:

> Dr William Watson Cheyne … condemns the absence of independent transport for the Army Medical Department. He declares that the New South Wales field hospital which was dispatched to South Africa under Dr Fiaschi has done the greatest

amount of good work of any field hospital during the campaign. The Army Medical Department would, he says, often have 'been in serious difficulties without it.'[25]

The *Bendigo Advertiser* added that Cheyne had referred to "the excellent arrangements of the New South Wales field hospital."[26] He had good reason to praise the New South Wales unit, but his account to the Commission would show that this was not to be taken as blanket acceptance of Burdett-Coutts' criticisms. He gave evidence on the second day of the hearings, on 25th July, and dispelled any claims that the RAMC was to blame for the conditions in the hospitals during the typhoid epidemic, or that patients had been treated in any way inhumanely. He largely restricted his criticism to what he considered the main shortcomings involving inadequate transport and army bureaucracy, and placed everything else within the context of the specific conditions they had encountered on the South African veld on their march to Bloemfontein. He admitted there was overcrowding in the field hospitals, but added that the wounded were nevertheless attended to and were, where possible, sent back to the base hospital.[27] As we have seen, however, Cheyne had been critical of inadequate transport in ox wagons, which could actually be detrimental to the healing of wounds, and orders had been given after Driefontein not to move the worst cases.[28] He gave an honest account of the surgeons, noting that there were likely to be good and bad surgeons in the RAMC just as in civilian surgery, but that the surgeons and non-commissioned officers in South Africa were generally very good. He diplomatically suggested that the St. John's Ambulance men were in need of training, but that all concerned were doing their best, and that, "if their best was not good, that was another matter."[29] He pointed out that the orderlies had also been subject to fever, and that everyone had been battling unusually adverse circumstances.

He responded to Burdett-Coutts' accusations that patients were left outside by noting that "being out in the open did not affect the patients a bit, for it was a beautiful time of the year,"[30] except during the heavy rain at Paardeberg, where hospital tents were flooded. He said he had seen men lying on the ground, "but never in the mud."[31] He said he had been aware of no instances where men had not received their rations. He was nevertheless critical of how an "unnecessary amount of the time of the medical staff was taken up in administrative and clerical work,"[32] which he thought clerks should be employed to do. It took around two hours a day to clear the medical reports. He noted that excessive bureaucratic arrangements not only held things up for the RAMC but were also the source of inconvenience for the civil surgeons. He hinted, however, that he was happy to go along with it to avoid misunderstandings. He recalled how he had started out without a uniform, but had ultimately found this a great inconvenience, "as he was constantly being arrested for going to places where it was thought he ought not to have gone."[33]

Cheyne had barely been home a month when he was called on to give evidence to the Commission, which welcomed his statement, as did the press. It has been noted that Cheyne's testimony was taken more seriously than many others because of his senior status, that "he was heard with great respect and several of his views went into the final report."[34] Crowther and Dupree have cited the rather different experience of Calder, "one of the few doctors of minor status to give evidence",[35] whose complaints about conditions in a requisitioned police hospital at Tuli were condemned as "loose criticisms of the medical services."[36] He was admonished by the Commissioners for leaving "a most false impression" and using inaccurate language.

Cheyne's testimony was treated with much more reverence, but perhaps not only because of his seniority. He was careful with the language he used, no doubt recognising that if he were to make blanket or inflammatory criticisms, the most important points would not be addressed. He was also a tremendously loyal man, as we have seen in his relationship with Lister. He was aware, as was the press, that he had been specifically recommended by Lord Roberts to represent the authorities' case against the charges, but overall, it can probably be said that it was his objectivity, not his loyalty, that won the day. He had been an eye-witness, and his well-argued views were already well known in the medical press. *The Times*' attempt to take his words out of context had backfired, and the Parliamentary inquiry had given him an opportunity to make it quite clear where he thought the difficulties lay and where praise was due.

In fact, a number of newspapers were careful to refer directly to Cheyne's comments in *The Times*, not the editor's interpretation of them. One paper found them "entirely free from the intemperate expressions which have characterised the utterances of some of his predecessors in writing or speaking upon the subject, and this adds to its weight. His remarks upon the necessity for providing the Royal Army Medical Corps with its own transport will be applauded far and wide."[37] Furthermore, as the mood in the press shifted, 28 Canadian soldiers who arrived in London about the same time as Cheyne enthused about the conditions to *The Daily*

Telegraph: "all gave emphatic denials to the charges that the field hospitals were incompetently and badly managed, asserting that the arrangements were the best possible under the circumstances. They unite in saying that they could not have been better treated."[38]

Nevertheless, Cheyne's account was a brave one, particularly given the responsibility Lord Roberts had clearly placed on his shoulders and the duty he equally felt in delivering an objective account. The Commission had stressed that there would be no censorship of statements, but in reality very few came forward to give a full and honest account of all the issues, and the Commission clearly did not live up to its promise if they did, as in Calder's case. The Commission published its report in January 1901. It cautiously dismissed Burdett-Coutts' charges, but recognised isolated cases of neglect. With regard to the charge that officers of the RAMC were "too military in their habits and conduct"[39] they found that "there has been a tendency on the part of some of the officers and men of the RAMC to treat the hospitals too much as if they were barracks, and to regard the patients in the hospitals too much as soldiers and not sufficiently as patients."[40] On the whole, however, the officers of the RAMC were considered to have done their job admirably in unpredictable circumstances, and deserved "great praise".[41]

The commissioners noted that

> ... as a rule, the civil surgeons have worked well with the officers of the RAMC. A few cases of friction occurred between them in certain hospitals. These arose from faults on both sides. Some civil surgeons not unnaturally felt a difficulty at first in becoming acquainted with the working of military hospitals, but as rule, they soon became familiar with their duties.[42]

When it came to the all important transport, Cheyne's main criticism, the Commission recognised the shortcomings of arrangements for stationary and general hospitals, but in considering whether the RAMC field hospitals should have their own transport, presented what may be seen as a rather lame objection that the staff were not experienced in handling the mules.[43] They regretted that some of the transport difficulties during the war had "caused unnecessary suffering."[44]

While Cheyne was away, his family had grown, and was also growing up. On 25th April 1900, while Bloemfontein was reeling under the epidemic, Margaret had given birth to a daughter. She had called her Julia Millicent[45] after her own mother, Julia, who had died when Margaret was young. By this time, Cheyne's eldest son, Joseph Lister Watson Cheyne, was away at boarding school, in East Grinstead,[46] and within a year, if not already, his second son Hunter would be away at school in Bray in Berkshire.[47] Margaret now only had her own children at home, Meta, Basil and baby Julia.

Fully a year before the protracted conflict in South Africa was finally brought to a close with the signing of the terms of the Boer surrender, the assessment began at home of the many issues the conflict had raised, and the government and medical establishment began to reflect on how organisational arrangements could be improved in the event of another war. On June 5th 1901, these thoughts were reflected at a dinner at the Cecil Hotel in London for the civil surgeons who had served in South Africa. Mr Brodrick, Secretary of State for War, was present, and noted in his speech that the military authorities and, he suggested tentatively, the RAMC, recognised that organisational changes were necessary.[48] He informed that he would shortly be asking civil surgeons their opinion on how this could best be undertaken to enable the authorities to attract "the best candidates for the [army] medical service,"[49] and "perfecting the scheme which would give to the country the best medical organisation in the world."[50] Treves, in a speech reminiscent of the days when antiseptic surgeons were presenting their fight against bacteria as a struggle against a ubiquitous foe,[51] lightened the mood of the dinner by comparing the protective power of the Navy to a perfect antiseptic dressing. Though there was a "difference in size between a bacillus and a battleship, and there was of course a difference in texture between torpedo netting and carbolic gauze ... the function was the same,"[52] he said.

Cheyne was asked by the *British Medical Journal* to "make some remarks on the suggestions of the [Hospitals] Commission"[53]. He confined his comments to a discussion on the Commission's remarks about "the best way of supplementing the Royal Army Medical Corps in the case of a great war, where the existing establishment is insufficient."[54] He suggested that, while it was relatively easy to procure materials, the question of staffing needed to be considered more carefully. He had no fears about the availability of doctors, whether army or civilian, but was more concerned about the availability of trained nurses, and noted that the army's almost exclusive use of male nurses in times of war was inefficient. Firstly, as there were no real openings for male nurses in civilian life at the time, they had to be specially trained, and then were in danger of losing their skills again during peace time. His solution was to suggest taking on female nurses, and perhaps he was recalling his first wife when he said they were

> ... in every way far superior to men, and that if matters could only be so arranged that they could be employed in military base hospitals as fully as in civil hospitals in peace, it would be a great advantage to the sick and wounded, while there would be no difficulty in getting an ample supply. [55]

He added that

> ... while women naturally cannot move with the army at the front, I cannot see any possible objection to their being as fully employed at the base as in civil hospitals in London, and if this were done there would be a large number of fully-trained orderlies available to nurse the sick at the front, and there would be but little necessity to recruit the orderly staff from partially or entirely untrained men.[56]

This suggests that he at least partially agreed with some of Burdett Coutts' allegations when it came to general medical arrangements in the military.

Where the Commission had insisted that there were no serious tensions between the army and civilian surgeons, Cheyne suggested that the issue of organisation was more relevant, and that it was a question of apportioning skills appropriately. He said he did not think

> ... the mixture of the two both at the base and at the front work[ed] well ... To put a good civil surgeon at a base hospital under a less efficient Royal Army Medical Corps man (and the Royal Army Medical officers are selected by seniority rather than by professional ability) is to lead to friction or to breed contempt of his superior officer in the mind of the civil surgeon. Neither is favourable to good work. On the other hand, to put a civilian at the front where he is not acquainted with the various duties which may be required of him - with the method of obtaining transport, equipment, rations, etc. - is apt to irritate the army medical men in charge owing to the civilian's apparent incapacity, but in fact to his want of knowledge of the arrangements.[57]

He suggested there were more strategic ways of making the best use of the skills of both. He recommended that stationary hospitals, which emulated more closely the conditions of civilian hospitals, be manned by civilian surgeons, and that the very specific skills of the RAMC, more suited to the conditions of battle, be used in the field hospitals. It was an eminently sensible suggestion, and calculated to reduce tension as well as maximise skills. The place of the RAMC was "at the front"[58], he said, for which it was trained.

On the question of whether this would detract from the experience of the RAMC practitioners, Cheyne noted emphatically that "during war the object of hospitals is not to serve as teaching institutions or as places for gaining experience, but for the treatment of the sick and wounded, and for that alone."[59] He proposed instead that "the army medical officer, when first sent out to the seat of war, could, unless pressure existed for his immediate services at the front, be temporarily attached to the base hospital for some weeks,"[60] and that "hospital classes might be held [at the base] for the benefit of those not at once drafted to the front, and for those on leave."[61]

He suggested that the base hospitals not only be staffed by civilian surgeons, but also that they be managed on civil lines, and linked to the army through an army medical officer, as had been the case at the Portland Hospital and other civil hospitals in South Africa. He believed young civilian surgeons were adequately experienced to take on this task, which would avoid the need to send out so many consulting surgeons and physicians, as had been the case in South Africa.[62] There was perhaps a hint here that this had been an avoidable extravagance. Whilst Cheyne seems to have believed himself quite prepared for the experience of war surgery when he left for South Africa, it most probably marked the beginning of a gradual deterioration in his health. When he left for the war, he was already weakened to some extent by a gruelling workload. As soon as he returned, the workload resumed.

Cheyne had already published detailed information on his experience of wounds and operating conditions during the war, and had sent back his first letters before his return. A week after his account of the march from Modder River to Bloemfontein, on May 12th 1900, the *British Medical Journal* published his observations on wound treatment, and he used these impressions a year later to organise his conclusions, which he set out in an address to the Midland Medical Society in November 1901.[63] His talk was a contribution to a debate on whether modern aseptic surgery was responsible for the success of medical services during the war. He disagreed with those who suggested the improved treatment compared to previous wars was the result of the advances made in modern surgery,[64] and focused instead on the lessons that could be learned from the specific conditions in which surgery had been conducted on the South African veld.

He partly took *modern surgery* to mean methods of achieving asepsis, and detailed why the hospitals could in no way be said to have benefited from it in the war in the same way as Listerism was understood in civilian surgery – as a careful process rather than a specific set of chemical antiseptics or modern equipment. He attributed the success instead to advances in the understanding of healing processes as a whole. Although aseptic conditions were almost impossible to achieve at the front, the dryness of the air on the veld favoured scabbing, as long as medical personnel understood Listerian principles sufficiently to avoid tampering with the wound with dirty hands and instruments, or relying on wet dressings which invited the growth of bacteria.[65]

First of all, Cheyne was not convinced the levels of antiseptics used were in any way helpful. Soldiers had been issued with a package of field dressings soaked in corrosive sublimate, an antiseptic containing mercury, and had been instructed to apply the dressing as best they could if they were struck down in the field. The logic was that, if infection could be prevented until the wound could be properly treated at a field hospital, limbs were less likely to be amputated. Dressings of this ilk had been issued to soldiers as early as the Crimean War,[66] and Cheyne himself had noted the details of how "[T]ampons of salicylic cotton, wrapped in salicylic gauze"[67] had been proposed in 1869 by Esmarch, the German authority on military surgery and first aid on the battlefield. His methods had been applied successfully by Reyher during the Russo-Turkish war of 1877 to 1878[68]. In South Africa, Cheyne was in favour of the principle, but dubious of the practice. He said the bandages contained too little antiseptic to be effective.

> Let us look first at the conditions under which a man is wounded in the field ... He goes into the fight provided with what is called a 'field dressing,' which essentially consists of a small piece of antiseptic gauze, a piece of wool, and a bandage. The gauze is quite small in amount, and, if there is any bleeding going on, it very soon becomes saturated with blood, and, to begin with, contains such a small quantity of antiseptic material in it that the blood

passing through it is not in any way deprived of the possibility of putrefying, so that this little dressing, as arranged in the packet which accompanies each soldier, is more a sort of clean rag than any actively antiseptic dressing.[69]

More significantly, however, emphasis on the field dressing missed an important point. Esmarch and Reyher had stressed that early attempts to treat a wound could only be effective if it were not probed, if no fingers touched any part of it, and if no attempt were made to extract the bullet or examine the wound at that stage [70], avoiding contamination as carefully as possible. Ogston would later note that this was specifically and logically a Listerian principle, but that it had largely been adopted and publicised by continental surgeons. He said in 1904 that abandoning probing in *practice*,

> like most of the advances on Listerian lines, came from the Continent. The recognition of the fact that bullets were usually more harmless than the means employed to remove them, particularly on or near the battle field, is unquestionably a gain which is to be ascribed to the influence of Lister and his methods.[71]

Cheyne noted that the effect of the so-called "antiseptic gauze", already dirty by this time, was in defiance of this very principle:

> ... under the circumstances in which it is employed it is very far from being even a clean rag. If the soldier or his comrade attends to the wound, they open this packet; their hands are covered with mud and sand and grease; they take out this piece of gauze with their dirty hands; very probably they lay it down on the ground while they are getting the clothes opened, and then they stick this piece of dirty gauze on the wound. And even if the dressing is made by the doctor the conditions are practically no better.[72]

He deduced that any success in wound treatment in South Africa could be attributed to the specific, dry climatic conditions of the veld, and if any modern surgical understanding were involved it was what surgeons now knew about the natural healing processes of the body. "The antiseptic dressing and the surgical treatment adopted had little or nothing to do with the result in these cases," he said, "and the only way in which we can bring in modern surgery as of value is because it has taught us the meaning and value of healing under a scab."⁷³ He was convinced that more moist conditions would have impeded the process of scabbing, and recalled that

> ... in the early stages of the war the field dressing which was provided contained a piece of mackintosh which the soldiers were instructed to apply outside the gauze. When this was done, however, it was found that sepsis and suppuration frequently occurred along the track of the wound and I have seen a number of cases brought in only a few hours after the infliction of the injury dressed in this way, and have found the gauze and discharge stinking, and of course, entire absence of scabbing. This was observed quite early in the campaign, and from that time orders were issued that the mackintosh was not to be used. But this very fact shows that the drying up of the discharge and not the antiseptic dressing was the important factor. It was also observed that those cases did best where the wounded lay out in the sun for a long time before they could be brought in, and where, therefore, a rapid and very dry crust formed.⁷⁴

Cheyne had taken careful case notes under difficult conditions, and Makins later had cause to be grateful to him for his documentation of "the abdominal wounds observed after the fighting at Karree [sic] Siding, on March 29"⁷⁵. Makins used the notes in his study on the effects of small bullet wounds, a feature of Boer War armaments which had differed from previous campaigns and which had achieved an impact on surgical results. These were the notes Cheyne had published in the *British Medical Journal* on 12th May.⁷⁶ He agreed that the Mauser bullets had produced smaller wounds than in previous wars, which limited the chances of sepsis and promoted healing, but later added a word of caution that this did not mean all wounds from these bullets remained aseptic.⁷⁷ The decision on whether to operate to remove the bullet depended on a number of factors, including the angle of entry, the general condition of the patient on arriving at the field hospital, and how much, if any of the clothing of the soldier the bullet took with it in its passage into the body.⁷⁸

Cheyne recommended that the principle lessons to be learned from the campaign for future wars were that, since approaching aseptic conditions in the field was clearly impossible, medical teams should attempt to replicate on the wound the drying effects of the South African veld. He supported procedures "to promote scabbing and prevent decomposition by the free use of antiseptic powders, and by the free exposure of the wound to air and sun. The object of these powders would be partly to absorb moisture, and therefore tend to make a cake, and partly also, being antiseptic, to impregnate the blood with such a quantity of antiseptic material that bacteria cannot grow in it ... I think that powder would be better than pastes, especially if these contained gummy materials ... I think if, in addition to the cyanide gauze, there was in the field dressing a packet of antiseptic powder, which the soldier was instructed to open and sprinkle over the wound before applying the gauze, or if the gauze itself were full of loose powder, considerable help would be obtained. Further, the regimental surgeon, although he could not carry antiseptic lotions with him, might quite well carry a quantity of antiseptic powder and dust the wounds with it before applying the dressings."[79]

Cheyne had, in fact, conducted his own experiments when he had first arrived in South Africa. Lister had been interested in the concept of antiseptic use as first aid for many years, and had published a paper [80] during the Franco-Prussian war, advocating the early treatment of gunshot wounds with carbolic. He was anxious to test antiseptic use in the field, and before Cheyne and Cheatle left for South Africa, he supplied them with double cyanide of zinc and mercury in powder form, and also in the form of "a paste made with a watery solution of carbolic acid."[81] Cheyne had "furnished various bearer companies with pepper boxes containing ... antiseptic powders of different strengths and kinds, but, owing to the ever-changing position of the bearer companies, I never knew what the result was, or indeed if they had ever been used. This is a point which, I think, might well be looked into by the authorities."[82] Here was another lesson Cheyne had learned about war. The controlled conditions of the laboratory and ward could not easily be emulated on the march and the battlefield, and he would have to be inventive if he wanted the type of evidence he had been used to. On the one hand, it had been a new experience for him, and an unprecedented opportunity to experiment in interesting conditions. On the other, it had given him an opportunity to reiterate that chemical antiseptics were worse than useless if used inappropriately, but that, with some inventive adaptations, they could provide a useful supplement to the natural healing processes of the body by emulating the dry conditions of the South African veld. Once again, it was about process and adaptation.

Finally, he turned to the location of the field and base hospitals, suggesting that the ideal would be to leave the field hospital on the field of battle and keep the patients there until it was safe for them to be transported to the base, assuming they could be adequately supplied with medical materials and food. He recognised that this would require an agreement between nations and relied heavily on the integrity of the enemy not to fire on the hospital, even if they took prisoners. He had absolute faith that this could be achieved,[83] despite mentioning on a number of occasions his experience of hospitals having been deliberately targeted and hit by Boer troops.[84] He was nevertheless recalling the harrowing conditions in which the wounded were transported away from the battlefield in South Africa, and the consequent effect on their chances of recovery, saying, "no one who has realised the sufferings of the patients in this campaign from the early transport, and the loss of life and limbs which have consequently ensued, will doubt as to the advisability of some plan of this kind."[85] He nevertheless recognised that his suggestion would involve addressing the understaffing of the RAMC.[86]

He chose not to mention under the heading "modern surgery" an innovation which had considerably helped assess bullet wounds during the war. Röntgen (or X-) rays had been discovered 1895, and this was the first conflict in which they were used by the military authorities. They were first discovered by Wilhelm Röntgen in Germany, but in Britain, Mackenzie Davidson, originally a student of Lister and Ogston, developed them to help locate small foreign bodies in humans.[87] As far as I am aware, Cheyne also chose at this stage not to enter publicly into the debate on another important innovation in medicine, though he would later engage in a "heated dispute"[88] with Almroth Wright on other aspects of the matter. Wright had developed inoculation against typhoid at the Army Medical College in 1896 and it was tested, largely in India, in the last years of the century, amid a split in medical opinion about how effective it was. In India, there was a general feeling it had worked, but evidence of its success came from clinical experience only, and the government was sceptical about adopting it.[89] Moreover, it was not considered to prevent the disease, only to lessen its effects. Lord Lansdowne at the War Office suggested it be put to the test, using public funding, on any British soldiers who were prepared to submit voluntarily to it. When the Boer War began, the inoculations were made available on a voluntary basis to soldiers leaving for South Africa. The opinions of individual Medical Officers, however, affected whether it was taken up, and under five percent were inoculated overall.[90] When the fever at Bloemfontein was at its height, the government came under fire for not making the inoculations compulsory. Arthur Conan Doyle criticised the policy in the *British Medical Journal*, writing from Bloemfontein:

> There is one mistake which we have made, and it is one which will not, I think, be repeated in any subsequent campaign. Inoculation for enteric was not made compulsory. If it had been so I believe that we should (and, what is more important, the army would) have escaped from most of its troubles ... We have had no death yet (*absit omen*) from among the inoculated, and more than once we have diagnosed the inoculation from the temperature chart before being informed of it. Of our own personnel only one inoculated man has had it, and his case was certainly modified very favourably by the inoculation.[91]

Cheyne's first active experience of treating wounds in war had been a lesson in adaptation of his methods, and an eye-opener as far as the organisation of medical staff was concerned. The experience would serve him well during the next major conflict in 1914. Moreover, he had made a favourable impression with Lord Roberts, the press and government commissioners. He had, in effect, acquitted himself well in the debates which ensued. In February 1901, Lord Roberts had commended him for the advice he had been able to offer, and he was mentioned in despatches, which were published in *The Times* on 10th February.

> Mr Watson Cheyne, FRS, and Mr Kendal Franks, FRCSI, consulting surgeons, who accompanied the Army, have rendered invaluable service by their advice and assistance to the medical officers. They have been unwearying in their work among the wounded and sick, and, humanly speaking, many a valuable life has been saved by their skill.[92]

Lister was delighted, and in a letter to Margaret Watson Cheyne the following day congratulated her husband on "the honourable mention of his name by Lord Roberts".[93] On 12th June 1901, Cheyne was invited to the presentation of medals at Buckingham Palace, to be made a Companion of the Order of the Bath, an honour bestowed by the King.[94] Once again, Lister had sent congratulations, succinct and to the point, "Although it does not do justice to the value of your work, yet it is a recognition of it."[95]

On his return, Cheyne gave a number of talks, not only to medical societies but to hospitals and the general public, detailing his experiences and the general difficulties of surgery under conditions of war. He was

Microbes and the Fetlar Man

approached by a Dr Percy Lewis, for example, to give a talk to patients and staff at the Folkestone Hospital, which he delivered with his characteristic sense of humour. For a shy man, he never lacked presence during his talks and lectures. The *Folkestone Herald* reported enthusiastically on the talk, saying that "some will have it that a Scotsman is devoid of a sense of real humour, but in Professor Cheyne we surely have an exception."[96] Cheyne would not have agreed with the rather stereotypical image of his countrymen, but he was apparently able to make light of his shyness, saying he "did not possess the qualifications for a popular lecture."[97] The talk gave him one of many opportunities to show lantern slides of photographs he had taken, detailing important events and difficulties he had experienced. It seems Cheyne had hardly been idle a moment in the short time he was in South Africa, and the *Folkestone Herald* report mentioned the lantern slides as a particular highlight:

> This great man – and he is great in his profession – was attached to Lord Roberts' staff, and during the few spare hours at his disposal he endeavoured to "snap-shot" many excellent and exciting scenes. They were duly made into lantern slides, and these were thrown onto the screen with the aid of Mr Arnold H. Ullyett's powerful lantern.[98]

When Cheyne gave the lecture in Shetland, it was reported in the press at the other end of the country, in the *Western Mercury* in Plymouth, which referred to the slides as in some cases "quite unique".[99] Cheyne was one of the first medical men to use lantern slides to illustrate his lectures in general.[100] These particular slides - some of which illustrate the previous chapter - appeared not long ago in his house in Fetlar.[101] They show horrific scenes of dirty rivers, horses struggling to cross temporary bridges and haul wagons up steep, muddy inclines, and collections of the dead and wounded, probably in Cronje's laager. He also illustrated cases where the Boers abused the Red Cross flag, showing the "effects of the shells poured on"[102] a hospital where he was operating. He had probably been the first to raise the matter with the authorities, and it was established that in calm weather, the flag tended to be limp and could not be distinguished from any other. It led to the invention of a new metal apparatus by a Dr Charles Muggs, which could theoretically make the cross more distinguishable.[103]

His lecture in Folkestone had apparently created "roars of laughter"[104] when he "described the various methods resorted to by the soldier in mending his often ragged clothes. If a piece was torn off there was often no cloth ready to make a patch."[105] He recalled how one

> ... Tommy on the march was going about in a deplorably tattered state, and in order to make himself respectable he conceived the idea of utilizing a square out of a biscuit tin. He made in this four holes, and with the aid of string fastened it on his trousers. When, added the professor, that man was marching up the hills, he resembled a heliograph on two legs.[106]

At home, however, tragedy had struck once again, and this time fate must have seemed particularly cruel. Cheyne's daughter, Julia, died on 22nd February 1901,[107] and had lived exactly three days fewer than his first daughter, Mary Frances, just under ten months. Once again, his skill as one of the country's leading surgeons had been of little help in saving his own children. Cheyne's own health must also have been beginning to suffer with the rigours of the campaign in South Africa on top of his ambitious work timetable, publications and public appointments. He was not alone. Sir William MacCormac's death at the end of 1901, at the age of 66, was widely attributed to a weakening of his constitution during the war,[108] and a close colleague and friend of Cheyne's, A.W. Hughes, Professor of Anatomy at King's, died of the typhoid he had contracted just before his return. He was only 39.[109] Cheyne had been remarkably lucky, considering his ill-health early in life and his exposure to typhoid and exhaustion in South Africa, but the experience probably marked the beginning of a gradual decline in his health which would force him to take more care of himself in the new century.

12

Picnics and concert parties
Building Leagarth

Even after Cheyne tried to slow down, his reputation continued to grow, which led to inevitable public and professional responsibilities. Whatever the "old trouble"[1] was that his wife Mary had mentioned to Lister as early as 1887, it clearly never quite went away, and his war experience would have been taxing even for a younger man who had not had a history in his youth of "incipient tuberculosis"[2]. Furthermore, now that Lister had largely withdrawn from public debate, Cheyne had assumed the responsibility of standard bearer for chemical antiseptics in the new world of dry aseptic techniques.

Even in the 1890s, Cheyne's thoughts had been turning more seriously to a concept of 'home'. All the time he had been with Lister in Edinburgh and London, Fetlar had remained a significant part of his identity. His students and colleagues all knew he still identified himself with the island, and some of them may already even have been there with him. He had been returning to Fetlar as often as he could during his years in London, and had been increasingly associating himself with whatever he could that reminded him of it. In 1892, he had become a founder member of the London Orkney and Shetland Association, which was broadened a couple of years later to include 'northern' culture in general. It was renamed the Orkney, Shetland and Northern Society (sometimes 'The Viking Club', or the 'Viking Society for Northern Research'), and was reconstituted as a "Social and Literary Society for all interested in the North".[3] Mainland Scotland was represented in its membership, as well as Orkney and Shetland, and it included the Duchess of Sutherland and the Marquis of Bute.[4] Cheyne was its President, and at least one of the meetings was held at King's College, his place of work.[5]

The society held discussions and lectures, particularly on themes related to Norse traditions and culture. Today, these are a strong part of the popular Shetland identity, and relate back to the fact that the islands were settled by emigrants from Norway around the 9th century. The settlers brought with them their language, which developed in Shetland and Orkney into a form of its own, known as Norn, and which resembled modern day Icelandic. In 1468, however, Orkney came under Scottish rule, handed to Scotland as part of a dowry from the Danish King when his daughter married James III of Scotland. Shetland followed suit in 1469. The islands were subject to a variety of cultural influences thereafter, both in terms of language and custom, and Scots English became the standard language. The Norse element nevertheless remained a popular part of the identity. The Shetland diaspora had been around for a long time, because so many people left the islands to find work, and societies of this type provided a basis for 'exiles' from the islands to meet and celebrate their culture.

There were developments in Shetland literature around the 1890s, when the Viking Club was established. A number of works, many of them in dialect such as those of Haldane Burgess, contributed towards centralising a Norse element in Shetland culture, not only for Shetlanders themselves, but in influencing the way they were viewed by others. Cheyne's childhood acquaintance and neighbour in Unst, Jessie Saxby, also an exile from the islands, published a number of works on folklore and custom, focusing particularly on the North Isles of Unst, Yell and Fetlar, where Cheyne had grown up. She would later contribute lectures to the London Viking Club.

The value of local languages, dialect and custom in national and regional identity was not a new one. All over Europe, since the end of the 18th century, stories and tales had been gathered, revised, renewed and sometimes even invented, and were often used to contribute to a collective identity. The poems collected by Elias Lönnrött in Finland, published as *Kalevala* in 1835-6, were instrumental in forming Finnish identity after centuries of rule by Sweden and then Russia. In Scotland, Campbell's *Popular Tales of the Scottish Highlands* had appeared in 1860-62, translated from the Gaelic. In Shetland, the revival of the Norse element in Northern Isles culture, including the establishment of the Viking fire festival *Up Helly Aa* in its current form from the 1880s, was at its height in the 1890s and early 1900s, just as Cheyne was beginning to turn his thoughts to home.

The inaugural lecture at the newly-constituted society in 1894, by Mr F. York Powell, an Oxford Don, was entitled, "Some Literary and Historical Aspects of Old Northern Literature."[6] "In order to give character to the Club"[7], the members and office-bearers were given titles borrowed from Icelandic. Cheyne, as President, was the Jarl (pronounced Yarl) or leader. The council of the Club was called the Law-Thing (from the Old Norse and Icelandic 'ting', or council), and before Mr Powell began his talk, he instructed everyone present that ladies in the audience were to be referred to as Skiald Majar, or Shield Maidens. This may sound very Wagnerian today, but for Cheyne, it was a very tangible way of consolidating his Shetland identity.

The consequences were twofold. His increasing involvement in Shetland cultural matters may have fuelled his desire to retain his links with home in a more solid way, and it influenced very strongly how he projected his identity in London. The eventual use of the words 'Viking blood' in his obituaries is almost certainly an image he cultivated himself, glossing over the fact that his Cheyne ancestors were Norman and his grandfather Watson had come from Scotland. This may sound a moot point to outsiders, but not in Fetlar, where some families could trace their ancestry in an unbroken chain to Norse landowners, and the community preserved a custom where incomers were buried on a different side of the graveyard to local people. Even today, 'incomers' buried on this side of the cemetery include people whose family arrived in Fetlar from another Shetland island in 1863 and have remained there ever since. Cheyne's final resting place, needless to say, was on this side, which is now becoming a little overcrowded as fewer and fewer people meet the requirements for the other area.

His ongoing involvement with Shetland culture probably had a bearing on Cheyne's decision to build himself a house of his own in Fetlar, and ultimately to retire there. He had regularly spent summer holidays on the island since he left, and until the early 1880s, he is likely to have stayed with his aunt and uncle Christian and David Webster who were still resident at the Manse. On Webster's death on 31st May 1881[8], however, Christian and her sister Grace would have had to vacate the Manse where they had spent their entire adult lives. A new minister, Mr Campbell, took up residence there, but he does not appear to have been the first choice of the parishioners, in Fetlar at least. A petition was raised on the island, possibly at the instigation of Christian and Grace, to ask for David Webster's son as replacement minister for the parish.[9] This would almost certainly have left him with an obligation to care for Christian and Grace at the Manse. Christian, was, after all, his step-mother, and had brought him up to all intents and purposes. She and Grace were clearly fond of the Manse, and had known no life outside the Church of Scotland, so this may well have been the driving force behind the petition. John McKessor Webster, who had grown up alongside Cheyne, was by this time Minister of Row (now Rhu) in Dumbartonshire.[10]

The petition might well have succeeded had he not refused. Whether it was because he was settled and happy in Row (he had married there in 1882) or whether he entertained no special love for Fetlar, he was unwilling to move back. Nor might he have relished the thought of the responsibility for his two elderly charges, and they seem to have fallen to Cheyne. By the terms of his grandfather's will, both women had their own means in terms of a third share of the estate, but Cheyne's and John Webster's studies had swallowed a good deal of the family funds, and they were now living on whatever was left. The upshot was that, when the new minister moved in, the two old ladies were forced to move not only from the Manse, but from Fetlar itself.

Shortly after Webster's death, they may have lived for a while in Unst, the neighbouring island, and if so, Cheyne would have visited them there. He also knew members of the Edmondston family in Unst, a landowning family with more than a few intellectuals of its own, and with whom he had spent time as a young man. Laurence Edmondston had been a medical practitioner, and also a naturalist, with a particular interest in ornithology. He had corresponded with many of the country's leading naturalists, and often played host to visiting expeditions at a time of intense activity in the field in the mid-late 19th century. His son Thomas became a botanist and had produced the first systematic catalogue of Shetland plants at the age of 16 before his untimely death on a Pacific expedition.[11] Cheyne had known the family well as a child, and had visited their house, as Edmondston's daughter Jessie Saxby recalled. As I noted in a previous chapter, he had gone there in particular for tuition to help him pass the university entrance exams, and they had known him as "Willie Watson",[12] recalling his prolific memory for facts. He had no lack of places to visit in the North Isles of Shetland, whether in Fetlar or otherwise, but no longer had a family home of his own.

Within two years of moving to Unst, however, Christian was also dead, leaving Grace on her own by 1883.[13] By the time of the 1891 census, and very probably much earlier, Grace was back in Fetlar,[14] living at the former Methodist Chapel, which the authorities had sold into private hands. It was officially known as Hillside Cottage, and this may even have been the name Grace gave it in order to expunge the memory of her father's rivals, but to the Fetlar community it was known simply as *The Chapel*.[15] Her father would have considered it a supreme irony, given his trials with the Methodists after they set up in competition so close to his Manse,[16] and might even have smiled to see them falling from favour with the Fetlar community and selling up. Grace also had reason to be grateful for the fact. At the Chapel, she was close to the Manse where she had lived so many years, and the graveyard where her family was buried. The whole family had an abiding passion – even obsession - with the Manse. For Cheyne, in all his years in Edinburgh and London, it was most probably the focus for his vision of home.

He certainly visited Grace at The Chapel when he made his annual pilgrimage to the island. He is pictured outside the house in one photograph with two of his sons, probably Lister and Hunter, given their apparent ages, sitting at his ever-present camera which is perched on a tripod, pointing towards the sea.[17] A lady in the background is likely to be Margaret. Grace was an old lady by now,[18] and she died in 1906. A woman called Mary Brown may have attended to her needs, probably from the time Grace moved back to Fetlar from Unst. Mary Brown was 40 when the census was taken in April 1891, and had previously been with Grace at the Manse, as a member of the staff there.[19]

Sir Watson with his camera outside The Chapel, where his aunt Grace lived. The woman in the background is probably Margaret, his second wife, and the two boys are probably Lister and Hunter Cheyne, his eldest sons

Cheyne had often been called on for medical help on his visits in the 1890s, though there were, by now, doctors in the two other North Isles of Unst and Yell. While he was visiting in September 1899, he was called out to attend to a shooting accident. John Smith, originally from Yell, but now a merchant in Edinburgh, was visiting his summer residence when he went rabbit shooting, and paused for a moment to rest the gun on his foot. The powerful Shetland wind had wrapped his coat around him and caused him to shoot himself in the foot, which Cheyne had to amputate.[20]

An incident recalled by Fetlar residents indicates that Cheyne was also taking colleagues, friends and perhaps even students with him to Fetlar on some of these summer visits. In 1894, a Mrs Catherine Coutts was heavily pregnant and digging a field over with a spade. One of Cheyne's guests is reputed to have stopped and asked her how she could be working so hard in her condition. She replied that she had already had six children and worked through the pregnancy of each one. They talked a while, and for some reason, before he left she asked for his name, saying she would name the child after him. He told her his name was Bevis, and the child was given Bevis as a middle name (as were subsequent generations of male children in the family, often to their horror). It is possible the name was actually spelled Beavis.[21]

In short, Cheyne's experiences during the Boer War, and the relentless pace he had set himself with his work, had ultimately turned his thoughts to a more permanent solution of 'home', and he too set his sights on somewhere as close to the Manse as he could find. The land around the Manse was owned by the Nicolson family, the main lairds on the island, who lived in an eccentric, and not always functional, neo-gothic mansion further to the west, called Brough Lodge. It is an extraordinary building, thankfully under renovation as I write, but one which grew organically with its different residents, and unfortunately suffered in the popular mind from the reputation of the man who built it. Sir Arthur Nicolson, from a well-to-do merchant family in Lerwick, had acquired the land in Fetlar in payment of a debt. He had travelled extensively in Europe and taken back with him ideas – and perhaps antiquities – for the house he wished to build himself in Fetlar.[22] It is almost as though he built it more as an experiment than a functional building. It had a chapel of red brick, not common building material in Shetland, and on the site of an Iron Age broch (a round tower) in the grounds he constructed a folly, a tower which looks half built or half ruined, depending on how you look at it.

Nicolson had also been the perpetrator of the Clearances in Fetlar in the early part of the 19th century, evicting people from their homes as part of his agricultural experimentation to clear land for sheep grazing. In Fetlar, the Clearances had not been much of a success economically, but the estate remained a central influence on people's lives on the island. Sir Arthur was rarely there himself – he installed a factor and lived elsewhere,

Brough Lodge, Fetlar, the home of the Nicolson family.

The tower at Brough Lodge, Fetlar, a mid-19th century folly.

corresponding with people like Cheyne's guardian David Webster on island matters.[23] His experiments had left a large proportion of the island depopulated, with most households concentrated into a relatively small area. Fetlar's economy never fully recovered from the misfortunes of the 19th century, compounded by agricultural disasters such as horse disease (sarcoptic mange) at the end of the century, which temporarily decimated the stock of ponies, the island's principle working animal. Nicolson's widow and her factor were considered responsible for the misery resulting from this too.[24] Throughout the 20th century, Fetlar was to suffer from the lack of a suitable pier, which precluded any effective involvement in the fishing industry, and since the 1980s, it has supported fewer than a hundred inhabitants.[25] In the collective memory of islanders, Sir Arthur Nicolson's clearances and his family's management practices on the estate represented the start of a gradual and permanent decline in the island's fortunes.

On his death, Nicolson left the house and the income from the land to his wife for her lifetime. He had resurrected a dormant baronetcy, a process which was probably a product of the same vivid imagination he applied to building his house. It nevertheless had important consequences when his wife died, as the land and house passed to the next in line for the title (Sir Arthur had no living children). Thankfully for the island, the new incumbent behaved rather more responsibly, but would unfortunately never quite shake off the poor reputation his ancestor had given the estate. This was the man who was in residence at Brough Lodge at the turn of the 20th century when Cheyne was thinking of building his house there. Sir Arthur T.B.R. Nicolson had spent most of his life before Fetlar in Australia, as a leading light in the state of Victoria, and had taken up residence on the island in the early 1890s, though he had inherited the baronetcy several years before he inherited the estate. He set about clearing up the mess left by the previous incumbents. The house was almost derelict and had to be restored before his wife and children could join him, and he spent a good deal of time putting the estate back on its feet.

They were an interesting family in their own right. Sir Arthur brought up four sons and a daughter at the restored Brough Lodge. One son, Eardley, was born with an intellectual disability of some form,[26] but was brought up alongside his brothers and sisters rather than being sent to an institution or hidden from view like George V's youngest son, Prince John (who had epilepsy). The life of the Nicolson family at Brough is documented in an extraordinary diary written daily by his wife, Annie, herself from a colourful colonial family associated for a time with the composer of *Waltzing Matilda*.[27] Her diary chronicled daily activities at the house, the events of the island community and the coming of the modern world. She was devoted to her disabled son Eardley, and was devastated by his death at the age of 24. She allegedly dabbled quietly in Christian Science, gave her sons everything they asked for, and was in some quarters considered a saint. Her family, however, lived under the shadow of their predecessor's misdeeds. By all accounts, her husband took the rehabilitation of the estate some way towards success, and is even said to have instructed his sons not to emulate the capricious behaviour of the original Sir Arthur. It was an uphill struggle, and reputations thus earned take more than a few generations to shake off.

Lady Annie Nicolson, whose diaries have given us so much information about life at Brough Lodge, Fetlar, in the early 1900s, and about visitors to Leagarth.

Sir Arthur T.B.R. Nicolson of Brough Lodge, who renovated the house and put the estate back on its feet when he inherited.

Moreover, on one issue, Sir Arthur and his heirs remained steadfast. They believed implicitly that the relationship between laird and tenants, where the laird was responsible for managing the livelihood of those living on the estate, was unequivocally in the best interests of all concerned. The family held that any change to the system would so alter the social fabric as to bring about its complete collapse. Sir Arthur T.B.R. Nicolson seems to have taken his own responsibilities within this arrangement seriously. He even, as noted earlier, pulled teeth and lanced abscesses where a dentist or the doctor was unavailable to come in from the neighbouring islands.[28]

Flit boats coming into Brough, Fetlar, showing Brough Lodge in the background.

The Nicolsons organised local events at Brough Lodge to coincide with national occasions such as the Coronation of Edward VII (which they celebrated too early, not having heard the news of the King's operation for appendicitis).[29] In short, because they owned and managed the bulk of the land on the island, they thought of themselves as an administrative centre, unofficial health and welfare establishment, and a source of formal entertainment. The old family chapel, probably originally built as a sort of folly on the whim of their ancestor, was never used as a religious building in their own time, and had become a storeroom for island goods arriving by boat at the pier below Brough Lodge. In bad weather, people waited in their kitchen for the steamer to arrive. Its arrival was never something that could be predicted with much accuracy. It depended on the weather, and in bad conditions it would lie off at another spot around the coast, wherever it could. Word would filter through and everyone would rush over there.[30] The Nicolsons owned the flit boat, or large six-oared boat which ferried people and goods out to the steamer, as the waters around the shore were too shallow for it to come closer, and there was no pier.

The flit boat transporting people and goods to (or from) the Earl of Zetland. *Sir Arthur T.B.R Nicolson is standing in the boat (left).*

When Cheyne asked to buy some land from the Nicolsons to build his house close to the Manse, they declined his offer. Cheyne's grandfather, the Reverend William Watson, had made no secret of his dislike for their ancestor,[31] but this had little to do with the decision. Sir Arthur T.B.R. Nicolson had no personal memory of this relationship. He did, however, have reason to wonder whether Cheyne might pose a threat. First of all, he was a potential professional rival, a successful son of the island who had been away, learned the ways of the modern world, and returned to propagate them in Nicolson territory. Secondly, he was an intellectual, and a potential challenger to their position of knowledgeable authority on legal and economic matters.

Moreover, Cheyne was by now a wealthy man, one of a new generation of self-made professionals. The Nicolsons were landowners in a much more traditional sense, and though they were the beneficiaries of a Trust fund arranged by Lady Nicolson's father,[32] most of their wealth was tied up in land they would find themselves gradually selling off in order to maintain their lifestyle. In fact, the eldest son, Arthur, succeeded in marrying

Microbes and the Fetlar Man

Golf at Brough Lodge.

a wealthy New Jersey heiress who was in search of a title. The young lady, Dolores Cubbon, got cold feet on reaching London and returned home without ever seeing Fetlar. The marriage was annulled.[33] Finally, but perhaps most importantly, the Nicolsons were serious sportsmen. At various times they installed at Brough tennis courts and a 9-hole golf course, and closely guarded all angling rights on the freshwater loch close to the Manse. As a member of the family once pointed out [34], their game also supplied their kitchen, and the almost daily trips to fish or hunt were not entirely for leisure. They were nevertheless prolific entertainers, played host to regular shooting parties, and stocked and fished the very loch which happened to be in full view of the place Cheyne wanted to build his house. It was clearly out of the question.

They were nevertheless dealing with a man who did not give up easily, and Cheyne stumbled on a remarkable piece of luck. About the same time, the land owned by the Earl of Zetland in Fetlar, a much smaller and less significant estate than the Nicolson one, came up for sale, and Cheyne snapped it up, finalising the purchase on April 1st 1901.[35] It was not his first choice of location, but he would make the best of it, and he selected a position for his house with exceptional views of the sea, albeit at a different angle from the Manse. For the Nicolsons, the matter had backfired in a much more important respect. Cheyne had intended to buy just enough land near the Manse to build his house and grounds.[36] He had never intended to become a landowner with an estate to manage. Suddenly, he found himself just that, and the Nicolsons found themselves with a rival who clearly intended to spend a good deal of time there.

The estate was small compared to the Nicolsons' Fetlar lands, and the Nicolsons also had extensive property elsewhere in Shetland and the Scottish mainland. Perhaps the size of Cheyne's estate made it more manageable and gave him the opportunity to run things slightly differently. The Earl of Zetland had owned the lands on an economic basis only, as an asset, and was rarely, if ever, there. He took no direct part in the community and the estate was managed by a factor. People did not know him personally and had probably never seen him. It was not uncommon for owners of lands to live elsewhere, and the original Nicolson, the builder of Brough Lodge, was very often in Edinburgh, leaving his affairs on the island in the hands of a factor. The Nicolsons at the turn of the 1900s, however, were very much resident, and many people on the island interacted with them on a daily basis. Cheyne was not only a regular visitor to the island, he made no secret of his intention to build a retirement home there. In the management of his house and land, he was to differ from the Earls of Zetland in that he interacted personally with local people, to whom he was already a familiar face. His residence there meant considerable local employment, and he was to differ from the Nicolsons in that he employed Fetlar people. The Nicolsons had always brought their staff in from other islands.

Cheyne employed as general factors for his estate Hay and Company, prominent merchants in Lerwick, Shetland's capital.[37] A representative was appointed on the island. George Garster [38] lived at Beala, not far from where Cheyne would be building his house. As was probably inevitable with newly-purchased estates, the affairs were not altogether in order, and Hay and Co. set about trying to collect back rents. They noted in a letter to Garster,

> I suppose you understand that Mr Watson Cheyne took over the whole of the arrears when purchasing the Fetlar lands from Lord Zetland and is therefore entitled to the payment of these arrears in full along with the current rent.[39]

The man at Hay and Co. was of the opinion that

[t]here are a few tenants in Fetlar who seem to think it is the correct thing always to keep the rent of at least one year in arrears, and these tenants are in arrears more than a year.[40]

To be fair, if rents were not systematically chased up, the tenants could hardly be blamed for holding off. On the other hand, lack of money for rent no doubt also played a part. It is important to remember that, after the main period of the haaf fishing in the 19th century, when fish had been cured and sold on by merchant enterprises at stations on the island, there had been little or no fishing industry there. This was in contrast to other areas in Shetland, and may sound surprising, given that the island was surrounded by some of the best fishing waters in the area. It was entirely owing to the lack of a suitable harbour or pier for larger vessels, such as steam drifters, and meant that Fetlar's economy remained largely agricultural.

The lack of a suitable pier or natural harbour even affected the way goods were brought into the island. People, livestock and staple goods were shipped in by steamer to a short distance from the island, and then unloaded onto a "flit boat", a large rowing boat with four or six oars, and brought in to shore. Larger cargo such as livestock was hoisted on and off the steamer by a sort of crane. The issue was never resolved in time to build an effective industry, and the lack of a suitable pier for the island remained the main point of discussion throughout the 20th century. In other words, it made the island's economy dangerously dependent on agriculture and seasonal employment on the estates. In time, a few public sector jobs provided by the local authority became an important part of the economy.

Garster managed to collect quite a number of the unpaid rents, £90/2/4d.[41] The full quota of annual income from his rents was around a £131,[42] but it is not clear how much of these were actually in arrears. Hay noted, "I have no doubt that you have done your very best to collect all you could, and I am still hopeful of you getting some more money from these in arrears before the end of the year."[43] They suggested that any tenants reporting difficulty in paying the remainder be allowed at least "to reduce the arrears"[44] until they were paid off.

In the meantime, Cheyne busied himself with the task of building Leagarth House. He contracted a Shetland architect and engineer, John Morgan Aitken, who had been involved as a subcontractor to the main architect for Lerwick Town Hall in 1881.[45] The design for Leagarth was architecturally pleasing, in a delicate, almost colonial style, with wooden panelling and facades, fortified against the Shetland weather by an inner structure of run-concrete, not an easy proposition in itself.[46] The small-grade stone for the concrete had to be brought across by boat from the peninsula known as Lambhoga, across the bay. Before it could be shipped, however, Cheyne had to arrange for a small jetty to be built at Houbie close to Leagarth. Once the jetty had been built, the stone for the house could be unloaded, and was used to mix the concrete for the walls.

Leagarth House in its original state, prior to 1920. The hall is on the back of the house, the low roof to the left of the photograph.

Leagarth House in its original state had a balcony and open porch facing seaward. Cheyne brought an elegant driveway around to the seaward side of the house, and had a small, private landing stage (different to the one for unloading the stone) built at the shore at the edge of the grounds, with steps leading down the steep face. As well as its pleasing appearance, the house was designed from the first to be functional. Unlike Brough Lodge, which the Nicolsons had needed to patch up and adapt to their purposes, and which always famously suffered from interminable damp and leaks, Leagarth began life with a modern kitchen with a large range, and an anteroom for washing and drying, with modern boilers. It had an airy, sunny drawing room on the first floor, overlooking the sea, and a wooden balcony for the less stormy days. Underneath was a wide porch for summer evenings. An integrated, purpose-built coach-house sat on the edge of the complex, in towards the land. An important feature of the house was a large recreation hall, with markings used for games. It had a beautiful gable window looking out across the garden to the sea. It was here that the many dances and concert parties were held to entertain the Fetlar community and guests from 'south'.

Cheyne furnished the house in 1901 with items bought at Maple & Co. in Tottenham Court Road in London [47], furnishers to the Victorian and Edwardian élite. Its transit to Fetlar in 1901, presumably just after the house was completed, was far from straightforward. It was initially shipped to Lerwick, where Hay and Co. arranged for it to be transported to Fetlar by the Zetland Steam Company. It would cost £12, for which they apologised in case he considered it excessive, adding that there were distinct advantages to sending it by steam instead of in a smack. As it turned out, there was not enough cargo space on the intended shipment, and the items would have to be shipped in two lots, so arrangements were made for it all to be shipped on the small local steamer, *Sunflower*. They also arranged for John Hughson, the merchant at Houbie in Fetlar, to land and house the furniture when it arrived.[48]

When the steamer arrived however, it would have been unable to land Cheyne's furniture directly onto the shore. There was now Cheyne's small pier at Houbie, but none of the steamers had ever been able to come right into land, either at Brough or at Houbie. They were dependent on the flit boat which was rowed out to the steamer anchored in the bay. It collected the goods, and deposited them onshore until everything was unloaded. It is likely that Cheyne's furniture was strapped to the boat, and the bigger items may have had to be strapped to two boards, laid cross-wise over the vessel. A motor car belonging to the Nicolsons was unloaded this way in the early 1930s, as photographs show,[49] and it was apparently more stable than it looked. The new activity at Houbie contributed to the largely good-natured rivalry between the two landed families. With more than a slight hint of irritation, Lady Nicolson commented in her diary on July 12th 1907 when the steamer went around to Houbie on an unscheduled stop,

> "It was a Brough day, but the steamer went to Hubie [spelled Houbie today; this is an older spelling too, but around 1900 it was often spelled "Hubie"] with Mrs Cheyne, Miss Clapperton, the children and servants.[50]

Car being brought ashore by flit boat. This was in the 1930s, but may also be how some of Cheyne's furniture was brought in.

Building Leagarth had been an organisational achievement on everyone's part. At this stage, however, Cheyne was still resident in Harley Street, and generally only spent his summers at Leagarth. He gradually built up the house and its functions, as he had every intention of retiring there. He reserved many of the adjustments, extensions and innovations for his move in 1920, perhaps to oversee them himself as a retirement project, and to accommodate the increasing need for more space. In the early 1900s, he contented himself with his summer visits, keeping certain members of staff on a retainer to oversee the place in the winter and prepare the house for his annual arrival.[51]

He always invited a party of colleagues, friends and students for the summer, most of whom signed a visitor's book, which contains an impressive array of names. Some of them even gained a degree of immortality in island anecdotes. On one occasion, one of the party was sent over to assist at a difficult birth (Cheyne himself had probably been asked to go but could not attend). The baby was delivered safely, and the family insisted on showing their gratitude. They asked the man's name and promised to bear him in mind when they named the child. If he ever knew the consequences, he must have been mildly amused. The family was apparently true to its word and called the child Algy. The problem was that the baby was a girl, and went through life with the name.[52]

Microbes and the Fetlar Man

In the absence of a doctor based on the island (Dr Taylor came in when required – weather permitting - from the neighbouring island of Yell), Cheyne regularly treated, and even operated on local people. His operating table remained for many years in the kitchen at Leagarth, though it is unclear whether this was its original location. Somewhere in the house was a Queen Anne style chair which had been made from some of the wood of one of Lister's operating tables. It was one of the items Cheyne loaned to the Lister Centenary exhibition at the Wellcome in London in 1927.[53]

Vera Nicolson of Brough Lodge.
She later married Lord Herschell, Lord-in Waiting to King George V

The Nicolsons probably did not consider Cheyne their full social equal, yet in a changing world where achievement was bringing substantial recognition to members of the medical profession who had come up through the ranks, this is precisely what he was. If the Nicolsons indeed harboured fears of changing practices, these were most realised when it came to employment. Cheyne did not differ significantly from them in terms of how he viewed the economic relationship between household staff and their employer. Both he and the Nicolsons saw it as something of a paternal relationship with responsibilities on both sides. Both families were also fond of their staff, and kept in touch with them when they left their service to marry. Lord Herschell, son of Vera Nicolson, the daughter of the family in residence when Cheyne was building his house, once told me that, when they visited the island in later years, the very first thing his mother did was visit Agnes Thomason, who had been her late brother Eardley's nursemaid at Brough. In one important aspect, however, Cheyne and the Nicolsons differed significantly. Cheyne's household staff consisted almost entirely of Fetlar people, and this was quite different to the Nicolson tradition of bringing in staff from Yell and other islands. It is not clear why they did not employ Fetlar staff. They may have thought staff from elsewhere less likely to have conflicting interests with the island community, though a number of them married into Fetlar families.[54]

The difference in Cheyne's style of management was sufficiently noteworthy to be the subject of comment. Many years later, in 1921, when Cheyne moved to Fetlar permanently and took on a full household staff, Laurence Williamson noted in a letter to his cousin that "[a]t Fetlar Cheyne is very active & friendly, employing Fetlar men. He is South [i.e. in England, Scotland or anywhere south of Shetland and Orkney] just now, & is settling down in Fetlar, building a big new house."[55] The house was probably Feal, built for the manager of the home farm. Williamson was a resident of Yell, the neighbouring island from which the Nicolsons tended to draw their household staff. He was nevertheless of a Fetlar family, and was fond of the island, so his remark was clearly intended in a positive way. The fact that he mentioned it at all means it was considered noteworthy.

Whatever lurked below the surface with the Nicolsons, relations nevertheless remained amicable, and were even seen by the islanders (from the outside, looking in) as a sort of good-natured rivalry, and source of considerable amusement. The Sunday *chariot race* to church quickly found its way into island lore. The Nicolsons and Cheynes had to approach the church from opposite ends of the island, and when they saw each other coming, raced their carriages to be the first to arrive. On one occasion, the

Nicolson carriage apparently came flying around the corner, and the horse slipped, depositing Lady Nicolson in a pool of mud, at which point she was forced to return home and forego the Sunday service.[56]

The Kirk on Sunday

Picnics and concert parties

The Kirk on Sunday

After Leagarth was built, Cheyne arrived each summer with a lively group of guests and a well-planned programme of entertainment. No guest was ever bored. Local people were respectful, but privately rather bemused by the groups arriving from London, largely because they tended to be genuinely interested in island life. Islanders had enjoyed plenty of contact with outsiders, but largely on maritime or agricultural business, and Cheyne's visitors were quite different. They tended to ask a battery of questions about things the islanders took for granted, and were greeted with bemused replies. Their questions have occasionally gone down in island history. They were fascinated in particular by the use of ponies and boats to collect peat from the peninsula called Lambhoga. In other parts of Shetland, peat, used for household fuel, tended to be cut closer to where people lived, but the workable Fetlar peat was found on a remote headland, which meant it had to be transported to other parts of the island by pony and boat. It was a well-established annual process of cutting, drying and transporting, and the transportation period took place in summer, just as Cheyne was there for his holidays with large parties of visitors from London. The islanders' bemusement occasionally turned to irritation when the questions lasted too long and they lost time at their work, or the brightly coloured clothes of the guests frightened the ponies.[57]

Microbes and the Fetlar Man

Flittin peats by boat from Lambhoga, Fetlar

These excursions to the hills were generally combined with picnics. The guests would be ferried around to remoter parts of the island, or to small, uninhabited islands in Bluemull Sound, in one of Cheyne's boats, complete with picnic and camera equipment, and these occasions were recorded for posterity in the copious photograph albums, visitors' books and logs Cheyne and his family kept at Leagarth. He had at least some of his boats built locally. One of them, which may have been the *Neesik* (the Shetland word for a porpoise) was built in 1922 by Thomas Walter Scott, who originally came from Scalloway in the Shetland Mainland. The boat had a 10-foot keel, indicating it was an "eela" boat, which people took out to the fishing relatively close to shore. In other words, Cheyne was not particularly interested in a fast sailing boat or competitions. He was interested in fishing and ferrying his guests around the islands. There is some indication that he had a boat built for him by a Fetlar man, Walter Shewan, not long after he had moved to Leagarth. Shewan's particular expertise was in sailing boats, and Sir Watson did not accept this one, possibly the *Smugga*, most likely because he preferred sturdier vessels specifically for fishing.[58] Walter Shewan, the son of the Nicolsons' factor at Still, was a talented man, and also something of an artist and sculptor. A whalebone carving he made of a seal survives.[59]

Picnics and concert parties

Flittin peats by pony in Lambhoga, Fetlar.

On one occasion, the guests took photographs of the events that ensued when a group of whales was beached in Basta Voe, on the neighbouring island of Yell. Whales had come close to shore, an important source of food if it could be got, and as the men of the district were away working, the women took out boats and drove around 60 whales ashore. Sir Watson and his 'house-party' went over there the next day in one of his boats to take photographs.[60]

Thankfully for us, guests and family recorded these summer stays at Leagarth in all manner of detail, so it is easy to give an example of how it all tended to unfold. In 1907, the "Leagarth party"[61], the men at least, collected "the Professor"[62] (Cheyne), his son Hunter, known familiarly as "Ha"[63], possibly his older brother Lister, and guests called "Booky"[64] (real name Brisket), Johnny and "the Welshman"[65]. They took the 8.15 train from King's Cross (later comments indicate this was the night train) and were seen off by "the one and only Arthur (surname unspecified) in a frock coat and deep dejection"[66], presumably because he was not going. Cheyne apparently arrived "in a state of hilarity"[67] in tweed suit and cap and, to everyone's surprise, "minus a beard"[68]. He had shaved off his long-standing beard in 1907, and it was such a noteworthy event for all who knew him that Sir Harold Stiles later commented that it coincided with his "conversion"[69] to 'aseptic' surgery, an interpretation Cheyne himself would almost certainly have refuted, as he did not consider himself by any means a convert. The author of the account of the journey north in 1907 restricted his comments to the fact that the great event revealed a "hitherto unsuspected double chin."[70]

One of Sir Watson's boats.

Following their arrival in Aberdeen, two hours late, they had breakfast, and joined other members of the party, many of whom are referred to by nicknames. Miss Darby, however, retained her real name, and was accompanied by her violin, the purpose of which would become clear later in the stay. The party now included "the Whuff encumbrance"[71], which we can only assume was Cheyne's dog. They made final purchases and headed for the "powerful and commodious"[72] *SS St. Giles*, temporarily thrown into confusion when reports circulated that she would not be departing until

1.30 pm. They consequently set off for an "amble"⁷³ around Aberdeen, and were jolted into action by reports it would leave at 11.30 after all, so everyone went rushing off to join the ship, arriving on time "more by good luck than good management."⁷⁴ They proceeded to accommodate themselves in their bunks, and discovered they were two bunks short. The consequence was that 'the Welshman' was later forced to spend an uncomfortable night "balancing on a 4ft (by 1ft 9 inches) sofa"⁷⁵, where he could hear Cheyne snoring away in his much more comfortable bunk. The vessel finally set off at 12 o'clock into choppy seas, which sent Miss Darby and the dog away "to seek temporary oblivion".⁷⁶ No amount of coaxing would persuade the seasick creature to indulge in his "epoch-making and world-wide known balancing act."⁷⁷

Sir Watson in one of his boats.

Cheyne, on the other hand, was quite at home on the sea, and he and the remaining party, hereafter referred to as "the dauntless three ... attacked a hearty meal."⁷⁸ Those not confined to sofas passed a reasonable night, and they docked in Lerwick, Shetland, at 7am. They were about to disembark when Cheyne threw the party into panic by announcing he had lost his tobacco pouch. It was discovered under his bunk, along with a whole box of cigars under his pillow. Cheyne clearly guarded his tobacco supply and his dog with his life.

Sir Watson, Sir Arthur T.B.R. Nicolson, Mr Campbell, the Church of Scotland Minister in Fetlar, standing by the flit boat.

They bought in further provisions, and then boarded the *Earl of Zetland*, the steamer which delivered passengers and goods to the North Isles of Shetland. They off-loaded goods on the island of Whalsay, and then at Burravoe in Yell before they finally saw the familiar shape of Lambhoga, the long peninsula on the west side of Fetlar. The steamer came in at Houbie, probably to the Nicolsons' irritation, as Sir Arthur's sister was on board and had to be fetched by road.[79] Cheyne's party was met by his motor boat, propelled by Mr Campbell, the Minister. However, when the boat was later noted to be in need of some repair, the following passage indicates the extent to which the party was composed of medical men:

> ... Johnny covered himself with glory (and incidentally with lubricating oil and petrol) and performed a brilliant operation by extracting the carburettor (without an anaesthetic). Some amoebic foreign bodies were extracted therefrom that the fons et origo mali were thereby discovered.[80]

Cheyne, at this point finally exhausted, apparently gave a short choral rendition and retired to bed. Throughout the stay, his snoring is mentioned more than once under the euphemism of "distant thunder."[81] His wife

Picnics and concert parties

Margaret, Meta, Basil, the servants and a Miss Clapperton had all arrived on 12th July, and had presumably made preparations for the arrival of Cheyne's party. They had already received a visit from Sir Arthur and Lady Nicolson, along with one of their guests, less than a week later, followed by a visit from their son, Arthur, the following day. Mrs Cheyne repaid the visit on 31st and went over the island to Brough in the afternoon, just before Cheyne's party set off from London. Relations were clearly amicable.[82]

A Leagarth party disembarking into the flit boat from the steamer Earl of Zetland, *to complete their journey to Fetlar.*

Microbes and the Fetlar Man

A Leagarth party, early 1900s.

The day after the arrival of Cheyne's party, apart from catching an early "mixed bag"[83] of fish with Basil, Meta and some of the party, his first activities on the island were to pay "a professional visit."[84] Unless this involved land-related matters, he was almost certainly responding to a call for medical help. It would not have been unusual. In the afternoon, they played tennis, and in the evening, set haddock lines and caught four score. Everyone who did not already have an alias was assigned one, in fine medical tradition, including the very appropriate *Microbe Militant*.[85]

And so it continued. On Sundays they visited the Kirk *en masse*, and on one occasion, a guest at Leagarth, Dr Hughes, played the organ.[86] The party occasionally took a boating excursion in Cheyne's yachts *Skua* and *Neesik*. Cheyne was famously an unrivalled handler of boats. They were sometimes accompanied on their excursions by members of the staff, particularly 'Jimmy Young' (James Young Robertson) and Jimmy Petrie. Apart from the usual provisions, an excursion in August was equipped with "the innovation of thermal flasks which were voted a great success"[87]. The concept of vacuum flasks had been developed much earlier, but they had not long been available commercially in 1907 at the time of this excursion. On another trip in the yachts to Moowick, "much interest was shown in Peats and Ponies"[88] as usual, Sylvia sketched, and Johnny shot a four and a half foot long seal. One of the Hughson men "removed the pelt"[89] of the seal in the evening. The writer of the account commented (rather flippantly perhaps for today's environmentally-conscious audience) on the wealth of things Johnny was going to make with the seal-skin, including a sporran, an operating apron, a pair of slippers and binding for his medical and surgical note books, indicating that there would be more than enough to cover the number of lectures he had attended.

Meanwhile, Mrs Cheyne resumed her visits to Brough Lodge. She went with one of her guests, Mrs Broomhead, on 6th August, after the men had arrived, and on 10th, took the Nicolsons' only daughter Vera back to spend the day at Leagarth. On 15th, one of the Leagarth guests, a Miss St. George, fell off her pony at Brough [90] (an incident recorded with a colourful drawing in the notebook made by the Leagarth guests),[91] and was taken in for tea in the drawing room in the absence of Mr and Mrs Cheyne, who failed to arrive as planned in their motor boat. On 17th, Mr Cheyne went by Brough in his motor boat, causing Lady Nicolson to remark on the fact that he was towing two other boats behind him. He relieved her curiosity by turning up for tea.[92]

The culmination of the summer activities was a lavish concert party, to which the community was invited, and which had become an annual feature of Cheyne's visits. In 1907, it was detailed in the *Shetland Times*, which referred to it by now as an "established custom",[93] and noted that the audience agreed "surely this was the best concert we have had yet."[94] Cheyne sang. Meta danced, and her "youthful grace and beauty of motion elicited enthusiastic applause."[95] Miss Darby's violin finally came into its own, and she and Miss St. George gave the audience a fine recital. Mr Reid punctuated the proceedings with a series of amusing songs, and the Leagarth party presented a "very laughable little farce, 'Box and Cox,'"[96] to their Fetlar audience. Finally, the two resident Ministers, Mr Campbell of the Established Kirk and Mr McAffer of the Free Kirk, led the thanks to Mr and Mrs Watson Cheyne. Everyone apparently deeply regretted that Sir Arthur and Lady Nicolson were unable to attend, suggesting that they had done so other years. Her diary does not indicate anything which might have prevented them. Their daughter Vera went along however, with her friends, thus avoiding a full-scale embargo.

Three days later, on 19th, Cheyne motored over to Mid Yell to take Miss Clapperton to the steamer in his launch, avoiding Brough when the steamer called there. Lady Nicolson noted his *faux pas* in her diary. The following day when Mrs Cheyne invited Vera and her brothers over

Microbes and the Fetlar Man

Images from the guest book at Leagarth, illustrating the 1907 summer party. The man in the Admiral's hat is clearly intended to be Sir Watson.

for music and supper, "they did not care to go."[97] Lady Nicolson crossed out the subsequent passage in her diary, presumably regretting having elaborated on the matter. On 21st, all this was forgotten, and she drove over to see the Cheynes, only to find them out. Lister (Cheyne) and Mr Briscoe went over to Brough on the following day, however, to meet two more Leagarth visitors, Drs. Attenburgh and Hughes, from the steamer, and there were clearly no ill feelings. Nor were there any complaints when the oldest Nicolson boy, Arthur, took down his books from the folly in their grounds where he studied, and had to cart them all the way over to Houbie in the portmanteau, as the steamer was calling there to collect seven of the Leagarth party, leaving at the end of their stay. This time it was Mrs Cheyne's turn to give them lunch.

Sir Watson (left) in his motor boat.

On 6th September, the dentist had called "to stop Mrs Cheyne's teeth"[98], apparently without ill effects, so, unlike the postman, Mrs Cheyne does not appear to have entrusted her teeth to Sir Arthur. Unfortunately, a few days before, Cheyne's motor boat had met with an accident on one of their

Microbes and the Fetlar Man

picnics. On 28th August, he had been towing the Nicolson boys in their own boats to the picnic spot, but when the weather turned, he had to abandon all boats on the wrong side of the island and return on foot, maybe an hour's journey. The ladies, having arrived by carriage, were thankfully able to return in comfort, but it was 3rd September before Cheyne's motor boat became more accessible when it drifted ashore after it had sunk two feet under water. Mr Campbell, the Minister, helped him remove the engine. He was apparently, amongst his many other talents, a gifted amateur mechanic.[99] On 6th, the steamer arrived to pick it up for repairs. It was returned by the steamer on 13th, when Dr Taylor came in from Yell to help Cheyne and Mr Campbell put the motor back in. The party finally left on September 21st, leaving only Lister and Hunter at Leagarth for a little while longer.[100]

Left: Meta Cheyne, right: Basil Cheyne (Sir Watson's two surviving children from his second marriage). Outside Leagarth House, early 1900s.

Cheyne's visits to Leagarth in the early 1900s were nothing if not eventful. His colleagues were given a taste of how different life in Shetland was, not to mention a better idea of where it was on the map of Britain. Between the two families and their guests, Fetlar was a lively place in the summer of 1907. Cheyne's ultimate retirement to the island in 1920 would involve major extensions and projects at Leagarth, and was recalled by islanders as a social heyday.

He continued to be called on in a medical capacity. In an earlier chapter, we saw how Lister's patient Margaret Mathewson visited him when he was home in the 1870s, to check on her shoulder, but by the time he had built Leagarth, there was a resident GP, Dr Taylor, in the neighbouring island of Yell, who also served Fetlar. However, he was not always able to come into Fetlar for reasons of weather, or perhaps he was needed elsewhere. Nor was he a surgeon. As a matter of course, it seems Cheyne refused payment for any medical work he did in Fetlar and other communities. The most often quoted anecdote involved an old lady in Fetlar who insisted on paying for her treatment. Cheyne replied that there was no need, and that, if she insisted, he would be delighted if she would knit him a pair of socks.

"No," she continued. She was determined not to be put off.

At this point, Cheyne quoted his Harley Street fees, around 50 guineas [101], many times more than the average family on the island could afford, and the old lady, mulling the matter over for a few seconds, replied, "And what colour would you like your socks, Sir Watson?"[102]

In the early 1900s, Cheyne received a letter from Laurence Williamson of Mid Yell, a man with close family connections in Fetlar, asking him if he would consent either to operate on a relative, or to advise whether she should attempt to go to an infirmary. Williamson was more than aware of what he was asking. Cheyne's national reputation had not gone unnoticed in Shetland, but people were also clearly aware that he was apt to look kindly on medical cases in Fetlar and the North Isles. Laurence Williamson wrote:

> I write this in my own name rather than direct from herself as it enables me to put the matter more fully & clearly. But it is done with great reluctance, as we have a deep inward desire to impose on no one, & to deal straightly with all. And viewing your high professional position it is asking very much. It is only considering your Fetlar leanings, the reality of human need & suffering, & that you are a man of understanding, that we write.[103]

He reminded Cheyne that he had already examined the patient while she had been visiting a relative, a Mrs Robertson, at Tresta in Fetlar in August 1900. Cheyne had discovered she had a tumour, and Williamson was now following up on the matter in the hope Cheyne would agree to operate. He offered a number of options for a consultation, either in Fetlar or in Yell:

> The alternatives as to seeing her are, if it fell in with your views & arrangements, we can fetch you from Hubie, or Brough, to Vatsie [pronounced Vatchie] near her home, on any Day fixed, & put you back there – weather permitting. Or we can bring her on any fine day to Hubie or Tresta. If an operation were made in her own house Dr Taylor would be near afterwards, if made in Fetlar arrangements would be made for her stay there afterwards. In either case her son would attend her. He is cool, determined, with a very clear head & would carry out directions with rigid precision, & has been long used with nursing.[104]

Cheyne would go on acting as unofficial consultant and surgeon in Shetland's remoter islands well into his retirement, and guests at Leagarth were also often sent to households who had asked for help. It is astounding to imagine how this tiny place, so often considered marginal in 21st century Shetland, let alone in Britain as a whole, had at its disposal throughout the 1920s the skills of one of history's most eminent surgeons, and would count two resident baronets among its population. Cheyne, still by all accounts a modest man who never forgot where he had come from, clearly saw nothing odd in it.

13

Getting older
1900-1914

When Cheyne was awarded the Companion of the Order of the Bath (C.B.) for his work during the Boer War, Lord Lister considered that it had not quite "done justice to the value"[1] of his work, though it had, at least, recognised it. On the one hand, this would be rectified by events in the first two decades of the 20th century, and on the other, the image of Cheyne as Lister's 'diehard' standard-bearer would become entrenched. This chapter will look at how Cheyne's own achievements were increasingly recognised and how this, ironically, coincided with his progressive relegation by many within the profession to the ranks of the grand old men of medicine. He was not beyond fighting back.

His diehard image was accentuated more than anything else by his continued crusade against the 'aseptic' surgeons throughout the period and beyond, and he rarely missed an opportunity to level criticism at procedures he considered lax in their attention to Listerian principles. He did not confine these comments to the major debates, but expounded them in his many talks to provincial medical associations and even to non-medical organisations. In addressing the opening meeting of the Torquay Medical Society in October 1903, he attributed the advances in modern surgery to "thoroughness",[2] by which he was implying that detailed Listerian procedures and attention to the natural healing capacities of the body were the most assured way of achieving asepsis, rather than what he considered the trappings of the new aseptic methods, or even chemical antiseptics *per se*.

On the one hand, he began to be seen as questioning innovation. Ironically, as someone who had spent his life as a pioneer, he was now increasingly portrayed as a kindly but rather docile old man with quaint, outdated views. On the other hand, his success and reputation as a cautious surgeon had helped build a lucrative private practice, and his reasoned approach to innovation generally won the approval of the medical establishment. His public roles also increased, and his lifelong campaign to promote a properly-funded system for pathological research led him into leading roles in new organisations. Whether as a result of these activities, or simply as a recognition of his overall achievements, he was showered with honours, and he always seemed to be mildly taken aback by it.

By the early 1900s, he counted some of the country's leaders among his patients, and his higher profile cases during this period include some of the foremost households in the country. At King's in December 1895, he had operated on Gladstone's house maid, Margaret Cheyney[3] for a tumour in the nose, and she seems to have made a full recovery. He also operated on Marcel Dessouter, an aviator who went on to become a pioneer in the manufacture of prosthetic limbs.[4] The cases which probably had the greatest public impact, however, involved two of the highest profile women in the country: Lady Curzon and Elizabeth Asquith.

Mary Victoria, Baroness Curzon of Kedlestone, was the much-loved wife of the Viceroy of India. Lord Curzon had been an MP when they met, and accepted the position of Viceroy of India in 1898, encompassing an important period of social events, including the coronation of Edward VII. The occasion gave Lady Curzon an opportunity to make use of her organisational talents and her taste for fine dress. The Delhi Durbar of 1903, held to mark the celebrations for the coronation, has been well publicised, and in some quarters criticised for its extravagance and cost. Lady Curzon commissioned a dress made of gold cloth decorated with peacock feathers, and was allegedly warned not to wear it, as the bird's feathers were considered bad luck.[5]

The explanation for her illness at the end of the summer of 1904 was no doubt a medical one, rather than the peacock feathers, but it is nevertheless an unfortunate anecdotal coincidence. She became seriously ill, and on returning to England for treatment, it was discovered she was suffering from peritonitis. It is unclear when Cheyne was first called to Walmer Castle to treat Lady Curzon, but her husband's correspondence indicates that by mid-September, she was suffering from an unusually high temperature.[6] Just prior to this, at the end of July 1904, Cheyne himself had been sufficiently ill to cancel appointments.[7] He was unable to deliver a paper in Oxford at the annual meeting of the British Medical Association.[8] Frederick Burghard, his co-author in the six-volume *Manual of Surgical Treatment* and a fellow confirmed Listerian, gave the paper on his behalf. By Saturday 24th September, Cheyne was at Walmer Castle, and Lady Curzon underwent an operation for peritonitis.[9]

Her condition was reported daily across the Empire, particularly when it took a turn for the worse on September 28th, and the *Sydney Morning Herald* printed starkly, "Lady Curzon worse. Her condition is exceedingly critical."[11]

On October 4th 1904, he was still in attendance, and her condition was by now dangerous. He missed the dinner for the opening of the winter session at King's, an occasion he usually attended with great enthusiasm. His presence at the Curzon residence was sufficiently significant for the speeches at the dinner to note that he, and two other King's men in attendance with him, were nevertheless bringing honour to their institution. The speaker was "proud to know that their staff and old students [were] taking such prominent part in the great struggle against disease which [was] going on in Walmer Castle."[10] It was announced as though the health of the nation were in peril.

Lord Curzon, who considered his wife's life to be for him "the most valuable on Earth"[12], was beside himself. Cheyne apparently spent long hours at her side, and did not leave until he considered her out of danger. He nevertheless had his own duties and practice to attend to, and no doubt took breaks in between to return to London. He was summoned again on 7th October, when Lady Curzon "was seized with rigors".[13]

Cheyne left Walmer Castle around 31st October,[14] apparently confident that Lady Curzon had turned a corner. After he left, however, she weakened, her temperature rose, and she was subject to "some wandering" until around 6th November when her temperature began to fall to normal levels "for the 2nd time only in 7 weeks."[15]

This time, Lord Curzon was confident his wife was on the road to recovery, and was prompted to write Cheyne an effusive letter of thanks: "I am hopeful therefore that the era of acute anxiety is coming to an end and that we may not have to call for you much in the future … Your coolness, composure and courage, your readiness of resource & clear decision, above all your absolute dedication to your patient have equally impressed both of us and will always leave in my mind sentiments of the warmest gratitude & regard."[16] He concluded by noting that he had mentioned Cheyne in a letter to the King. Curzon's policies did not always put him in the best political position with the government of the day, but he nevertheless remained Viceroy of India until his resignation in 1905, and his recommendation of Cheyne most probably influenced future events.

By this time, Lady Curzon's mother had arrived from America, and was staying at the South-Eastern Hotel (it is not clear why she was not staying at the Castle), being fed bulletins about her daughter's condition. Her sister was there too, but appears to have visited the Castle personally.[17] Sir Thomas Barlow, physician to the King, had also been in attendance during this time.[18] Her condition continued to fluctuate for some time, but she eventually recovered sufficiently for them to return to India. A year later, however, they returned permanently when Lady Curzon became ill again. She died in 1906, and her inconsolable husband had a chapel built to contain her body. He too, in time, was buried there, and the peacock feather dress is currently displayed at Kedleston Hall, the family residence.[19] Cheyne was presented with a photograph of Lady Curzon wearing the dress, and it later hung in his library at Leagarth.

Framed photograph of Lady Curzon in her peacock feather dress, presented to Cheyne by Lord Curzon.

In 1913, he was also to perform an emergency operation on Elizabeth Asquith, the daughter of the Prime Minister, Herbert Asquith. She had begun to feel ill while she was studying in Germany, so returned home, and was met at Victoria station by her mother and father. Dr Parkinson saw her, and called in Cheyne to perform an appendectomy at 10 Downing Street on November 30th.[20] She made a good recovery and went on to

Getting older

have a colourful life as an author, marrying a Rumanian diplomat, Prince Antoine Bibesco, and living in Rumania until her death in 1945.

Sir Lister Cheyne, Sir Watson's eldest son.

There is some suggestion that Cheyne knew Asquith personally, perhaps as a result of operating on his daughter. His descendants recall a tale in the family involving Cheyne's eldest son, Joseph Lister Cheyne, who was a Lieutenant in the 16th Lancers, based at the time in Ireland in March 1914, at the Curragh in County Kildare. Irish Home Rule was pending, but the Ulster Volunteers were not prepared to accept it. The British Government contemplated action against the Volunteers, and although the events which took place are complex, Ulstermen in the British army are alleged to have been given the option of unofficially disappearing while the action took place, in order not to have to fight fellow Ulstermen. It became known as the Curragh Incident. At some point in the proceedings, a message had to be taken from the Curragh to the Prime Minister, to ensure he did not receive conflicting messages, and someone had to be found whose word the Prime Minister would trust on a personal level. Lieutenant J.L. Cheyne's sons recall having heard that their father was asked to deliver the message to Cheyne, who was to take it to the Prime Minister, as they were personally acquainted and Asquith would trust him.[21] If this account is accurate, and did indeed take place during the Curragh incident, it is an indication that Cheyne's operation on Asquith's daughter only a few months before this had brought him into the confidence of the family.

After Cheyne returned from the Boer War, he began a number of long-term public commitments, many of them connected with his desire to see medical research well funded. He had continued to develop his work on treatment for cancer, specialising particularly in operations for cancer of the breast, and in 1902, he was asked to sit on a committee to work on an "investigation into the causes, treatment and prevention of cancer"[22], a joint initiative by the Royal College of Surgeons and the Royal College of Physicians. The King had apparently been taking "great interest in the prevention and treatment of cancer and [was] glad that the Royal Colleges … [were] taking up the question of cancer research."[23] He hoped for "important results"[24]. Shortly afterwards, a fund was established to pay for the research, with the Prince of Wales as President and an impressive array of Vice Presidents including Lord Lister and the Prime Minister, Balfour.[25] Cheyne served on the Executive Committee of the fund and became Honorary Treasurer in 1912.[26] It was to contribute to the cost of new laboratories at the Royal Colleges of Surgeons and Physicians, and despite all his commitments, Cheyne was finally in a position to play a direct role in helping to fund medical research.

His work on medical committees had not diminished much since before the Boer War, when he was never still. He was President of the Pathological Society of London when Queen Victoria died in 1901,[27] and was at the same time President of the Harveian Society of London.[28] He presided over most of the discussions of these societies and took his presidency seriously. In June 1901, he announced he would stand for re-election to the Council of the Royal College of Surgeons, and in July was elected with the second highest number of votes.[29] In July 1902 he served on a committee which was subsidising fares for people wishing to attend the Egyptian Medical Congress in Cairo in December.[30] In 1903 he was also involved in discussions on the decision to move King's College Hospital from Lincoln's Inn Fields to Denmark Hill, where it stands today, a process which was to be long and protracted. The foundation stone for the new buildings was not laid until 1909, but the hospital was moved in order to serve a district where people were more in need of treatment and did not have access to it, a rationale Cheyne supported wholeheartedly when he pointed out in 1904 "that the institution ought to be near its poor patients".[31] An initial sum of £30,000 was needing to be raised for the move, and a removal fund was set up, to which Cheyne put his signature in the appeal.[32]

Cheyne also became fairly heavily involved in editorial work about this time as a member of a number of committees. On October 1st 1906, he was elected President of the New Sydenham Society, with which he had remained involved after his more prolific period of translation in the 1880s and 90s [33], and by July 1913 he was a member of the editorial committee of a new periodical, the *British Journal of Surgery*.[34] In 1907, he had also become a key member of a committee which formed to produce a fitting tribute to Lord Lister on the occasion of his 80th birthday.[35] Lister, by this time, had long withdrawn from the public eye, and was reluctant to be drawn back into it, even for celebratory occasions. Despite the importance of his work, however, his papers had never been collected together, and were scattered throughout publications and journals. At a meeting on April 4th 1907 at the Royal College of Surgeons, an editorial committee was appointed to commemorate Lister's 80th birthday by "publishing in quarto form a complete collection of his scientific papers."[36] The committee consisted of Sir Hector Cameron, Professor Watson Cheyne, Mr R. J. Godlee, Dr Dawson Williams, and Dr C. J. Martin, Director of the Lister Institute. When he received a letter from the Committee asking his permission to publish the papers, Lister characteristically replied that he found the proposal "almost overwhelming in its kindness".[37]

However, this concealed the fact that the idea had been negotiated with Lister by those at the heart of the committee two months before it was announced at the RCS meeting. The suggestion had effectively taken Lister by surprise,[38] as he had felt weighed down by the need to publish the papers, yet knew he was too ill to attempt it himself. He had provisionally made arrangements for the papers to be published after his death by a nephew, Dr A.H. Lister of Aberdeen.[39] He had even set aside £1,000 in his will for the work involved, which he knew would be considerable, and had written guidelines for what was to be included.

In a frank letter to Sir Hector Cameron on February 19th 1907, Lister noted that he was delighted for the Committee to take this over, but he needed to be convinced that the exact content the Committee had in mind would meet his own guidelines. He was concerned that they were more likely to concentrate too heavily on his work on antiseptic surgery, and he wished the publication instead to show the full development of how it had come about. He needed to be clear that his early work would be included, particularly on suppuration, which was the logical antecedent to the development of antiseptics. Lister also indicated his willingness to engage in "not too prolonged"[40] discussions with the members of the committee who lived in London, including Cheyne, in order to help produce the work, but he no longer considered himself "capable of much mental effort"[41]. These were his most closely trusted disciples, but he still needed to oversee it personally to make sure his work was not misrepresented or important aspects understated. Such was the experience of a lifetime.

The arrangements were well underway before the public announcement, and Dr Lister is likely, according to Godlee, to have been relieved, as Lister thought at the time that he "was becoming more and more engaged with his work as a physician"[42]. Lister himself had noted that publication by the committee of colleagues "would be far better done than it could possibly be done by any nephew after I am gone".[43] He was confident that they would be more likely to understand the finer issues if the work were done in consultation with him, and that if he were still alive to advise, albeit if and when his illness allowed, it would be more likely to meet his approval. Godlee, in his biography of his uncle, quotes Lister as saying, "This new plan will involve full as much work on my part as my poor brain can do. But if I can do *anything*, it will be better than leaving things in the utter confusion in which they would have been left to A.H."[44]

The editorial committee was also keen to publish Lister's work as he intended it, and now had the benefit of his 'active' involvement. They went about the detailed task of collecting the papers, and Godlee notes that Dr Martin and Dr Dawson Williams undertook the main share of editing, interviewing publishers and discussing the work with Lister himself, "for he had very decided views to which attention had to be paid."[45] The committee had also been "instructed to prepare a note of a biographical character showing the sequence and interrelation of the various researches and observations made by Lord Lister and recorded in the papers to be reprinted."[46] For Cheyne, this was another opportunity to indulge his

Getting older

very detailed recollections of Lister's experimental work. Each member of the committee contributed a section of the introduction, and Martin and Dawson Williams again did the bulk of the work in collating it into a "connected whole."[47]

The papers were finally published in June 1909 to critical acclaim. Had Cheyne been younger and with fewer commitments, he might have played a more intensive role in the publication, but he too was getting older. His energies were not what they had been even in the early years of the century when he had completed the many volumes of his *Manual of Surgical Treatment* with Burghard. His writing during this period and into old age often involved fond recollections of his time with Lister. He was not alone in this – many of Lister's ex-students published their memoirs, including Leeson and Stephenson. Cheyne only really published papers and transcripts of his addresses on the subject. The difference is that he had remained with Lister throughout his career, while Lister was still active. It is reflected in the number of times Cheyne was present at tributes to Lister, whether the great man himself was there or not. When Lord Lister was presented with the freedom of the City of London in June 28th 1907, Cheyne was prominent among the guests.[48]

His respect for Lister was boundless, but there was also a general recognition within the medical establishment that Cheyne was high on the list of people who knew Lister's work better than anyone. Rickman Godlee, Cheyne, and Sir Hector Cameron in Glasgow were generally associated with Lister in the public imagination, and each represented a relationship with a different aspect of Lister's work. Sir Hector Cameron had worked under him during his years in Glasgow, and Godlee, as Lister's nephew, assistant and biographer, was probably in many ways a substitute for the son Lister never had, although Cheyne could also be said to have found himself in this position, particularly in a professional capacity. This was where Cheyne occupied a special place in Lister's circle, both in reality and in the public mind. He had been with Lister through the most difficult periods of his experimental and surgical work, particularly the move to London with its uncertainties and difficulties. He had been privy to Lister's own speculations about the implications of experimental results for his antiseptic principle. He had become Lister's very vociferous standard bearer in the face of opposition, particularly after the master withdrew from the fray, a period which just happened to coincide with an increasing reliance on 'aseptic' methods of sterilisation in surgery. Where Godlee was the biographer of Lister's life, Cheyne was at one point described

Microbes and the Fetlar Man

as more or less "the biographer in England of microorganisms",[49] the bacteriological essence of Lister's achievements.

On the other hand, an era was drawing to a close, and many of the iconic figures of 19th century science had either died or were old men. In October 1901, Cheyne and Lister had both travelled to Berlin to Rudolf Virchow's 80th birthday celebrations, Cheyne representing the Pathological Society of London, of which he was President at the time.[50] Lister's own 80th birthday was celebrated six years later, though he himself had withdrawn from the public stage. He was aware he could be receding into the background, and his correspondence indicates that he was grateful for the efforts of men like Hector Cameron and Cheyne in keeping his principles on the medical agenda. He had written a letter to Cheyne on 27th December 1902, "Wishing Mrs Cheyne and you and the bairns a prosperous and happy New Year"[51], thanking him for keeping the issues alive, and noting, with some satisfaction, "Now that I am on the shelf, my professional brethren seem to grow in kindly feeling to me."[52] Cheyne himself clearly felt a responsibility to keep the flag flying, in case the great achievements of the era were fudged, misunderstood or worse, misinterpreted, as not only Lister but most of the great leaders of his familiar battlefields had become less actively involved. Pasteur had died in 1895 and Robert Koch in 1910. Cheyne's tribute to Koch in the *British Medical Journal* had been as personalised as his tribute to Lister would be, and illustrated how well he felt he knew him. He still considered himself something of a spokesman for Koch's methods in Britain:

> ... I first ... made his personal acquaintance at the International Congress in 1881, when he brought forward his great work on the method of cultivating micro-organisms. Since that time I have had the great pleasure of his friendship, and the opportunity of working for several weeks on two occasions in his laboratory in Berlin. Dr Koch appeared somewhat retiring to strangers, but when one got to know him intimately one found that he was a most genial companion, and he was ever ready to help in one's work in every way possible. Like every one who has been intimately associated with Professor Koch, I conceived the greatest admiration and affection for him.[53]

He was not afraid to resurrect the tuberculin episode of 1890, and – bravely for an obituary where less palatable recollections tend to be glossed over – confronted it head-on, absolving Koch of responsibility for the premature release of the treatment.⁵⁴ This was Cheyne once again reading the minds of detractors and pre-empting any comments. He dwelt instead on Koch's efforts to improve the treatment which had now become the basis of new vaccination methods being developed largely by Almroth Wright, a point not without a certain ironic significance, as we shall see in Cheyne's work during World War I.

Cheyne's vociferous public deference to Lister and Koch means it has not always been easy for history to distinguish Cheyne's own achievements. Coincidentally, however, the period was also one in which he was appropriately showered with honours in his own right. Many of them brought him intense pleasure. On December 21st 1906, on a Friday as the week's work was drawing to a close, he was summoned to the Board Room at King's College, where his old house surgeons and dressers presented him with the portrait they had commissioned by the artist W.C. Penn. It was considered an "excellent likeness"⁵⁵ of him. Mr Keyton Beale, who had been Cheyne's first house surgeon, made the presentation, and Cheyne responded by "saying that he should never forget how fortunate he had been in having such excellent workers in his wards".⁵⁶ He had indeed been lucky enough to attract many students who would remain as devoted to Listerian principles as himself. He had enjoyed a glittering career at an extraordinary moment in the history of medicine. The portrait hung for many years in Shetland, at Cheyne's house at Leagarth in Fetlar, until it was donated to Surgeons' Hall in Edinburgh by his son J.L. Cheyne and his wife in 1953.⁵⁷

Cheyne had never lacked academic honours. On 7th April 1905 he was made an honorary Doctor of Law at Edinburgh,⁵⁸ and on June 26th 1907 an honorary Doctor of Science at the University of Oxford.⁵⁹ He was a member of the Court of Examiners ⁶⁰ at the Royal College of Surgeons from 1902 to 1907, and in 1909 took part in a committee which looked at joint exams between the Royal College of Physicians and the Royal College of Surgeons.⁶¹

In 1908, however, Cheyne's achievements were finally recognised with one of the greatest honours the country could bestow. King Edward VII "recognized [Cheyne's] conspicuous services ... to bacteriology, to pathology, and to practical surgery"⁶² when he made him a baronet in the list of honours he customarily issued for his birthday. The baronetcy was hereditary and would pass to his eldest son. He was one of only eleven medical baronets created by Edward VII, and only three of these were surgeons.⁶³ Sir Lauder Brunton, who had also studied at Edinburgh, was made a baronet at the same time.

Microbes and the Fetlar Man

Lister had received his baronetcy in 1893 and subsequently, in 1897, became a baron (Lord Lister of Lyme Regis).[64] When he heard of Cheyne's announcement, he sent immediate congratulations, and his subsequent correspondence clears up a long-standing mystery. Cheyne had to make a decision about how he would be addressed. As Sir William Watson Cheyne, a logical shortened form of address would be Sir William, but he was never addressed this way. He decided instead on "Sir Watson", and this is how he was known henceforward, to his colleagues, friends and tenants, in London, Edinburgh, Lerwick and Fetlar.

During the time I lived in Fetlar, Cheyne's ultimate home, he was only ever known locally as "Sir Watson", and it always sounded odd if visitors called him "Sir William". It appears that the decision on the name was influenced, if in fact not suggested, by Agnes Lister's sister, Lucy Syme, who had managed Lister's London household for him in the early years of the decade. Lister wrote to Cheyne on 15th June 1908 to let him know that "Miss Syme much prefers Sir Watson to Sir William for you, as so much more distinctive. I thought I might as well let you know this."[65] So Sir Watson it became, and in 1927 it was to provoke a rather strange anecdote in the press:

> At a reception in London, Sir William Watson Cheyne was introduced to Frederick Bramwell recently as "Sir Watson Cheyne". Sir Frederick did not catch what was said, but seeing a large albert asked, "Watch and chain? Yes, but who presented him with a watch and chain?[66]

He was apparently known quite differently to his grandchildren, however, as "Nono", because he was forever saying, "No".[67] Congratulations on the baronetcy poured in. The Royal College of Surgeons congratulated him on July 9th,[68] and at the same meeting elected him Vice President along with Rickman Godlee. The *BMJ* noted he "had [a] large share - by experiment, by his writings, and by his example in extending the practice of antiseptic surgery, and has attained to the distinguished position he holds by sheer force of character and work."[69] The baronetcy was noted with pleasure, the same day, at the Annual Meeting of the Edinburgh Royal Infirmary Residents' Club, of which Sir Watson was a long-standing member.[70]

On the evening of Friday, July 11th, his colleagues at King's, and his students past and present, held a complimentary dinner in honour of the baronetcy at the Waldorf Hotel in London. It proved so popular an occasion that invitations had to be limited to old King's staff and students.[71] Everyone present (at least it appears so), signed their name on a notice of the dinner, which pictured an explosive scene on one of Sir Watson's Shetland boats and the unmistakable Fetlar peninsula of Lambhoga in the background. The scene is interesting, and must have been drawn by one of the many colleagues or students who had accompanied him to Leagarth in the period before 1908. The fact that they chose this scene to mark such an important occasion in his life is also indicative of the extent to which he was associated with Fetlar by his colleagues in London. The notice was framed and has hung in Leagarth ever since. For many years it sat, appropriately, in the dining room.

Complimentary dinner given for Cheyne by his students, ex-students and colleagues at King's College Hospital.

Further recognition was to follow in June 1910, when he was appointed Honorary Surgeon in Ordinary to George V, not long after the King's accession to the throne in May.[72] Another moment of personal gratification, however, involved his election as President of the Edinburgh Royal Infirmary Residents' Club on June 24th 1910.[73] His speech at the annual meeting on June 16th 1911 at the Caledonian Station Hotel provided him with another opportunity to reminisce about his days as Lister's house surgeon, and the excitement of his early bacteriological work.[74]

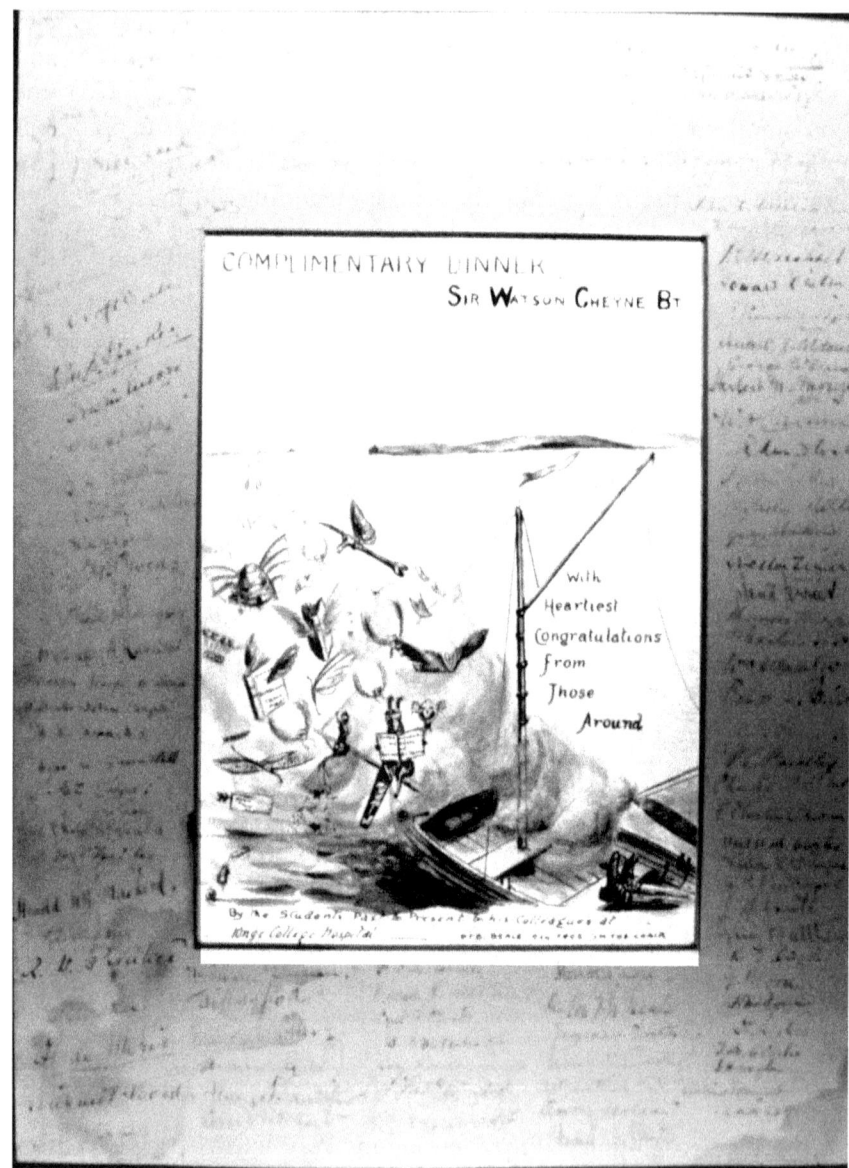

The notice contains signatures of the guests. These hung in the dining room at Leagarth House.

In Sir Watson's 60th year, he faced a tragedy which he must have known for some time was on the horizon. Lord Lister died on 10th February 1912, the nation mourned and Sir Watson lost a man who had become a surrogate father. Lister had long been hindered in walking and his ability to concentrate for any period of time was severely diminished, but Godlee doubted Lister's own belief that it was a "very slight paralytic stroke".[75] His standing in British science was equivalent to that of Pasteur in France and Koch in Germany. He was, moreover, recognised by the nation and the profession as a deeply principled man with extraordinary integrity, and this had helped make him a national icon. His standing with the nation was clear when the Royal College of Surgeons and the Royal Society obtained permission for him to be buried in Westminster Abbey. Lister himself, however, had expressed a wish to be buried alongside his wife in Hampstead cemetery, so although the funeral service was held at the Abbey, he finally came to rest with Agnes, whose own death had affected him so deeply.[76] The honour of the Abbey alongside his own wishes would have pleased him. As his nephew Godlee said, Lister always considered himself a Londoner at heart.[77]

Cheyne was one of eight pall-bearers at the funeral on 16th February. As a group, they represented many of the institutions connected with Lister: London, Edinburgh and Glasgow Universities, the Royal College of Surgeons, the Lister Institute, the Royal Society and King's College.[78] Although Godlee and Cheyne were representing the Royal College of Surgeons and King's College respectively, they were clearly also representing his closest friends and colleagues, some of whom were no longer around. Sir Hector Cameron, Lister's other 'old friend', was ill himself and was forced to cede his representation of Glasgow University to Sir Donald MacAlister. In asking Cheyne to represent King's College, Mr A.W. Howlam said that the College considered him the most suitable "both on personal and professional grounds", and added, "… you will I know be glad to pay this last tribute of respect to your best teacher and friend."[79]

Sir Frederick Treves represented the King, and a wreath was sent by the German Emperor, who had apparently known Lister personally.[80] All the other Royal Colleges and medical associations were represented, as well as other British universities. The Pasteur Institute in Paris sent a delegate, and had also sent a public notice to the *BMJ* on the announcement of Lister's death, calling him the "rénovateur de la chirurgie"[81]. The Chinese representation led the non-European contingent. A number of memorial services were held around the country, and in Glasgow, the 81 year-old Nurse Bell attended, who had been Lister's nurse at Glasgow Royal Infirmary when he had begun working on antiseptic treatment.[82]

As well as being a pall-bearer, Cheyne was also invited to sing in the choir at the funeral, as he had done at the Coronation of King George V in 1911. His invitation detailed how the procession was to take place, and clearly had "choir" written along the side,[83] so we can only presume that, having accompanied the coffin with the funeral procession, Cheyne then took his place in the choir. The memories of Lister which Cheyne chose to air in public were deeply personal. His comments in the *BMJ* recalled his first encounters with Lister, as would many of his later talks. In a world which by now took Lister's achievements for granted, Cheyne tried his best to convey just how different his teachings had been from anything else available to students at the time. He said that Lister had taught them to think, which we should understand in the context of 19th century education, where a teacher's relationship with students tended to be an authoritarian one, and students were expected to regurgitate their teacher's words. He said:

> It was not a case of being taught things which we could find in the books; everything was new, and the treatment of the various cases was always looked at from the fresh point of view, which resulted from the abolition of the danger of septic diseases. The treatment which he advised, which he carried out before us, and which ultimately proved completely successful, was often diametrically opposed to what we were taught by the other surgeons ...[84]

Cheyne needed no incentive to eulogise about Lister, but his role on the Memorial Committee after Lister's death may help explain how often his public talks and published articles during this period involved such extensive tributes to his mentor. Many of these talks coincided with important occasions such as Lister's 80th birthday, his funeral, and the Lister centenary in 1927. However, he was not simply making sure no one forgot the long and arduous process it had been to convince the world of Listerian principles. He was also promoting the Lister Memorial Fund as joint Treasurer. After the Fund was established to create a national memorial to Lord Lister in 1912,[85] Cheyne took on the work of Honorary Treasurer along with Lord Rothschild. In November, he even donated £100 of his own money to the fund,[86] and in October had opened the winter session of the London Medical Society with an address "on the early history of the late Lord Lister."[87]

In March of the following year, he was present at a meeting in Oxford where the Royal Society's plans for the fund were announced, and these were particularly dear to his own heart. He explained the proposals for establishing a "real living monument in the form of a fund for the advancement of surgery"[88], and hoped to encourage the formation of a committee in Oxford and district to help raise the money. They hoped to be able to offer a prize "not so much for the advancement of work in the early days of a man's life, but as a recognition of good work completed."[89] In March 1914, Sir Thomas Brock was commissioned to make a medallion portrait of Lord Lister, to be placed as a permanent memorial in Westminster Abbey.[90]

Sir Watson by now had his own, very committed involvement in the work to fund British research. He had become a founder member of the Cambridge Research Hospital for Special Diseases, set up to investigate conditions which, perhaps because they tended to be long and protracted, and not immediately or necessarily fatal, did not attract the interest of research and were ill-funded. It was felt that these diseases were neglected, but that they nevertheless "incapacitate the sufferer from active work, cause great pain and misery, and finally leave the victim helpless and entirely dependent on the kindness and charity of friends or of strangers."[91] They included conditions such as rheumatoid arthritis, and "tragedies of the sick-room"[92] which "arouse no sympathy outside because life is not directly threatened; nevertheless, they are as distressing to those who understand them as the more sensational dramas of disease in which the battle is waged against death."[93] They recognised the chief difficulty of a hospital in treating conditions like this – its constant battle to raise funds in competition with the more obvious causes which appealed to the public mind. Any funds they could raise came largely from within the medical community[94], and they made a number of appeals for donations in the medical journals. These led to plans, early in 1911, for a purpose-built hospital which would include laboratories, where the hospital and labs had hitherto been two miles apart. The work of charities for marginal and unpopular causes is not new, but it was typical of Cheyne, particularly in this phase in his life, to recognise the combined need to fund medical research in general, and specifically in areas where there was unlikely to be much accolade for the researchers.

Cheyne was politically conservative (he would later stand for Parliament on a Unionist ticket), and on a number of issues, he supported a philanthropic approach to funding medical institutions, rather than a government-led one. In medical politics, he was not alone in opposing some of the organisational changes of the early 1900s, but it no doubt contributed to his growing image as a diehard. He was not, in reality, any more sceptical about many of the changes than others in the profession, young or old. In 1904, he joined a widespread protest to plans to introduce registration of nurses. The move intended to provide a national standard for nurses, and to bring in qualifications by examination. Until this time, individual hospitals tended to be responsible for training nurses, and gave them a certificate after their probation. Cheyne's own first wife, Mary, had been a probationer at St. Bartholomew's in London. Importantly, during this training and probationary period, nurses would generally be subject to the views of presiding surgeons and physicians. This may be why Lister and Cheyne felt the existing system worked well. It allowed them to ensure nurses received a thorough grounding in Listerian antiseptic procedures in the hospitals under the tutelage of surgeons who adhered to them. Given the growing popularity of dry aseptic methods, which avoided the use of chemical antiseptics where possible, Cheyne can perhaps be forgiven for thinking that this standard was likely to become the norm.

An important focus of the public protest, however, was an objection to the examinations, and this is also where the Nightingale Fund joined the opposition to the proposals. Florence Nightingale was still alive, and had resisted earlier attempts to bring in nurse registration, along with many London and provincial nurse training schools. The letter of protest published in the *British Medical Journal* announced, "We know that today Miss Nightingale's opinion remains the same as it then was".[95] Her own objection was that the personal qualities required to make a good nurse were not proved by examination, and the proposed system was likely to produce nurses who were technically competent only. These qualities included "good temper, manner, tact, discretion, patience, and unselfish womanliness."[96] We can assume such personal requirements were not brought forward when formal incorporation had been introduced for doctors and physicians. Oddly, registration for midwives had been introduced in 1902 with much less fuss. The massive protest about registration in nursing had its desired effect, and successive governments hesitated about it until after World War I, when a bill finally saw its way through Parliament at the end of 1919.[97]

Cheyne had chosen surgery over bacteriology for practical reasons, but had never really left the latter behind. The result was that, as a practical surgeon with years of experience of both success and failure, and as one of the country's first experimental bacteriologists in medicine, he had developed a combination of skills, and he brought this experience to bear in an argument which arose with Almroth Wright around 1906. Wright had read a paper at the Royal Medical and Chirurgical Society which, to Cheyne, appeared to suggest that a vaccine treatment for tuberculosis, using a modified version of tuberculin, would render surgical interventions largely unnecessary in treating the disease. Wright had been widely credited with developing the first anti-typhoid vaccine in 1896, though Richard Pfeiffer was working on the matter simultaneously[98]. In addition, Metchnikoff in France had identified the significance of phagocytes, or cells which ingest other cells, and Wright had picked up on this and developed a method for measuring the strength of these cells in the blood. He considered that, in order to function properly, they required another substance in the blood he termed 'opsonins' and he devised what he called an opsonic index for measuring them. It effectively indicated the extent of a person's immune response, and could help indicate whether treatment had been successful. He considered that raising the incidence of opsonins could not only boost immunity, it could also inform the development of treatment for diseases like tuberculosis.

Cheyne's objection to this was Wright's implication that this new treatment completely negated the need for surgery or any other form of intervention, and he doubted whether opsonins and phagocytes were entirely responsible for levels of immunity in the body. Cheyne considered Wright's assertions rash and premature, but unlike his experience with Koch's tuberculin, where he had taken the opportunity to replicate the experiments himself, he could only make theoretical arguments on the basis of Wright's published bacteriological results. He published a lengthy word of caution against adopting Wright's treatment without investigating some important questions first. Given the acrimonious developments his arguments with Wright would undergo within a few years, he was conspicuously careful to congratulate Wright on his work before he launched his initial objections.

In Cheyne's article in *The Lancet* in January 1906, he tried to show that moving to a full reliance on Wright's methods, and removing the option of surgery altogether for tuberculosis, deprived practitioners of a choice of treatment depending on the particular details of the case. He said that, in the case of the different manifestations of tuberculosis, a surgeon operated only rarely, particularly in bones and joints, and would try a number of other options first. Only when other methods failed would he consider operating. He took particular exception to what he considered Wright's sweeping statement that he "would, in view of this new asset in medicine, fain induce the surgeon to abate something from his conviction that extirpation and the application of antiseptics offer, in connexion with bacterial infections, the only possible means of cure."[99] In plain English (Wright seems to have been fond of over complex sentences), he was implying that surgeons not only commonly operated in diseases with a bacteriological origin, but that they were convinced that these operations, along with the use of antiseptics, were the only way of treating the disease.

This had touched a raw nerve in Cheyne which he could hardly let pass without public comment, given his very central role in the debate on antiseptics in the early 1900s, not to mention his comprehensive earlier work on tubercular bones and joints. He argued first of all that a surgeon took a range of possible treatments into account before considering an operation, and only after these had failed would he resort to intervention. As a bacteriologist, he also drew attention to his own earlier work, particularly the relative significance of local factors involved in the disease, such as the number of bacteria and where the disease develops in the body.

Cheyne's responses endeared him to the more cautious element in the surgical sector. His written arguments often read as though he was thinking aloud or delivering a lecture, talking himself through the evidence as he wrote. He began his analysis of Wright's arguments by saying only, "... we must remember that, though from time to time Professor Wright speaks of his conclusions as new facts, we are really not dealing with facts but with theories, with excellent working hypotheses, no doubt, but still with theories – which are certain to undergo considerable modification as the investigation goes on and experience accumulates."[100]

He nevertheless welcomed a renewed look at what he called "the value of tuberculin in tuberculous disease"[101], and reiterated his earlier conclusions that aspects of the treatment could be valuable with further research and modification. He said he had been "shocked"[102] at its wholesale dismissal when it had failed to live up to original expectations. He had felt at the time that "thorough investigation as to the best method of using it would have overcome many of the objections,"[103] and he now hoped that "Koch's discovery may be replaced on its proper footing".[104] His issues with Almroth Wright would nevertheless be revived during World War I.

In the purely surgical sphere, on the other hand, on April 16th 1909, he delivered at a meeting of the Ipswich Clinical Society an address on an operation he had devised for 'moveable kidney'[105], or nephroptosis, a condition in which the kidney is displaced. His account began with a consideration of what the statistics told him about the possible causes. He concluded from the fact that the majority of cases occurred in younger women that their clothing could have a considerable bearing on the matter. The tendency of fashionable women to wear tightly-laced corsets could, he reasoned, push the kidney down. Where intervention was considered likely to be effective, the profession in general was unable to report much long-term success, either because it had been inappropriate in the first place and did not address the root of the condition, or because the kidney, once fixed, began to move again.

A standard operation at the time had involved making a lumbar incision and attempting to fix the kidney back in place. Cheyne had nevertheless found that it was difficult to replace the organ into its correct position here, and it tended to be fixed too low. He devised an operation which allowed the kidney to be moved with greater ease, and which, importantly, allowed the surgeon to explore whether other organs, such as the liver or gall bladder, could be involved in causing the symptoms. He claimed considerable success for the procedure. Oddly, however, he does not mention the value of x-rays in the article, and these had been invaluable in progressing exploration of this type. They had been introduced in 1895, and his case notes at King's mention Roentgen rays, as they were known then, as early as 1896.[106] They were invaluable during the Boer War in locating bullets; by 1903, X-ray photographs were attached to the notes for one of Cheyne's hospital cases.[107]

They involved the sad case of an attempted suicide of a 69 year-old widowed Master Mariner called John Chaffer Brennand. He had been found in the lavatory at Temple Station, close to King's College, where he had tried to shoot himself in the head, and was taken to King's where he underwent an operation. The bullet "had passed behind the right eye and had divided the optic nerve and the third nerve. Passing across to the other sis [sic] of the face the fifth nerve had been divided in the floor of the left orbit."[108] An x-ray photograph, however, had shown a second bullet, which he was unlikely to have fired himself if the other attempt had been so successful. Remarkably, he made a full recovery, and was able to give them an explanation for the second bullet as soon as he could speak again. Twenty years ago he had been shot with a revolver during an attack by Malay pirates, and doctors in the hospital in Singapore had thought it wise not to attempt to remove the bullet. If Cheyne was present when this was revealed, I can hardly help wondering if a fleeting image of the tales of his father came to mind. He was clearly very familiar with the use of X-rays when he wrote the article on moveable kidney, yet he does not mention them when others were lauding the value of x-rays in explorations of this type. An article in the *Canadian Journal of Medicine and Surgery* in 1909 noted that, "The introduction of the Roentgen rays as a means of diagnosis marks the most important advance made in the surgery of renal and ureteral calculi in modern times."[109]

His Bradshaw Lecture to the Royal College of Surgeons on 4th December 1908 had been on much more familiar ground, too familiar, some might have said, and he took it as an opportunity to stand once again on his antiseptic soapbox. He anticipated the objections of those wishing to hear something new, and any other objections, by noting that the wonderful benefit of lectures of this type was that they were not only useful for presenting innovation, but also for considering whether innovation was heading along the right road by placing it in the context of past developments.[110] In other words, he was suggesting once again that many had forgotten the bacteriological basis for asepsis, and he not only had every right to remind them, it did not make him an anachronism for doing so. He said:

> In a subject such as medicine, where knowledge is still very imperfect, there is a constant oscillation in the views which are current from time to time, and a constant swing of the pendulum of medical opinion in one or other direction, sometimes very extreme and far beyond what is justified by the facts of the case. One has constantly to be on one's guard not to be carried too far by this swing – on the one hand not to disregard too hastily the results of previous research and experience, and on the other not to overlook what seems to be sound in the new work. Nowhere is oscillation of opinion more marked at the present time than in the views founded on experimental pathology."[111] He noted that he was more than willing to take up the invitation to contribute to the debate on the treatment of wounds at that time, given that they were in the middle of one of these "extreme oscillations"[112] and his main concern in the lecture was to consider whether it had "swung too far.[113]

He could argue his way through familiar scientific ground, but he had not been quite so thick-skinned when his personal integrity was questioned. In February 1902 a publication called *What's What* had been brought to his attention which contained a favourable mention of his services as a surgeon, along with similar mentions of some of his colleagues. There was some implication that this could be taken as personal advertisement in blatant disregard of medical ethics, and to make matters worse, the *Medical Times and Hospital Gazette*, of whose existence Sir Watson had apparently been unaware until then, had suggested in "a most offensive paragraph"[114] that it was not certain that those whose names had been mentioned in *What's What* were entirely innocent.[115] He reacted swiftly and violently. He wrote a letter to the editor of *What's What*, a Mr Harry Quilter, and one to the editor of the *Medical Times and Hospital Gazette*, and the replies he received appeared alongside his own letter in the *British Medical Journal*. He stated his case in no uncertain terms, protesting that he had never been consulted on the mention of his name, and would never have condoned it if he had. His bringing it to public attention alerted others whose names had appeared in the publication, and they were similarly outraged. This prompted on the one hand a public apology from Harry Quilter [116], whose response to the *BMJ*'s assessment of his publication in January had not been quite so restrained.[117] On the other hand, an outraged counter-claim was issued from the editor of the *Medical Times* implying that the vehemence of Sir Watson's criticism, whose "three verbose epistles"[118] he regarded as "abuse",[119] was an attempt to taint his reputation as a journalist.

Sir Watson had spent so many years defending Lister that perhaps no one had really expected him to react so violently to allegations against himself. Had they known anything of him, they would have realised that the 'three verbose epistles' were simply his way of making sure he was covering every possible angle to make sure no one missed the point. He also continued to defend the integrity of others where he felt it was warranted. On July 12th 1913 he was signatory to a letter of resignation from a group of honorary medical staff at Mount Vernon Hospital for Consumption and Diseases of the Chest. They felt the Management Committee of the hospital had acted illegally and immorally in disregarding the decisions of the medical board, and in dismissing two physicians.[120]

Sir Watson was clearly at a crossroads in terms of his active presence in the surgical and research world by the second decade of the 1900s. He nevertheless continued to play a leading role at medical meetings and congresses, and Lady Cheyne had, by now, established a role for herself in the social side of these events. During the 17th International Congress of Medicine in London in 1913, while her husband led the Section of Surgery, she directed "an influential ladies' committee".[121] They organised a trip to the botanical gardens, and arranged for a party of a hundred surgeons and their wives to be shown around the University of Cambridge and Barbers' Hall, the nine hundred year-old "cradle of English surgery".[122] Sir Watson and Lady Cheyne also "gave a river party to a large number of the members of the Surgical Section" on the banks of the Thames."[123]

Sir Watson had consolidated his career and his family had consolidated its social position. Lady Cheyne had been presented at court. Sir Watson was a baronet and recipient of almost all the honours open to his profession. In 1911, he had made a journey to Canada with his son Hunter, who was by now following him into the medical profession.[124] The same year, Lady Cheyne had taken a house in Eastbourne, presumably to be near Meta, who was at school there.[125] We know little about his son Basil, though he may already have been developing signs of the old family problem of tuberculosis. Sir Watson had also become known as a particularly fine singer, and was more than once called upon to exercise his talents. On Thursday, 22nd June, 1911, he had sung in the choir at the coronation of King George V and Queen Mary at Westminster Abbey, at the direction of Frederick Bridge. Their music included works by Merbecke, Tallis, Orlando Gibbons, Sir John Stainer and Händel.

In the instruction booklet for participants, Händel was cautiously described as "an Englishman in all but birth",[126] a reference, perhaps, to more than the provenance of the composer, playing to what would become acute sensitivities about the German origins of the ruling House of Saxe-Coburg and Gotha. In fact, only a few years later, as Cheyne approached thoughts of retirement, a war was declared against Germany which would change for ever the way war was viewed in the public imagination, and would force King George V to change the name of his dynasty to the House of Windsor. In 1908, Cheyne had become a consultant surgeon in the Royal Naval Reserve, finally fulfilling his early maritime ambitions, and was once again to see active service. He was 61 years old when the war began, and by the time it ended, would be almost 66.

14

Antiseptics at the Dardanelles
World War I

In July 1914, war was declared against Germany, and men (at first volunteers, but later conscripts) were mobilised to the front for what all considered would be a short conflict. It was to last over four years, and would give new meaning to the horrors of war, with gas poisoning and long-term shell-shock for many who returned. For Cheyne, it became on the one hand a renewed focus for addressing the issues of wound sepsis, and on the other, a major opportunity to return briefly to the laboratory. Late in life (he was nearly 62 when the war began), he also finally found himself in a position to go to sea, as he had intended as a student. This strange combination of circumstances culminated in a renewed battle over antiseptics.

In 1908[1] he had become a member of the Medical Consultative Board to the Royal Navy under the presidency of the Medical Director-General. In July 1915[2], he became temporary Surgeon-General, based at Chatham Naval Hospital, and was ultimately given the rank of Surgeon Rear-Admiral[3] when the ranks were reorganised. It was to put him in an important position to address the subject of the treatment of wounds in war, and bring him into contact with some of the country's leading military personnel. On May 23rd 1914, just prior to the outbreak of war, he attended a dinner of the Royal Navy Medical Club at Princes Restaurant in Piccadilly. Among the other guests were Vice-Admiral Sir John Jellicoe and Prince Louis of Battenberg.[4] Another simultaneous, and perhaps even more important appointment was to put him in a strong position within the medical profession. From 1914-1916, for the first two years of the war, he was President of the Royal College of Surgeons.[5] It gave him a central platform from which to make recommendations on the ways in which wounds should be treated at sea and on land.

Sir Watson was a patriotic man, and his initial reaction to the outbreak of war placed him in a difficult position with regard to his German colleagues, for whom in most cases he still had the highest regard. In

355

the early months of the war, he put his signature to a protest by British scientists about a manifesto issued by ninety-three of Germany's leading intellectuals. In Britain, the manifesto was considered to be openly "glorifying and exulting Prussian militarism"[6], and inappropriate for men of their intellectual standing. It clearly saddened Sir Watson, who counted some of the signatories as his friends, as well as valued colleagues. When he delivered his Hunterian Oration in 1915, he addressed the dilemma many scientists must have faced when he said that "whatever may be thought of German Kultur in general,"[7] Professor Ehrlich, one of the signatories of the manifesto, was still "the greatest medical asset"[8] the world had at that time. He hoped he would "still be able to go on with his work"[9] on infective disease, and that he would not "be crippled in health or opportunity by the war."[10] In the climate of war, it was perhaps not the wisest thing to bring his personal feelings into the matter, as it was not, strictly speaking, necessary to his argument. However, Ehrlich himself is said to have been deeply affected by the outbreak of war, and suffered a stroke shortly afterwards.[11] The military conflict saddened Cheyne with respect to the way in which he was now forced to view his colleagues, and he was making a point. Perhaps he wished he could rise above the divisions of war, and continue to see colleagues simply as colleagues.

In November 1914, he was given an opportunity to do so when he joined the British medical profession in jumping to the aid of Belgian doctors. With the country in the front line of fighting, many of them had been "deprived of their laboratories, instruments, and their medical stores".[12] An emotional appeal had been published by Professor C. Jacobs of the University of Brussels to raise money among the medical profession in Britain to help. Belgian doctors and pharmacists were absolutely poverty-stricken and many of them and their families were homeless. Some, said Professor Jacobs, "have had to work as navvies in order to have a few pence in their pockets; others have told me that they had not seen bread for a fortnight."[13] The response in Britain was to form a committee to raise funds, chaired by Rickman Godlee, and Cheyne was one of eight initial signatories to their appeal in the medical journals. They intended to send packets of medical and pharmaceutical items, supplied by Burroughs, Wellcome and Co., and conveyed to Belgium through intermediaries. They also appealed for surgical instruments such as scalpels, syringes, forceps and scissors, and for funds to help the refugee Belgian doctors who had managed to reach Britain, and had lost everything. Sir Watson remained active in the search for funds, and found this easier than the dilemmas he had found himself facing over his German colleagues.

Sir Watson's eldest son, Lister, was an officer in the 16th Lancers, and had embarked at Dublin for the Continent on 16th August. They were part of the first Expeditionary Force of the war to land in France in September, and were to fight in many of the major battles, including Mons, Ypres and the Somme. At the end of October, Cheyne's second son Hunter, who had followed him into the medical profession, had been appointed a temporary surgeon to the Fleet [14], and by 1st January 1915, he was working on the hospital ship *Soudan*.[15]

By November 1914, Sir Watson had already been making his views known on wound treatment – to which, of course, he was no stranger. He had acquired considerable experience of gunshot wounds, both in civilian life at King's and as Consultant Surgeon to Lord Roberts during the Boer War. His case notes at King's show treatment of bullet wounds from hunting accidents and attempted suicides.[16] War wounds were characterised by a number of factors, some of which depended on where the wound was sustained. Men on ships were unlikely to be as exposed to complications such as tetanus and gangrene, for example, as the soldiers fighting in the fields of France, which were crawling with bacilli. Naval wounds tended instead to involve fragments of shell, or parts of the ship itself, and could often be problematic as they left a lacerated wound and jagged edge, which did not easily scab over or make use of the body's natural healing capacities.[17] Sir Watson, in his simultaneous position as President of the Royal College of Surgeons and consultant to the Royal Navy, was well-placed to make recommendations about the treatment of these wounds, which he hoped would filter their way into policy. When wounded men began to be sent home to Britain with septic wounds, he – and many others on both sides of the argument - seized on this as a focus. How could wounds, a difficult enough issue in themselves, be prevented from turning septic, given the organisational challenges doctors were facing in the field? Cheyne believed that, despite the obvious differences in the environment of war, the wounds themselves were not necessarily different to those encountered in civilian practice and that, if other factors could be overcome, there was no reason why they should become septic.

The sepsis of wounds may seem an obvious focus, but it was particularly highlighted as other potential medical risks , such as typhoid, were already beginning to be addressed through vaccination. In 1896, Pfeiffer and Koller had developed an anti-typhoid vaccine in Germany, and almost simultaneously, Almroth Wright in Britain published his own work on a vaccine. Wright departed from Pasteur's method and used killed typhoid bacilli. Only a small proportion of troops had received the vaccine during the Boer War, and it remained a voluntary decision on the part of the soldier. Calls to make the process compulsory were rejected, prompting high-profile criticism from commentators such as Arthur Conan Doyle in the wake of the major typhoid outbreak at Bloemfontein.[18]

When World War I was declared in 1914, Wright went out to Boulogne with the British Expeditionary Force at the suggestion of General Sir Alfred Keogh, Director General of Army Medical Services at the War Office. Lord Kitchener, Secretary of State for War, had been heavily against the involvement of bacteriologists at the front,[19] so Wright had gone out to France as a Consultant Physician. Kitchener supported vaccination, and despite initially low figures, a sizable proportion of troops were ultimately vaccinated, though still voluntarily.[20] This helped deflect attention onto the high incidence of wound sepsis, and there were marked differences of opinion in how this should be approached.

Cheyne was sceptical about the ability of Wright's 'physiological methods' to treat septic wounds, and preferred instead to take antiseptic measures to prevent wounds from turning septic in the first place. Wright, on the other hand, was opposed to the use of antiseptics, and had conducted tests he said showed that they inhibited the natural healing processes of the body by killing macrophages and encouraging the growth of anaerobes. Cheyne was concerned that abolishing the use of antiseptics altogether afforded the wound too little protection against sepsis, though he agreed that antiseptics were of little use if they were used too long after a wound had been sustained. It was once again the methods of using antiseptics which fuelled the argument on Cheyne's part, but as he was so well-known for his antiseptic stance, this once again tended to present him as a die-hard, obsessed with their use per se, and he was forced to enter into detailed explanations of his views.

His campaign focused on preventing infection taking hold by advocating that soldiers carry in their kit an effective antiseptic for immediate use in the field. The time factor was his main concern, not the quantity or use of antiseptics for their own sake. He also pointed out that surgeons in the navy faced different problems in treating the wounded.[21]

Four months into the war, Cheyne was asked to give a lecture on treatment of wounds to the Medical Society of London. He delivered the talk on November 16th 1914, as the introduction to a discussion on the subject, and open a discussion it certainly did. It would continue for much of the period of the war, and revived many of the old animosities.

At the Medical Society, he was in his element. He first reminded his audience of the causes of sepsis by returning to Lister's original work with compound fractures. His listeners could have been in no doubt where this was leading. Why, he asked, when Lister had been so successful at avoiding sepsis in wounds, were they "not having the same experience today in this war?"[22] Clearly, wounds in the army and navy were sustained under very different circumstances from those presented in civilian surgical practice, but as commentaries had shown after the Boer War, the bullets or the quantity of them were not necessarily responsible for any worse effects. This argument was, however, driven by the fact that conditions in South Africa were relatively dry, and some types of bullet tended to make a relatively clean wound,[23] though even here, Cheyne had cautioned on surgeons becoming complacent about the need to disinfect.[24] Dirt could be introduced when the bullet dragged pieces of clothing with it and when dirt entered the wound from the ground where an injured man lay, and wounds arriving in civilian hospitals from accidents were often filled with dirt and debris. So what was the difference? In an accident in the street, in a civilian context, the injured tended to reach the hospital relatively quickly, and could be treated antiseptically before sepsis was given much opportunity to set in. In the fields of war, and on the decks of vessels in combat at sea, on the other hand, it could be hours and even days before the wounded could reach a place of primary treatment, let alone be transported to hospital.

He had experienced this problem in South Africa, when transport difficulties prevented surgeons from accessing the wounded in the field in good time. This time, he was aware of the difficulties faced in the fields of France and Belgium, but also in the navy. He noted that "in the old days the wounded were carried down from the deck to the surgeon's quarters as soon as they fell, but nowadays the surgeons and non-combatants on a battleship are all kept below … consequently the wounded must lie where they fall till the battle is over."[25] From here, they would be transported to a hospital ship, and from there to a base hospital, but so far they had been transported on destroyers, which were able to move quickly, and destroyers tended not to carry surgeons or medical teams. In other words, in both the field and at sea, injuries were not attended to for some considerable time, and sepsis had plenty of opportunities to set in. He advocated measures to ensure medical teams were nearer the wounded, such as carrying them on board destroyers, for example, but this would not really strike at the root of the problem, particularly if there were logistical difficulties. If the time to first aid, hospital ships, field hospitals and bases could not be reduced, something had to be done to prevent sepsis before the wounded arrived there.

For Sir Watson, there was only one way this situation could be alleviated effectively – by the early, preventative, use of antiseptics. At this point in the lecture, he was ready to launch a renewed attack on aseptic, or non-chemical, methods of disinfection. He was once again in his element. He began by recalling that when he had suggested that suppuration in a wound made by a surgeon through unbroken skin could only be due to an error of disinfection on the part of the surgeon, his remarks had caused "great offence"[26], but he was sticking to his guns. Aseptic methods of disinfection, for Cheyne, left too many loopholes for error, and only chemical antiseptics, used properly and at the right time, could act as a safeguard. As he got into his stride, he told his audience horror stories of nurses who kept disinfected towels in chamber pots and how ineffective steam disinfectors must be if instruments were only placed in them for a few minutes when the apparatus would take "hours to boil a potato"[27].

> I think the man who annoys me most is the boiled water man!" he said. "Some surgeons seem to take a particular pride in emphasizing their contempt for antiseptics and the extreme simplicity of their methods. A surgeon comes to an operation and finds a dish containing some fluid. He asks what that is, and the nurse, who has been carefully trained in real aseptic work, says, in fear and trembling: "That is carbolic lotion for your instruments." It is most instructive to see the look of contempt on the surgeon's face as he says: "Carbolic lotion! Who on earth uses antiseptics nowadays? I thought that no one out of an asylum ever thought of them. Take it away and bring me a bowl of boiled water." He does not disinfect his hands, but trusts to the protection of boiled gloves, which, however, are soiled at the very beginning by being put on with unsterilized hands. He thinks that he is no end of a great man, and the unfortunate thing is that the nurses and students think so too and follow his fatal example.[28]

The budding politician then came out when Cheyne appealed to everyone's Achilles heel:

> I wonder whether, if such a man had a son at the front at the present time, he would feel happy in the thought that the only thing that the

surgeon had at hand for the treatment of wounds in war was a basin of boiled water! The futility and littleness of it all makes me sick![29]

The focus of his argument, however, was that, in war conditions, because of the time delay, wounds were already septic on arrival at full medical care, and the natural healing processes of the body, such as the clotting of blood, were already impeded. Cheyne therefore advocated the *early* use of a chemical antiseptic in the field or at least as soon as possible after injury, to stave off infection until the wound could be properly treated. Furthermore, he provided evidence that wounds should ideally be disinfected within 24 hours, but that after 48 hours it was not only useless to use antiseptics, it could be detrimental, as they could damage any "actively granulating tissues"[30] This is the area where Sir Watson's argument has sometimes been misunderstood in retrospect, and why he was often regarded as having been one of the last "die-hard" antiseptic surgeons, advocating the use of antiseptics at all costs. On the contrary, he said:

> My own view is that if a wound is aseptic, while I do not willingly put antiseptics into it, their entrance will not interfere with healing at all, so long as no bacteria have been let in by the surgeon. But on the other hand, if bacteria are already established in a wound, I think it is the worst treatment possible to meddle with it more than one can help, and more especially to syringe it out with antiseptics and to poke gauze and other things into it. The general practice at present is just the reverse, namely, to avoid antiseptics when the matter is of no consequence, and when their use outside the wound may help to prevent the reinfection of instruments and materials used; and on the other hand, to use them when the wounds are septic, when they cannot kill the bacteria, and when the only effect they can have is to damage the natural defences of the tissues.[31]

He also worried about relying too heavily on the protective power of injecting anti-tetanic serum, and hoped that "those who have to treat these wounds (in war conditions) will not take it into their heads that, after making these injections, they have done all, or indeed anything, of any real importance, for the safety of the patient."[32]

On the other hand, he agreed with Wright in considering good drainage an essential element in cleaning the wound, and deplored introducing gauze to try and soak up the pus. He illustrated the dangers of this with another of his anecdotes, this time about an experience when he was building his house at Leagarth in Fetlar. Sir Watson always told a story much better than most:

> Some years ago I built a house in the country and we had some difficulty in getting a proper water supply. At length I found a spring on the top of a hill in the neighbourhood where the earth had been somewhat hollowed out and which was used by a crofter not only for household purposes but also to water the animals. I got leave from the landlord to make a proper well there and to get the overflow for my own use. This I did, making a nice clean well for the crofter, with an overflow to water the animals, and this left ample water for my house. The crofters are very conservative people, though they always vote for the Radical candidate, and this particular old lady had a great affection for the old well and resented very much the new one. To relieve her feelings she put all sorts of things – rotten fish etc. - into the well. We did not notice this at first, but thought the water was extra good and tasty, but presently the flow of water through our pipe began to fail and ultimately stopped altogether. On examining the well the spring was all right, and therefore there was nothing for it but to dig up the pipe. About a hundred yards from the well I found a piece of dirty linen, evidently a piece of the old lady's chemise, which was very rotten, and completely prevented the flow of the water. Now this is exactly what you do when you fill a drainage tube with gauze – you prevent the free exit of the pus, and the gauze becomes very rotten and poisonous.[33]

His contemporaries were split on the matter of antiseptic use. Some considered the whole idea of antiseptics outdated and pedantic, on both a theoretical level and a practical one. Others applauded it, and it brought any remaining fellow Listerians back out of the woodwork. In the discussion which followed his lecture, Sir Rickman Godlee said that he "agreed with every word"[34] of Cheyne's address "from beginning to end"[35], and that "old fogey"[36] or not, he also believed that people would soon rediscover Listerian principles and that antiseptic surgery would "come back into its own,"[37]

On 28th November Albert Wilson, who had also been a pupil of Lister's, wrote to the *BMJ* expressing his support for Cheyne's "so clearly placing before the profession our responsibilities to the wounded,"[38] saying "Aseptic surgery spells disaster,"[39] and is "a frightful waste of time and clean linen."[40] At the end of the debate, Sir Watson had suggested suppositories of antiseptic be carried by soldiers to be applied in the field if they were injured.

Recent research had identified a powder of equal parts boric acid and salicylic acid, which he called borsal, and he estimated that the chances of disinfection were most likely to be successful if the powder could be used eight to twelve hours after an injury had been sustained, a timescale based on Reyher's experience during the Russo-Turkish war.[41] This timescale was considered too short by those currently at the Front, however. At the suggestion in December 1914 of Sir Arthur May, Director-General of the Naval Medical Service, Cheyne assembled a small investigative committee consisting of himself, Fleet-Surgeon Bassett-Smith, a bacteriologist in the Royal Naval Medical Service, and Arthur Edmunds. He considered himself "most fortunate in the choice of his colleagues"[42] on the basis that they were "men with great scientific knowledge"[43].

Arthur Edmunds had been Sir Watson's House Surgeon, and defended Cheyne in a similar way to the one in which Sir Watson had defended Lister. His background had similarities to Cheyne's in that his pathway into medicine had not been particularly straightforward. He had needed to follow his father into business to make a living, and had studied at night school. After he had served his time as Cheyne's House Surgeon at King's, he had become Demonstrator in Physiology in 1902, but was more interested in surgery, and was appointed Sambrooke Surgical Registrar, just as Sir Watson had been. Cheyne had also made him his private assistant, charging him with laboratory testing of sections from his cases, which were stained for microscopic examination. He contributed to a number of Cheyne's publications, including the second edition of the *Manual of Surgical Treatment*. He had been appointed Assistant Surgeon at King's in 1913, and, perhaps with Cheyne's help, had become a consultant to the Royal Navy at the onset of war. Edmunds was known for his ingenuity, and made many of his own instruments.[44] In fact, Cheyne commended him during the trials at Chatham for devising a method for making uniform slabs of agar, which he called Edmunds' cell, as a medium for testing the growth of bacteria with different antiseptic substances.[45]

On February 15th 1915, Sir Watson took advantage of his Hunterian Oration at the Royal College of Surgeons to deliver the results so far. He even apologised for the fact that the oration had very little to do with John Hunter, in whose honour the annual lecture was held, though he was sure the great man would have understood had he been alive to hear it.⁴⁶ Sir Watson was completely in his element, and made no attempt to conceal his excitement to be working with experimental bacteriology again. He returned to the military-type terminology of the old days when antiseptic surgeons were standing at the ready in battle mode to wage an endless war against microbes. To illustrate the problem of how to treat wounds antiseptically in time to inhibit the growth of bacteria, he used an appropriately warlike analogy of soldiers rooting around in woods looking for the enemy. The gunners might fire all over the place without hitting anything, because the enemy was hidden. He therefore advised opening up the wounds if necessary and applying antiseptic across the whole surface.⁴⁷ He went further and dreamed of a time when medicine could outwit enemy commanders altogether without leaving a scratch on their men.

> Now I suppose we all dream about various things at times – at any rate, I know that I do – and I have often thought that if I were writing a sixpenny romance on war I would make our gunners use shells containing some anaesthetic substance heavier than air which would diffuse along the ground and search out and anaesthetise the enemy, however well they were hidden. They and their guns could then be carted away at leisure. How furious it would make my noble and illustrious namesake [Kaiser Wilhelm] if he found that his best troops were carted off the field and woke up to find themselves prisoners of war without a scratch on them!⁴⁸

His main point remained the early use of antiseptics to prevent sepsis in wounds. He, Bassett-Smith and Edmunds developed soluble bougies containing an antiseptic substance, designed to be used at the front as a form of first aid, until the injured man could be brought into suitable hospital conditions and the wound thoroughly disinfected. The experiments were conducted partly at Chatham and partly at Bassett-Smith's laboratories at the Royal Naval College in Greenwich. They even learned "a good deal [more] about antiseptics"⁴⁹ than they had known before. The object was to look for a suitable antiseptic substance which would prevent the growth of bacteria for several hours. They were not needing it to be any more effective than that, just enough to get the injured man to a hospital where the wound could be more thoroughly disinfected. They discovered a number of antiseptics suitable for the purpose, but in the course of tests, also claimed that borsal powder actually disinfected a wound more thoroughly, to the point where it prevented tetanus and gangrene. This theoretically opened up the possibilities for testing beyond the first aid idea as well, the only problem being that, because it was a powder, it would entail a complex procedure of anaesthetising the patient and opening up the wound to make sure the powder reached all the recesses.

They had conducted trials on guinea pigs and also on some human cases of gunshot wound, but the time had clearly come for trials in real conditions, at the front and on board the naval vessels at sea. Their recommendations were already misunderstood in the medical press, which was something of a taste of things to come.[50] Sir Watson said, perhaps a little half-heartedly, that he did not fancy "knocking about the North Sea on the off-chance of having a scrap with the Germans,"[51] (under normal circumstances, he would probably have jumped at the chance to knock about the North Sea). Instead, Arthur Edmunds found himself stationed on a vessel in the Dardanelles, and was able to conduct the tests, sending Cheyne back his observations. Cheyne trusted Edmunds implicitly.

Tests in the army, however, were carried out in France, where Sir Watson did not have quite so much influence, and those asked to undertake the tests largely misunderstood the point. The doctors in France to whom he had sent the trial materials had for some reason assumed that Cheyne had sent them the antiseptics so that they could be used in the field hospitals, and had completely misunderstood the point of his experiment. He and Edmunds were probably right in thinking their general attitude towards antiseptics, and Sir Watson's reputation, had made them assume he was simply pushing the same old antiseptic argument again. The idea had been that a small section of the front be set aside for trying out this early treatment, as soon as possible after a soldier was wounded, and that the doctors involved could follow each case and keep data to see whether it made it any difference to the wound when it arrived at the hospital, and whether it prevented sepsis in the long-run. He hoped that, this way, the data could be followed by a single surgeon who had understood the point of the trials. Surgeon A. however, assumed he was asking him to compare different antiseptic preparations in the hospital, not on the front lines, and sent word asking for cresol instead of borsal, which he already had, saying that "owing to the press of wounded"[52], he had not had time to do the trials. He proceeded to send the antiseptic pastes on to various hospitals. Sir Watson was understandably frustrated. Apart from anything else, it was precisely at a point when there was a press of wounded that this could be useful at the Front. If a wound had been treated with antiseptics almost immediately, it could perhaps stave off infection until the surgeons had time to treat the wound properly. Surgeon A, moreover, was clearly not a great fan of antiseptics, as he had intimated at an early stage that he did not think they would work at all on the wounds they were dealing with in France.[53]

Sir Watson's only success was with another surgeon, B., who had called in to see him personally before he left for the Front, and Cheyne was able to explain fully the rationale behind the trials. The results he had sent back had been "quite good".[54] Heartened by this, Sir Watson then made an attempt to go out to the Front himself, to ascertain the situation he was dealing with, and presumably also to explain fully what he meant. He therefore asked A and another Surgeon D if they would recommend to the Director-General that he be invited to the Front, and he would set off immediately. A did not reply on the matter, and D was vociferously against it, saying that disinfection of wounds at the front was not possible with antiseptics, and that if Sir Watson disagreed, "no useful purpose"[55] would be served by his presence there. Cheyne was sufficiently frustrated to suggest that some might ask why he had not used his own influence to obtain permission. He had hoped, however, that if he had been invited, the surgeons might have been more disposed to assist him, and anyway, he had "never tried to get anything in that way" and was "not going to begin now."[56]

The catalogue of misunderstandings went on, and would dominate Sir Watson's work during the first two years of the war. Perhaps more than anything else, it is an illustration of the extent to which his reputation as a die-hard Listerian had become so entrenched that people were not even reading his instructions properly, assuming he was simply on another antiseptic soapbox. He had a good deal more success with Edmunds, not only because he had been involved in the experiments before he went out to the Dardanelles and understood the instructions and the rationale behind them, but also because he had not formed a prior opinion that the use of antiseptics at the front would be a useless exercise.

Edmunds had been appointed Staff-Surgeon on the hospital ship *Soudan*, which was also where Sir Watson's second son Hunter (W.H. Watson Cheyne) was based at the time. He and Edmunds were given joint charge of a ward, and Sir Watson may have engineered this to optimise the conditions for Edmunds to conduct the trials. He needed to trust the people involved in conducting them, given what had happened in France. Edmunds reached the ship in the Dardanelles on 30th April, but a contingent of wounded had just passed through, and he had to wait several days before the next. Even when this happened, he was not able to trial the concept of the wounded receiving antiseptic first aid at the front, as they were already on board the hospital ship before he saw them. What he *was* able to test, however, was whether the antiseptic procedures they had developed were effective within 8-12 hours, or whether the surgeons in France had been right in their assumption that antiseptics would not work at all.[57]

Edmunds' first communication reached Sir Watson at Chatham on May 24th, along with a report of 17 sample cases where he had been able to test the paste and powder. The vast majority of the cases were treated within the first 8-12 hours after their injury had been sustained, and Edmunds and W.H. Watson Cheyne proceeded according to Sir Watson's instructions. They disinfected the wound, dusted it with borsal powder and introduced some cresol paste into it. They then anaesthetised the patient, scrubbed the skin around the wound with 1- in-20 carbolic lotion, opened up the wound, cleaned it out, washed all recesses with "Lister's strong mixture"[58] (1-in-20 carbolic with corrosive sublimate, a compound containing mercury) and clipped away badly soiled tissues. Following this, they dried the wound with borsal powder, put in some cresol paste, inserted drainage tubes and applied antiseptic dressings. They had a good deal of success, and very few cases of sepsis, though Edmunds noted, "We have simply lived in the theatre; one day from 5 A.M. to 2 A.M. next morning".[59] Results from other surgeons on the ship proved equally satisfactory, but the lengthy procedure was to become a focus of criticism.

In June, Sir Watson himself was able to spend some time on another hospital ship, *Rewa*, in Malta.[60] They took back to Chatham 548 wounded soldiers and sailors, some from the *Soudan*, and he was impressed by how many compound fractures and other wounds from the *Soudan* had healed compared to similar cases from other ships.[61] When the *Soudan* returned to the Dardanelles after depositing its wounded in Malta, Edmunds found that, where the wounded had been transferred almost directly to the hospital ship previously, dressing stations had now been established on shore. Cases were reaching the ship much later, and most of them were already septic. Sir Watson did not have the same level of influence with the army and land-based operations as he had with the Navy, and any hopes of trials for his first-aid idea had been dashed in France, so Edmunds had to be content with the data he had managed to gather so far. On 19th June, Edmunds wrote:

> This trip has not been good for trying the powder and paste as we got all our cases from a base and they were septic on arrival. I am

getting fairly convinced that the real trouble is that most men have been so impressed by their teachers with the dangers of antiseptics that they are timid of using any. I think we can claim reasonable success in cleaning up wounds here, and I am absolutely sure that we have done a lot of good to them… I am more and more convinced that military wounds are like other ones, and that the injection of bacteria by the bullet is nonsense … The fact is that practically no one ever sets to work to clean up wounds under an anaesthetic. They dress wounds, but that is all, and when cleaned they pass through such a number of hands that they are pretty certain to go wrong."[62]

On board, at least, he was able to influence matters. He noted that "the men on board were much impressed with the antiseptic you gave them and want more of it."[63] Sir Watson added sardonically at the end of his talk, "I omit the name of this antiseptic for the present in case they should "investigate" it in France.[64]

Cheyne had returned from the Dardanelles unable to test his ideas fully. He did feel, however, that Edmunds had produced enough evidence to show that antiseptics had a positive effect if the wounds were treated within 8-12 hours. Surgeons nevertheless considered it difficult to go through the cumbersome procedures involving anaesthetics with such a press of wounded, so Sir Watson continued to believe that, if he could only trial the use of antiseptics as a first-aid measure, before the wounded man even reached preliminary treatment, it would solve this problem by staving off infection until the surgeon had more time to disinfect the wound properly.

Edmunds' finding that war wounds were no different to wounds received in civilian practice was an important one in terms of the ongoing debate. One of the foundations for the scepticism about antiseptic use at the front was the belief that bullets took with them into the body a range of dangerous organisms, and this complicated the treatment of wounds in comparison with civilian practice. Sir Watson contended that this was not the main problem. The problem was that if a wound was not treated antiseptically soon enough, bacteria could enter it from all manner of sources, whether in civilian or war conditions, and that it was within the power of the surgeon to prevent this happening. Almroth Wright did not consider this approach in the least effective.

On the assumption that bullet wounds sustained in war carried in virulent bacteria, Wright advocated treating wounds by what he termed 'physiological methods', irrigating them with a salt solution to encourage the release of fluid containing phagocytes (which absorbed harmful bacteria in the body), and to stimulate the natural defences. He considered that antiseptics killed these phagocytes as well as the bacteria. Sir Watson questioned Wright's understanding of how the body produced natural defences. He was concerned about the significance of Wright's opsonins. Wright contended that these developed in blood serum and weakened bacteria, enabling the phagocytes to absorb them. He considered that, by injecting minute quantities of dead bacteria (the same ones causing the diseases) into the bloodstream, the opsonic index, or strength of opsonins in the blood, would increase, and fortify the defences of the body. Sir Watson did not consider this sufficient protection, if indeed protection at all, and suggested it was too general a solution to take account of individual differences in each person or the various ways in which the body defended itself against bacteria.[65]

Nor did everyone at the British hospitals in Boulogne agree with Wright, particularly Surgeon-General Sir George Makins, who had been assigned a ward at No. 26 General Hospital, close to Wright's 13th. Makins had written extensively on bullet wounds on the basis of his experiences during the Boer War. In France, he was given the services of two surgeons and a bacteriologist to study a method of antiseptic wound treatment which consisted of inserting drainage tubes and irrigating with hypochlorites (a type of antiseptic), which they called Dakin's Solution.[66]

In mid-1915, these tests were brought to Sir Watson's attention [67], and he met one of the researchers, Dr Dakin, in mid-September at Chatham. As they had no opportunity to test it on cases arriving at Chatham, it was decided that he and Dakin would go out to the Dardanelles on the hospital ship *Rewa*, to do trials there. By this time, Edmunds had been appointed Surgeon on the *Rewa*, and there had been quite a number of wounded in August and September. They hoped to be able to test the hypochlorites and some other antiseptics Cheyne took out with him, and to treat the cases on the homeward voyage. He hoped this would give them a fortnight to observe the cases without having to move them.

They left Plymouth in September and landed in Malta in October, moving on from there to Mudros Bay and elsewhere, returning to Mudros with a number of cases, mostly of dysentery. They then returned to the Peninsula, and anchored off Cape Helles for five days, after which they returned, once again, to Mudros. There was next-to-no fighting when they were at the Peninsula, and they could only find 28 cases which were suitable for treatment with hypochlorite. Unfortunately, Cheyne had been taken ill on the outward journey, and was unable to observe the treatment first hand. When they returned to Mudros the third time, he decided to return home on the *Aquitania* and leave the observation of the cases on the *Rewa* to Edmunds. He also left him material for further work. Once again, however, his plans were thwarted, this time by an ironic coincidence. All the hypocholorite cases, instead of remaining with Edmunds on the *Rewa*, were transferred to the *Aquitania* on which Sir Watson was returning home. Because he was ill and still unable to look after them, Edmunds was sent on board to do so, and the investigation was disrupted to the point where they had to be content with the results they had so far. Trials in periods of war were clearly not easy. Nevertheless, on the basis of the results they had, Cheyne was able to write in January 1916 that he believed the method was "well worth following up."[68]

By this time, however, Almroth Wright had initiated a major campaign against the use of antiseptics on wounds at the Front, and directed his criticism heavily at Cheyne in the medical press. He had also misunderstood Sir Watson's concept of early first aid, and based some of his opposition on the fact that the wounded had to wait for some time after injury before their wounds could be treated. This was specifically the problem Cheyne was attempting to address.

Wright was a person of strong convictions and, by all accounts, apt to be dogmatic. In September 1916, he published in *The Lancet* an exhaustive eleven-page vitriol, the first two pages of which were entirely devoted to a justification for controversial argument, which he intended to use in no uncertain measure against Cheyne. His objections to the bacteriological evidence pale alongside the language he used. At one point he called Sir Watson "an amateur [in] matters of the laboratory,"[69] and ridiculed his commendation of Edmunds' apparatus for making uniform blocks of agar by saying

> "Some day we shall be told that the difficulty of carrying sugar into one's tea has been overcome in a very ingenious manner by the device of sugar-tongs, and that in connexion with it this or that reinventor's name ought to go down in posterity."[70]

After referring to other statements of Sir Watson's as "imaginative fiction"[71] (though, to be fair, Sir Watson had referred to Wright's theories as a "curious mixture of fact and fantasy"[72]), he then launched his attack on an even more personal level: "Let us now, rising to a more abstract and general point of view, deal with those three intellectual requisites of a scientific worker in which Sir W. Watson Cheyne comes, as it appears to me, hopelessly short."[73] He accused him of not paying attention to quantitative data and measurement, dealing only in "qualitative expressions and nothing in the form of figures".[74] He went too far when he accused him of writing with no clarity of expression, saying "the reader has, of course, been forewarned that we must not expect from Sir W. Cheyne precise or logical phrasing, or clear and continuous mental images".[75] The eleven-page denunciation by Wright used anything but clear and logical language itself, regularly digressed from the point in order to pillory Sir Watson and must have been difficult for an expert to follow, let alone a person with no understanding of the subject. No wonder he prepared his readers in the first two pages with prior justification for his methods of attack. Sir Watson's own language, filled as it was with images of battalions of antiseptics routing the bacterial enemy, may have been taken as rather naïve or even flippant in some quarters, given the fact that they were at war, but it could hardly be said that it was unclear, or that his language was "entangled".[76]

Wright did not confine himself to firing missiles at Sir Watson as a bacteriologist, but moved on to question his intellectual abilities as a surgeon. Once again, he based his criticism on the fact that Sir Watson was speaking in general of wounds where bacteria were on the surface, but that he considered that bullets in war took bacteria to a much deeper level. Beyond this, he resorted to direct insult, which he presumably considered amusing, implying that he had drawn wrong conclusions from the evidence, and that "confused cerebration and deficient logic may deduce a wrong lesson from the data of clinical observation."[77] Where Sir Watson apparently drew plausible conclusions, Wright considered them to be irrelevant. He also took some of Sir Watson's statements out of context. Cheyne had been particularly disappointed in the responses he had from France, and had pointed out that there was little point in a surgeon trialling something in which he did not believe - in this case antiseptics. He had simply meant that an investigator should keep an open mind as to the possibilities. Wright turned this on its head and accused Sir Watson of being subject only to belief and not to scientific evidence.[78] In short, he accused him of deficiency in all the things Sir Watson had spent a lifetime defending, and even suggested that he was incapable of seeing Lord Lister's work "in any broad perspective."[79]

Whatever the rights or wrongs on both sides, the vehement language of Wright's attack did not do his arguments justice. Sir Watson, with admirable restraint, replied briefly in a letter to the editor of *The Lancet* the following week:

> You will hardly expect me to reply to Sir Almroth Wright's personal attack on me either by defending my own or dissecting his mentality. It is a great pity that he has made this such a personal matter; so long as it remained impersonal any criticisms of methods or views, however trenchant, could have been answered, and possibly in course of time the two lines of thought would have more nearly approximated one another.[80]

In June 1917, two wards, an operating theatre and a pathological laboratory were set up in Etaples under Makins to compare the treatment of severe wounds using hypochlorite solutions with aseptic methods, and the clinical trials initially suggested that the antiseptic method was more effective, though this was later questioned.[81] Whichever side history would prove right or wrong, these episodes show how polarised the antiseptic/aseptic debate had become, and how entrenched was the view in many quarters that Sir Watson's advocacy of antiseptics was held not on the basis of evidence, but through an unnecessary loyalty to Listerian principles. Sir Watson, on the other hand, genuinely believed that the new methods were not thorough enough to account for all eventualities, and his views of antiseptics were not as simple as his detractors claimed.

In 1916 his Presidency of the Royal College of Surgeons came to an end, depriving him of a high profile platform for his views. He continued at Chatham, however, and in the King's New Year's Honour's List, was made a Knight Commander of the Order of St. Michael and St. George (K.C.M.G.) for his services to the Royal Navy. From 1917, he found himself representing war-related issues in a quite different way, one which was relatively new to him. In August, with retirement from active medical work looming and perhaps slightly worn out by the arguments his brief return to the laboratory had engendered, he accepted an invitation to stand for Parliament.

The Order of St. Michael and St. George (K.C.M.G.), presented to Cheyne in 1916 for services to the Royal Navy.

15

A very interesting lecture

Cheyne and politics

When Christopher N. Johnston K.C. was elevated to the judicial bench in August 1917,[1] it left a vacancy in his parliamentary seat for the Universities of Edinburgh and St. Andrews. A peculiarly Scottish institution, the university seats had been extended to the whole country in the 17th century, but remained something of an anomaly as they allowed eligible people to vote twice. The seats were elected by graduates of specific universities, whether or not they were still resident there, and voters were allowed a second vote in their geographical constituency. The seats were generally held by Conservatives, and successive Liberal and Labour administrations attempted to modify the system to ensure an individual only had one vote. Ramsey MacDonald was famously elected to the Combined Scottish Universities many years later, after he had spent years objecting to the university seats as Labour Prime Minister.[2]

In 1917, a "representative committee of the constituency"[3] invited Sir William Watson Cheyne to stand for the University seat of Edinburgh and St. Andrews, and he accepted. He was still a consultant surgeon for the Royal Navy at Chatham, and the country was still in the throes of war. He was also still in private practice when he stood, though he noted in his election speech that he intended to retire very soon, and would continue his Royal Naval duties until the end of the war.[4] Sir Watson took all these offices and duties seriously, and was once again committing himself to work which could easily become overwhelming. He was sixty-four.

Active political involvement may seem an unlikely move for a man who always described himself as shy, and who was increasingly seen by younger medical scientists as dyed-in-the-wool. His views were nevertheless strong ones, whether they were considered old-fashioned or otherwise, and his energies in expressing them had not diminished in the slightest. Moreover, he had for many years been an active member of Conservative organisations. In 1885 he was present at a dinner given for Conservative graduates of Edinburgh and St. Andrews Universities who were resident in London.⁵ He described his political leanings and intentions when it was announced that he would stand, by saying that he had "always belonged to the Conservative and Unionist party in politics,"⁶ and although he would "continue to associate himself with that line of political thought ... at the present time, when we are struggling for our very existence [he was referring to the war], party politics must be put completely in abeyance, and all must devote their energies to the task of defending the country from an utterly unwarranted attack, and of retaining our most cherished freedom."⁷ He would, in other words, give his support to the coalition government, under the Liberal Lloyd George, at least in terms of "the noble struggle"⁸ in which it was engaged. Lloyd George had taken over as leader from an earlier coalition under Asquith, whose daughter had been one of Sir Watson's patients.

Cheyne's public speaking had been well tested in the medical sphere, and it was largely through a growing involvement in the political side of medical matters that his candidacy came about. As early as 1911, he had demonstrated a side of his character which at first seemed out of kilter with the sober and carefully-reasoned lectures he gave on medical matters. In late 1911, the medical profession had been divided in its reaction to proposals for a National Insurance Act, and many had opposed it. The Act had been proposed by Lloyd George, who was, at the time, chancellor with Asquith's Liberal government. In return for a small contribution each week, topped up by employers and by the government, workers would have for the first time a right to subsidised sick leave for up to 26 weeks. The clauses over which many in the medical profession expressed concern made provision for workers to receive free treatment for tuberculosis and the free right to see a doctor appointed by a panel. It led to heated debate over how the role of the medical profession would change as a result, and how the new duties would be adequately funded. Political representation within the medical world also focused on the fear that it would restrict the choice of patients about which doctor they could see, and perhaps some local doctors would lose their patients and their living.

In December, 1911, Sir Watson was persuaded to chair a meeting of the medical profession at the Queen's Hall in London, the purpose of which was to support medical associations around the country in opposing the bill in the form it appeared at the time.⁹ This was, as far as I can see, his introduction to politics, and probably the start of a gradual move towards standing for Parliament. If his speech as chairman at the Queen's Hall is anything to go by, Sir Watson began his involvement in a sphere which was largely unfamiliar to him with more than a slight bang. He was, in fact, urging simply that the profession unite solidly under the BMA to push for clarification and modifications to the bill, but the atmosphere and language portrayed in the report is unusually rousing for a man who was not essentially given to excess. The meeting was apparently well attended, and the stewards wore rosettes. While they scrutinised the tickets and people took their seats, the audience joined in "strains of 'Rule Britannia'".[10]

Sir Watson opened the proceedings by saying that they "had met under extraordinary and serious circumstances."[11] He went on to say that

> A very able and astute statesman had conceived the idea of national insurance against sickness and unemployment, and had produced a scheme which undoubtedly contained many points of the most admirable nature and which deserved the most careful consideration. This statesman had said to the public, "If you will subscribe a small sum weekly I will arrange that if you are taken ill you shall have a great variety of benefits - among them free medical attendance." *(Laughter.)* ... It was a very extraordinary thing - hardly conceivable, indeed - that up to the present no definite arrangements had been made with the medical profession to provide the attendance that had been promised ... *(Loud applause)* ... What were they to do under those circumstances? *(Cries of "Fight," and applause)* They had met to say to the Chancellor and anyone else it might concern that they would have no more specious promises (applause) and that until the necessary and most limited demands of the medical profession were definitely acceded to, and until they were put in such a form as prevented any possible inequitableness or misunderstanding, they would refuse to carry out medical treatment as the Chancellor desired it under the Act. *(Loud cheers.)* ... This was not a strike of the medical profession. They were not refusing to continue to treat the poor as they had always done. *(Applause.)* They were only saying to the Chancellor of the Exchequer, "We do not like your terms; they are not sufficiently good or definite, and we prefer to go on as before in our relations with the sick poor."[12]

He raised loud applause when he suggested the poor were better off with the current conditions than they would be under the provisions of the Act, because they had a broader choice of doctor,

... men of the most kindly disposition, who did an amount of charity towards the poor of which the public had not the slightest conception. The poor had also open to them great charities, provident dispensaries, and great hospitals. Would diminished hospital accommodation and smaller charitable contributions be compensated for by the sort of medical attendance this Act proposed to provide? ... *(Cries of "No")*. He would make one appeal to the meeting, and that was if they were to carry their point they must be united. *(Applause.)*[13]

He believed the poor were not likely to be easily fooled, and would not accept it. Moreover, the result of the Act would be "that hundreds and thousands of their medical brethren would be reduced to a state of penury."[14]

He advocated that the whole profession stand united behind the British Medical Association, and somehow managed to anticipate any close examination of his own past relations with the organisation. He had resigned after an altercation with them on hospital issues some years ago, and was quick to note that this was a time for unification, where personal differences should be put aside. "Do not on any account follow my example,"[15] he said. "It would be a disastrous example under present circumstances."[16] He distanced himself from the unpopular Council of the BMA by suggesting that they "must do as other employers did - get fresh servants who would serve them better."[17] This was greeted with loud applause and cries of "New Council"[18]. He ended by expressing a hope "that any speeches that were made would avoid anything approaching personalities in any reference to the Council or the Association, or in any way lead to controversy or disunion."[19] This was wishful thinking.

His demands were simply for the profession to remain united, but it was rousing language for Sir Watson, and he would rarely, if ever, use it again in his political career. It is hardly commensurate with the comment that his eventual speeches in Parliament sounded more like a lecture.[20] Either the writer in the *BMJ* had exaggerated the atmosphere, or Sir Watson was – for a moment at least – a radical speaker. He had set the tone for an enthusiastic protest, but the crowd rapidly began to take matters too far, and before long, it was out of hand. When Sir Victor Horsley, who was known to be in favour of the bill, rose to speak, he could say nothing for the clamour of "cheers and groans"[21] from opposing factions. Sir Watson invited him to the stand and appealed for order, saying, "Gentlemen, you will spoil the meeting if you persist; I must ask you to hear Sir Victor Horsley,"[22] but he was no longer in control of the meeting, and it was only with a great deal of effort he calmed it sufficiently to allow Horsley to say a few words. During the attempt to restore order, however, Horsley reputedly delivered some insulting remarks to Cheyne: "Why can't you keep your meeting in order? I want to contradict your lying statements."[23]

Sir Watson calmed the meeting sufficiently to allow Horsley to speak, and resisted the temptation to respond to his remark. Horsley noted that the "six cardinal points" to be represented to the government by the BMA did not represent what members had agreed.

> If he had begun with the Representative Meetings of May and June last, he would have found that the Representatives ordered the Council to divide the six points, such of the points as could be incorporated in the bill to be incorporated, leaving others to be settled by the Insurance Commissioners.[24]

At this juncture, the noise was so uncontrollable that Horsley gave up. The meeting was eventually restored to sufficient order to discuss the points and come to some resolutions about how it wished the case to be represented to the government.

The subsequent reaction of the medical profession to the debacle was one of horror that its members could have acted in so undignified a way. A letter to the *BMJ* a week later called for an apology to Horsley. Dr Herbert L. Carre Smith said, "I am opposed to Sir Victor Horsley's political opinions, but I felt a sense of shame, when it could go forth to the public that we would not hear one of the leaders of our own profession."[25] Horsley's behaviour towards Sir Watson, however, was severely criticised by another victim of his wrath, Fred Smith.[26] The British Medical Association ultimately sent a request for modification of the Bill, to ensure it represented the six 'cardinal points', including that patients would have a choice of doctor, and that remuneration for doctors be commensurate with the duties and commitments they were asked to take on.[27]

Subsequent remarks on Sir Watson's political career have stressed that he kept his involvement to medical matters, but in these early stages of involvement in national issues he seems to have taken an active interest in a broader sense. It indicates that his political commitment extended beyond medicine in these early days, and that he was also loyal to a party line. The National Insurance Act had been passed in 1911, but resistance to its provisions continued into 1912, and there were moves to make certain sectors exempt from paying their contribution. A meeting was held in the Albert Hall in June 1912 to organise a campaign to exempt domestic staff on the grounds that the staff themselves objected to having to "submit to compulsory deductions"[28] from their wages. Their employers objected to becoming tax collectors for Lloyd George, to whom the speakers variously referred as a "tyrant" and a "benevolent busybody."[29] *The Spectator*, which was critical of the protests, was nevertheless relatively convinced the staff seemed to be there of their own accord and had not been coerced by their employers,[30] but other newspapers lampooned the meeting in rollicking style.

The audience consisted of both staff and employers, and Sir Watson and Lady Cheyne were not only present, they were on the platform of speakers. Members of some of the country's leading households were there, including Hilaire Belloc,[31] the author, who had been outspoken against the Act in the first place. The meeting was ostensibly organised by the Servants' Tax Registers' Defence Association, but if the reports in the press are anything to go by, most, if not all the speakers were employers. A lone butler, a Mr Littler, was probably the only staff representation on the platform. Moreover, the political sympathies of the majority of the employers at the meeting were Conservative or Unionist, parties which had opposed the National Insurance Act in the first place, leading to suspicions that they could be encouraging the continued protest.[32]

The press went to town on the meeting, and some reports belonged more in the society columns than the political. The servants were apparently all in their best dresses and jewellery, and joined "spiritedly in the popular airs kindly pumped out by an anti-tax gentleman at the grand organ"[33] until the real business started. The chairperson, the Countess of Desart, then made her grand entrance to the accompaniment of Pomp and Circumstance, before "dealing exhaustively with the Norman Conquest and the refusal of John Hampden to pay ship money."[34] A Mrs Dockerell reminded the audience of the indignation of the medical profession at the Act, and envisaged a future with "Lord and Lady Limehouse taking their domestic tax-gatherers in charabancs to Epping Forest and entertaining them to rare and refreshing fruits."[35] When they fell ill on the fruit, they would be attended to by National Insurance doctors at no cost. After what appears to have been a short and nominal speech by the token butler, the motion was carried to continue the opposition.

This gives us something of an idea of Sir Watson's views, and where he stood socially and politically, but also indicates that, when he was finally persuaded [36] to stand for Parliament in 1917, his Unionist political affiliations were as defining a factor in his political career as his duty to the medical profession. He nevertheless determined to restrict himself on the whole to medical subjects, an area on which he considered himself on fairly solid ground. He was officially nominated by Harvey Littlejohn, Dean of the Faculty of Medicine at the University of Edinburgh, and Dr Barry Dow, Assessor of the University of St. Andrews.[37] He was returned unopposed on August 10th, and noted that he intended to give up his private practice shortly, to "devote his energies to learning a new science, the business of politics."[38] There were later suggestions that he treated his contribution to the nation's politics as an extended academic exercise. The Royal College of Surgeons of England commented that, "The House always listened to him with attention not unmixed with amusement, for he addressed it as though he was lecturing to a class."[39]

His speech of thanks singled out issues involving education and the fact that many medical men returning from the war had found themselves unemployed. He returned to his long-standing soapbox advocating the value of a strong, early, scientific component in education and, given the patriotic fervour of the war, risked more than a little of his credibility with the comment that German attention to this very matter had contributed to their making "such a great fight."[40]

He later claimed that his maiden speech on the proposed employment of bone-setters in the Royal Army Medical Corps was the only one he made "of any value".[41] It took place only a day after his election, and he began apologetically by saying, "he should not have dreamt of making a speech so soon but for the nature of the matter before the House".[42] He also noted that he had made a decision to retire from private practice when he was elected, as "he wished to speak with complete independence from suspicion of personal gain."[43] This statement is interesting in itself, as it demonstrates that, although he was approaching 65 years of age when he entered Parliament and had amassed a small fortune from private practice, he retired because of his election, not the other way round. It was not that he had taken up politics as a sort of retirement activity. On the contrary, he spoke little, but when he did, took it very seriously. On this first occasion, he set out to object to proposals that bone setters be employed by army and navy medical corps without raising questions of registration or legality. The issue had arisen over whether the RAMC should accept the services of a Mr Barker, a bone setter. Sir Watson objected to the suggestion that an unqualified practitioner be employed, as it could only open the flood gates to the employment of other unqualified men and women:

> If bonesetters were admitted, where was the thing to stop? Were they to refuse faith healers who gave wonderful accounts of their work? Then there were cancer curers, and plenty of herbalists and all sorts of unqualified persons who had asked to come in. His own experience of bonesetters, as a rule, was that they were not educated for their job. They were usually brought up in places where a considerable number of accidents occurred, and had acquired a reputation for facility in dealing with them. He did not say that there were not some men who, having become bonesetters, had set to work to try to learn something about bones and joints; but the average bonesetter did not. ... The human frame was a very delicate organization. It should not be meddled with by people who did not know it as intimately as it was possible to know it after years of study. In this country any man who said he was a bonesetter could be employed, and would have a clientele who believed very strongly in him. Would any business man like to employ a man who had never spent any time in learning that particular business, but had in some way or other acquired a reputation for knowing something about it?[44]

We can only assume the Members of the House had not expected to be treated to a medical lecture when they took their places that day, but Sir Watson proceeded to entertain them with a potted lesson on the treatment of bones and joints.

> It happened often that if a joint was injured a certain amount of inflammation occurred in the lining membrane of the joint, and if it was kept at rest sufficiently long the two folds of the lining membrane stuck together and sometimes stuck so firmly together that in the course of time the joint could not be worked; not only did the two bones stick together but the muscles and other structures, and the patient got a stiff joint. What happened when the bonesetter operated on what he thought was a dislocation was that he tore through these adhesions and released the joint.[45]

This knowledge, he said, was the result of scientific investigation, and medical men had learned to be on the alert for signs of intervention by bone setters.

He had seen many such cases and more than once had had to amputate. A still more tragic thing was that a tumour sometimes developed in the bone, and he had known cases in which the bone setter, still thinking it was a case of dislocation, had violently broken up the part, and the disease, which was local, had then spread throughout the whole body.[46]

MPs present that day were somewhat bemused by their introduction to medicine. "Mr Noel Buxton spoke of the charm of the speech of Sir Watson Cheyne, and hoped the House would not be carried away by it."[47] He went on to remind Sir Watson that osteopaths (which had something more of a professional ring to it than bone setters) underwent a four-year training.[48] Sir Watson nevertheless carried the day, and the proposals were dropped.[49]

When it came to the question of returning servicemen, Sir Watson also had a good deal to say, though not necessarily always in his role as MP. In May 1917, prior to his election to Parliament, he had been appointed by the Minister for Pensions to a Medical Advisory Committee. It was to consider, for example, remuneration for general hospitals which were to take on some of the care of disabled servicemen returning from the war, where places in military hospitals were not available.[50] It was to be one of a number of roles where he highlighted the economic consequences of the war for medical practitioners. The Royal Colleges had welcomed the suggestion of a Committee of Inquiry at a joint meeting with the Central War Committee on August 15th. They had been seriously concerned that "no more medical men from England and Wales could be called upon to take commissions in the army without seriously endangering the supply of doctors for the treatment of the civil community."[51] They pointed out that there was "widespread feeling … among medical men, both inside and outside the army, that something could be done to relieve the situation by better organization in the use made of the personnel of the R.A.M.C."[52]

Cheyne formed part of the Committee of Inquiry into the organisation of army medical personnel, and spent the whole of September in France. The committee's remit was to look into the use of medical officers in the British army there, and there was a suggestion that they should extend the enquiry to the army based in Britain. Rickman Godlee and Harold Stiles were also involved in the committee, which was composed of leading members of the medical profession, not medical members of Parliament (except for Cheyne). They interviewed army medical personnel about their work, their experience and whether, under certain circumstances, they might be employed differently.

Some of the notes taken by the Committee from 11th -17th September, in minute writing on small notepaper, have a curious tale to tell. After Cheyne's death, many of the contents of Leagarth House were auctioned in Fetlar in 1933, and afterwards, some remaining papers were burned in the grounds of the house. A crofter, J.J. Laurenson (known to everyone as "Jeemsie") had been at the sale, and was walking along the beach. He was known as a story-teller and local historian, and had an interest in Sir Watson's papers. He saw a wad of them blowing along the beach and picked them up. They looked like a diary and reports, or notes of some kind, and had clearly blown onto the beach from the sale. Thankfully, the tide must have been out. He ultimately sent the papers to London where they found their way to the Wellcome Library,[53] and there they remain today, a loose set of notes detailing some of the results of the Committee of Enquiry in France in September 1917. They also include a report on hypochlorite solution Cheyne made when he was testing it with Dr Dakin, and his personal copy of his Hunterian Oration, with corrections and additions.[54]

J.J. ('Jeemsie') Laurenson, who found some of Sir Watson's papers in 1933.

Microbes and the Fetlar Man

The notes from the interviews in France are incomplete. Not all the papers were rescued. They nevertheless contain the remarks made by the various medical staff interviewed. The notes from an interview with Colonel Percy Woodhouse, for example, said, "Very short of dentists. Would like a dental corps but would prefer a dental advisor under himself. Would not shift a man unless he applies but if he applies & has good reason it will be carefully considered."[55] He was against women path[ologists?]. Colonel Firth, on the other hand, agreed to women anaesthetists and "might find use for one woman doctor."[56]

The notes sometimes give a brief history of the person concerned. Captain Hebb of the 107th Field Ambulance had joined in 1915 and had gone out to France in August 1916. He had previous experience in general practice, and had been at the Field Ambulance around "13 months, chiefly at main dressing station, does not go to advanced posts. Only 2 officers as C.O. is away - administered ambulance - does no surgery. May have 30 to 40 patients at a time enlarging to 120."[57] Another interviewee reported,

> Day's work morning sick parade 7.30 where men report sick & if any man in billets is ill must see him first - had not had such a case. Probably ailments so trifling that he can handle. If at all severe sent to ambulance, runs from ½ hour to 1½ hour. Then no one except sanitation, does a round with his junior normally twice a day. Does not go over top. Nothing else to do. In winter has to inspect & treat trench feet. No instruction in duties. One picks it up from friends. In motor ambulance convoy an M.O. is not required. No treatment required.[58]

The committee assessed its findings and submitted its report to the War Office, but there were delays in the Government publishing it as "a copy had not been sent to France, and consideration of the Army Council had … occupied two months."[59] At the opening of Parliament in February 1918, Sir Watson criticised this and other delays, fearing that the country's interests were being jeopardised by "too much yielding to expediency, undue delay in making decisions and in taking action on them".[60] He criticised the prevalence of "parochial instead of wide views"[61] and the fact that "underhand intrigue, whether in favour of an individual or a section of the community, was not put down with a firm hand."[62] He went so far as to suggest that incompetence was condoned, and in some cases appeared to be rewarded.[63] Since his Boer war experiences, Cheyne had been openly critical of the wasteful and cumbersome bureaucratic procedures involved in the work of the military medical authorities, and he reiterated it during this debate. The *BMJ* also highlighted the delays caused by unwieldy bureaucracy, and cited the case of one hospital where 4,000 signatures were said to be required on a daily basis. They calculated that "the signatures for a year, if arranged in a row, would extend from London to Aldershot".[64] It was clearly a great waste "of medical man power at a time when there is a great shortage of it."[65]

Around the same time, in early 1918, Sir Watson had chaired a committee to look into a dedicated medical service for the Royal Air Force. Until the Royal Air Force officially came into being as a separate entity in 1918, medical services for air corps had come under army services, and there was no specialised provision. It was argued that Air Force injuries and medical requirements needed specialist provision of their own. The Watson Cheyne Committee, as it became known, had first set about trying to avoid in its plans for the Air Force the well-publicised and debilitating levels of bureaucracy in the Army Medical Corps. They recognised specifically that "the maintenance of health, the treatment of disease, and the direction of research, while associated to a certain extent, should be controlled by officers who are experts in the branch of work that they direct - not merely adepts in the routine of administration."[66] They recommended graded pay for the service, and promotion on the grounds of merit alone, which was considered "a praiseworthy departure from the precedent."[67]

Once again, however, the response to the Committee's recommendations was delayed, not helped by the fact that the Medical Officer of the RAF had resigned over a difference of opinion about the roles and responsibilities of the Medical Administrative Committee. Under the recommendations of the Watson Cheyne Committee, the new role of the Administrative Committee was to be purely advisory, which may have gone some way to resolving problems. One of the conditions of the post offered to his replacement, Colonel Fell, was that he "would be guided by the principles laid down in the report of" the Watson Cheyne Committee.[68]

The other important issue which developed in these latter stages of World War I was the question of compulsory recruitment of medical practitioners. In a parliamentary debate on April 12th 1918, Sir Watson "stated the position of the medical profession"[69] on moves to raise the maximum age for compulsory military service for medical practitioners to 56 in contrast to the maximum age for others, which was by this time 51. Conscription had been introduced for the first time in Britain two years after the start of World War I, in 1916. He argued that, while the profession was not opposed to the idea *per se*, he asked Parliament to take account of the circumstances of older medical practitioners. It was in many cases difficult enough for young men to return from the war to find they could walk back into their previous work or practice, whether because of injury or because their practice had been taken over by others. For older men, this was more complex still, as they were unlikely to be able to move on to other work in the same way as younger men might. He also warned against the physical consequences of expecting a man of 50 or 55 to go into the trenches, and hoped those responsible for selection would have the common sense to recognise that his experience may be of more use at home. He considered it a waste of resources to send out men who might be busy at home but find themselves with little to do for whole periods of time abroad.

The question ultimately turned to how best to use the country's medical practitioners abroad without detriment to medical provision at home.[70] His suggestion that there was no real opposition to the move *per se* was contested by R.A. Lundie in Edinburgh, who said he had yet to find medical men who approved of it, and that where Sir Watson had suggested it was an honour, most considered it a slight.[71] He suggested that, instead of raising the compulsory age, the government might enjoy more success if it were to offer commissions to medical men aged up to 60-65 for a limited period. There was also a suggestion that medical students "of three years' standing"[72] could take on first-aid duty and relieve more experienced practitioners. The motion of raising the maximum age, however, was carried by 249 votes to 95.[73]

Cheyne addressed the issue of medical practitioners returning from the war on a separate occasion, however, in October 1918, when he proposed a fund be set up to guarantee loans to practitioners returning to find themselves without a practice. In some cases this was because their existing practice had been taken over while they were away, and in others, because they had been conscripted or volunteered before they had initially set one up. He suggested that, although the government was theoretically responsible for helping to reinstate practitioners it had forced off to war, "the time for action would probably have passed before any tangible result could be got in that direction"[74], and that instead, a fund be set up for loans to help the practitioners back onto their feet.

In a similar vein, even before his election to Parliament, he had been an active supporter of the Medical Benevolent Fund's efforts to raise emergency money to "afford assistance to members of the profession who, in consequence of having joined the Army Medical Service, find themselves in temporary difficulties."[75] He also took up the cause of non-medical men who stood to lose their war pensions on a technicality. In October 1918, he objected to a clause in the War Pensions Bill where pensions could be withdrawn if a serviceman refused to submit himself to medical examination. He said there might be a number of reasons why he would refuse to do it – including cases where treatment had not previously worked.[76]

Another early domestic parliamentary task for Sir Watson was to secure an amendment to the new Education Act in mid-1918. The power of local education authorities was extended to provide medical treatment as well as medical inspection for all children up to the age of eighteen in schools or educational institutions. The British Medical Association had raised concerns that this could detract from the work of existing practitioners, and they lobbied for an amendment which would remove the education authorities' powers to provide treatment, restricting them to inspection. When this failed, Sir Watson proposed alternative amendments to try and protect the existing livelihoods of private practitioners who might be affected by the new arrangements. His amendments were not carried in the form he had intended, but he nevertheless secured a proviso "that the local education authority was not to establish a general domiciliary service, and should consider how far it could avail itself of the services of private medical practitioners."[77]

Sir Watson had been in Parliament just over a year when a general election was called in November 1918. Under the new Representation of the People Act, his constituency of the Universities of Edinburgh and St. Andrews was amalgamated with the constituency of Aberdeen and Glasgow Universities, to become the Combined Scottish Universities. It carried three seats at Westminster. Only a few days before Armistice Day on November 11th, which marked the end of the War, he had been ill with influenza. This was the height of one of the deadliest pandemics to hit Europe, and it was known euphemistically as "Spanish 'flu", though other countries were as hard hit as Spain. Sir Watson was no longer a young man, but ironically, this may be what saved him from the pneumonia and septicaemia into which the 'flu developed for the millions of people across the world who lost their lives. It seemed to hit younger people particularly hard. By the beginning of December, he was clearly on his feet again, and set about canvassing. He had an advantage in Aberdeen, which had now been added to his constituency, as he had been at school and university

Microbes and the Fetlar Man

there, and had kept in constant contact with the alumni of his old Arts Faculty. He visited Aberdeen in early December to meet the local electors, and was "entertained at dinner by his class-fellows of the 1868-72 Arts Class."[78] He was elected on 14th December with the greatest share of the votes (3719), along with two other pro-coalition candidates, one Liberal and one Conservative. There had been six candidates, three of whom were medical men, which reflects the degree of medical interest involved and the influence in particular of the medical Faculties.[79] Sir Watson was nevertheless the only one of the medical candidates to be elected.

For much of his parliamentary career, he restricted his input to active representation on medical committees, and one of these involved issues over how the medical profession was represented in Parliament. In October 1918, a controversial meeting had been held at Steinway Hall to urge the formation of a group to help elect more medical men to Parliament, as there were concerns that the views of the profession were not being heard, particularly in the wake of the Ministry of Health and National Insurance issues. Sir Watson, already an MP by this time, suggested that, however many medical MPs were in Parliament, a more pressing issue involved the establishment of a mechanism by which he and other medical Members of Parliament could be informed and coached on matters of medical interest. There was a degree of opposition from BMA members to the formation of a new mechanism, who felt the body already existed in its own Medical Parliamentary Committee [80], which was in the process of co-opting members from other bodies. Many felt, however, that a new mechanism was required. Sir Watson had resigned his membership of the BMA a few years beforehand, over what he called its "tendency to trade unionism"[81], though he was quick to add that he did not feel the same degree of antagonism towards it as some were showing at the meetings at the time, and was delighted that its Medical Parliamentary Committee was making moves to become more representative of the profession. His chief interest was that the profession should have more influence in Parliament and that he should be appropriately informed in order to do what he could towards it. It was ultimately agreed to establish a representative body, and that a provisional committee be set up to organise elections for it.[82]

In February 1919, Sir Watson became Chairman of a new Commons Medical Committee, with strong representation from the Royal Army Medical Corps, formed to discuss medically-related matters and reach consensus on them in order to present a united front when they came up in Parliament. Their object was to represent the consensus of the medical profession, rather than the views of any political party, as they felt the profession's views were not sufficiently taken into account in legislation, though this seems to gloss over the fact that views within the profession itself could be polarised at times. To this end they were hoping to attract communications from doctors on matters of interest. They began by appointing a sub-committee, involving Sir Watson, to keep an eye on the bill proposing a Ministry of Health, which was due to find its way through Parliament in 1919.[83]

At a three-hour [84] meeting on May 2nd, at which nursing and pharmacy services were also represented, moves were made to form a Parliamentary Medical Committee of a more permanent nature. Not everyone was convinced that existing mechanisms were strong enough. Sir Henry Morris noted "that the proper course would be to have the medical profession as a body directly represented in Parliament just as the universities were. He hoped the British Medical Association would send representatives to the proposed committee, and that those representatives, so far as their membership of that body was concerned, would detach themselves from all the other interests and activities of the Association."[85] The representative from the BMA hinted that the Association may wish instead to concentrate parliamentary representation of the medical profession within its own ranks, but the new committee nevertheless went ahead.[86]

In 1919, a Ministry of Health was established "to bring together the medical and public health functions of central government"[87]. Sir Watson's chief concern was to ensure its remit extended to research, without which he "could not contemplate a Ministry of Health,"[88] and that the Minister should have an advisory body which could initiate proposals as well as advise on them.[89] When it came to the management of hospitals, his opinions were governed by his loyalty to the existing concept of voluntary hospitals. He was against the idea of government involvement in management. In 1923, after the Ministry of Health had been in operation for four years, he said in a speech at the opening of a bazaar at Kirkwall, Orkney, in aid of the funds of the Balfour Hospital, that "a feather in the cap of British people was the provision of hospitals on a voluntary basis."[90] He said the future of these hospitals "hung in the balance"[91] and he would consider it "an immense pity if they became state institutions".[92] He perhaps feared the levels of bureaucracy he had seen in the RAMC during his war service in South Africa.

When it came to research, he was also active in raising money privately, partly, perhaps, because he had become disillusioned with the extent to which government would commit to it. However, he had become aware of the difficulties of private funding when it came to causes like the Cambridge Hospital for Special Diseases, which catered for lower profile illnesses which may not have been life-threatening, but nevertheless caused considerable discomfort. A similar issue arose in March 1920, when he led a delegation to Lord Balfour at the Privy Council to deliver proposals for state awards for scientific research. The delegation, composed of representatives of the British Medical Association and the British Science Guild, proposed that the state should grant awards to medical scientists after they had completed an investigation. Much of this work tended to be done by medical men on modest salaries, and whilst they were often supported by scholarships while they were conducting the research, they were left thereafter with no means of support. As they generally published their research immediately, and were governed by ethical standards, they derived no personal gain from it and were left with nothing. Balfour, in response, raised the difficulties of administering awards of this type, given that those who initiated great discoveries were often not those who brought them to fruition. He was effectively recognising the fact that discoveries are rarely brought about by one person alone. He preferred the idea that the state should contribute more enthusiastically to research whilst it was being undertaken. The delegation resisted the idea, and proposed that a fund of £20,000 be subdivided between a number of researchers.[93]

Sir Watson was described as speaking in the House "with authority and without dogmatism on what he knew, and only when his particular knowledge was likely to be of service to the members of the House."[94] His speeches were said to be "short, pithy and to the point"[95] and "have often been commented on as models of what Parliamentary speeches should be."[96] On the other hand, he was often reticent to speak, possibly because of the number of times the Speaker had brought him up on his 'lectures'. In fact, on one occasion, despite strong feelings against a bill, he waited for the third reading before he opened his mouth, until he felt it was absolutely appropriate to speak, even if he risked the bill going ahead before he taken a chance to object. He said, "I have not as yet spoken in this House on this Bill. On the last occasion I did not speak partly because it was not necessary and partly because the time was short, but chiefly because I did not approve of the Amendment proposed, and I could not have given it cordial support."[97]

Nevertheless, many of the amendments he proposed were successful, and he most definitely triumphed in his ability to give MPs a sound lecture, which tried their patience at times. On one occasion when he treated the House to an account of discoveries and experiments on the circulation of blood, he was brought up on it by the Speaker, who said, rather sarcastically, "This is a very interesting lecture, but we are dealing with the Bill"[98]. Sir Watson was undeterred, and retorted, "This is a very intricate question. I am a new Member of the House, and it is somewhat difficult to keep within the rules of order in a case like this, but I am sure Mr Speaker is perfectly fair. All I was trying to do, however, was to point out that delay might make many experiments impossible. If that is wrong, I am afraid I must leave out all I was going to say."[99]

The above 'lecture' refers to his attempts to secure an amendment to a Bill which would require an extra certificate for scientists and doctors using dogs, as opposed to other animals, in laboratory experiments. The use of laboratory animals for medical research had been one of his key issues since the days when he had been working in his makeshift laboratories. The Cruelty to Animals Act of 1876 had amended an earlier act of Parliament and was intended to prevent suffering to animals if they were used in laboratory experimentation. It set limits on how they could be used and had introduced a requirement for researchers to obtain a licence. Experiments were only allowed to proceed if they were essential to saving or prolonging human life. Animals had to be anaesthetised before procedures took place, used for only one process of experimentation and had to be killed at the end of the study, to prevent them from suffering any painful consequences of the work.

During this period, the animals used in laboratories included dogs. There was a strong antivivisection lobby in the country which raised protests in particular about the use of dogs. A high-profile case in which a researcher was accused of publicly dissecting an unanaesthetised dog led to violent protests, and a statue was erected in London as a memorial to the Brown Dog, which had been reported in severe pain. Anti-vivisection groups waged an extensive campaign to abolish the use of dogs in laboratories, which led to a bill introduced by Sir Frederick Banbury in 1919. Sir Watson and his chief ally, William Whitla, headed the opposition to the proposals, which were initially known as the Dogs' Protection Bill. Medical opinion against the bill was strong, and there was by now fairly entrenched antagonism between the medical profession and anti-vivisectionists. Sir Watson focused his own attack largely on bureaucratic issues, noting that it would create additional red tape and prove a further hindrance to researchers.

Microbes and the Fetlar Man

With ongoing sensitivity about the use of animals in laboratories, it is worth looking briefly at the issues in Sir Watson's own context. Where he was concerned, perhaps two main issues were at stake. Firstly, the medical press and the profession were generally in favour of experimentation on animals which would contribute to research on human disease and medical conditions. A large proportion of the people Sir Watson was representing in Parliament were members of the medical profession. It was they who had nominated him, and his personal pro-vivisection views were well-known. They had been well-known throughout his professional life, and he saw it as part and parcel of the laboratory experimentation he had attempted to put – and keep – on the medical agenda. Secondly, the focus of protests was on whether or not the animals suffered during the experimentation, and this is also where Cheyne and his group concentrated their response. Leeson's lengthy description, which I detailed in another chapter, of Lister's general humanity towards animals despite his pro-vivisection stance, is a good example of the sensitivity of the debate. It was such that Leeson felt the need to defend Lister's views, given that in all other respects he was an outspokenly humane man.[100] The issue, therefore, revolved around the degree of cruelty involved, and how this should be interpreted.

Sir Watson had been directly targeted by the anti vivisection lobby, given his public opinions on the matter, and he made a lengthy attempt to justify his views lest he be considered inhumane. He was not entirely insensitive to this type of criticism. He wished to divert attention to what he saw as the necessity of using animals in medical research, and objected that further restrictions and red tape would hold this back. As always, he employed personal anecdotes to illustrate his points, which no doubt left the House of Commons once again bemused.

> Only the other day a lady whom I had not met before came to me, and, when she heard that I was opposed to this Bill, she expressed great disgust and astonishment that I should oppose such humanitarian ideas, and that I should advocate such a terrible thing as vivisection. I tried to get to know what her idea of experiments upon animals was. She told me that vivisectionists had cellars at the bottom of their houses where they kept the animals so that their cries could not be heard and that when the house was quiet they went down to those cellars to see the results of their experiments. The public as a whole think that you experiment upon animals to see what happens. That is not the spirit in which experiments are

conducted ... I was only trying to show by a very simple illustration what an experiment means. I wanted to show that experiments on animals were the continuation of research, and that they were not merely carried on in order to see what might happen. It is a very important point.[101]

In an opportune move, Sir Watson and his allies were actually given the chance to demonstrate their support for anti-cruelty measures with a bill which was finding its way through Parliament at roughly the same time as the Dogs' Protection Bill. During the second reading of the Animals Anaesthetics Bill in May 1919, both Cheyne and Whitla contributed to the debate. They expressed their wholehearted support for the bill, and even suggested areas in which it should be extended. It was a private member's bill, introduced by Lieutenant Colonel W. Guinness, who said it aimed to correct an omission in the existing law. Although anaesthesia was compulsory for investigation conducted on animals in laboratories, "in a far larger number of cases in which research was not the object of operations - for they were carried out merely for the convenience or profit of the owner - such care was not enjoined nor was any humane device prescribed to prevent pain."[102] Sir Watson said "he had been horrified to learn what had been done without the use of anaesthetics"[103] and William Whitla added that he thought anaesthesia ought to be extended to the docking of terriers' tails.[104]

On May 9th, however, Sir Watson introduced a deputation of the Royal Society of Medicine to the Home Office. They presented their case against the Dogs' Protection Bill, noting in particular the difficulty of using any other animals for certain experimentation. Sir Hamar Greenwood, Parliamentary Secretary to the Home Office was sympathetic to their case, and promised the government would do its utmost to prevent the passing of the Bill. He was clearly on their side, and suspected that the Bill had only managed to get through to a second reading because it had been scheduled on a quiet Friday afternoon where there was little discussion.[105]

Greenwood had drafted an amendment, which was presented at the second reading, to allow experimentation on dogs if the researcher could show that the experiment could not be done on any other animal[106]. During this reading, William Whitla had focused on rabies (hydrophobia) as an example of where he considered experimentation on dogs benefited humans, but could also contribute towards protecting dogs themselves. Rabies is a classic zoonosis, a disease which can be transmitted from animals to humans. Dogs are the main transmitters of rabies to humans, though all mammals are theoretically susceptible. It follows, therefore, that by protecting dogs from rabies, human exposure to the disease could diminish. It is used today as an argument for vaccinating dogs against rabies to control the disease in general.[107]

During the debate in 1919, Whitla – no doubt coincidentally – may have pre-empted modern views when

> [he] saw in the distance a vision of the absolute annihilation of hydrophobia and rabies by the protection of inoculation, and this could be achieved only by experiments properly carried out. No dog should be allowed to be kept unless protected by inoculation, and in time probably every dog would be saved from rabies."[108]

In other words, vaccinate sufficient dogs effectively, and it may be possible to eradicate rabies in humans. He turned the argument about protecting animals on its head by advocating experimentation on them to save the animals themselves, which would have the convenient consequence of protecting humans at the same time.

At the third reading on 27th May 1919, Sir Watson finally had his say. He proposed a further amendment which would reject the bill altogether. He noted that

> The bill in its present form implied that cruelty was being practised, and that the medical profession delighted in torture and could not be trusted to deal with animals ... At the time that he sought licences he found it difficult to get the second signature, and when the certificates were obtained they had to be taken to the Home Office, and used to lie there for some considerable time before they were gone through. Under the bill ... it would be necessary not only to persuade the informed people but also the Home Secretary, who perhaps knew little about this particular department of science, that the experiment was necessary.[109]

He delivered his coup by illustrating why he did not think observations on human and post-mortem examinations were in any way a substitute for experimentation on dogs.

Sir Watson spoke of the situation which occurred when the Germans first used gas. Personally, he was much alarmed, for if the enemy had gas enough, it seemed to him they could easily destroy whole armies. How was that risk to be dealt with? Were the authorities to sit down and wait while the doctors watched gassed men and waited for a post-mortem examination? Had experiments on dogs not been made much valuable time would have been wasted and many lives would have been lost. But certain experiments on animals - dogs and goats - were made and complete protection was quickly found against the gas.[110]

His amendment was carried by 101 votes to 62 and the measures were rejected.[111] As an old man, when he had retired to Fetlar, a number of photographs show him with a small dog under his arm. The little animal was sent up to Shetland from the Army and Navy Stores in Aberdeen, and "Brandy" as he was called, became his constant companion. By today's standards, Sir Watson's views on vivisection may be questioned by more than anti-vivisectionists. Within the context of his profession at the time, he seems to have separated his use of animals in a laboratory from his relationship with them outside it. He was, after all, a pupil of Lister, who was considered one of the greatest humanitarians in the history of medicine, and also separated his personal views on animals from his professional ones.

A final, rather charming anecdote about Sir Watson's parliamentary career involves an incident in March 1919 regarding the proposed wording of provisions for consultation between the Minister for Health and the medical profession. He wished to change the wording of the clause: "It shall be lawful for His Majesty by Order in Council to establish Consultative Councils" to "It shall be the duty of the Minister to set up advisory councils"[112]. Apart from the fact that his suggestions will be applauded by those in favour of plain English, he did not wish the appointment of the councils to be "subject to the passing whim or fancy of a Minister"[113], adding quickly that, though they trusted the present incumbent, Mr Addison, who knew who might succeed him? He was forced to withdraw his suggestion on a point of protocol, which was patiently explained to him, but he got his way when it came to a further discussion on the composition of the councils. He managed to secure an improvement in the English, changing "every council shall consist of persons of both sexes" to "every council shall consist of men and women".[114] His mini-crusade to bring plain English to the House of Commons had on one occasion, at least, been successful, and could perhaps serve as a lesson today, well beyond parliamentary texts.

Sir Watson was in Parliament for around five years, and retired at the 1922 general election, as he was becoming subject to more frequent heart attacks.[115] By that time, he was also firmly established at Leagarth, with other things to attend to.

16

Coming Home

The Fetlar years

Leagarth House and gardens from the sea.

Cheyne's retirement from medicine had been almost as busy as his working years. He was still MP for the Scottish Universities, and always spent at least a month in Fetlar in summer. At some point after 1915, however, his health clearly began to deteriorate more substantially. He later noted that he had only served five years in Parliament because he "began to have heart attacks" and "had to be careful."[1] He retired from his practice in Harley Street and his son Hunter, who had by now qualified in medicine, may have kept it on for a while, though he also resided in Crewkerne, Somerset.[2] Sir Watson, on the other hand, had major plans for Leagarth.

In the interim, in 1919, he took up temporary residence in a house called Beechgrove, at 111 Sydenham Hill, South London. It was an impressive, sprawling house with a long garden close to Sydenham Hill

Coming Home

Woods, originally built in the 1860s by an East India merchant[3]. The house was not too far from King's College Hospital at Denmark Hill, and it was clearly a good temporary solution after Harley Street. He also began to enjoy some of the things he may not have had too much time for before. He became something of a patron of the arts. In March 1919, he supported a performance of the Greek drama *Antigone*, by Aberdeen University students. Other patrons included his colleague Treves, who was a former Lord Rector of the University, and the Archbishop of Canterbury.[4] In 1922, he was to subscribe to an inaugural concert for a British Empire Music Festival dedicated to the works of composers from the Empire.[5]

At least one Fetlar member of staff was with him at Beechgrove. Christian Anderson, daughter of a family at Tresta in Fetlar, had been with him at least since 1911,[6] when she worked alongside staff from other parts of the islands. One of these others, Andrina Slater, had previously been a servant in the household of Lady Cheyne's brother, John Scott Smith, in Lerwick before she was taken down to London.[7] Lady Cheyne also had a Fetlar maid with her in Eastbourne, Christina Bain, when she was resident there in 1911. Meta was at school in Eastbourne.[8]

Christian Anderson's status at Beechgrove took a major leap forward when she was woken one night by a noise, and went to investigate. She disturbed a burglar at his work, woke the household, and the burglar fled, though we have no idea whether he got away or was arrested. Sir Watson presented her with a small, engraved silver dish, inscribed "Beechgrove, 1920,"[9] and it remained in Christian's family for many years before they donated it to the museum in Fetlar. Sir Watson was fond of his Fetlar staff and would ultimately leave them a year's wages in his will.[10]

The silver tray engraved "Beechgrove, 1920", which Sir Watson presented to one of his staff, Christian Anderson, when she disturbed a burglar.

While he was residing at Beechgrove, however, he received word from the Palace that he had been appointed Lord Lieutenant for Orkney and Shetland. It was a popular appointment in Shetland, and one publication referred to it as "a step in the right direction."[11] He was to succeed Malcolm Alfred Laing, an Orkney resident who had held the post from 1892 until he died in 1917. Prior to this, the post – though not hereditary, seems to have been largely the preserve of the Dundas family, Earls of Zetland, who resided in Yorkshire, and Sir Watson was the first Shetlander to be appointed. This fact, and the previous absentee status of the Earls of Zetland, may have reinforced his decision to return permanently to the islands. The press was delighted. One newspaper suggested that "there are none in Shetland but will heartily congratulate Sir Watson on his new appointment."[12] Shortly afterwards, in 1920, he was also made President of the Zetland Territorial Force Association.[13] The Lord Lieutenant is the British monarch's representative in the counties, and Cheyne attended events and occasions on the King's behalf. He was representing the King and the Imperial War Graves Commission, for example, when he performed the unveiling ceremony of the memorial at the Naval Cemetery at Scapa Flow, in Orkney, where many of those who had served in the Grand Fleet were buried.[14] He also opened the Lerwick Seaman's Mission in Charlotte Street in December 1926.[15] On these occasions, he appeared in full ceremonial dress.

Whether or not his appointment as Lord Lieutenant was influential in his decision to return to Leagarth permanently, or whether it simply reinforced it, he and Lady Cheyne left Beechgrove and returned to Shetland around 1920. For most of the 1930s and 40s, Beechgrove was to become home to Lionel Logue, another Harley Street doctor, perhaps best known for his role as speech therapist to King George VI. The house was demolished in 1983,[16] after standing derelict for some time, but ironically, had been purchased in 1952 by the King Edward's Hospital Fund for London, and developed as a Home for the Aged Sick. For a number of years before it was demolished, the site was "squatted by an elderly West Indian who built a shack on the steep slope below the boundary wall of the former house. He gave the wall itself a lively colour scheme of lilac and terra-cotta and ensured the Royal Mail knew his presence by posting up the house name and number on a tree where it remains today."[17] I cannot help but think Sir Watson would have been mildly amused had he lived to see it. If I have learned anything about him during the course of writing his biography, I almost suspect he would have enjoyed the man's company on occasion as a welcome retreat from the bustle of his household.

Coming Home

The Leagarth staff.

In 1920, he moved himself, his family, and all the possessions and baggage of a glittering career to Leagarth. His Boer War memorabilia adorned some of the walls, and he had to apply for an exemption certificate at the end of 1920 for a Mauser Rifle 6546 and Martini Henry Rifle, trophies of the war in South Africa.[18] In pride of place were his photograph of Lady Curzon in her peacock feather dress, and a signed photograph of Queen Victoria, ostensibly presented to him for attending to Brown, her Highland servant, though this must remain speculation.[19] Brown died in 1883 at Windsor Castle. The notice of one of the complimentary dinners given to Cheyne by his students, signed by all present, could be seen in the dining room above the huge, solid dining table, and elaborately-worked notices of appreciation from his Fetlar tenants were hanging framed in the hallways. In the library, successive guests would progressively leave behind the books they had read during their holiday, until the shelves of Sir Watson's medical works swelled with variety.

He had built a set of apartments for his own use at the back of the house, where he could live apart from the many guests. It may also corroborate family suggestions that he and Lady Cheyne spent increasing amounts of time apart. Jessie Saxby, one of his old friends in Unst, also commented rather enigmatically in an article about Cheyne's achievements that he "needed the quiet and rest of home."[20] This coincides with local memories of his life in Fetlar. In his later years, he rarely attended the regular dances and parties at Leagarth, but withdrew to his apartments

Microbes and the Fetlar Man

Leagarth staff, 1920s, standing outside the steps to Sir Watson's apartments.

until the very end, when he would appear with his dog Brandy under his arm to say goodnight. As I have noted elsewhere, one of his dogs, probably Brandy, had been sent up from the Army and Navy Stores in Aberdeen[21], and Sir Watson carried him around the house and gardens with him, a fairly constant companion who perhaps understood him better than anyone.

Sir Watson's private apartments had V-lined wooden panels, and consisted of living rooms, a bedroom and bathroom. Alongside them was a small laboratory, in which he continued to experiment. It has long since disappeared, but it was used by visiting scientists and naturalists who had cause to be grateful for the facility. It was useful to Colonel H.H. Johnston, for example, when he stayed at Leagarth to gather botanical specimens around 1919, and subsequently published additions to the Shetland flora.[22] No doubt the laboratory contained microscopes, shelves full of chemicals, and presumably – though none have ever been found – volumes of notes. He made careful notes on everything, from his very early days as a student, and it is hardly conceivable that he did not continue into old age. One of Sir Watson's grandsons [23] recalled that his grandfather had a manservant. The man's name has not survived, but he apparently looked after him in the private apartments.

The entrance to the laboratory next to Cheyne's apartments at Leagarth House.

Sir Watson was still asked to operate on local cases or advise on medical matters. Jessie Saxby wrote that, during his retirement, "he was often called to help the doctors of neighbouring Isles when some desperate case was

Microbes and the Fetlar Man

on hand."[24] She said, "his courtesy and kindness to the local "medicos" won their hearts."[25] Another Shetland writer, T.M.Y. Manson noted that "he was always ready to travel to Lerwick to perform a serious operation in the Gilbert Bain Hospital, as Shetland had no resident surgeon-consultant in those days."[26] A Mrs J. Robertson recalled how he operated on her father in Lerwick around 1922 or 1923. Her father was an engineer, and Sir Watson subsequently invited him to Leagarth as a guest [27].

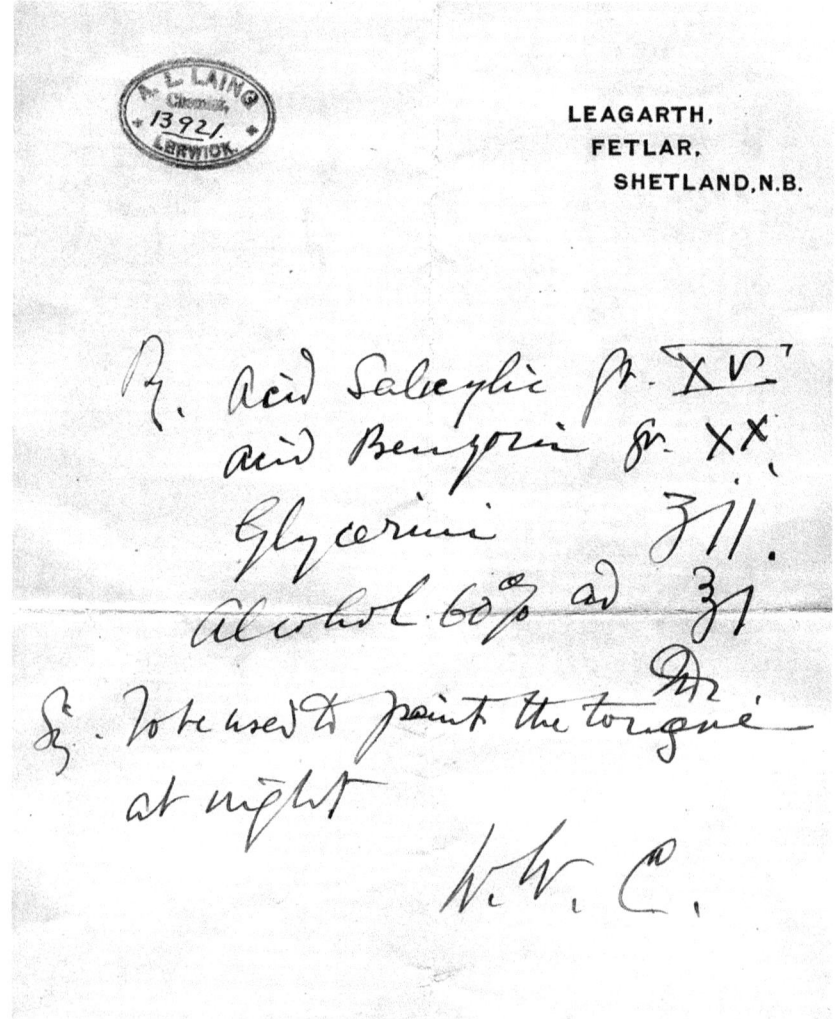

Prescription written by Sir Watson for Christian Anderson, a child in Fetlar, 1921.

Once again, he charged little or nothing for his services when he knew a patient could not afford it, particularly in Fetlar. Manson wrote that if Sir Watson was asked (and presumably only if he was asked) he quoted his Harley Street fees at fifty guineas, adding, "If you can pay that, very good. If not, it will be nothing."[28] In Fetlar, knowing full well his fees far exceeded incomes, there is no record of him ever charging anything. He also prescribed medication, presumably when it was impossible for Dr Taylor in Yell to visit, or if it proved difficult for a family to take a child out to the doctor. One of the prescriptions he wrote in Fetlar survives. It was for a little girl called Christian Anderson, a niece of the Christian Anderson on his staff who had disturbed the burglar at Beechgrove. It was directed to A.L. Laing, a Lerwick chemist, for a mixture "to be used to paint the tongue at night."[29] The little girl lived into her seventies.

Sir Watson had been only 48- or 49-years-old when he started building Leagarth, but it was not until he retired there permanently that he started to develop it as the full-scale working hub of his estate. An early development was the Home Farm.[30] Opposite Leagarth House, slightly further inland, he built a new house in 1921, subsequently known as Feal (pronounced Fyell), intended to house a Home Farm manager who would organise produce for the household.

A young man on the island, Frank Coutts, had been invalided out of the Royal Naval Reserve in 1916 with an injury which had cost him an eye. He was also, his descendants have speculated, traumatised by his experiences during the war, and by an earlier experience where he had been a member of one of the crews sent to pick up remaining bodies from the Titanic.[31] Sir Watson clearly recognised the psychological, as well as the physical effects of war, and when Frank Coutts married and began to bring up what ultimately became a very large family in a tiny two-roomed croft house, Sir Watson built Feal and employed him to manage the Home Farm. The family moved into Feal in 1922.

It may have been a coincidence that he decided to plan the workings of his estate this way, but other evidence suggests he was deliberately trying to create local employment. Feal was built around 1921, and in 1920, in his capacity as Lord Lieutenant, he had received a letter from the King asking him to promote employment for servicemen returning from the war.[32] The Home Farm may have been a project in this vein. The Leagarth project itself had been a source of employment for many years, from the early 1900s when Cheyne first built the house. His various smaller projects may have provided temporary employment too, however. At some earlier

Frank Coutts, manager of Cheyne's Home Farm, at Feal, the house Sir Watson built for the Home Farm in 1921. The dog was called Popeye and the cat was probably one called Old Jane Puss.

point he built a 'trout ladder' to encourage trout to swim up the burn from the sea to spawn, and there has been some speculation that the gradient of the burn was sufficient to allow the trout to reach their destination without help. One conclusion is that Sir Watson saw it as an opportunity for a small project which would have temporary employment spinoffs.[33]

Piecing together what Frank Coutts told his family about the period gives something of an insight into his work. From the start, Cheyne seems to have encouraged in the Leagarth staff some systematic rules for bacteria-free working practices. Frank Coutts' eldest daughter, Catherine, remembered that her father was instructed always to wear a white coat when milking the cows,[34] and other family members mentioned they were told never to drink out of cracked cups, which could harbour bacteria.[35] The white coat was not for cosmetic effect. Sir Watson's views on fashion for fashion's sake had been made quite clear on more than one occasion,[36] and there is every indication that at least some of his employees were aware of the reasons for the procedures. He may, in fact, have been promoting a clean environment, taking precautions against infection and diseases such as tuberculosis. The issue of systematic testing for tuberculosis in cattle, a prime source of infection in children, was not satisfactorily resolved until the 1950s and 60s, using tuberculin as a diagnostic tool, but Sir Watson was clearly aware of the possibility of infection in his own livestock. In 1899, he had contributed correspondence to *The Lancet* suggesting the importance of experimentation to determine whether the tubercle and diphtheria bacilli could be destroyed by boiling milk. He had said,

> It would be of even greater importance to determine whether tubercle bacilli, when present in milk, can be destroyed by boiling; because it is very probable that, in children at any rate, tubercular disease is frequently induced by the milk from tuberculous cows.[37]

It is not out of the question that Sir Watson explained to Frank Coutts the potential dangers from the milk and the science behind them. The manager of his Home Farm retained a lifelong respect for his employer. He brought up nine children, including my husband, the youngest. Some predeceased him, through sad accidents, but none of his offspring died from childhood disease.[38] The Home Farm delivered the produce to the Leagarth kitchen, which, in summer in particular, was at full capacity with all the visitors.

Overview of Leagarth gardens, early 1920s. when they were being landscaped. The glasshouses are under construction.

Then came Sir Watson's most creative project of all, his gardens. He employed a gardener and an assistant gardener, and probably others during the development stage. In their heyday, from 1920 until he died in 1932, the landscaped gardens defied even the Shetland weather. They were celebrated throughout the islands, and in summer were open to the public on Sunday afternoons. Well-attended tea parties were held on the long lawn overlooking the sea, and the gardens were a talking point throughout the islands and beyond. At the entrance to them was a wooden plaque with the Gurney poem:

"The kiss of the sun for pardon,
The song of the birds for mirth,
One is nearer God's heart in a garden
Than anywhere else on earth."[39]

Sir Watson in the completed gardens at Leagarth. The Lodge is visible just behind the gardens.

They were a planning achievement in their own right, adjacent to the house. They included a potting shed, greenhouses and a hothouse whose temperature was maintained by hot water pipes, heated by a boiler. There was even a frog pond, and the frogs seemed to breed reasonably well. In a sunken alpine garden, protected by a wooden fence, guests could sit in relative shelter from the sharp winds. In the high-walled kitchen garden, Sir Watson tried to recreate the Manse at Tresta where he had grown up alongside some of the few trees in the windswept landscape. Most parts of Fetlar, and indeed Shetland as a whole, are visibly devoid of any trees taller than bush size, but a minister in the 1700s, long before Sir Watson's family's time there, had planted an impressive array of trees at the Manse, most of which had survived into maturity. The result, on one of the most windswept coasts in Britain, was a miniature wood overlooking the beach, where Cheyne no doubt remembered playing as a child. It was probably this scene he was attempting to recreate, to compensate for not being allowed to build close to the Manse garden. There was a court for outdoor sports, and a small summer house perched close to the sea where, it was rumoured, his son Basil used to spend time to fend off the symptoms of his tuberculosis.[40] It is not out of the question that Basil's health was another reason for the family's permanent move to Leagarth, to be by the sea. Sir Watson was a great believer in the sea as an aid to recovery from tuberculosis, as he had advocated in detail in his lectures in 1892.[41]

In 1920, he made some serious additions to the house, removing the upper balcony and building an enormous glass verandah around the side facing the sea. Its purpose was to house his vast collection of geraniums, but it also, over the years, protected the house from the gales and provided a haven for guests to sit out of the wind. Around the same time, he built a small lodge in the grounds of Leagarth to house his head gardener, James Young Robertson, known locally as "Jimmy Young". The Lodge, as it is still known, was a wood-panelled cottage on the edge of the grounds, and it has remained in fairly constant use ever since.

In the sunken alpine garden at Leagarth, protected from the wind.

Importantly, in a move which would set a precedent for the family's relations with the Fetlar community, Sir Watson opened his recreation hall regularly for dances. The islanders had never had an official or substantial place where they could all gather as a community. Impromptu dances and wedding entertainment were held in people's byres, or anywhere suitable, but there was no community hall as there is today. In the 1920s, Sir Watson addressed the situation by opening the hall he had built for games and recreation. On a regular basis, families trooped in with their children for dances and entertainment. On occasion, the adults dressed up

The head gardener at Leagarth, James Young Robertson ("Jimmy Young"), on the left, and his assistant, Willie Garriock.

Sir Watson in the verandah at Leagarth, with his collection of geraniums.

in home-made costumes and took part in concert parties, a home-grown equivalent of music hall entertainment with local jokes and songs that parodied events and people everyone knew well. Local people wrote the material, and some of the lyrics and scripts even survive. The 1920s were, all in all, a social heyday in Fetlar, and Sir Watson's hall became its focus. The community of some 236 souls[42] had good reason to remember Sir Watson's retirement at Leagarth with a certain fondness.

Coming Home

The hall was also a focus for entertainment for visiting ships, and Sir Watson was given to inviting the officers and crew ashore. He still gave the occasional choral recital on some of these occasions. On the whole,

The Fetlar community at one of the many events held in the old hall at Leagarth..

Fetlar people dressed for a concert party, probably 1920s.

however, as the decade progressed, he became more withdrawn during events in his house. He would appear only at the end of the monthly dances, an event which became a local tradition, just as Sir Watson had become a local institution. Everyone may have had a reasonable understanding of

One of the many Leagarth concert parties, 1920s.

Sports on the lawn at Leagarth.

A picnic on the lawn outside the verandah at Leagarth, 1920s.

how great his achievements were in the world, but this was not why he was popular. He was remembered with affection principally because he became part of the community. He seems to have enjoyed a respectful relationship with most of his tenants, (though legal issues occasionally came into play over the course of managing the estate), and they sent him a number of letters or tokens of their esteem over the years,[43] generally organised by one of the ministers. When Sir Watson's son, J.L. Cheyne reached the age of 21, the two ministers, Mr Campbell and Mr McAffer, sent a letter of congratulations on behalf of the tenants, along with a gift of silver plate.[44]

After Sir Watson's death, J.L. Cheyne was to follow in the tradition of making parts of Leagarth available to the community by leasing his father's old apartments on the side of the house to convert into a public hall. The announcement appears to have been made during the celebrations in Fetlar for his son's marriage in 1938.[45] The hall remained a focus for island entertainment until the community raised the money to buy the nearby Free Kirk buildings and convert them into a hall in the 1980s. Many a couple still living had their wedding reception in what became known as Sir Lister Hall. It was managed by a local committee, who organised the entertainment, and was also used for badminton sessions, with matches against neighbouring islands. There were some good players amongst them, and this was long before the days of leisure centres. This 1920s generation was formally taught to dance, and they were known as fine dancers. I witnessed their dancing myself, albeit when they were older men and women, and it was, indeed, impressive. They passed on their skills to their children, and even in the late 20th century, Fetlar dances were recalled throughout the islands as memorable occasions. The move to the former Free Kirk in the 1980s marked the end of an era, in many ways, and Sir Lister Hall, by now needing repairs beyond the scope of the community, reverted to part of the house, though it was used as a development office in the mid to late 1980s.

The new hall in the former Free Kirk was the result of major fundraising in Fetlar, and an extraordinary achievement. When fundraisers went around to collect contributions, the same Jeemsie Laurenson who had sent the remains of Sir Watson's World War I notes to London is said to have been particularly interested in contributing. He was by now an old man, and had little more than his pension, but wanted to live long enough to see dancing in the Free Kirk which, of course, would never have been allowed when it was a working church building.[46] Unfortunately, he did not live to see the building opened.

Apart from the general community events, when Sir Watson first retired to Leagarth, he continued to throw his annual concert party, which had become such an institution that it was reported in the Aberdeen newspapers, as well as the Shetland ones. Perhaps the one he held in 1921 to celebrate the installation of electric lighting at Leagarth was the best of all,[47] and may even have been one of the last of the old style. The event was chaired by Mr Campbell, the Parish Minister, and the programme had been entirely organised by two local people, Walter Shewan and his sister. Mr Doig from Aberdeenshire, the electrician who had installed the lighting system, played the banjo. The Nicolsons were out in force with their jazz band, and the three brothers, Sir Arthur – the son of the Sir Arthur who was in residence when Cheyne had bought the estate –, Stanley and Lionel played mandolin and guitar. Vera, their sister who had spent many days at Leagarth in the early 1900s, had married a Lord-in-Waiting to the King, Lord Herschell, and moved away. The Nicolsons' jazz band was joined by Mr Campbell's daughter on piano and Basil on the bells. Walter Shewan had written a sketch, probably lampooning all manner of local personalities, which had everyone in stitches, and Sir Watson was in fine fettle with his rendition of The Admiral's Broom. The star of the evening, however, was the "electric illumination,"[48] which was "a complete revelation to many of the islanders". The whole system, with an engine and dynamo, had been "fitted up by the firm of Messrs Bell and Robertson Ltd., 66 Spring Garden, Aberdeen,"[49] and designed by Mr James Cobban from Aberdeen.

It was, sadly, coming to the end of an era. Sir Watson had put an enormous effort into setting up Leagarth for his retirement. Not long after they moved to Fetlar, however, his wife Margaret died suddenly on April 9th, 1922, while she was away at Ravensbourne, Melrose.[50] She had barely had a chance to enjoy her husband's retirement. He had been a busy man during his second marriage, and while his children were small, but Leagarth had always been a place where he could spend more time with them. From the early 1900s, his two youngest children, Basil and Meta, are pictured in photographs on Shetland ponies, or on picnics on the small islands around Fetlar. In one photograph, Sir Watson is teaching Basil to fish in the Houbie Burn. Lady Cheyne was spared the second tragedy of the year, but Sir Watson was not. Basil had contracted tuberculosis, and probably laboured for some time under the condition. It worsened despite the sea air at Leagarth, and in desperation he was sent to Switzerland for treatment, but he died there, in Leysin, on July 2nd 1922 at the age of 23,[51] less than three months after his mother.

Watson teaching his son Basil to fish in the Houbie Burn, Fetlar.

Microbes and the Fetlar Man

The trout ladder Sir Watson built on the Houbie Burn.

Coming so soon after Margaret's death, the loss of Basil perhaps made Sir Watson realise it was finally time to slow down the pace of his life. He always gave his own health as the reason he stood down from Parliament in the 1922 election, which was held in November, but perhaps the deaths of Margaret and Basil in quick succession earlier in the year contributed to the decision. Losing Basil must have been devastating, though given his understanding of the disease, Cheyne was probably aware of how serious Basil's health was before he considered sending him to Switzerland. For all his ground-breaking work on tuberculosis, and all the people on whom Sir Watson had operated with varying degrees of success, he could not save his youngest son. Basil is commemorated by a plaque in the Fetlar Kirk, and this and the many photographs of holidays at Leagarth are all that remain of his short life.

The family tragedies of 1922 were a turning point. Sir Watson began to travel extensively again, by land and by sea. He is pictured in one photograph standing by a motor car, holding up a spent tyre while the driver fitted a new one. Tyres in those days did not last long. Each winter, he set off on a cruise to far-flung places, a habit which seems to have begun around 1922, just after Margaret and Basil died. In an address in 1927, he mentioned a "pretty long sea voyage …some five years ago."[52] On this first cruise, either people were aware he was a medical man (though not necessarily how important he was), or he was called on to attend to a serious case, because he clearly spent a good deal of time discussing the causes of sepsis with people.

> [T]here were a considerable number of people on board who had been operated on but hardly with success, and it was interesting to hear them discussing the troubles through which they had passed, and praising their surgeon for the great attention to them in disagreeable circumstances, and blaming everything but the real cause.[53]

It was yet another opportunity for him to criticise the aseptic surgeons of the new generation.

His daughter Meta accompanied him on these early cruises. On 6th December 1923, they were due to set sail from Liverpool in the *Ortega*, belonging to the Pacific Steam Navigation Company. For some reason, however, they cancelled, and sailed instead on the *Oropesa*, with the same

Sir Watson on one of his many journeys, dealing with motor problems.

company, on 10th January 1924.[54] Sir Watson was 70 years old when they cancelled, and may have been ill, so the voyage had to be postponed. When they finally sailed, the passenger lists show that the vessel called at Havana and South American ports including Callao, Valparaiso, Antofagasta,

Talcahuano, Punta Arenas, and Montevideo.[55] The last time Sir Watson had made this journey, he had been travelling home to Fetlar from Callao with his parents, and he had not even been two years old. He cannot have remembered it very well, if at all, but had probably had to piece it all together from what people had told him. In short, Sir Watson was retracing his childhood journey, and this time he was not travelling on a convict ship, but in first class accommodation. We can only imagine what he was thinking as he passed through all the ports, whether he remembered anything or just enjoyed the journey. As he travelled through the Panama Canal, which had been officially opened in 1914, he may have speculated how his parents' journey could have been shortened if they had had the option of the canal, and perhaps his mother would never have arrived in Fetlar in such a poor state of health. His life would have been quite different.

The following year, on December 22nd, 1924, he and Meta set sail from Bristol for the Caribbean, for Kingston, Jamaica in the *Camito*.[56] Interestingly, Sir Watson's former political adversary, the Labour MP Ramsey MacDonald, was on the same cruise.[57] No doubt politics were laid aside if they ran into each other on deck, particularly as Mr MacDonald's party had recently lost a general election; he had to step down as Prime Minister. Something much more worrying to Sir Watson is likely to have happened on the voyage, however. It may be on their return from this trip that Meta met her future husband, a Mr Herbert M.T. Browne, originally from London. The indications are that they became acquainted on one of these cruises, and that Sir Watson did not approve of the match. She nevertheless married him, by all accounts against her father's wishes,[58] in 1926. There is some evidence, though unconfirmed, that Sir Watson attempted to set them up in business in coffee in Tanganyika,[59] (now Tanzania). On 17th December 1925, Sir Watson and Meta boarded the *Usaramo* of the Deutsche Ost-Afrika Linie in Southampton bound for Beira, Mozambique.[60] Meta was 29 years old. They were clearly bound for East Africa, and he may have been accompanying her out there to begin, or at least prepare for, a new life. Browne may already have been there. They were ultimately to live on an estate called Mavrimi, at Amani, Tanga,[61] in Tanganyika.

However, assuming Sir Watson was still in East Africa in February 1926, he added a codicil to his will. By the terms of his existing settlement, Meta had been going to receive on his death £4,000, and £300 a year to be paid to her by Sir Watson's heir, J.L. Cheyne, but only so long as she remained a spinster.⁶² This was amended so that the £4,000 was to be held instead in trust. Meta's annual allowance would automatically become void if she married, but Sir Watson was also clearly concerned about the capital sum. It may be further evidence that he did not trust Browne, and was ensuring he could not access Meta's inheritance. Judging by the signatories, the amendment was signed in Nyasaland (now Malawi). Meta returned to marry Browne in Marylebone, close to their old home, in the summer of 1926.⁶³ On 27th November 1926, not long after Meta's marriage, Sir Watson sailed from London for Kenya. He was 73 years old and noticeably, this time, was travelling alone.⁶⁴ The ship's name was the *Ormuz*,⁶⁵ and it travelled the route between Britain and Australia, though Sir Watson was only travelling part of the way.

These were his few moments of leisure. Even in retirement, and apart from all the projects and voluntary medical duties, Sir Watson found himself with a host of engagements, many of them outside Shetland. As part of the ongoing commemoration of Lister's work, the Lister Memorial Fund committee, for which Sir Watson had hitherto been Honorary Treasurer, decided in July 1920 that the Royal College of Surgeons would administer the fund, as the original committee could not, "in view of its constitution,"⁶⁶ be permanent. Besides, they were all getting older. Sir Watson had, at least, managed to divest himself of one long-standing commitment. It had nevertheless been decided that a sum of £500 and a bronze medal was to be awarded every three years for distinguished contributions to surgical science, irrespective of nationality,⁶⁷ and the recipient would be required to deliver a lecture in London "under the auspices of the Royal College of Surgeons."⁶⁸ The awarding committee consisted of an illustrious collection of representatives of the Royal Society, the Royal College of Surgeons of England, the Royal College of Surgeons in Ireland, the University of Edinburgh and the University of Glasgow.

In July 1924 they announced that they would be awarding the first Lister medal to Sir William Watson Cheyne, to whom the *BMJ* referred as Lord Lister's "chief disciple."⁶⁹ It was a recognition not only of Sir Watson's loyalty to Lister, but also of his contribution to Lister's achievement, and the fact that, "for many years, he was intimately associated with Lister in his work."⁷⁰ He delivered the lecture at the Royal College of Surgeons, Lincoln's Inn Fields, on Thursday, May 14th, 1925, at 5 o'clock in the afternoon, but true to form, did not feel he could say all he wished about Lister in a single lecture. The upshot was that he wrote his thoughts out in full, condensed them for the benefit of the lecture itself, and then published the whole narrative in book form, adding his remaining text as an appendix.[71] The lecture itself covered thirty-six pages and his appendix ninety-two. He called the book, *Lister and His Achievement*,[72] and published it with Longman, Green and Co., but paid for the publication himself. He later noted that he published it this way in order to be "able to give away copies gratis to those whom I wished to do so."[73] Apart from being an appropriate tribute to Lister, the lectures contain a good deal of autobiographical material on the development of Sir Watson's own professional relationship with him, and have been used ever since as a useful source of information on a man who normally gave little away about himself.

By 1926, he was engaged in correspondence with the Wellcome Museum in London, which was gathering material for a major exhibition to celebrate the centenary of Lister's birth in the next year. Sir Watson called in to the Museum on his way out to Kenya in November 1926, and dropped off some of his Lister memorabilia for the exhibition. He asked Mr Malcolm, curator at the Museum, to keep hold of the items until he returned from Kenya, and only then to send them on to him at Leagarth, an arrangement which was to be the source of some confusion later.[74] He left a forwarding address with the staff at the Wellcome, who could write to him at Tuloa Farm, Molo, Kenya Colony.[75] He would be staying with Robert Pringle [76], who had a plantation in Kenya and was a brother of his daughter-in-law, Nelita, his son Lister's wife.[77] He may have been travelling on to visit Meta.

The Wellcome Museum sent a telegram to Sir Watson at Tuloa Farm on 24th January to remind him they had asked his permission to print copies of his lectures *Lister and His Achievement*.[78] They had written in December of the previous year, presumably just as he arrived in Kenya, and Sir Watson had not responded. He finally replied from the New Stanley Hotel in Nairobi to say that his response had warranted more than a telegram, as the matter was not simple. He had published the work himself, with Longman and Co., and did not think it fair to the publishers to authorise a reprint by the Wellcome. He had only published the work himself to give to friends, and he offered them instead the opportunity to give away or sell any remaining copies from this batch.[79]

He was nevertheless back from his travels in Kenya in time to play a leading role in the event. Mr Malcolm had also asked him for a photograph, and in March, only a month before the centenary celebrations, Sir Watson was forced to admit he had asked Elliott and Fry (his London photographers) if they had any, but the only ones they had given him

Microbes and the Fetlar Man

had "no moustache or no hair at all."[80] By this time, he was writing from somewhere near Ffestiniog in Wales, and was proving a hard man for the Museum to track down. He sat for new photographs and wrote to say he would send them when they were ready.

By 6th April, he was staying at the Langham Hotel in Portland Place in West London, and Mr Malcolm sent him a "programme of the proceedings for tomorrow evening".[81] Mr Wellcome could not be present, so Sir Watson was to open the proceedings with "a cablegram ... from Mr Wellcome, which he wishes to be read by you to the assembly."[82] After the long-winded preparations, the first of Sir Watson's engagements for the Lister centenary around the country went according to plan on 7th. In his address, he "described certain original pieces of apparatus devised by Lister which were on the table before him. They were for the reception and preservation of sterile liquids and for the purpose of holding liquid culture-media, Lister having begun his bacteriological studies before the discovery of solid culture media."[83]. The museum, moreover, had managed to obtain one of Lister's specimens of urine from 1878. It was still sterile.[84]

Sir Watson gave another address at the McEwan Hall for the Edinburgh celebrations for Lister's centenary on July 20th 1927, and contributed to a small book of reminiscences.[85] The major publication to come out of the centenary, however, involved the three addresses he gave on Lister and his work on April 8th at King's College Hospital, which were published as *Three Orations* by John Bale, Sons and Danielsson in London. Coincidentally, or perhaps because of the publicity from the centenary celebrations, he was also awarded an honorary fellowship of the Royal College of Physicians of Edinburgh on July 22nd, along with Sir Hector Cameron, another of Lister's leading disciples.[86]

Sir Watson was still at the Langham the day after his speech at the Wellcome, when Mr Malcolm wrote to thank him.[87] He even wrote to the Wellcome again in June 1927, offering an "imperfect sample of Lister's silver wire needles"[88] he had come across for their Lister exhibition. He also offered to sit for a new portrait during a small window in his schedule when he would be "in London off & on the week 4th July to 12th July"[89] before going "to Edinburgh for an oration".[90] Nevertheless, he noted that "On July 2nd I get the L.L.S. Birmingham & then have to hang about ... at my son's [J.L. Cheyne] at Tidworth till the 11th. Perhaps if you wrote to me at the Langham (to await arrival) about the 2nd or 4th it would be better than replying up here as I may leave before the next post comes."[91] Sir Watson's life was as frenetic as ever, and Mr Johnston-Saint, Secretary of the Wellcome Museum, suggested he only sit for the colours

& tones and that the portrait, by Saloman, be done from a photograph.[92] Presumably by this time he had somehow taken time out to commission Elliott and Fry to produce one with hair and a moustache.

However, he had admitted in a letter to the Museum on 6th July that he was not well, hardly a surprising revelation given the schedule he was setting himself. He noted that he had left his room at the Langham Hotel, as he had been feeling unwell on

> ... Saturday & Sunday after the function at Birmingham University & I gave up my room ... they told me that they would be quite full for some days so I was arranging to go straight through London for Edinburgh on the 12th or 13th [July]. I shall be in London later ... it depends on what I choose for my next winter's cruise."[93] He said that it would be best to wait for the sitting until he reached London, unless ... he [the artist] lives in this neighbourhood when he or I could motor over.[94]

The portrait sitting was still unresolved in November 1927, however, and the long-suffering Mr Malcolm had cause to write to let him know that "Mr Saloman is very anxious to arrange a sitting or an interview with you with regard to painting your portrait. I have told him that we will let him know later as to when you are in London."[95]

On his return from Kenya, something more serious was becoming apparent. Sir Watson was starting to show signs of forgetfulness, and his subsequent correspondence with Mr Malcolm provides us with written evidence of the tales his family later told of the beginnings of a gradual decline in his health. Back in Fetlar after the celebrations, he sent an irritated letter to Mr Malcolm admonishing him for not returning his Lister memorabilia after the exhibition, and noting that he had called in to the Museum, just missed Malcolm, and could find little trace of Lister.

> ... no one knew where the Lister area was & I could only see a picture of Lister himself & of Hector Cameron. Evidently at the present time they are not the interesting parts. Would you please send at once by post my medal and class certificate which I value more than any other of my possessions & send the large picture, chain & instruments as soon as possible so that I can replace them for visitors to see.[96]

Poor Mr Malcolm responded with admirable restraint, apologising for missing him and reminding him that, not only was Lister amply represented

Microbes and the Fetlar Man

in their displays, but that Sir Watson might recall that he had asked him to withhold shipment of the goods to Fetlar until he informed him of his return from Kenya.[97] Sir Watson regained his composure, admitted his error, and thanked Malcolm for holding on to the other material, as the weather had been particularly bad that winter, so sending it on may not have been wise.[98] Mr Malcolm dispatched the memorabilia forthwith, along with a list of items: a large photograph taken at the Banquet to Lord Lister on May 26th, 1897, a framed photograph of Lister on his arrival in Edinburgh in 1852, the certificate of merit awarded to Sir Watson for Clinical Surgery by Lister from the life-changing 1872-73 session, the case of eleven catheter needles Lister had presented to him as first prize in Clinical Surgery in 1873, two cases of bougies which had belonged to Lister, the Queen Anne chair made from the wood of Lister's operating theatre and 34 reprints of Lister's work from medical journals.[99]

It all arrived safely on May 26th, 1928.[100] Shortly afterwards, Sir Watson was due back in London at an engagement for the King's birthday[101], but unfortunately for poor Mr Malcolm, the question of the returned items did not end there. Sir Watson insisted he had loaned the Museum his Lister medal, which had not been returned with the other artefacts, but Mr Malcolm had no record of it.[102] Thankfully, Sir Watson remembered he still had it, admitting in a moment of lucidity on May 26th, "You are quite right about the medal. I remember deciding that I need not send the medal as well as the certificate so I chose the latter."[103] Mr Malcolm may have been mildly alarmed when Sir Watson added, "I may be looking you up about the end of the week or beginning of the following week,"[104] but at least this time he said he would phone ahead so as not to miss him. It also meant they might be able to resolve the issue of the portrait once and for all. It had been completed by now, but the Museum's policy was not to display portraits until the subject had approved it, and Sir Watson's approval was still pending.[105] We must assume he was happy with it on his visit, and that Mr Malcolm was finally free to file the matter away for ever.

Whether this was indeed the early signs of decline, or simply the confusion of a man whose life never stood still for more than a few weeks at a time, we can only speculate. Family recollections place the beginnings of this phase of his illness slightly later, with what was to become his final cruise.[106] His correspondence suggests that he was by now accustomed to spending at least part of each winter out of the country, and had covered much of the globe by this time. In 1928, he set off on a world tour, the highlight of which was a visit to New Zealand. It was hardly his first world

The woman on the right is possibly Ella Davis, which means this was taken on Sir Watson's New Zealand trip in 1928.

tour, but at least this was in more leisurely circumstances than the first. He set off from Southampton in the *Tamaroa* [107] on 9th November 1928 with a family friend, Ella Davis. On the passenger list, Sir Watson, strangely, gave his address as Mrs Davis' home at Bourne in Lincolnshire, unless the crew member filling in the forms had simply assumed their addresses were the same.[108] On the outward journey, he apparently operated on a passenger for appendicitis [109] and took advantage of his stay to give an address to the New Zealand branch of the British Medical Association, but this is likely to have been coincidental. Much more important events had been planned. They landed in Wellington, and were met by members of the city's Shetland Society.[110]

New Zealand had always had a strong Shetland presence. Some had emigrated in search of a better life, or because they were evicted from their homes, and some had gone in search of wealth during the gold rush. Many more were to emigrate with the assisted passages in the 1950s, and a good proportion of New Zealand surnames have a Shetland origin. In fact, the person who had persuaded him to visit New Zealand was its ex-Prime Minister, Robert Stout, another Shetlander and an old friend. He originally came from Lerwick, and had been Prime Minister of the country on two occasions in the 1880s. By this time, he was in his 80s. At the end of 1909, he had returned to London with an illness, and Sir Watson had performed a successful operation on him.[111] He was patron of the Shetland Society of Wellington, and members of the Society were there to greet Sir Watson when he disembarked.[112]

On the Saturday evening, he was entertained by Sir Robert Stout and the Shetland Society at New Century Hall. He even consented to becoming joint patron,[113] along with Stout, which is something of an indication of the esteem in which he was held. A local newspaper referred to him as "one of the greatest, if not the greatest of [Shetlanders]".[114] There was "an unmistakable community spirit,"[115] and 300 [116] people of Shetland extraction had come from all over the country to see him. A romantic eulogy followed on the subject of Norsemen, and a greeting was delivered in Maori by Mr Kingi Tahiwi, with a translation by the Society's president, Mr Isbister.[117] The address was handwritten in an elaborate folder with illuminated letters, which was later presented to their guest, in English on the left, and in Maori on the right.[118] The occasion also just happened to coincide with Sir Watson's 76th birthday, and a huge birthday cake with candles was borne into the room to a violin accompaniment. Sir Watson was visibly moved, and responded by saying, "I am utterly dumbfounded.

Coming Home

I have never been at a gathering like this before."[119] He added that he felt a year younger rather than a year older.[120] He also remarked that he had been particularly interested to hear the address in Maori, as no doubt his father would have been. There followed an evening of "native music and song"[121] with Auld Lang Syne, against the background scenery of a Shetland croft house.

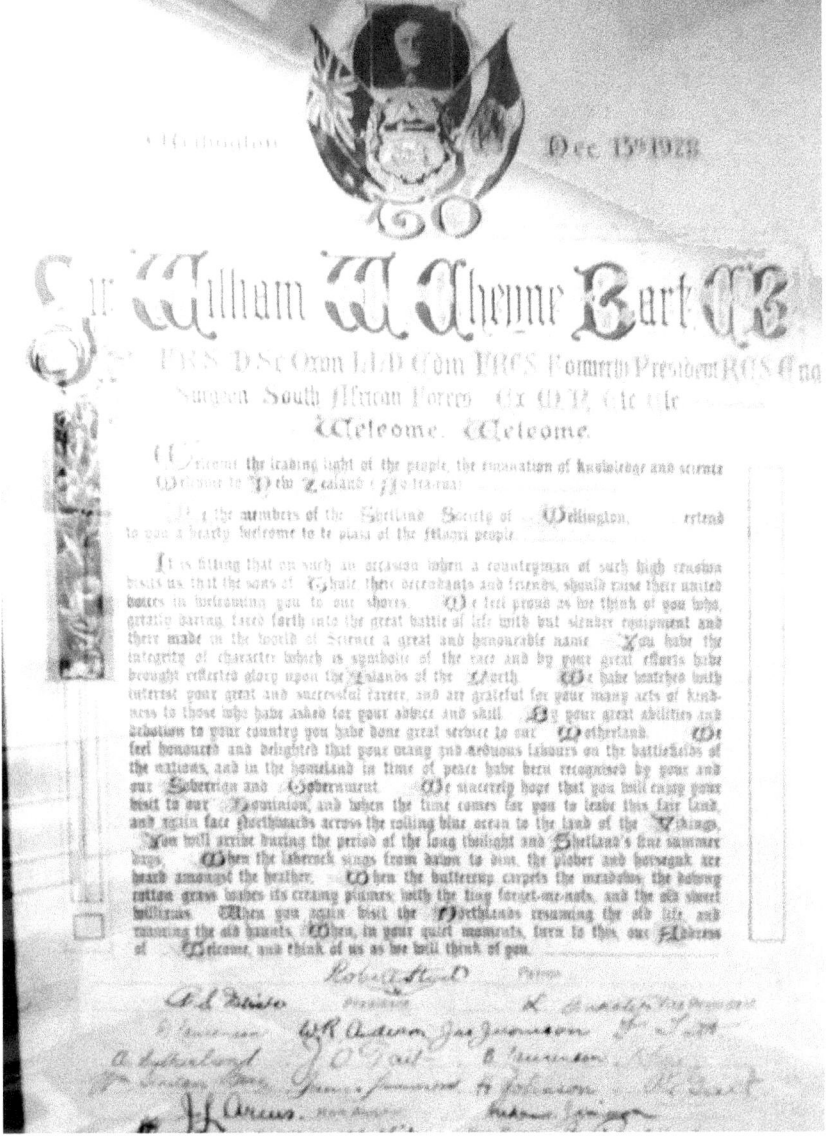

Translation

Haere mai! Haere mai!

Haere mai te tauira o te Iwi, haere mai te mangai o te matauranga. Haere mai ki Ao-tea-roa.

Ko matou ko nga morehu o te Shetland Society i Poneke, takiwa o Ao-tea-roa e powhiri atu nei. Haere mai ki te marae o te Iwi Maori.

Ko tou mana, ko nga mea i oti i runga i tou matauranga kau whakawhanuitia ki nga wahi katoa o te ao. Na reira e to matou huanga, haere mai. Ko enei o nga uri a Thule e mihi atu nei ki a koe, e powhiri atu nei. Eo ora mai te waka, toia mai ki uta.

E tino hari ana matou ki a koe mo tou kaha me to manawanui i a koe i haere nei ma runga i te ara whaiti, i te ara ururua ki te whakahoki mai i nga kete o te matauranga hei painga mo nga Iwi o te ao. I runga i au mahi pai kua rangatira tou ingoa, kua tino rangatira rawa atu hoki te ingoa o to tatou Iwi me nga moutere o te tai ki te raki ara nga moutere o o tatou tipuna. I whai haere tonu o matou whakaaro i a koe i te wa i haere nei koe ki te whakatoa i tou ingoa a e tino mihi atu ana matou ki a koe mo ou arohatanga ki nga tangata e inoi atu ana kia whakatuwheratia mai kia ratou au kete matauranga.

I runga i tou kaha me tou matauranga kua whakawhanuitia nei ki te ao, ka oti i a koe nga mahi e hoki ai te painga ki te kainga o nga tipuna e ngakaunuitia nei e koe. E koa ana, e whakanui atu ana matou ki a ko mo tou toa i te wa o te Pakanga Nui me au mahi pai i te kainga kua whakanuharotia nei e to tatou Kingi me tana Kawanatanga.

E tumanako ana kia tae mai he painga, he hakoakoatanga ki a koe i te wa e noho ai koe ki tenei moutere, a, a te wa e hoki ai koe, e mangai ai te ihu o to waka ki te tai ki te raki, ki te kainga o nga Vikings e tumanako ana me tae koe i te wa e rangona ai te waiata a te Laberock mai ano i te ata tae noa ki te ahiahi, i te wa ano hoki e pa mai ana ki te taringa te reo o te Plover raua ko te Horseguk, i te wa e whakapaipai ana a Papa i tona uma ki nga pua rakau o te Raumati.

Ka tae koe ki te kainga o nga tipuna a i ou kopikopiko haere tanga i ena moutere, mahara mai ano kia matou e noho atu nei e pupuri nei i tou ahuatanga i roto i o matou ngakau.

Kati. Haere mai! Haere mia ki Ao-tea-roa.

The illuminated copy of the welcome speech given to Cheyne by the Shetland Society of Wellington, New Zealand, on his final voyage. The right-hand side is written in Maori.

Sir Watson intended to return via Australia and spend time in Italy before going home.[122] There is no reason to suspect he did not, but in the meantime he made a two-month tour of New Zealand.[123] Some glass slides at Leagarth survive, probably from this tour, and show enigmatic scenes of Ella Davis with raincoat and pickaxe alongside a wooden hut. On the hut is a barely legible sign, which with modern digital techniques can be magnified to be able to read the lettering more or less accurately. It reads, "Supply depot. Civilized only."[124] Close by, if the order of the photos is anything to go by, are three graves, at least two of which appear to belong to victims of a shipwreck. These two are marked by wooden signs, but the third is marked by an (unfortunately illegible) inscribed headstone. All the graves have what seems to be recently-dug soil. The significance of the photographs must, for the moment, remain a mystery. I can find no evidence that he went to Palau on this journey, though it is interesting to speculate whether he ever returned to see the place where his father had died. If, as the newspaper suggests, he was returning via Australia, it is not out of the question he revisited Hobart, the place of his birth.

On the return journey, however, something catastrophic happened. He met with an accident, and hit his head. His grandchildren recall being told that this was generally considered the beginning of a rapid decline in his health, and other sources note that he never fully recovered from the accident.[125] He became more and more forgetful. He was still resident at Leagarth, but was not always himself. When he returned from New Zealand, he corresponded at some length with E.S. Reid Tait, who was compiling a list of his and Captain Cheyne's publications, and who may have visited him at Leagarth in April 1929.[126] He was no longer there by November, when someone had given the Minister, Mr Carson, permission to go through the house to look for papers on Captain Cheyne. He was accompanied by Jimmy Young (Robertson), who was probably taking care of the house by this time.[127] Mr Reid Tait was still compiling information about Andrew, and Sir Watson was clearly no longer there to look for the information at Leagarth. Sir Watson's family ultimately arranged for him to be admitted to Holloway Sanatorium in Virginia Water in Surrey.

If his memory had gone, he may at least have been spared the pain of Meta's death. According to the plaque erected to her memory in the Fetlar Kirk, she died in Tanganyika (now Tanzania) in 1929,[128] but oddly, official records suggest she died slightly later.[129] She was only in her early thirties, and her marriage, however it turned out to be, had been short-lived. Sir Watson's grandson thought she died of cancer,[130] and if so, it is yet another irony, given the volume of work Sir Watson had done on the subject.

Probably Ella Davis holding a pickaxe outside a hut, on Sir Watson's last journey to New Zealand.

Sir Watson died at Virginia Water on 19th April, 1932.[131] His remains were cremated, and his ashes brought home to Fetlar, where he was finally laid to rest at Tresta, in the cemetery close to the Manse. It was where he had begun his long journey at the age of two, and he was, fittingly, as close to the sea as anyone could be buried on land.

His legacy was profound. Materially, he left an estate in England of £58,000.[132] He had "disposed of his Scottish property by a trust disposition."¹³³ He left £4,000, his medical instruments and the copyright of his books to his son Hunter, and the remaining property, including Leagarth, to his oldest son Joseph Lister Cheyne, who also inherited the baronetcy and £4,000. Sir Watson left a year's wages to the indoor and outdoor servants who had been with him for at least two years.¹³⁴

Hunter Cheyne, who followed his father into medicine.

Sir Lister Cheyne, as Sir Watson's eldest son now became, had inherited Leagarth, but his life was elsewhere. He did not intend initially to keep it, and intended instead to keep a boat on which the family had been used to spending their summer holidays. He could not afford to maintain both, and initially decided against Leagarth. When it was valued, however, it was clear a sale would not raise much and it seemed easier to keep the house on and sell the boat.[135] It was a fortunate decision in many ways. Lister Cheyne's children, his grandchildren and great-grandchildren have spent summers in Leagarth ever since, keeping a visitors' book just like the ones kept during Sir Watson's time. His Boer War photographs showing Cronje's surrender still hang on the stairway. The gardens, sadly, deteriorated when the wooden paling around them eventually came down, and his laboratory ultimately crumbled. The remainder of the house, however, was protected to a degree by the glass verandah he built, and I had the privilege of caring for it for some years over the winters when the various branches of the family were away. It has recently undergone extensive renovations, and the Cheyne family is committed to maintaining it.

On April 25th 1932, his friends and colleagues gathered to honour him at King's College Hospital, where Dr Wilfred Trotter represented the King. A Shetland service was held for him, as former Lord Lieutenant of the County, at St. Columba's Church in Lerwick on Sunday, May 1st at 3 in the afternoon.[136] Tributes poured in when news of his death reached London, and glowing obituaries appeared in the medical and scientific press, as well as a number of local and national papers. He was described as "a true man of science as well as a great clinician,"[137] with "a warm heart and a tender nature, which he hid deliberately, or from a sort of innate shyness."[138] None of the writers seemed able to resist highlighting his opposition to 'aseptic' surgical methods, but *The Lancet* came perhaps the closest when it said, "He could not suffer fools but none could want for a truer or stauncher friend, and few men who have reached the high position to which he attained were more deservedly trusted and honoured."[139]

The Royal Society noted that "overwork in the Great War and in the House of Commons aged this otherwise strong and active man rather quickly",[140] but that he was fortunate enough to be able to "spend his remaining years in simplicity and tranquillity, delighted that he was able to go to bed and leave the door open."[141] The writer called him "a straight fighter", who "bore no ill-feeling in defeat,"[142] and added, in a moment of light-heartedness, that his well-known "worship of My Lady Nicotine"[143] would no doubt have upset his grandfather, who had expressed strong views on the subject.

Sir Watson's grandchildren on Tresta Beach, Fetlar. Left–right: Andrew (Sandy), Joseph (later Sir Joseph) and William (Bill).

Sir Watson would have been particularly gratified to learn that he was remembered as a Shetlander. *The Lancet* noted that "he was at his best when on holiday at his island home in Shetland where he was greatly beloved; lucky were those who were privileged to be his guests, and they were many."[144] He would have been equally delighted to hear that his colleagues associated him unequivocally with the sea, his abiding passion, and that he "handled boats with consummate skill."[145] The *British Medical*

Microbes and the Fetlar Man

Journal summed it up by saying, "The last active years of his life were largely devoted to sailing on his beloved ocean around his native island."[146]

Sir Watson's final resting place in the Fetlar Kirkyard

Coming Home

After Sir Watson's death, in 1933, many of the original contents of the house were sold at an auction, which took place at Leagarth. People came from all over Shetland and from further afield, and Sir Watson's effects were scattered to the four corners of the earth. Sadly, any papers which were not considered important were burned in the grounds of the house, but as I have noted, a number of them were rescued by J.J. Laurenson and sent to London.

In terms of medical science, his legacy was extraordinary. He had dedicated his life to defending and adapting Lord Lister's antiseptic principle, and had ultimately been recognised for his own contribution to the bacteriological revolution in surgery. To the island of Fetlar, and the seas around it, he had dedicated what was left. Knowing how the wind can whistle around the verandah on a stormy day at Leagarth, I suspect that somewhere in the eaves of the house Sir Watson can still sometimes be heard, with Brandy the dog under his arm, coming into the hall to say goodnight.

Sir Watson and his dog, outside the verandah at Leagarth.

Notes and References

Introduction
1 Worboys, M., *Spreading Disease: Disease Theories and Medical Practice in Britain, 1865-1900*, Cambridge University Press, 2000:151.
2 Ibid:8.
3 *The Third King's College Hospital and the Medical School 1913-1918*, Surgeon-Rear-Admiral Sir William Watson Cheyne Bt.:339.
4 Ibid.
5 Cheyne, W.W., *Lister and His Achievement*, London (Longman, Green and Co.), 1925:24.

Chapter 1
1. Worboys, M., *op. cit.*,:151.
2. Shineberg, D., *They Came for Sandalwood: a study of the sandalwood trade in the south-west Pacific, 1830-1865*. Melbourne University Press, 1967:2.
3. Shineberg, D. (ed.), *The Trading Voyages of Andrew Cheyne 1841-1844*, Canberra (Australian National University Press), 1971:5.
4. Ibid.
5. Ibid:4.
6. Fenton, A., *The Northern Isles: Orkney and Shetland*, East Lothian (Tuckwell Press), 1997. Chapter 66: Haaf Fishing.
7. Johnson, R., From a Minister's Diary, Lerwick, *Shetland Life*, May 1987:21.
8. Heritors were generally local landowners who were expected to contribute financially to parish obligations such as the maintenance of the Manse, and funds for the poor. They often preferred to make voluntary contributions to a poor fund rather than be assessed through tax on the basis of their property. Website of the National Archives of Scotland, Records of the Poor, http://www.nas.gov.uk/guides/poor.asp accessed 26/08/2013.
9. Ibid.
10. Shineberg, D. (ed.), *op.cit.*, 1971:4.
11. Ibid:3.
12. Cheyne family records, and Manson, T.M.Y., Transcript of a lecture given to the Northmavine History Group at the Sullom Public Hall on 21st March 1989, copy at Fetlar Museum Trust, Fetlar Interpretive Centre, Fetlar, Shetland.
13. Mack, J., Thomas Irvine of Midbrake (II), *Shetland Life*, June 1988, Volume 92:37.
14. Cheyne family records, and Manson, T.M.Y., Transcript of a lecture given to the Northmavine History Group at the Sullom Public Hall on 21st March 1989, copy at Fetlar Museum Trust, Fetlar Interpretive Centre, Fetlar, Shetland.
15. Shineberg, D. (ed.), *op.cit.*:16-17.
16. Ibid:16.
17. Ibid.
18. Cheyne family records, and Manson, T.M.Y., Transcript of a lecture given to the Northmavine History Group at the Sullom Public Hall on 21st March 1989, copy at Fetlar Museum Trust, Fetlar Interpretive Centre, Fetlar, Shetland.

19. Letter to Sir Joseph Cheyne from Dorothy Shineberg, dated 10th August 1992, Shetland Archives, Lerwick. D1/542/13/22.
20. For example, entry for Sunday 26th August 1849: "Capt. & Miss Cheyne came in after breakfs.", diary of the Reverend William Watson, 1849-1854, Shetland Archives, Lerwick. D1/542.
21. Manson, T.M.Y., 1968, in Shineberg, D.,(ed.) *op.cit.*:3.
22. Shineberg, D. (ed.), *op.cit.*:4.
23. Ibid.
24. Obituary of Harry Cheyne, *The Scotsman*, Edinburgh, Monday, February 29th, 1915. "... the fourth of the five stalwart sons of Mr Henry Cheyne, W.S., of Tangwick, Shetland ... The eldest son, John, destined to become Procurator of the Church ... Mr Cheyne was admitted to the Society of Writers of the Signet in 1868 ... to the present generation he was known as a partner in the firm of Mackenzie and Kermack."
25. Shineberg, D. (ed.), *op.cit.* 1971:3.
26. Ibid:3. Also: Cheyne family records, and Manson, T.M.Y., Transcript of a lecture given to the Northmavine History Group at the Sullom Public Hall on 21st March 1989, copy at Fetlar Museum Trust, Fetlar Interpretive Centre, Fetlar, Shetland.
27. Letter from William Watson to a friend, 27th November 1848, Shetland Archives, Lerwick. D1/542/13/3.
28. Shineberg, D. (ed.), *op.cit.*:24.
29. Catalogue to the Manse Library at Hillswick, Shetland Archives, Lerwick. D25/32.
30. Estensen, Miriam, The Life of Matthew Flinders, New South Wales (Allen and Unwin), 2003:5.
31. Shineberg, D. (ed.), *op.cit.*:4.
32. Shineberg, D., *They Came for Sandalwood: a study of the sandalwood trade in the south-west Pacific, 1830-1865*, Melbourne University Press, 1967:87.
33. Ibid. Chapter 6: The Jobs, the Ships and the People;82-97.
34. Cheyne, A., in Shineberg, D. (ed.), *op.cit.*:5-7.
35. Shineberg, D., *They Came for Sandalwood: a study of the sandalwood trade in the south-west Pacific, 1830-1865*, Melbourne University Press, 1967:203.
36. Cheyne, A., in Shineberg, D. (ed.), *op.cit.*:83-84.
37. Cheyne, A., in Shineberg, D. (ed.), *op.cit.*:64.
38. Ibid.
39. Ibid.
40. Ibid.
41. Cheyne, A., in Shineberg, D. (ed.), *op.cit.*:7.
42. Shineberg, D. (ed.), *op.cit.*:1.
43. Ibid.
44. Shineberg, D. (ed.), *op.cit.*
45. Ibid:58.
46. Cheyne, A., 1855:21 in ibid.
47. Shineberg, D. (ed.), *op.cit.*:53.
48. Ibid:13.
49. Ibid.
50. Ibid.
51. Shineberg, D., *They Came for Sandalwood: a study of the sandalwood trade in the south-west Pacific, 1830-1865*. Melbourne University Press, 1967: 168-169.
52. Ibid.
53. Cheyne, A., 1855:21 in Shineberg, D. (ed.), *op.cit.*:13.
54. Ibid:16.

Notes and References

55. Squires, N., The Man who lost a "coral kingdom", BBC news, June 7th, 2007. http://news.bbc.co.uk/2/hi/programmes/from_our_own_correspondent/6730047.stm Accessed 26/8/2013.
56. Darwin, Charles, *The Voyage of the Beagle*, 1845 second edition, Hertfordshire (Wordsworth Editions Limited), 1997:429.
57. Squires, N., The Man who lost a "coral kingdom", BBC news, June 7th, 2007. http://news.bbc.co.uk/2/hi/programmes/from_our_own_correspondent/6730047.stm Accessed 26/8/2013.
58. Cheyne, A., 1855:16 in Shineberg, D. (ed.), T *op.cit.*:58.
59. Ibid.
60. He was born in 1770, according to Scott, H., *Fasti Ecclesiae Scoticanae, The Succession of Ministers in the Church of Scotland from the Reformation*, Edinburgh (Oliver and Boyd), 1928:297.
61. Reverend William Watson, in Shineberg, D., *op.cit.*:17.
62. Reverend William Watson, in Shineberg, D. (ed.), *op.cit.*:17.
63. Letter from William Watson to a friend, dated 21st February 1852, Shetland Archives, Lerwick, Shetland. D1/542/13/10.
64. Scott, H., *Fasti Ecclesiae Scoticanae, The Succession of Ministers in the Church of Scotland from the Reformation*, Edinburgh (Oliver and Boyd), 1928:297.
65. Ibid.
66. Australian Government website http://australia.gov.au/about-australia/australian-story/convicts-and-the-british-colonies Accessed 26/08/2013.
67. Loch, John D., Immigration Agent, Hobart Town, letter dated 31st January 1853, as part of Despatches from Lieutenant-Governor Sir W. Denison:Van Diemen's Land, sent 10th January 1853, Hansard, Parliamentary Papers, 1854:56 (436) (436-I) Emigration (Australia).
68. Ibid:56.
69. Ibid.
70. Ibid:57.
71. Ibid:56.
72. Ibid:56.
73. Ibid:63.
74. Ibid:56 and 63.
75. Shetland Museum and Archives.
76. Johnson, Robert L., The deserted homesteads of Fetlar, *Shetland Life*, Vol. 13, November 1981:26.
77. Loch, John D., Immigration Agent, Hobart Town, letter dated 31st January 1853, as part of Despatches from Lieutenant-Governor Sir W. Denison:Van Diemen's Land, sent 10th January 1853, Hansard, Parliamentary Papers, 1854:63 (436) (436-I) Emigration (Australia).
78. Letter from William Watson to a friend, dated 21st February 1852, Shetland Archives, Lerwick. D1/542/13/10.
79. Diary of William Watson, Saturday 14th February, 1852, Shetland Archives, Lerwick. D1/542.
80. Diary of William Watson, Thursday 19th February, 1852, Shetland Archives, Lerwick. D1/542.
81. Diary of William Watson, Monday, 29th March 1852, Shetland Archives, Lerwick. D1/542.
82. Letter from Eliza Cheyne to Christian Webster, Edinburgh, 8th April 1852, Shetland Archives, Lerwick. D1/542/13/11.
83. Ibid.

84. Contract signed by Captain Andrew Cheyne and others to transport 290 named convicts to Van Diemen's Land, National Archives, Kew, TS18/498.
85. Ibid.
86. Ibid.
87. Johnson, R.L., From a Minister's Diary, *Shetland Life*, May 1987:21.
88. Letter from Eliza Cheyne to Christian Watson, April 8th 1892, Shetland Archives, Lerwick. D1/542/13/11.
89. Mortality on Board the Ship Lady Montague, *The Courier*, Hobart, Wednesday, 3rd November 1852, http://trove.nla.gov.au/ndp/del/article/2958264 Accessed 17/03/2013.
90. Letter from the Home Office to the Office of the Committee of the Council for Trade, 10th June 1854, National Archives, Kew. HO 45/5814 7721.
91. Contract signed by Captain Andrew Cheyne and others to transport 290 named convicts to Van Diemen's Land, National Archives, Kew, TS18/498.
92. Passenger Arrivals, Tasmania, Australia, December, 1852. Accessed through www.ancestry.co.uk 17/03/2013.
93. Shineberg, D. (ed.), *op.cit.*:2.
94. Shipping Intelligence, Port of Hobart Town: Departures, Colonial Times, Hobart, Tasmania, Friday, 21st January 1853, "Jan. 19th – Lady Montague ... Passengers - Mrs Cheyne, child and servant." http://trove.nla.gov.au/ndp/del/article/8772820?searchTerm=Captain%20Andrew%20Cheyne&searchLimits Accessed 23/03/2013.
95. Passenger Arrivals, Tasmania, Australia, December 1852. Accessed through www.ancestry.co.uk 17/03/2013.
96. Journal of Samuel Donnelly, Surgeon on the Lady Montague, National Archives, Kew: ADM 101/254, 1A-1H.
97. Letter from Mr Addington at the Foreign Office, dated September 8th, 1853, National Archives, Kew, HO45/4725.
98. Ibid.
99. Ibid.
100. Ibid.
101. Journal of Samuel Donnelly, Surgeon on the Lady Montague, National Archives, Kew: ADM 101/254, 1A-1H.
102. Passenger Arrivals, Tasmania, Australia. December 1852. Accessed through www.ancestry.co.uk 17/03/2013.
103. Journal of Samuel Donnelly, Surgeon on the Lady Montague, National Archives, Kew: ADM 101/254, 1A-1H.
104. Ibid.
105. Chapman, T.D. and 8 other signatories, Southern Tasmanian Council of the League Protest Against the Convicts per Lady Montague, 11th December, 1852, Colonial Times, Hobart, Tasmania, Tuesday 14th December 1852. Accessed: 17/03/2013 via http://trove.nla.gov.au/ndp/del/article/8772544
106. Headlam, Charles, Letter to The Right Hon. Sir John S. Pakington, Bart., printed in Launceston Examiner, Tasmania, February 25th, 1854:2. http://trove.nla.gov.au/ndp/del/article/36288467 Accessed 16/03/2013.
107. Manson, T.M.Y., Transcript of a lecture given to the Northmavine History Group at the Sullom Public Hall on 21st March 1989, copy at Fetlar Museum Trust, Fetlar Interpretive Centre, Fetlar, Shetland.
108. Shineberg, D. (ed.), *op.cit.*:18.

109. Loch, John D., Immigration Agent, Hobart Town, letter dated 31st January 1853, as part of Despatches from Lieutenant-Governor Sir W. Denison: Van Diemen's Land, sent 10th January 1853, Hansard, Parliamentary Papers, 1854:56 (436) (436-I) Emigration (Australia).
110. Manson, T.M.Y., Transcript of a lecture given to the Northmavine History Group at the Sullom Public Hall on 21st March 1989, copy at Fetlar Museum Trust, Fetlar Interpretive Centre, Fetlar, Shetland.
111. Ibid.
112. Journal of Samuel Donnelly, Surgeon on the Lady Montague, National Archives, Kew: ADM 101/254, 1A-1H.
113. Diary of William Watson, 13th April 1853, Shetland Archives, Lerwick. D1/542:324.
114. Shineberg, D. (ed.), *op.cit*.18.
115. Loch, John D., Immigration Agent, Hobart Town, letter dated 31st January 1853, as part of Despatches from Lieutenant-Governor Sir W. Denison: Van Diemen's Land, sent 10th January 1853, Hansard, Parliamentary Papers, 1854:56 (436) (436-I) Emigration (Australia).
116. Shineberg, D., *They Came for Sandalwood: a study of the sandalwood trade in the south-west Pacific, 1830-1865*, Melbourne University Press, 1967:31.
117. Diary of William Watson, Thursday June 9th 1853, Shetland Archives, Lerwick. D1/542.
118. Shineberg, D. (ed.), *op.cit*. :18.
119. Ibid:19.
120. Cheyne, A., *Sailing Directions from New South Wales to China and Japan, Including the Whole Islands and Dangers in the Western Pacific Ocean and Coasts of New Guinea and Safest Route through Torres Strait*, London (J.D. Potter), 1855:174-175.
121. Shineberg, D. (ed.), *op.cit*.:19.
122. Law Report: U.S. District Court, Daily Alta, California, 6th October, 1853, Volume 4, No.267. http://cdnc.ucr.edu/cdnc/cgi-bin/cdnc?a=d&cl=search&d=DAC18531006.2.16&srpos=1&e=06-10-1853-06-10-1853--en--20-DAC-1--txt-IN-Cheyne---- Accessed 17/03/2013
123. Shineberg, D., *They Came for Sandalwood: a study of the sandalwood trade in the south-west Pacific, 1830-1865*. Melbourne University Press, 1967:91.
124. Shineberg, D. (ed.), *op.cit*.:25.
125. Howden Smith, Arthur D. (ed.), *The Narrative of Samuel Hancock, 1845-1860*, New York, 1927, http://web.uvic.ca/~qxm/hancocksmallpoxonly.pdf Accessed 03/06/2012.
126. Letter from John Davey, loose page inside the log books of the schooner Acis 1863-66, Shetland Archives, Lerwick, Shetland. D1/542.
127. San Francisco Earthquake History 1769-1879, The Virtual Museum of the City of San Francisco, http://www.sfmuseum.org/alm/quakes1.html Accessed 17/03/2013.
128. Diary of the Reverend William Watson, 22nd December 1853, Shetland Archives, Lerwick. D1/542.
129. Shineberg, D. (ed.), *op.cit*.:19.
130. Loch, John D., Immigration Agent, Hobart Town, letter dated 31st January 1853, as part of Despatches from Lieutenant-Governor Sir W. Denison: Van Diemen's Land, sent 10th January 1853, Hansard, Parliamentary Papers, 1854:56 (436) (436-I) Emigration (Australia).

131. Darwin, Charles, *The Voyage of the Beagle*, 1845, second edition, Hertfordshire (Wordsworth Editions Limited), 1997:350.
132. Ibid.
133. Ibid.
134. Shineberg, D. (ed.), *op.cit.*:19.
135. Ibid.

Chapter 2

1. Shineberg, D. (ed.), *op.cit.*:19.
2. Fetlar tradition, mentioned in i.a. Manson, T.M.Y., Transcript of a lecture given to the Northmavine History Group at the Sullom Public Hall on 21st March 1989, copy at Fetlar Museum Trust, Fetlar Interpretive Centre, Fetlar, Shetland.
3. Loch, John D., Immigration Agent, Hobart Town, letter dated 31st January 1853, as part of Despatches from Lieutenant-Governor Sir W. Denison:Van Diemen's Land, sent 10th January 1853, Hansard, Parliamentary Papers, 1854:56 (436) (436-I) Emigration (Australia).
4. Shineberg, D. (ed.), *op.cit.*:18.
5. Fetlar tradition, mentioned in i.a. Manson, T.M.Y., Transcript of a lecture given to the Northmavine History Group at the Sullom Public Hall on 21st March 1989, copy at Fetlar Museum Trust, Fetlar Interpretive Centre, Fetlar, Shetland.
6. The Courier, Hobart, Wednesday, 3rd November 1852, http://trove.nla.gov.au/ndp/del/article/2958264 Accessed 17/03/2013.
7. Cheyne, A., *Sailing Directions from New South Wales to China and Japan*, London, (J.D. Potter), 1855. Copy in the British Library, London. 1500/220, inscription on inside front cover: "To Henry Cheyne, Esq., N.S., 6 Royal Terrace, Edinburgh."
8. Shineberg, D. (ed.), *op.cit.*: 19.
9. Ibid.
10. *Nautical Magazine and Naval Chronicle*, 24 (September 1855), Cambridge University Press, 2013.
11. Shineberg, D. (ed.), *op.cit.*:19.
12. Ibid: 2.
13. Fetlar Parish Records, Shetland, in Shineberg, D., (ed.) *op.cit.*:19.
14. Johnson, R.L., unpublished account in possession of the Cheyne family. Johnson confirms that the story was told him by his own aunt and uncle, who were from Fetlar, and who had been told the story by people who were old enough to have known the young William Watson as a boy.
15. Shineberg, D. (ed.), *op.cit.*:19.
16. He was born in 1770, according to Scott, H., *op.cit.*:297.
17. Inventory of the personal estate and extract trust disposition and settlement of the Reverend William Watson, minister of Fetlar and North Yell, Shetland Archives, Lerwick. SC12/36/4:42.
18. Ibid.
19. She was born 6th May 1811, according to Scott, H., *op.cit.*:297.
20. She was born 27th April 1813, according to Scott, H., *op.cit.*:297.
21. Inventory of the personal estate and extract trust disposition and settlement of the Reverend William Watson, minister of Fetlar and North Yell, Shetland Archives, Lerwick. SC12/36/4.
22. Ibid.
23. Personal communication, Sir Joseph Cheyne to author, 2000. See also Shineberg, D. (ed.), *op.cit.*:19.
24. Ibid.

25. Inventory of the personal estate and extract trust disposition and settlement of the Reverend William Watson, minister of Fetlar and North Yell, Shetland Archives, Lerwick. SC12/37/4:42.
26. Manson, T.M.Y., Transcript of a lecture given to the Northmavine History Group at the Sullom Public Hall on 21st March 1989:5, copy at Fetlar Museum Trust, Fetlar Interpretive Centre, Fetlar, Shetland.
27. Shineberg, D., *They Came for Sandalwood: a study of the sandalwood trade in the south-west Pacific, 1830-1865*. Melbourne University Press, 1967.
28. Cheyne, A., February 1844, in Shineberg, D. (ed.), *op.cit.*: 308-9.
29. Shineberg, D. (ed.), *op.cit.*:25.
30. Ibid:13.
31. Ibid:22.
32. Tetens, A., *Among the Savages of the South Seas: Memoirs of Micronesia*, Oxford University Press and Stanford University Press, 1958:4-6.
33. Ibid.6.
34. Ibid:5.
35. Shineberg, D. (ed.), *op.cit.*:22.
36. Stevens, Charles E., 1867, Report of Proceedings at the Pelew islands in the Matter of the Murder of Andrew Cheyne, National Archives, Kew, FO72/1155.
37. Shineberg, D. (ed.), *op.cit.*:25.
38. Ibid:24.
39. Stevens, Charles E., 1867, Report of Proceedings at the Pelew islands in the Matter of the Murder of Andrew Cheyne, National Archives, Kew, FO72/1155.
40. Ibid.
41. Letter from Dorothy Shineberg to Sir Joseph Cheyne, 29th December, 1985. Shetland Archives, Lerwick, Shetland. D1/542/15/14.
42. Shineberg, D., *They Came for Sandalwood: a study of the sandalwood trade in the south-west Pacific, 1830-1865*, Melbourne University Press, 1967:139-40.
43. Ibid:140.
44. Ibid:139-40.
45. Hansard, Parliamentary Papers, 1854 (436) (436-I) Emigration (Australia).
46. Personal communication to author by John Coutts, 23/03/2013.
47. Personal communication to author by Sir Joseph Cheyne, 2000.
48. Correspondence between David Webster and Sir Arthur Nicolson, 1856-1863, Shetland Archives, Lerwick. D.24, Box 38. Item 7.
49. Letter from the Rev. William Watson to Sir Arthur Nicolson, dated 7th August 1839, copy in possession of Cheyne family.
50. Fenton, A., *The Northern Isles: Orkney and Shetland*, Tuckwell Press, 1997:606.
51. i.a. Collier, A., *The Crofting Problem*, Cambridge University Press, 1953:30.
52. Personal communication to author by John Coutts, 2010, who heard the story from J.J. Laurenson, Fetlar, as a child.
53. Johnson, Robert L., The deserted homesteads of Fetlar, *Shetland Life*, Vol. 13, November 1981:26.
54. A daguerreotype of Williamson, probably taken in California, forms part of the collection of Fetlar Museum Trust, Fetlar Interpretive Centre, Fetlar, Shetland. F-2513.
55. *Second Statistical Account*, 1834-45: United Parishes of Fetlar and North Yell, Volume 15:25.
56. Johnson, Robert L., unpublished manuscript on the life of Sir William Watson Cheyne, dated April 1982:8, copy belonging to the Cheyne family.
57. Manson, T.M.Y., Transcript of a lecture given to the Northmavine History Group at the Sullom Public Hall on 21st March 1989:4, copy at Fetlar Museum Trust, Fetlar, Shetland.

58. Johnson, R.L., From a Minister's Diary, *Shetland Life*, April, 1987:21.
59. Ibid.
60. Personal communication to author, John Coutts, on a Fetlar tradition, 27/04/2013. He had heard the story from J.J. Laurenson in Fetlar as a child.
61. Ibid.
62. Letter from David Webster to Sir Arthur Nicolson, dated 20th January 1857, Shetland Archives, Lerwick. D24, Box 38, Item 7.
63. Ibid.
64. Scott, H., *op.cit.*:297.
65. 1861 Census: Fetlar and North Yell, ScotlandsPeople: http://www.scotlandspeople.gov.uk/.
66. Ibid.
67. Fenton, A., *The Northern Isles: Orkney and Shetland* Tuckwell Press, 1997:67.
68. Manson, T.M.Y., Transcript of a lecture given to the Northmavine History Group at the Sullom Public Hall on 21st March 1989, copy at Fetlar Museum Trust, Fetlar, Shetland.
69. Rules for the Management of the Fetlar Parochial Library, September 18th, 1828, Shetland Archives, Lerwick. D24, Box 37, Item 1. The rules are unlikely to have changed very much by Cheyne's time.
70. Catalogue of the Parish Library of Fetlar, 1831, Shetland Archives, Lerwick. Box 37, Item 1.
71. Cheyne, Sir William Watson (1852-1932), Plarr's Lives of the Fellows Online. http://livesonline.rcseng.ac.uk/biogs/E000222b.htm Accessed 24/03/2013.
72. Manson, T.M.Y., Transcript of a lecture given to the Northmavine History Group at the Sullom Public Hall on 21st March 1989:7, copy at Fetlar Museum Trust, Fetlar, Shetland.
73. John Webster graduated from the University of Aberdeen in 1869: http://www.ebooksread.com/authors-eng/university-of-aberdeen/roll-of-the-graduates-of-the-university-of-aberdeen-1860-1900-hci/page-35-roll-of-the-graduates-of-the-university-of-aberdeen-1860-1900-hci.shtml Accessed 24/03/2013.
74. Johnson, R.L., unpublished manuscript on the life of Sir William Watson Cheyne, dated April 1982:9 copy belonging to the Cheyne family.
75. Letter from William Watson to a friend, dated 27th November 1848, Shetland Archives, Lerwick. D1/542/13/3.
76. Johnson, R.L., unpublished manuscript on the life of Sir William Watson Cheyne, dated April 1982:9, copy belonging to the Cheyne family.
77. Letter from Reverend David Webster to Sir Arthur Nicolson, dated 7th April 1857, Shetland Archives, Lerwick. D24, Box 38, Item 7.
78. Letter from Reverend David Webster to Sir Arthur Nicolson, dated 12th May 1857, Shetland Archives, Lerwick. D24, Box 38, Item 7.
79. Letter from Reverend David Webster to Sir Arthur Nicolson, dated 7th April 1857, Shetland Archives, Lerwick. D24, Box 38, Item 7.
80. Letter from Reverend David Webster to Sir Arthur Nicolson, dated 12th May 1857, Shetland Archives, Lerwick. D24, Box 38, Item 7.
81. Letter from the Rev. William Watson to Sir Arthur Nicolson, dated 7th August 1839, in possession of Cheyne family.
82. Letter from Reverend David Webster to Sir Arthur Nicolson, dated 12th May 1857, Shetland Archives, Lerwick. D24, Box 38, Item 7.
83. Letter from Reverend David Webster to Sir Arthur Nicolson, dated 17th July 1857, Shetland Archives, Lerwick. D24, Box 38, Item 7.
84. Ibid.
85. Ibid.

Notes and References

86. Ibid.
87. Letter from Reverend David Webster to Sir Arthur Nicolson, dated 17th July 1857, Shetland Archives, Lerwick. D24, Box 38, Item 7.
88. Letter from Reverend David Webster to Sir Arthur Nicolson, dated 7th August 1857, Shetland Archives, Lerwick. D24, Box 38, Item 7.
89. Ibid.
90. Letter from Reverend David Webster to Sir Arthur Nicolson, dated 7th September 1857, Shetland Archives, Lerwick. D24, Box 38, Item 7.
91. Ibid.
92. Ibid.
93. Ibid.
94. Letter from Reverend David Webster to Sir Arthur Nicolson, 29th March 1858, Shetland Archives, Lerwick. D24, Box 38, Item 7.
95. Letter from Reverend David Webster to Sir Arthur Nicolson, 19th April 1858, Shetland Archives, Lerwick. D24, Box 38, Item 7.
96. Ibid.
97. Letter from Reverend David Webster to Sir Arthur Nicolson, 2nd May 1859, Shetland Archives, Lerwick. D24, Box 38, Item 7.
98. Letter from Reverend David Webster to Sir Arthur Nicolson, 11th May 1859, Shetland Archives, Lerwick. D24, Box 38, Item 7.
99. Letter from Reverend David Webster to Sir Arthur Nicolson, 13th May 1859, Shetland Archives, Lerwick. D24, Box 38, Item 7.
100. Shineberg, D. (ed.), *op.cit.*:20.
101. Ibid.
102. Shineberg, D., *They Came for Sandalwood: a study of the sandalwood trade in the south-west Pacific, 1830-1865*. Melbourne University Press, 1967:145.
103. Cheyne, A., in Shineberg, D. (ed.), *op.cit.*:59.
104. Cheyne, A., in Shineberg, D., (ed.) *op.cit.*:21.
105. Ibid:22.
106. Ibid:23.
107. Hezel, Francis X., *The First Taint of Civilisation, a history of the Caroline and Marshall Islands,* University of Hawaii Press, 1994:307.
108. Tetens, A., *op.cit.*:9.
109. Cheyne, A., in Shineberg, D. (ed.), *op.cit.*:23.
110. Shineberg, D. (ed.), *op.cit.*:23.
111. Ibid.
112. Anderson High School website: http://www.anderson.shetland.sch.uk/wordpress/ahs/history-of-the-school/ Accessed 26/03/2013.
113. Manson, T.M.Y., Transcript of a lecture given to the Northmavine History Group at the Sullom Public Hall on 21st March 1989, copy at Fetlar Museum Trust, Fetlar, Shetland.
114. MacCarthy, F., *Byron: Life and Legend*, Faber and Faber Ltd., 2005:11.
115. Obituary, Sir William Watson Cheyne, Bt., K.C.M.G., C.B., F.R.S., F.R.C.S., *BMJ*, April 30th, 1932. 1932:821. (*BMJ* 1932;1:821)
116. Inventory of the personal estate and extract trust disposition and settlement of the Reverend William Watson, minister of Fetlar and North Yell, Shetland Archives, Lerwick, Shetland. SC12/36/4:42.
117. http://www.measuringworth.com/ukcompare/relativevalue.php), accessed 26/03/2013.
118. Manson, T.M.Y., Transcript of a lecture given to the Northmavine History Group at the Sullom Public Hall on 21st March 1989:7, copy at Fetlar Museum Trust, Fetlar, Shetland.

119. Aberdeen Grammar School website, http://grammar.org.uk/visitors/history.php, accessed 26/03/2013.
120. Obituary, Sir William Watson Cheyne, Bt., K.C.M.G., C.B., F.R.S., F.R.C.S., *BMJ*,April 30th, 1932. 1932:821. (*BMJ* 1932;1:821)
121. Manson, T.M.Y., Transcript of a lecture given to the Northmavine History Group at the Sullom Public Hall on 21st March 1989:7, copy at Fetlar Museum Trust, Fetlar, Shetland.
122. Shineberg, D. (ed.), *op.cit.*:24.
123. Stevens, Charles E., 1867, Report of Proceedings at the Pelew islands in the Matter of the Murder of Andrew Cheyne, National Archives, Kew, FO72/1155.
124. Ibid.
125. Ibid.
126. Ibid.
127. Spoehr, Florence M. (trans.) in Tetens, A., *op.cit.*:103.
128. Shineberg, D., *op.cit.*: 25.
129. Ibid.
130. Manson, T.M.Y., Transcript of a lecture given to the Northmavine History Group at the Sullom Public Hall on 21st March 1989:7, copy at Fetlar Museum Trust, Fetlar, Shetland.
131. For example, a letter from the British Consul in Manila, Mr Ricketts, to Henry Cheyne, 3rd June, 1867, which is responding to an enquiry made by Henry Cheyne in April that year. Shetland Archives, Lerwick. D1/542/13/12.
132. Power, Sir D'Arcy and Le Fanu, W.R., *Lives of the Fellows of the Royal College of Surgeons of England 1930-1951*, London, 1953:143.
133. Scott, H., *op.cit.*:297.
134. Crowther, M.A. & Dupree, M.W., *Medical Lives in the Age of Surgical Revolution*, Cambridge University Press, 2010:16.
135. Power, Sir D'Arcy and Le Fanu, W.R *op.cit.*:143
136. Ibid.
137. Cheyne, W.W., in Manson, T.M.Y., Transcript of a lecture given to the Northmavine History Group at the Sullom Public Hall on 21st March 1989:7, copy at Fetlar Museum Trust, Fetlar, Shetland.
138. Manson, T.M.Y., Transcript of a lecture given to the Northmavine History Group at the Sullom Public Hall on 21st March 1989:7, copy at Fetlar Museum Trust, Fetlar, Shetland.
139. Crowther, M.A. & Dupree, M.W., *op.cit.*:16.
140. Saxby, J., Shetland's Greatest Men, Shetland Archives, Lerwick. D11/21/14.
141. Scottish Census, 1871, ScotlandsPeople: http://www.scotlandspeople.gov.uk/.
142. University of Edinburgh alumni,http://www.ed.ac.uk/about/people/famous-alumni Accessed 28/04/2013.
143. McCrae, Morrice, *Simpson: the turbulent life of a medical pioneer*, Edinburgh (Birlinn), 2011:13.
144. University of Edinburgh, http://www.ed.ac.uk/about/people/millennial/ conandoyle Accessed 28/04/2013.
145. Crowther, M.A. & Dupree, M.W., *op.cit.*:65.
146. Godlee, R.J., *Lord Lister*, London (MacMillan & Co.), 1918: Chapters 8 and 9.

Chapter 3
1. Crowther, M.A. & Dupree, M.W., *op.cit.*:102.
2. Worboys, M., *op. cit.*:24.
3. Woodward, J., *To Do the Sick no Harm, A Study of the British Hospital System to 1875*, London and Boston (Routledge Kegan Paul), 1974. In Brunton, D. (ed.),

Medicine Transformed: Health, Disease and Society in Europe, 1800-1930, Manchester University Press, 2004:39-40.
4. Ibid.
5. Cheyne, W.W., Lister and His Achievement, being the first Lister Memorial Lecture delivered at the Royal College of Surgeons of England on May 14th, 1925, London (Longmans, Green & Co.), 1925:3-4.
6. Ibid:101.
7. Margaret Mathewson, whose Sketch forms the subject of Chapter 5, was in hospital for eight months. Mathewson, Margaret C., *Sketch of Eight Months a Patient in the Royal Infirmary*, Edinburgh, August 8th 1879, photocopy at Shetland Archives, Lerwick. SA2/340.
8. Margaret Mathewson in a letter to her brother Arthur, dated February 24th 1877, cited in Goldman, M., Lister Ward, Bristol and Boston (Adam Hilger), 1987:37.
9. Leeson, J.R., *Lister As I Knew Him*, London, 1927:4.
10. Ibid:5.
11. Ibid:6.
12. Winter, A., Ethereal Epidemic: Mesmerism and the Introduction of inhalation anaesthesia to early Victorian London, *Social History of Medicine*, 4, 1991. In Brunton, D. (ed.), *op.cit.*:104.
13. Ibid:105-6.
14. Ibid:110.
15. Ibid:109.
16. Ibid.
17. Ibid:110.
18. McCrae Morrice, *Simpson: the turbulent life of a medical pioneer*, Edinburgh (Birlinn), 2011:167.
19. Ibid: 197.
20. Ibid:200
21. Ibid:198.
22. Ibid:202.
23. Ibid:210-17.
24. Goldman, M., *Lister Ward*, Bristol and Boston (Adam Hilger), 1987:6.
25. Leeson, J.R., *op. cit.*,:2.
26. Goldman, M., *op.cit.*:150.
27. Ibid.
28. Nightingale, F., *Notes on Nursing*, London (Harrison), 1860: 120-121. Accessed 20/10/2012 via http://books.google.es/books?id=YxIDAAAAQAAJ&pg=PA120&source=gbs_toc_r&cad=4#v=onepage&q&f=false
29. King's College Collections, London. http://kingscollections.org/exhibitions/specialcollections/nightingale-and-hospital-design/florence-nightingale-and-hospital-design Accessed 20/10/12.
30. Nightingale, F., Notes on Nursing, London (Harrison), 1860:120-121. Accessed 20/10/2012 via http://books.google.es/books?id=YxIDAAAAQAAJ&pg=PA120&source=gbs_toc_r&cad=4#v=onepage&q&f=false
31. Worboys, M., *op.cit.*:4.
32. Ibid:39-41.
33. Ibid:76.
34. Ibid.
35. Crowther, M.A. & Dupree, M.W., *op.cit.*:5.
36. Waddington, I., *The Medical Profession in the Industrial Revolution* /Dublin, Gill and Macmillan, Humanities Press), 1984:138-143. In Brunton, D. (ed.), *op.cit.*:90-93.
37. Crowther, M.A. & Dupree, M.W., *op. cit.*,:50.

38. 1871 Scottish census, via ScotlandsPeople, http://www.scotlandspeople.gov.uk/
39. Crowther, M.A. & Dupree, M.W., *op. cit.*,:50.
40. Letter from Mr Ricketts, Acting British Consul in Manila, to Henry Cheyne, 3rd June 1867, Shetland Archives, Lerwick. D1/542/13/12.
41. Sir Patrick Cheyne, personal communication to author, 04/01/2013.
42. Letter from Mr Ricketts, Acting British Consul in Manila, to Henry Cheyne, 3rd June 1867, Shetland Archives, Lerwick. D1/542/13/12.
43. Ibid.
44. Shineberg, D., in a letter to Sir Joseph Cheyne, 24th April 1967, mentions that it was alluded to by the Acting Consul in Manila in writing about the disposition of Cheyne's papers. Shetland Archives, Lerwick. D1/542/15/1.
45. Obituary, Sir William Watson Cheyne, Bt., K.C.M.G., C.B., F.R.S., F.R.C.S., *BMJ*,April 30th, 1932. 1932:821. (*BMJ* 1932;1:821).
46. 1871 census: Fetlar and North Yell, ScotlandsPeople: http://www.scotlandspeople.gov.uk/.
47. Roll of the Graduates of the University of Aberdeen, 1860-1900.
48. Scott, H., *Fasti Ecclesiae Scoticanae, The Succession of Ministers in the Church of Scotland from the Reformation*, Volume III, Synod of Glasgow and Ayr, Edinburgh (Oliver and Boyd), 1928:366-7.
49. Sir D'Arcy Power, Le Fanu, W.R., *Lives of the Fellows of the Royal College of Surgeons of England 1930-1951*, London (Royal College of Surgeons), 1953:143.
50. Crowther, M.A. & Dupree, M.W., *op.cit.*:45.
51. Cheyne, W.W., *Lister and His Achievement,* London (Longmans, Green & Co.), 1925:23.
52. Cheyne's certificates from the University of Edinburgh, in possession of the Cheyne family (viewed December, 2012).
53. Crowther, M.A. & Dupree, M.W., *op.cit.*:67.
54. Ibid.
55. Ibid:66.
56. Ibid: 67; 382.
57. Stephenson, G.S. (under the pseudonym Alisma), *Reminiscences of a Student's Life at Edinburgh in the Seventies*, Edinburgh (Oliver and Boyd), 1918:6.
58. Ibid.
59. Crowther, M.A. & Dupree, M.W., *op.cit.*:67.
60. Cheyne's certificates from the University of Edinburgh, in possession of the Cheyne family (December, 2012).
61. Ibid.
62. Stephenson, G.S. (under the pseudonym Alisma), *op.cit.*:30.
63. Ibid:32
64. Crowther, M.A. & Dupree, M.W., *op.cit.*:78.
65. Stephenson, G.S. (under the pseudonym Alisma), *op.cit.*:30.
66. Ibid:31.
67. Ibid.
68. Ibid.
69. Crowther, M.A. & Dupree, M.W., *op.cit.*:78.
70. Cheyne's certificates from the University of Edinburgh, in possession of the Cheyne family (viewed December, 2012).
71. Stephenson, G.S. (under the pseudonym Alisma), *op.cit.*42.
72. Ibid.
73. Crowther, M.A. & Dupree, M.W., *op.cit.*:81.

74. Natural History Museum website, accessed 28/06/2013 http://www.nhm.ac.uk/nature-online/science-of-natural-history/expeditions-collecting/hms-challenger-expedition/voyage-preparations/index.html
75. Cheyne, W.W., 22 W. Preston St., Edinburgh, Natural History Notebook, C. Wyville Thomson Prof., Summer Session 1871, Edinburgh University Special Collections: MS2708.
76. Ibid.
77. Ibid.
78. Crowther, M.A. & Dupree, M.W., *op.cit.*:81.
79. Obituary, Sir W. Watson Cheyne, Bt., K.C.M.G., C.B., F.R.S., F.R.C.S., *BMJ*,30th April 1932:821. (*BMJ* 1932;1:821).
80. Cheyne's certificates from the University of Edinburgh, in possession of the Cheyne family (Viewed December, 2012).
81. Crowther, M.A. & Dupree, M.W., *op.cit.*:82.
82. Museums Galleries Scotland website: Edinburgh University Anatomy Museum, http://www.museumsgalleriesscotland.org.uk/member/edinburgh-university-anatomy-museum - accessed 28/06/2013.
83. Cheyne's certificates from the University of Edinburgh, in possession of the Cheyne family (viewed December, 2012).
84. Crowther, M.A. & Dupree, M.W., *op.cit.*:94.
85. Ibid:46-7.
86. Ibid:49-50.
87. Ibid:65.
88. Ibid.
89. Cheyne's certificates from the University of Edinburgh, original in possession of Mr A. Cheyne, London (viewed December, 2012)
90. Crowther, M.A. & Dupree, M.W., *op.cit.*:81.
91. Cheyne, W.W., Lister and His Achievement, being the first Lister Memorial Lecture delivered at the Royal College of Surgeons of England on May 14th, 1925, London (Longmans, Green & Co.), 1925:24.
92. Stephenson, G.S. (under the pseudonym Alisma), *op.cit.*:65.
93. Crowther, M.A. & Dupree, M.W., *op.cit.*:86.
94. Conan Doyle, A., in Crowther, M.A. & Dupree, M.W., *op.cit.*:83.
95. Ibid.
96. Stephenson, G.S. (under the pseudonym Alisma), *op.cit.*:20.
97. Crowther, M.A. & Dupree, M.W., *op.cit.*:91.
98. Ibid.
99. Stephenson, G.S. (under the pseudonym Alisma), *op.cit.*:20.
100. Ibid.
101. Cheyne, W.W., Lister and His Achievement, being the first Lister Memorial Lecture delivered at the Royal College of Surgeons of England on May 14th, 1925, London (Longmans, Green & Co.), 1925:23.
102. Ibid.
103. Crowther, M.A. & Dupree, M.W., *op.cit.*:92.
104. Ibid: 92-95.
105. Stephenson, G.S. (under the pseudonym Alisma), *op.cit.*:15.
106. Scottish Census, 1871, via ScotlandsPeople, http://www.scotlandspeople.gov.uk/
107. Maugham, W.C., *Annals of Garelochside, Paisley and London* (Alexander Gardner), 1897. http://www.electricscotland.com/history/garelochside/ Accessed 29/08/2013.
108. Cheyne, W.W., 22 W. Preston St., Edinburgh, Natural History Notebook, C. Wyville Thomson Prof., Summer Session 1871, Edinburgh University Special Collections: MS2708.

109. Cheyne, W.W., *Three Orations. The Lister Centenary*, London (John Bale, Sons & John Danielsson, Ltd.), 1927:11.
110. Cheyne, W.W., *Lister and His Achievement*, London (Longmans, Green & Co.), 1925:3.
111. Ibid:23-25.
112. Godlee, R.J., *Lord Lister*, London (MacMillan), 1918:11-12.
113. Ibid:12.
114. Ibid.
115. Ibid:18
116. Ibid:15.
117. Ibid:28.
118. Ibid:34.
119. Ibid: Chapter VIII The Glasgow Professorship:85-104.
120. Cheyne, W.W., *Lister and His Achievement*, London (Longmans, Green & Co.), 1925:5.
121. Worboys, M., *op.cit.*:76.
122. Cheyne, W.W., *Lister and His Achievement*, London (Longmans, Green & Co.), 1925:5.
123. Ibid:44.
124. Worboys, M., *op.cit.*:89.
125. Cheyne, W.W., *Lister and His Achievement*, London (Longmans, Green & Co.), 1925:9.
126. Ibid.
127. Ibid.
128. Worboys, M., *op.cit.*:79.
129. Ibid.
130. Cheyne, W.W., *Lister and His Achievement*, London (Longmans, Green & Co.), 1925:8.
131. Worboys, M., *op.cit.*:80.
132. Leeson, J.R., *Lister As I Knew Him*, London, 1927:23.
133. Ibid:94.
134. Cheyne, W.W., Lectures 1851-1877, Cheyne, Sir William Watson, 1852-1932, Royal College of Surgeons of England, London, MS0021/4/1:116-117.
135. Cheyne, W.W., Listerism and the Development of Operative Surgery, *BMJ*, December 13th 1902:1851. (*BMJ* 1902;2:1851).
136. Letter from Sir William Watson Cheyne to the Wellcome Museum, dated May 13th, 1928, Wellcome Library, London, WA/HMM/EX/A.11: "Would you please send at once by post my medal and class certificate which I value more than any other of my possessions & send the large picture, chain & instruments as soon as possible so that I can replace them for visitors to see."
137. Crowther, M.A. & Dupree, M.W., *op.cit.*:116.
138. Ibid:101.
139. Cheyne, W.W., *Lister and His Achievement*, London (Longmans, Green & Co.), 1925:12.
140. Ibid:19.
141. Ibid:45.
142. Worboys, M., *op.cit.*:80.
143. Cheyne, W.W., *Lister and His Achievement*, London (Longmans, Green & Co.), 1925:50.
144. Worboys, M., *op.cit.*95.
145. Ibid.
146. Ibid:91.

147. Cheyne published Lister's surgical statistics in 1879: Cheyne, W.W., Statistical Report of all Operations performed on Healthy Joints in Hospital Practice by Mr Lister from September 1871 to the Present Time, together with such accidental wounds of joints as occurred during the same period, *BMJ*,November 29th, 1879:859-864. (*BMJ* 1879;2:859).
148. Hutt, P., Heath, I., Neighbour, R., Confronting an Ill Society, Oxford (Radcliffe), 2005:102.

Chapter 4

1. Cheyne, W.W., Listerism and the Development of Operative Surgery, *BMJ*,December 13th 1902:1851. (*BMJ* 1902;2:1851).
2. Lectures 1851-1877, Cheyne, Sir William Watson, 1852-1932, Royal College of Surgeons of England, London, MS0021/4/1. (Cheyne's notes from Lister's lectures).
3. Ibid.
4. Ibid.
5. Cheyne, W.W., Listerism and the Development of Operative Surgery, *BMJ*,December 13th 1902:1851. (*BMJ* 1902;2:1851).
6. Ibid:1852.
7. Ibid.
8. Ibid.
9. Ibid.
10. Stephenson, G.S. (under the pseudonym Alisma), *op.cit.*:98.
11. Crowther, M.A. & Dupree, M.W., *op.cit.*:97.
12. Leeson, J.R., *op.cit.*, :81.
13. Cheyne, W.W., *Three Orations. The Lister Centenary*, London (John Bale, Sons & John Danielsson, Ltd.), 1927:10.
14. Cheyne's certificates from the University of Edinburgh, in possession of the Cheyne family (viewed December, 2012).
15. Ibid.
16. Cheyne, W.W., *Three Orations. The Lister Centenary*, London (John Bale, Sons & John Danielsson, Ltd.), 1927:10.
17. Ibid.
18. Sir D'Arcy Power, Le Fanu, W.R., *Lives of the Fellows of the Royal College of Surgeons of England 1930-1951*, London (Royal College of Surgeons), 1953:144.
19. Ibid.
20. Cheyne, W.W., *Three Orations. The Lister Centenary*, London (John Bale, Sons & John Danielsson, Ltd.), 1927:10.
21. Obituary, Sir W. Watson Cheyne, Bt., K.C.M.G., C.B., F.R.S., F.R.C.S., *BMJ*,30th April 1932:821. (*BMJ* 1932;1:821).
22. Crowther, M.A. & Dupree, M.W., *op.cit.*:67.
23. Reference written by Lister in favour of Cheyne, 10th February, 1880. Royal College of Surgeons, Edinburgh, GD5/138.
24. Leeson, J.R., L *op.cit.*:141.
25. Ibid:21-23.
26. Obituary, Sir W. Watson Cheyne, Bt., K.C.M.G., C.B., F.R.S., F.R.C.S., *BMJ*,30th April 1932:821. (*BMJ* 1932;1:821).
27. Leeson, J.R., *op.cit.*:141.
28. Ibid.
29. Ibid:140.
30. Ibid:144.
31. Ibid.

32. Lectures 1851-1877, Cheyne, Sir William Watson, 1852-1932, Royal College of Surgeons of England, MS0021/4/1:48.
33. Cheyne, W.W., *Lister and His Achievement*, London (Longmans, Green & Co.), 1925:27.
34. Leeson, J.R., *op.cit.*:81.
35. Cheyne, W.W., *Lister and His Achievement*, London (Longmans, Green & Co.), 1925:28.
36. Ibid.
37. Ibid.
38. Leeson, J.R., L *op.cit.*:146.
39. Ibid.
40. Cheyne's certificates from the University of Edinburgh, in possession of the Cheyne family (viewed December, 2012): class certificate of merit for General Pathology, Faculty of Medicine, University of Edinburgh, winter session, 1873-4.
41. Ibid: class certificate of merit for anatomy, Faculty of Arts, University of Edinburgh, winter session, 1873-4.
42. Ibid: class certificate of merit for Diseases of Women and Children, Faculty of Medicine, University of Edinburgh, winter session, 1874-5.
43. Ibid: class certificate of merit for Senior Surgery, Faculty of Medicine, University of Edinburgh, winter session, 1874-5.
44. Ibid: class certificate of merit for the Practice of Medicine, Faculty of Medicine, University of Edinburgh, winter session, 1874-5.
45. Leeson, J.R., in Goldman, M., *Lister Ward*, Bristol and Boston (Adam Hilger), 1987:141. Cheyne, along with the Conservator at the Wellcome Museum, expressed doubts about aspects of Leeson's memoirs. Cheyne described them as "in no sense a scientific production", though they pictured "Lister and his methods very well." Letter from Cheyne to Wellcome Museum, October 6th, 1926, Wellcome Library, London. WA/HMM/CO/AP.
46. Cheyne, W.W., *Lister and His Achievement*, London (Longmans, Green & Co.), 1925:25.
47. Ibid:25
48. Ibid:25-6.
49. Ibid:26.
50. Cheyne's class certificate of merit for Medical Juriprudence, Faculty of Medicine, University of Edinburgh, summer session, 1875, belonging to the Cheyne family.
51. Cheyne, W.W., *Lister and His Achievement*, London (Longmans, Green & Co.), 1925:74.
52. Ibid.
53. Crowther, M.A. & Dupree, M.W., *op.cit.*:67.
54. Ibid:130.
55. Ibid:138.
56. Ibid:135.
57. Ibid.
58. Ibid.
59. Cheyne, W.W., *Lister and His Achievement*, London (Longmans, Green & Co.), 1925:74.
60. Ibid.
61. Stephenson, G.S. (under the pseudonym Alisma), *op.cit.*:75.
62. Crowther, M.A. & Dupree, M.W., *op.cit.*:142.
63. Ibid:143.

64. Ibid:142-3
65. *The Lancet,* June 19th, 1875, in Godlee, R.J., Lord Lister, London (MacMillan and Co.), 1918:367.
66. Godlee, R.J., *op.cit.*:367.
67. Godlee, R.J., *op.cit.*:54-55.
68. Crowther, M.A. & Dupree, M.W., *op.cit.*:142.
69. Ibid.
70. Copy deeds relating to 24 merks land in Bouster, Yell, Papers of Captain Andrew Cheyne, Shetland Archives, Lerwick, Shetland. D/542.
71. Inventory of the Personal Estate of the Late Revd. David Webster, 1881, Shetland Archives, Lerwick, Shetland. SC12/36/7/p.194.
72. Little Reward for Brain Work, *Dundee Evening Telegraph,* August 26th, 1920:3, British Library, accessed through the British Newspaper Archive, 22/01/14.
73. Meldungsbogen für den ausserordentlichen Hörer der Universität in Wien (Cheyne's registration certificate for University of Vienna), in possession of the Cheyne family (viewed December, 2012)
74. Stepansky, Paul E., *Freud, Surgery and the Surgeons,* New Jersey, (Analytic Press), 1999:38.
75. Cheyne's lecture timetable from the University of Vienna, in possession of the Cheyne family (viewed December 2012).
76. Cheyne, W.W., *Lister and His Achievement,* London (Longmans, Green & Co.), 1925:74.
77. Cheyne's lecture attendance certificate from the University of Vienna, in possession of the Cheyne family (2012). Also: Cheyne, W.W., *Lister and His Achievement,* London (Longmans, Green & Co.), 1925:74.
78. Godlee, R.J., *op.cit.*:349.
79. Ibid.
80. Cheyne, W.W., *Lister and His Achievement,* London (Longmans, Green & Co.), 1925:74.
81. Godlee, R.J., *op.cit.*:332.
82. Ibid.
83. Ibid:333.
84. Ibid:334.
85. Ibid:347.
86. Cheyne, W.W., *Lister and His Achievement,* London (Longmans, Green & Co.), 1925:75.
87. Ibid.
88. Ibid:76.
89. "Small-Pox": public notice urging the residents of Lerwick to seek vaccination against smallpox and offering free vaccination, January 1877, Shetland Archives, Lerwick. 50.10.19.
90. Cowie, R., Shetland: Descriptive and Historical, 1874: "Memoir" section, digital transcript viewed on http://www.electricscotland.com/History//index.htm 20/11/2014.
91. Brunton, D., *Medicine Transformed: The Politics of Vaccination,* University of Rochester Press, 2008: Chapter 9: Vaccination in Scotland: Victory for Practitioners:141-162.
92. Notice to registrars issued by W.P. Dundas, with requirements of compulsory vaccination, and sample copies of schedules A-C, 1863, Shetland Archives, Lerwick. D23/156/1A.
93. Ibid.

94. Ibid.
95. Correspondence from Dr Loeterbagh, Lerwick, to A.D. Mathewson, registrar, Yell, Shetland, dated 7th June and 26th July 1865, protesting the terms offered him for expenses by the Pariochial Board and threatening to resign the post of Vaccinator, Shetland Archives, Lerwick. D23/151/31/724.
96. Cowie, R., Shetland: Descriptive and Historical, 1874: "Memoir" section, digital transcript viewed on http://www.electricscotland.com/History//index.htm 20/11/2014.
97. Letter from A.D. Mathewson, September 8th 1873, Shetland Archives, Lerwick. D23/150/32/23.
98. Letter from Margaret Mathewson to her brother Arthur, dated October 22nd 1876, A.D. Mathewson papers, Shetland Archives, Lerwick. D23/193/17/72.
99. Leeson, J.R., *op.cit.*:145.
100. Ibid.
101. Cheyne, W.W., *Lister and His Achievement*, London (Longmans, Green & Co.), 1925:75.
102. Reference written by Lister in favour of Cheyne, 10th February, 1880. Royal College of Surgeons, Edinburgh, GD5/138.
103. Cheyne, W.W., *Lister and His Achievement*, London (Longmans, Green & Co.), 1925:75.
104. Ibid.
105. Ibid.
106. Ibid.

Chapter 5

1. Mathewson, Margaret C., *Sketch of Eight Months a Patient in the Royal Infirmary, Edinburgh, August 8th 1879*, copy in Shetland Archives, SA2/340. A number of versions of the Sketch have now come to light, and are the subject of a work in progress by Mary Wilson Carpenter.
2. Goldman, M., *op.cit.*:18.
3. Fenton, A., *The Northern Isles: Orkney and Shetland*, Tuckwell Press, 1997:67.
4. Goldman, M., *op.cit.*:18.
5. Carpenter, M.W., Lister's relationship with patients: a successful case. Notes and Records of the Royal Society, August 2013:4 Notes Rec. R. Soc. published online May 29, 2013.
6. Goldman, M., *op.cit.*:147.
7. Abrams, L., *Myth and Materiality in a Woman's World: Shetland 1800-2000*, Manchester University Press, 2005:100.
8. For example: Photograph in Fetlar Museum Trust Photo Archive, Fetlar Interpretive Centre: F-2693.
9. Letter from Margaret Mathewson to her brother Arthur, dated October 22nd 1876, A.D. Mathewson papers, Shetland Archives, Lerwick. D23/193/17/72.
10. Mathewson, Margaret C., *op.cit.*
11. Ibid:7-8.
12. Hamilton, D., *The Healers: a history of medicine in Scotland*, Canongate Books, 1981:174.
13. Johnson, R.L., From a Minister's Diary, *Shetland Life*, May 1987:23.
14. Ibid.
15. The Rev. Mr Dishington, in Edmondston, A., *A View of the Ancient and Present State of the Zetland Islands*, London (Longman, Hurst, Rees and Orm) and Edinburgh (John Ballantyne and Co.), Volume II, 1809:88-89.

16. Anderson, Bruford, Henderson: School of Scottish Studies, SA1970.240 Tobar an Dualchais, http://www.tobarandualchais.co.uk/en/fullrecord/65344/3;jsessionid=1AFAADA422BFFF7EFACDC8230C99C628 - accessed 0/08/2013.
17. Diary of Lady Annie Nicolson, Brough Lodge, Fetlar, 25th January 1913. Shetland Archives, Lerwick. D24/Box82.
18. Times Past: 100 Years Ago, *Shetland Times*, October 28th, 2011. http://www.shetlandtimes.co.uk/2011/10/28/times-past-135 Viewed 06/03/2015.
19. Diary of Lady Annie Nicolson of Brough Lodge, Saturday, 18th January, 1902, Shetland Archives, Lerwick. D24/Box82.
20. Cowie, R, Shetland: Descriptive and Historical, 1874, Part I, Chapter 8: Health and Longevity, digital version accessed on http://www.electricscotland.com/History/shetland/index.htm 20/11/2014.
21. John Coutts, personal communication to author, 21st August 2013.
22. Goldman, M., *op.cit.*:33.
23. Mathewson, Margaret C., *op.cit.*
24. Goldman, M., *op.cit.*:20.
25. Letter from Margaret Mathewson to her brother Arthur, dated October 22nd 1876, A.D. Mathewson papers, Shetland Archives, Lerwick. D23/193/17/72.
26. Ibid.
27. Letter from Margaret Mathewson to her brother Arthur, April 30th, 1876, Shetland Archives, Lerwick. D23/193/17/27.
28. Letter from Margaret Mathewson to her brother Arthur, dated October 22nd 1876, A.D. Mathewson papers, Shetland Archives, Lerwick. D23/193/17/72.
29. Ibid.
30. Ibid.
31. Ibid.
32. Cowie, R, Shetland: Descriptive and Historical, 1874, Part I, Chapter 11: Rheumatism: "Dyspepsia", digital version accessed on http://www.electricscotland.com/History/shetland/index.htm 20/11/2014.
33. Goldman, M., *op.cit.*:20.
34. Mathewson, Margaret C., *op.cit.*
35. Mathewson, Margaret C., *op.cit.*
36. Letter from Margaret Mathewson to her father, dated 22nd February, 1877, Shetland Archives, Lerwick. D23/151/43/4.
37. Mathewson, Margaret C., *op.cit.*
38. Goldman, M., *op.cit.*:24.
39. Wilson Carpenter, M., 2012, The Patient's Pain in her Own Words: Margaret Mathewson's "Sketch of Eight Months a Patient, in the Royal Infirmary of Edinburgh, A.D. 1877", *Interdisciplinary Studies in the Long 19th Century*, no. 15 (2012):I(2)
40. Mathewson, Margaret C., *op.cit.*
41. Letter from Margaret Mathewson to her family, February 27th, 1877. Shetland Archives, Lerwick. D23/151/43/5.
42. Mathewson, Margaret C., *op.cit.*
43. Ibid.
44. Letter from Margaret Mathewson to her father, dated 27th February, 1877, Shetland Archives, Lerwick. D23/151/43/5.
45. Diary of the Reverend William Watson, 1846, Shetland Archives, Lerwick, Shetland. D1/542; also in Johnson, R.L., From a Minister's Diary, Shetland Life, May 1987:21.

46. Mathewson, Margaret C., *op.cit.*
47. Letter from Margaret Mathewson to her father, dated 27th February, 1877, Shetland Archives, Lerwick. D23/151/43/5.
48. Mathewson, Margaret C., *op.cit.*
49. Ibid:9.
50. Ibid.:10.
51. Ibid.
52. Ibid:9-10.
53. Leeson, J.R., *Lister As I Knew Him*, London (Baillière, Tindall and Cox), 1927:142-3.
54. Letter from Margaret Mathewson to her father and brother, dated April 30th 1877, Shetland Archives, Lerwick. D.23/151/43/12.
55. Mathewson, Margaret C., *op.cit.*:19.
56. Ibid:19.
57. Ibid:19-20.
58. Ibid:11.
59. Ibid:15.
60. Mathewson, Margaret C., *op.cit.*:6.
61. Margaret Mathewson in a letter to her brother Arthur, dated February 24th 1877, cited in Goldman, M., *op.cit.*:37.
62. Ibid:29.
63. Leeson, J.R., *op.cit.*:137.
64. Goldman, M., *op.cit.* 45.
65. Ibid:41.
66. Ibid:41.
67. Leeson, J.R., *op.cit.*:145.
68. Pringle, L., 11th November, 1872, in Goldman, M., *op.cit.*:41.
69. Ibid.
70. Mathewson, Margaret C., *op.cit.*:17.
71. Ibid:16.
72. Ibid:17.
73. Letter from Margaret Mathewson to her father, dated July 16th 1877, Shetland Archives, Lerwick. D23/151/43/30.
74. Leeson, J.R., *Lister As I Knew Him*, London, 1927, in Goldman, M., *op.cit.*:118.
75. Ibid:111.
76. Ibid:86.
77. Mathewson, Margaret C., *op.cit.*:29.
78. Ibid.
79. Ibid:41.
80. Ibid.
81. Ibid:42.
82. Letter from Cheyne to L.W.G. Malcolm, Conservator at the Wellcome Museum, dated October 6th, 1926, Wellcome Library, London, WA/HMM/CO/AP/123.
83. Cheyne, W.W., *Three Orations: the Lister Centenary*, London (John Bale, Sons and Danielsson, Ltd.), 1927:21.
84. Letter from Margaret Mathewson to her father, dated April 30th 1877, Shetland Archives, Lerwick. D23/151/43/12.
85. Mathewson, Margaret C., *op.cit.*:61.
86. Leeson, J.R., *op.cit.*7:142-3.
87. Mathewson, Margaret C., *op.cit.*:60.

Notes and References

88. Personal communication: John Coutts (05/04/2013) who was brought up in Fetlar and who was told this by his mother. The latter was born in 1906 and was taught at school not to speak in dialect to non-Shetlanders for fear of offending.
89. Robertson, T.A., Graham, John J.J., *Grammar and Usage of the Shetland dialect*, Lerwick (The Shetland Times), 1991:4.
90. Mathewson, Margaret C., *op.cit.*:61.
91. Obituary Notices, the Royal Society, 10.1098/rsbm.1932.0007 Obit. Not. Fell. R. Soc. 1 December 1932 vol. 1 no. 1 26-30. http://rsbm.royalsocietypublishing.org/content/obits/1/1/26.full.pdf+html
92. Obituary in the Times, quoted in Manson, T.M.Y., Profiles from the Past, No.XXXVI: Sir Watson Cheyne, A World Role in Surgery, The New Shetlander, Shetland Archives, SA4/591/5/83.
93. Smith, B., The Development of the spoken and written Shetland dialect: a historian's view, in Doreen J. Waugh (ed.), Shetland's Northern Links: language and history, Edinburgh, Scottish Society for Northern Studies:30-43.
94. Mathewson, Margaret C., *op.cit.*60.
95. Ibid:62.
96. Ibid:63.
97. Ibid:65.
98. Ibid.
99. Ibid:92.
100. Ibid:68
101. Ibid:73.
102. Ibid:71.
103. Mathewson, M.:28-29, in Carpenter, Mary W., The Patient's Pain in her Own Words, Margaret Mathewson's Sketch of Eight Months a Patient, in the Royal Infirmary, Edinburgh, *Interdisciplinary Studies in the Long Nineteenth Century*, No.15, (2012):II(9). http://www.19.bbk.ac.uk/index.php/19/article/view/636/848 Accessed 27/05/2013.
104. Mathewson, Margaret C., *op.cit.*:74-75.
105. Letter from Margaret Mathewson to her father, dated April 30th 1877, Shetland Archives, Lerwick. D23/151/43/12.
106. Letter from Margaret Mathewson to her father, dated May 10th 1877, Shetland Archives, Lerwick. D24/151/43/15.
107. Letter from Margaret Mathewson to her father, dated June 11th 1877, Shetland Archives, Lerwick. D24/151/43/22.
108. Ibid.
109. Letter from Margaret Mathewson to her father, dated April 30th 1877, Shetland Archives, Lerwick. D23/151/43/12.
110. Mathewson, Margaret C., *op.cit.*:99.
111. Letter from Margaret Mathewson to her father, dated May 24th 1877, Shetland Archives, Lerwick. D24/151/43/18.
112. Letter from Margaret Mathewson to her father, dated July 2nd 1877, Shetland Archives, Lerwick. D24/151/43/27.
113. Ibid.
114. Ibid.
115. Letter from Margaret Mathewson to her father, dated July 16th, 1877, Shetland Archives, Lerwick. D23/151/43/30.
116. Mathewson, Margaret C., *op.cit.*:109.
117. Ibid.
118. Leeson, J.R., *op.cit*1.
119. Mathewson, Margaret C., *op.cit.*:110.

120. Mathewson, Margaret C., *op.cit.*:111.
121. Ibid.
122. Ibid.
123. Cheyne, W.W., *Lister and His Achievement*, London (Longmans, Green & Co.), 1925:26.
124. Mathewson, Margaret C., *op.cit.*:110.
125. Carpenter, M.W., Lister's Relationship with Patients: "A successful case."", Notes and Records of the Royal Society, Notes Rec. R. Soc. published online May 29, 2013:11.
126. Mathewson, Margaret C., *op.cit.*:110.
127. Carpenter, M.W., Lister's Relationship with Patients: "A successful case."", Notes and Records of the Royal Society, Notes Rec. R. Soc. published online May 29, 2013:11.
128. Ibid:3.
129. Graham, J.J. in Johnson, L.G. *Laurence Williamson of Mid Yell*, Shetland Times, 1971:Introduction.
130. Carpenter, M.W., Lister's Relationship with Patients: "A successful case."", Notes and Records of the Royal Society, Notes Rec. R. Soc. published online May 29, 2013:2.
131. Mathewson, Margaret C., Holograph manuscript of part of Margaret Mathewson's Sketch of her treatment in Edinburgh Royal Infirmary in 1877, Edinburgh, 15th December, 1878, Shetland Archives, Lerwick, D1/557/4:5-76.
132. Mathewson, Margaret C., Holograph "Sketch of Eight Months a Patient in the Royal Infirmary of Edinburgh, 1877," July 26th 1879, Shetland Archives, Lerwick, D1/557/1.
133. Carpenter, M.W., Lister's Relationship with Patients: "A successful case."", Notes and Records of the Royal Society, Notes Rec. R. Soc. published online May 29, 2013:2. http://www.ncbi.nlm.nih.gov/pmc/articles/PMC3744355/
134. Margaret Mathewson, 'A Sketch of Eight Months a Patient in the Royal Infirmary of Edinburgh, A.D. 1877', dated 27 September 1879, first six pages holograph, the remainder copied by Laurence Williamson, Shetland Archives, D.7/77. Cited in: Carpenter, M.W., Lister's Relationship with Patients: "A successful case."", Notes and Records of the Royal Society, Notes Rec. R. Soc. published online May 29, 2013:8-9. http://www.ncbi.nlm.nih.gov/pmc/articles/PMC3744355/
135. Letter from Margaret Mathewson to Dr Cheyne on Fetlar, dated 4th August 1879, Shetland Archives, Lerwick. D/23/193/20/13.
136. Carpenter, M.W., Lister's Relationship with Patients: "A successful case."", Notes and Records of the Royal Society, Notes Rec. R. Soc. published online May 29, 2013:10.
137. Mathewson, Margaret C., *op.cit.*:110.
138. Ibid:17.
139. Ibid.
140. Lectures 1851-1877, Cheyne, Sir William Watson, 1852-1932, Royal College of Surgeons of England, London. MS0021/4/1 (Cheyne's notes from Lister's lectures).
141. Letter from Joseph Lister to Mr J.A. Collinson, dated 31st May 1870, in private ownership, photocopy at Fetlar Museum Trust, Fetlar, Shetland.
142. Ibid.
143. Mathewson, Margaret C., *op.cit.*:177.
144. Ibid:176-177.
145. Ibid:41.
146. Cheyne, W.W., *Lister and his Achievement*, London (Longmans, Green & Co.), 1925:127 (In discussing the merits of Listerian antiseptic versus aseptic methods,

Notes and References

Cheyne notes that the aim of both is to "prevent living septic organisms from entering wounds.")
147. Mathewson, Margaret C., *op.cit.*:171.
148. Ibid.
149. Ibid.
150. Mathewson, Margaret C., *op.cit.*:182.
151. Ibid.
152. Ibid.
153. Ibid.
154. Goldman, M., *op.cit.* 145.
155. Letter from Margaret's father, A.D. Mathewson, to his remaining children, dated 18 October 1880, Shetland Archives, Lerwick. D1/411/3/6.

Chapter 6
1. Letter of recommendation from Lord Lister for Cheyne, 10th February, 1880, Royal College of Surgeons (Edin.), RCS/5, GD5/138.
2. Cheyne, W.W., *Lister and His Achievement*, London (Longmans, Green & Co.), 1925:75.
3. Cheyne, W.W., *Three Orations. The Lister Centenary*, London (John Bale, Sons & John Danielsson, Ltd.), 1927:14.
4. Pathological Society of London, Tuesday, May 6th 1879, Micrococci in Antiseptic Dressings, *BMJ*,May 24th 1879:781.(*BMJ* 1879;1:776).
5. Cheyne, W.W., *Lister and His Achievement*, London (Longmans, Green & Co.), 1925:76.
6. Ibid:75.
7. Ibid:76.
8. Ibid:79.
9. Ibid:82.
10. Ibid:80.
11. Ibid:76.
12. Worboys, M., *op.cit.*:151.
13. Pennington, T.H., Listerism, its Decline and its Persistence: the Introduction of Aseptic Surgical Techniques in three British Teaching Hospitals, 1890-99, *Medical History*, 1995:39-43.
14. Cheyne, W.W., *Lister and His Achievement*, London (Longmans, Green & Co.), 1925:79.
15. Cheyne, W.W., List of his works in order, sent to E.S. Reid Tait, 1928-29, Shetland Archives, Lerwick, Shetland. D6/262/9/5.
16. Ibid.
17. Reference for Cheyne from Lister, dated 10th February, 1880, Royal College of Surgeons (Edinburgh), RCS/5; GD5/138.
18. Obituary: Sir W. Watson Cheyne, Bt., K.C.M.G., C.B., F.R.S., F.R.C.S., *BMJ*,April 30th, 1932:821. (*BMJ* 1932;1:821).
19. Worboys, M., *op.cit.*:175.
20. Reference for Cheyne from Lister, dated 10th February, 1880, Royal College of Surgeons (Edinburgh), RCS/5; GD5/138.
21. For example: Cheyne, W.W., *Three Orations. The Lister Centenary*, London (John Bale, Sons & John Danielsson, Ltd.), 1927:6.
22. Leeson, J.R., *op.cit.*:103.
23. Crowther, M.A. & Dupree, M.W., *op.cit.*:108.
24. Ibid:109.
25. Godlee, R.J., *op.cit.*:396.

26. Ibid:15.
27. Cheyne, W.W., T*hree Orations. The Lister Centenary,* London (John Bale, Sons & John Danielsson, Ltd.), 1927:9.
28. Ibid: 20.
29. Leeson, J.R., *op.cit.*24.
30. Cheyne, W.W., *Three Orations. The Lister Centenary,* London (John Bale, Sons & John Danielsson, Ltd.), 1927:5.
31. Ibid:4.
32. Ibid.
33. Cheyne, W.W., *Lister and His Achievement,* London (Longmans, Green & Co.), 1925:31.
34. Godlee, R.J, *op.cit.*:393.
35. Ibid:396
36. Ibid:397.
37. Ibid:396.
38. Baxter, Thomas C., The Antiseptic Trio; Pasteur, Lister and Stewart – Their Legacy for our Times, Proceedings of the 17th Annual History of Medicine Days, University of Calgary, 2008, http://dspace.ucalgary.ca/bitstream/1880/47474/1/2008_HMD_Baxter.pdf - accessed 31/10/12.
39. Cheyne, W.W., *Lister and His Achievement*, London (Longmans, Green & Co.), 1925:32.
40. Ibid.
41. Goldman, M., *op.cit.*:136.
42. Letter from Margaret Mathewson to her father, dated October 18th, 1877, Shetland Archives, Lerwick, Shetland. D23/151/43/39.
43. Letter from Margaret Mathewson to her father, dated November 6th, 1877, Shetland Archives, Lerwick. D23/151/43/42.
44. Ibid.
45. Ibid.
46. Goldman, M., *op.cit.*:136.
47. Spence, J., Surgical Statistics, *BMJ,*January 24th 1880 (119-121):120. (*BMJ* 1880;1:119).
48. Handwritten draft of Cheyne's Boylston Prize submission (Volume II):p.286 of transcription. There is no date, but he submitted his final version of the essay for the 1880 prize. The original is in possession of the Cheyne family, and a transcript by the author is available at http://www.watson-cheyne.com/handwritten-manuscript-volume2.pdf.
49. Ibid:p.288 of transcript.
50. Ibid.
51. Ibid: pp.288-290 of transcript.
52. Tirard., N., Watson Cheyne, W., Phillips, J., Halliburton, W.D. (eds.), King's College Hospital Reports, Volume I, London (Adlard and Son), 1895:398.
53. Ibid.
54. Ibid.
55. Stewart, J., in Obituary: Sir W. Watson Cheyne, Bt., K.C.M.G., C.B., F.R.S., F.R.C.S., *BMJ,*April 30th, 1932:822. (*BMJ* 1932;1:821).
56. Cheyne, W.W., *Lister and His Achievement,* London (Longmans, Green & Co.), 1925:34.
57. Sir D'Arcy Power, le Fanu, W.R., *Lives of the Fellows of the Royal College of Surgeons of England, 1930-1951,* London (Royal College of Surgeons), 1953:144.
58. Cheyne, W.W., *Three Orations. The Lister Centenary,* London (John Bale, Sons & John Danielsson, Ltd.), 1927:5.

59. Ibid.
60. Ibid.
61. Cheyne, W.W., *Lister and His Achievement*, London (Longmans, Green & Co.), 1925:34.
62. Ibid.
63. Cheyne, W.W., Listerism and the Development of Operative Surgery, *BMJ*,December 13, 1902:1852.
64. Leeson, J.R., *op.cit.*:141.
65. Crowther, M.A. & Dupree, M.W., *op.cit.*:114.
66. Ibid:113.
67. Letter from Professor Spence in favour of William Watson Cheyne, 10th May 1878, London, Royal College of Surgeons of England, MSS552(1).
68. Ibid.
69. McCrae Morrice, *Simpson: the turbulent life of a medical pioneer*, Edinburgh (Birlinn), 2011:151.
70. Cheyne, W.W., *Lister and His Achievement*, London (Longmans, Green & Co.), 1925:34.
71. Richardson, R., Inflammation, Suppuration, Putrefaction, Fermentation: Joseph Lister's Microbiology, Notes and Records of the Royal Society, May 29th, 2013: Abstract.
72. Santer, M., Joseph Lister: first use of a bacterium as a 'model organism' to illustrate the cause of infectious disease of humans, Notes and Records of the Royal Society, London, Notes Rec R Soc Lond. 2010 Mar 20;64(1):59-65.
73. Cheyne, W.W., *Lister and His Achievement*, London (Longmans, Green & Co.), 1925:33.
74. Lister, J., On the lactic fermentation and its bearing on pathology, *Transactions of the Pathological Society of London*, Vol. 29, London, October 1878 (425-467):426.
75. Santer, M., Joseph Lister: first use of a bacterium as a 'model organism' to illustrate the cause of infectious disease of humans, *Notes and Records of the Royal Society*, London, Notes Rec R Soc Lond. 2010 Mar 20;64(1):59-65.
76. Ibid.
77. Cheyne, W.W., *Lister and His Achievement*, London (Longmans, Green & Co.), 1925:33.
78. Santer, M., Joseph Lister: first use of a bacterium as a 'model organism' to illustrate the cause of infectious disease of humans, Notes and Records of the Royal Society, London, Notes Rec R Soc Lond. 2010 Mar 20;64(1):59-65.
79. King's College London, Archive Catalogues, Identity Statement, Cheyne, Sir William Watson (1852-1932), http://www.kingscollections.org/catalogues/kclca/collection/a-e/10ch30-1 Viewed 02/09/2013.
80. Obituary, Sir William Watson Cheyne, F.R.C.S., F.R.S., *The Lancet*, April 30th, 1932:963.
81. Obituary: Sir William Watson Cheyne, Bt., K.C.M.G., C.B., F.R.S., F.R.C.S., *BMJ*,April 30th 1932:822. (*BMJ* 1932;1:821).
82. Crowther, M.A. & Dupree, M.W., *op.cit.*:136.
83. Obituary: Sir William Watson Cheyne, Bt., K.C.M.G., C.B., F.R.S., F.R.C.S., *BMJ*,April 30th 1932:822. (*BMJ* 1932;1:821).
84. http://www.measuringworth.com/ukcompare/relativevalue.php Accessed 26/03/2013
85. Cheyne, W.W., List of Cheyne's works, sent to E.S. Reid Tait, 1928-29, Shetland Archives, Lerwick, Shetland. D6/262/9/5.
86. Ibid.

87. Wright, William, M.D., D.Sc., F.R.C.S., The Fellowship of the Royal College of Surgeons of England, The Canadian Medical Association Journal, May 1932:600. http://www.ncbi.nlm.nih.gov/pmc/articles/PMC402358/pdf/canmedaj00118-0083.pdf Accessed 04/11/12
88. Stewart, J., in Obituary, Sir William Watson Cheyne, F.R.C.S., F.R.S., *The Lancet*, April 30th, 1932:963.
89. Sir D'Arcy Power, Le Fanu, W.R., *Lives of the Fellows of the Royal College of Surgeons of England 1930-1951*, London (Royal College of Surgeons), 1953:144.
90. Cheyne, W.W., Handwritten, bound copy of Cheyne's essay submitted for the Boylston Prize, 1879: opening page. Transcript available at http://www.watson-cheyne.com/handwritten-manuscript-volume1.pdf. Original document in possession of Cheyne family.
91. Ibid.
92. Wylie, W.G., Hospitals: their history, organization and construction, Boylston Prize essay for 1876. https://ia801408.us.archive.org/8/items/hospitalstheirhi00wyli/hospitalstheirhi00wyli.pdf. Accessed 02/09/2013.
93. Ibid.
94. The Boylston Prize, *BMJ*, June 18th 1881:974. (*BMJ* 1881;1:972).
95. Antiseptic Methods, *BMJ*, June 26th 1880:984. (*BMJ* 1880;1:977).
96. Ibid.
97. Ibid.
98. Cheyne, W.W., *Lister and His Achievement*, London (Longmans, Green & Co.), 1925:40.
99. Leeson, J.R., *op.cit.*:16.
100. Ibid:17.
101. Ibid.
102. Goldman, M., *op.cit.*:39.
103. Obituary: Sir William Watson Cheyne, Bt., K.C.M.G., C.B., F.R.S., F.R.C.S., *BMJ*, April 30th 1932:822. (*BMJ* 1932;1:821).
104. Cartwright, F.F., Robert Bentley Todd's contributions to medicine, *Proceedings of the Royal Society of Medicine*, September 1974:895, Proc R Soc Med. 1974 September; 67(9): 893–897.
105. Cheyne, W.W., *Lister and His Achievement*, London (Longmans, Green & Co.), 1925:34-5.
106. Ibid:35.
107. Ibid.
108. Ibid.
109. Ibid.
110. Santer, M., Joseph Lister: first use of a bacterium as a 'model organism' to illustrate the cause of infectious disease of humans, Notes and Records of the Royal Society, London, Notes Rec R Soc Lond. 2010 Mar 20;64(1):59-65.
111. Cheyne, W.W., Statistical Report of All Operations Performed on Healthy Joints in hospital practice, by Mr Lister, from September 1871 to the present time, together with such accidental wounds of joints as occurred during the same period, *BMJ*, November 29th, 1879:859-864. (*BMJ* 1879; 2:859).
112. Spence, J., Surgical Statistics, *BMJ*, January 24th 1880:119-121. (*BMJ* 1880;1:119).
113. Worboys, M., *op. cit.*:152.
114. Godlee, R.J., *op. cit.* 331-332.
115. Ibid:332.
116. Worboys, M., *op. cit.*:8.

Notes and References

117. Richardson, R., Inflammation, Suppuration, Putrefaction, Fermentation: Joseph Lister's Microbiology, Notes and Records of the Royal Society, May 29th, 2013: Conclusions. Ruth Richardson notes that Lister had anticipated "the postulates before Koch had formulated them."
118. Nobel Prize website: Robert Koch: http://www.nobelprize.org/nobel_prizes/medicine/laureates/1905/koch-bio.html Accessed 21/04/2013.
119. Ibid.
120. Ibid.
121. Worboys, M., *op. cit.* :177.
122. Ibid.
123. Ibid.
124. Obituary, Robert Koch, M.D., *BMJ*, June 4th 1910:1384. (*BMJ* 1910;1:1384).
125. Watson Cheyne, W. (trans.), in Koch, R., *Investigations into the Etiology of Traumatic Infective Diseases.* Translated by W. Watson Cheyne, F.R.C.S., Assistant Surgeon to King's College Hospital, London. The New Sydenham Society, 1880: Preface.
126. Ibid.
127. Ibid.
128. Ibid.
129. Ibid.
130. Worboys, M., *op. cit.* :172.
131. Ibid:173.
132. Nova et Vetera: Alexander Ogston - Bacteriologist, *BMJ*, August 7th 1954:356. (*BMJ* 1954;2:355).
133. Ibid.
134. Worboys, M., *op. cit.* :173.
135. Ibid.
136. Ibid.
137. International Medical Congress: Discussion on the Relations of Minute Organisms to Unhealthy Processes occurring in Wounds, *BMJ*, October 1st 1881:546. (*BMJ* 1881;2:545).
138. The Jacksonian Prize – a Restrospect, *BMJ*, Vol. 1, No.1996, April 1st, 1899:818. (*BMJ* 1899;1:818).
139. Ibid.
140. Ibid.
141. Sir D'Arcy Power, Le Fanu, W.R., *op. cit.*,:144.
142. Cheyne, W.W., *Antiseptic Surgery, its Principles, Practice, History and Results*, London (Smith, Elder and Co.), 1882.
143. Cheyne, W.W., List of Cheyne's works, sent to E.S. Reid Tait, 1928-29, Shetland Archives, Lerwick, Shetland. D6/262/9/5.
144. Cheyne, W.W., *Antiseptic Surgery, its Principles, Practice, History and Results*, London (Smith, Elder and Co.), 1882:x.
145. Ibid:xi.
146. Ibid.
147. Cheyne, W.W, *Die antiseptische Chirurgie; ihre Grundsätze, Ausübung, Geschichte und Resultate*, Leipzig, 1883.
148. Cheyne, W.W., List of Cheyne's works, sent to E.S. Reid Tait, 1928-29, Shetland Archives, Lerwick. D6/262/9/5.
149. Worboys, M., *op. cit.* : 174.
150. Ibid.
151. Ibid.

152. Cheyne, W.W., *Antiseptic Surgery, its Principles, Practice, History and Results,* London (Smith, Elder and Co.), 1882:120.
153. Ibid:122-123.
154. Ibid:125.
155. Worboys, M., *op. cit.* 175.

Chapter 7

1. Worboys, M., *op. cit.* :212.
2. Ibid:175.
3. The Bacilli of Tubercle, *BMJ,*May 13th 1882:709. (*BMJ* 1882;1:706).
4. Ibid:710.
5. Ibid.
6. Reports of Societies, Royal Medical and Chirurgical Society, Tuesday May 23rd 1882, Micro-organisms in Disease, *BMJ,*May 27th 1882: 778. (*BMJ* 1882;1:777).
7. Ibid.
8. Ibid.
9. Micro-organisms, *BMJ,*May 27th, 1882:787. (*BMJ* 1882;1:785).
10. Minutes of the first meeting of the Council of the Association for the Advancement of Medicine by Research, April 20th 1882, Wellcome Library, London, MS5310:3.
11. Ibid:4.
12. Worboys, M., *op. cit.* :205.
13. Ibid:27.
14. Dogs' Protection Bill, House of Commons, Hansard, HC Deb 27 June 1919 vol. 117 cc511-53 http://hansard.millbanksystems.com/commons/1919/jun/27/dogs-protection-bill lines 519-520. Accessed 17/05/2013.
15. Cruelty to Animals Act 1876 (Amendment), House of Commons, Hansard, HC Deb 30 January 1968 vol. 757 cc1093-4
16. Cheyne, W.W., Lectures on Clinical Surgery by Joseph Lister, F.R.S. &c. Winter Session, 1872-73, Royal College of Surgeons of England, London, MS0021/4/1.
17. Vivisection, *Aberdeen Journal,* Friday, 1st May 1891:5, accessed through British Newspaper Archive, 13/01/2104.
18. Ibid.
19. Dogs' Protection Bill, House of Commons, Hansard, HC Deb 27 June 1919 vol. 117 cc511-53
20. Leeson, J.R., *op. cit.* :69.
21. Ibid.
22. Ibid:97-98.
23. Cheyne, W.W., Report to the Association for the Advancement of Medicine by Research on the Relation of Micro-Organisms to Tuberculosis, Handed in to the Association for the Advancement of Medicine by Research on February 1, 1883 by W. Watson Cheyne, M.B., F.R.C.S., Assistant Surgeon to King's College Hospital. Published in The Practitioner, April 1883, No. 178, (Vol. XXX. No. 4).
24. Mr Watson Cheyne, on the Bacillus of Tubercle, *BMJ,*March 17th 1883:527. (*BMJ* 1883;1:520).
25. Minutes of the Association for the Advancement of Medicine by Research, London (Wellcome Library), MS5310:13.
26. Ibid:11.
27. Ibid.
28. Cheyne, W.W., Report to the Association for the Advancement of Medicine by Research on the Relation of Micro-Organisms to Tuberculosis, Handed in to the Association for the Advancement of Medicine by Research on February 1, 1883 by

W. Watson Cheyne, M.B., F.R.C.S., Assistant Surgeon to King's College Hospital. Published in The Practitioner, April 1883, No. 178, (Vol. XXX. No. 4).
29. Worboys, M., *op. cit.* :203-204.
30. Minutes of the Association for the Advancement of Medicine by Research, Fifth meeting of the Executive Committee, July 12th, 1882, London (Wellcome Library), MS5310:12.
31. Ibid:12-13.
32. Ibid:13.
33. Ibid:13.
34. Association for the Advancement of Medicine by Research, *British Medical Journal* July 15th 1882:109. (*BMJ* 1882;2:105).
35. Cheyne, W.W., Report to the Association for the Advancement of Medicine by Research on the Relation of Micro-Organisms to Tuberculosis, Handed in to the Association for the Advancement of Medicine by Research on February 1, 1883 by W. Watson Cheyne, M.B., F.R.C.S., Assistant Surgeon to King's College Hospital. Published in *The Practitioner*, April 1883, No. 178, (Vol. XXX. No. 4).
36. Ibid.
37. Ibid.
38. Ibid.
39. Ibid.
40. Ibid.
41. Ibid.
42. Letter from Cheyne to the Committee of the Association for the Advancement of Medicine by Research, 7th November 1882, Wellcome Library, MS5310:4-5: "Since my return I have been engaged on experiments with the materials obtained from Messrs. Toussaint & Koch & also on other experiments which I thought it necessary to make."
43. Letter from Cheyne to the Committee of the Association for the Advancement of Medicine by Research, 7th November 1882, Wellcome Library, MS5310:2.
44. Ibid:5.
45. Ibid.
46. Ibid.
47. Minutes of the Association for the Advancement of Medicine by Research, February 22nd, 1883, London (Wellcome Library), MS5310:10.
48. Ibid.
49. Letter from Cheyne to the Committee of the Association for the Advancement of Medicine by Research, February 1st 1883, London (Wellcome Library), MS5310.
50. Minutes of the Association for the Advancement of Medicine by Research, February 22nd, 1883, London (Wellcome Library), MS5310:11.
51. Green, Henry T., An Address given at the Opening of the Section of Pathology at the Annual Meeting of the British Medical Association, in Liverpool, August, 1883, *BMJ*, August 4th 1883, (229-232):231. (*BMJ* 1883;2:229).
52. Wells, Spencer T., The Hunterian Oration, delivered February 14th, at the Royal College of Surgeons of England, *BMJ*, February 24th, 1883, (341-343):341. (*BMJ* 1883;1:341)
53. Reports of Societies, Royal Medical and Chirurgical Society, Tuesday, December 9th 1884, *BMJ*,Dec. 13th, 1884: 1193. (*BMJ* 1884;2:1193).
54. Cheyne W.W., The Bacillus of Tubercle, A Speech made during the discussion on Dr Kidd's paper at the Royal Medical and Chirurgical Society, London, on January 13th, 1885, *BMJ*,Jan 24th, 1885:169. (*BMJ* 1885;1:169).
55. Notice in the *BMJ*,March 10th, 1883:467. (*BMJ* 1883;1:464).
56. Ibid.

57. Ibid.
58. Notice in the *BMJ*,May 12th, 1883:921. (*BMJ* 1883;1:918).
59. Research Scholarships of the British Medical Association, Notice in the *BMJ*,October 27th, 1883:841. (*BMJ* 1883;2:828).
60. International Health Exhibition, 1884, South Kensington, *BMJ*,May 17th 1884:968. (*BMJ* 1884;1:959).
61. Ibid.
62. Working-Health Laboratories at the International Health Exhibition, *BMJ*,July 12th 1884:82. (*BMJ* 1884;2:71).
63. Ibid.
64. Ibid.
65. Hart, E., Abstract of a Lecture on the International Health Exhibition of 1884, its influence and possible sequels, delivered before the Society of Arts, Wednesday, November 26th, 1884, *BMJ*,Dec. 6th, 1884,(1115-1122):1120. (*BMJ* 1884;2:1115).
66. Ibid.
67. Ibid.
68. The Biological Library at the International Health Exhibition, *BMJ*,July 5th, 1884:27. (*BMJ* 1884;2:24).
69. Ibid.
70. Ibid.
71. Ibid.
72. Working-Health Laboratories at the International Health Exhibition, *BMJ*,July 12th 1884:82. (*BMJ* 1884;2:71).
73. Hart, E., Abstract of a Lecture on the International Health Exhibition of 1884, its influence and possible sequels, delivered before the Society of Arts, Wednesday, November 26th, 1884, *BMJ*,Dec. 6th, 1884,(1115-1122):1119. (*BMJ* 1884;2:1115).
74. Notice in *BMJ*,July 26th, 1884:180. (*BMJ* 1884;2:179).
75. Hart, E., Abstract of a Lecture on the International Health Exhibition of 1884, its influence and possible sequels, delivered before the Society of Arts, Wednesday, November 26th, 1884, *BMJ*,Dec. 6th, 1884,1115-1122. Proposal for Permanent Establishment of Health Laboratories:1121. (*BMJ* 1884;2:1115).
76. Ibid.
77. Ibid.
78. Ibid.
79. Ibid.
80. Cheyne, W.W., List of Cheyne's works, sent to E.S. Reid Tait, 1928-29, Shetland Archives, D6/262/9/5.
81. Cheyne, W.W., On Foulbrood, *BMJ*,Oct 10th 1885:697. (*BMJ* 1885;2:697).
82. Ibid.
83. Cheyne, W.W., List of Cheyne's works, sent to E.S. Reid Tait, 1928-29, Shetland Archives, D6/262/9/5.
84. Worboys, M., *op. cit.* :178, note 124.
85. Carlisle Poisoning Case, *Tamworth Herald*, Saturday, 18th September, 1886:6, accessed through British Newspaper Archive, 19/01/2014.
86. Cheyne, W.W., Reports to the Scientific Grants Committee of the British Medical Association, Report on Micrococci in Relation to Wounds, Abscesses and Septic Processes, *BMJ*,September 20th 1884, (553-556):553. (*BMJ* 1884;2:553).
87. Cheyne, W.W., List of Cheyne's works, sent to E.S. Reid Tait, 1928-29, Shetland Archives, D6/262/9/5: No.13.
88. Ibid.
89. Ibid.

90. Cheyne, W.W., *Manual of the Antiseptic Treatment of Wounds for Students and Practitioners*. London (Smith and Elder), 1885: Preface.
91. Worboys, M., *op. cit.* :178.
92. Cheyne, W.W., *Manual of the Antiseptic Treatment of Wounds for Students and Practitioners*. London (Smith and Elder), 1885:10.
93. Cheyne, W.W., "Antiseptic Surgery" in Heath, C. (ed.) *Dictionary of Surgery*, Volume 1, London (Smith and Elder) 1886:71.
94. Cheyne, W.W., List of Cheyne's works, sent to E.S. Reid Tait, 1928-29, Shetland Archives, D6/262/9/5: No.16.
95. Worboys, M., *op. cit.* :64.
96. Cheyne, W.W., Reports to the Scientific Grants Committee of the British Medical Association, Report on a Study of Certain of the Conditions of Infection *BMJ*,July 31st 1886:197-207. (*BMJ* 1886;2:197).
97. Worboys, M., *op. cit.* :179.
98. Cheyne, W.W., List of Cheyne's works, sent to E.S. Reid Tait, 1928-29, Shetland Archives, Lerwick. D6/262/9/5, no.15.
99. Cheyne, W.W. (ed.), Essays by Various Authors on Bacteria in Relation to Disease, London (New Sydenham Society), 1886: Prefatory Note.
100. Ibid: Editor's Preface:xii.
101. Ibid: vii-viii.
102. Richardson, R., Inflammation, Suppuration, Putrefaction, Fermentation: Joseph Lister's Microbiology, Notes and Records of the Royal Society, May 29th, 2013.
103. Flügge, C., Cheyne, W.W. (trans.), *Micro-organisms, with special reference to the etiology of the infective diseases*, London (New Sydenham Society), 1890.
104. Worboys, M.,*op. cit.* :113.
105. Ibid.
106. Brody, H., Rip, M.R., Vinten-Johansen, P., Paneth, N., Rachman, S., Map-making and myth-making in Broad Street: the London cholera epidemic, 1854, *The Lancet* 356: 64 - 68, 2000.
107. Worboys, M., *op. cit.* :247.
108. Worboys, M., *op. cit.* :248.
109. Preliminary Report of the India Cholera Commission, *BMJ*,3rd January 1885:43. (*BMJ* 1885;1:33).
110. Ibid.
111. Ibid:44.
112. Klein, E., Remarks on the Etiology of Asiatic Cholera, *BMJ*,March 28th 1885, (650-652):650. (*BMJ* 1885;1:650).
113. Ibid.
114. The Organisms of Cholera, *BMJ*,Jan 17th 1885:141. (*BMJ* 1885;1:140).
115. Ibid.
116. Worboys, M., *op. cit.* :251.
117. Ibid.
118. The Discussion on Cholera at the Royal Medical and Chirurgical Society, *BMJ*,March 21st, 1885:613 (*BMJ* 1885;1:606); also Johnson, G., On the Biology, Pathology and Treatment of Cholera, Royal Medical and Chirurgical Society, Reports of Societies, *BMJ*,March 28th, 1885:654-6. (*BMJ* 1885;1:654).
119. The Discussion on Cholera at the Royal Medical and Chirurgical Society, *BMJ*,March 21st, 1885:613. (*BMJ* 1885;1:606).
120. Johnson, G., On the Biology, Pathology and Treatment of Cholera, Royal Medical and Chirurgical Society, Reports of Societies, *BMJ*,March 28th, 1885:656. (*BMJ* 1885;1:654).

121. Klein, E., Some Remarks on the Present State of our Knowledge of the Comma-Bacilli of Koch, *BMJ*,April 4th 1885:693.(*BMJ* 1885;1:693)
122. Cheyne, W.W., The Cholera-Bacillus of Koch, *BMJ*,April 11th 1885:756-7. (*BMJ* 1885;1:756).
123. Atalić, B., Drenjančević-Perić, I., Fatovic-Ferenčić, S., Emanuel Edward Klein, a diligent and industrious plodder or the father of British microbiology, Medicinski Glasnik, Volumen 7, Number 2, August 2010:112. http://www.ljkzedo.com.ba/medglasnik/vol72/05_mgvol72.pdf Accessed 13/09/2013.
124. Cheyne, W.W., Report on the Cholera Bacillus, Reports to the Scientific Grants Committee of the British Medical Association, *BMJ*,April 25th, 1885:821-823 (*BMJ* 1885;1:821); May 2nd, 1885:877-879 (*BMJ* 1885;1:877); May 9th, 1885:931-934 (*BMJ* 1885;1:931); May 23rd 1885:1027-1031. (*BMJ* 1885;1:1027).
125. Ibid.
126. Cheyne, W.W., Report on the Cholera Bacillus, Reports to the Scientific Grants Committee of the British Medical Association, *BMJ*,April 25th 1885,(821-823):821. (*BMJ* 1885;1:821).
127. Ibid.
128. Report on the Cholera-Bacillus, Reports to the Scientific Grants Committee of the British Medical Association, *BMJ*,May 23rd, 1885,(1027-31):1031. (*BMJ* 1885;1:1027).
129. Stephenson, G.S. (under the pseudonym Alisma), *op. cit.* :75.
130. Worboys, M., *op. cit.* :250.
131. Illingworth, C.R., Cholera and Comma—Bacilli, *BMJ*,June 13th, 1885:1231. (*BMJ* 1885;1:1231).
132. The Organisms of Cholera, *BMJ*,Jan 17th 1885:141. (*BMJ* 1885;1:141).
133. Worboys, M., *op. cit.* :251.
134. Richardson, R., *Imperial Contagions. Henry Vandyke Carter and Indian Leprosy: Anatomy, Geography, Morphology, Medicine, Hygiene and Cultures of Planning in Asia*, Hong Kong University Press, 2013.
135. Ibid:11.
136. Hime, T.W., The Cholera-Bacillus from a Public Health Point of View, *BMJ*,May 23rd 1885:1080. (*BMJ* 1885;1:1079).
137. Ibid.
138. Ibid.
139. Ibid
140. Cheyne, W.W., The Comma-Bacillus: Proposal for a Commission of Inquiry, *BMJ*,May 30th 1885: 1127. (*BMJ* 1885;1:1126).
141. Ibid.
142. Ibid.
143. Klein, E., The Comma-Bacillus, Proposal for a Commission of Inquiry, *BMJ*,13th June 1885:1224. (*BMJ* 1885;1;1224).
144. Hime, W.T., The Cholera-Bacillus: Proposal for a Commission of Inquiry, *BMJ*,June 20th 1885:1270. (*BMJ* 1885;1:1268).
145. Koch, R., Further Researches on Cholera, *BMJ*,January 2nd 1886, 6-8:7. (*BMJ* 1886;1:6)
146. Worboys, M., *op. cit.* :251.
147. Obituary, Royal Society, Obit. Not. Fell. R. Soc December 1932, Vol. 1 no.1:30.
148. Worboys, M., *op. cit.* :251.
149. Maglen, Krista, 'The First Line of Defence': British Quarantine and the Port Sanitary Authorities in the Nineteenth Century," 0951-631X *Social History of Medicine* Vol. 15 No. 3 pp.413-428:414.

150. Ibid.
151. Ibid:413.
152. Obituary: Waldemar Haffkine, C.I.E., *BMJ*,November 8th, 1930:801. (*BMJ* 1930;2:801).
153. Obituary: Sir William Watson Cheyne, F.R.C.S., F.R.S, *The Lancet*, April 30th, 1932:963.
154. Ibid.
155. Ibid.
156. Worboys, M., *op. cit.* :69.
157. Ibid.
158. Obituary: Sir William Watson Cheyne, F.R.C.S., F.R.S, *The Lancet*, April 30th, 1932:963.
159. Worboys, M., *op. cit.* :69.
160. Lost Hospitals of London website http://ezitis.myzen.co.uk/paddingtongreen.html Accessed 14/09/2013.
161. Cheyne, W.W., Report on Micrococci in Relation to Wounds, Abscesses, and Septic Diseases, Reports to the Scientific Grants Committee of the British Medical Association, *BMJ*,September 27th 1884, (599-605):599. (*BMJ* 1884;2:599).
162. Worboys, M., *op. cit.* :254.
163. Cheyne, W.W., On Early Tracheotomy in Diphtheria, *BMJ*,March 5th 1887:504. (*BMJ* 1887;1:502).
164. Ibid.
165. E.J. Edwardes, Assistant Physician to the N.W. London Hospital, levelled an argument against the causative nature of the bacillus on April 30th, 1887 in the *BMJ,BMJ*, April 30th, 1887:961. (*BMJ* 1887;1:961).
166. Worboys, M., *op. cit.* :255-257.
167. Hardy, A., Tracheotomy versus intubation: Surgical intervention in diphtheria in Europe and the United States, 1825-1930, Bull. Hist. Med., 1992, 66: 536-559. http://discovery.ucl.ac.uk/3441/1/3441.pdf - accessed 05/08/2013.
168. Cheyne, W.W., On Early Tracheotomy in Diphtheria, *BMJ*,March 5th 1887:504. (*BMJ* 1887;1:502).
169. Ibid:505.
170. Crowther, M.A. & Dupree, M.W., *op. cit.* :152.
171. Ibid.
172. Ibid.
173. Lister, J., Admission of Ladies to the Meetings of the British Medical Association, *BMJ*,9th February 1878:212-213. (*BMJ* 1878;1:211).
174. Godlee, Sir R.J., *op. cit.* :476.
175. Stephenson, G.S. (under the pseudonym Alisma), *op. cit.* :22.
176. Woman Doctors Justified, *Aberdeen Evening Press*, 3rd October 1918, British Library, accessed through British Newspaper Archive, 19/01/2014.
177. Obituary: Frances Helen Prideaux, *BMJ*,December 5th 1885:1089. (*BMJ* 1885;2:1089).
178. Ibid.
179. Ibid.
180. Ibid.
181. Cheyne, W.W., *Lister and His Achievement*, London (Longmans, Green & Co.), 1925:96-98.
182. Ibid:98
183. Ibid.
184. Ibid.
185. Cheyne, W.W., Report on a Study of Certain of the Conditions of Infection,

    Reports to the Scientific Grants Committee of the British Medical Association, *BMJ*,July 31st 1886:197. (*BMJ* 1886;2:197).
186. Cheyne, W.W. Lectures on Suppuration and Septic Diseases, delivered at the Royal College of Surgeons, February 1888, *BMJ*,February 25th, 1888:404-409 (*BMJ* 1888;1:404), March 3rd 1888:452-458 (*BMJ* 1888;1:452) and March 10th 1888:524-529 (*BMJ* 1888;1:524).
187. Cheyne, W.W. Lectures on Suppuration and Septic Diseases, delivered at the Royal College of Surgeons, February 1888, *BMJ*,March 10th 1888, 524-529:529. (*BMJ* 1888;1:524).
188. Worboys, M., *op. cit.* :185.
189. Mathewson, Margaret C., *op. cit.* .
190. Cheyne, W.W., Handwritten submission for the Boylston Prize, Harvard University, 1879. In possession of the Cheyne family. Transcript made by author, available at http://www.watson-cheyne.com/handwritten-manuscript-volume1.pdf Accessed 08/06/2013.
191. Cheyne, W.W., *Antiseptic Surgery, its Principles, Practice, History and Results*, London (Smith, Elder and Co.), 1882:596.
192. 1901 Census for Fetlar and North Yell, ScotlandsPeople, http://www.scotlandspeople.gov.uk/
193. Mr K. Hughson, Yell, personal communication to author, 01/12/2013.
194. Marriage notices, Surrey Mirror, Saturday, 12th July 1884:1, accessed through the British Newspaper Archive, 13/01/2014.
195. Leeson, J.R., *op. cit.* :164.
196. Stephenson, G.S. (under the pseudonym Alisma), *op. cit.* :107.
197. 1881 Census, National Archives, Kew, accessed through www.ancestry.co.uk 13/01/2014.
198. Ibid.
199. Mr A. Cheyne, London, personal communication to author, December 2012.
200. Registry of Marriages, National Archives, Kew, accessed through www.ancestry.co.uk13/01/2014.
201. Papers of Sir Joseph Lister, 100 pages of notes and letters on Iodide Gauze, 1882-1888, Royal College of Surgeons of England, London. MS0021/4/4.
202. Ibid: Item 1, 22nd December 1882.
203. Ibid: Item 2, 6th September 1887.
204. Ibid: Item 3, 16th September 1887.
205. Registry of births, National Archives, Kew, accessed through www.ancestry.co.uk 13/01/2014.
206. Registry of births, National Archives, Kew, accessed through www.ancestry.co.uk 13/01/2014.

Chapter 8
1. Mathewson, Margaret C., *op. cit.* :9.
2. Some Open Scholarships and Prizes, *BMJ*,September 19th, 1903:667. (*BMJ* 1903;2:663).
3. Notice in the *BMJ*,January 12th, 1889:89. (*BMJ* 1889;1:86).
4. Cheyne, W.W., Astley Cooper Prize essay submission, King's College Archives, London, G/PP5/7/1-2.
5. Ibid:4.
6. Ibid.
7. Cheyne, W.W., Extracts from Three Lectures on Tubercular Diseases of Bones and Joints, Delivered at the Royal College of Surgeons of England, *BMJ*,Nov 29th, 1890, (1227-1229):1227. (*BMJ* 1890;2:1227).

8. Cheyne, W.W., Astley Cooper Prize essay submission, King's College Archives, London, G/PP5/7:5-6.
9. Ibid:6.
10. Ibid:7.
11. Bradbury, S., Galbraith, W. and Lyster, M.E., Cytological Laboratory, Department of Zoology, University Museum, Oxford. A Microscopical Finder Slide, Article first published online: 11th November, 2011. DOI: 10.1111/j.1365-2818.1872.tb02107.x http://jcs.biologists.org/content/s3-97/38/197.full.pdf - accessed 09/08/2013.
12. Koch, R., *Investigations into the Etiology of Traumatic Infective Diseases*. Translated by W. Watson Cheyne, F.R.C.S., Assistant Surgeon to King's College Hospital, London. The New Sydenham Society, 1880.
13. Crookshank, E.M., *Photography of Bacteria*, London (H.K. Lewis), 1887.
14. Cheyne, W.W., Astley Cooper Prize essay submission, King's College Archives, London, G/PP5/7:8.
15. Cheyne, W.W., Extracts from Three Lectures on Tubercular Diseases on Bones and Joints, *BMJ*,November 29th, 1890:1227. (*BMJ* 1890;2:1227).
16. Cheyne, W.W., Astley Cooper Prize essay submission, King's College Archives, London, G/PP5/7:9.
17. Ibid.
18. Ibid.
19. The Astley-Cooper Prize, *BMJ*,June 28th, 1890:1495. (*BMJ* 1890;1:1490).
20. The Annual Meeting [of the British Medical Association], Section of Pathology, *BMJ*,August 9th, 1890, (347-354):352. (*BMJ* 1890;2:347).
21. Harveian Society of London, *BMJ*,October 25th, 1890:958-9. (*BMJ* 1890;2:957).
22. Ibid:959.
23. Ibid.
24. Notice in the *BMJ*,December 20th 1890:1441-2. (*BMJ* 1990;2:1438).
25. Sir D'Arcy Power, Le Fanu, W.R., *Lives of the Fellows of the Royal College of Surgeons of England 1930-1951*, London (Royal College of Surgeons), 1953:145.
26. Cheyne, W.W., Extracts from Three Lectures on Tubercular Diseases of Bones and Joints, Delivered at the Royal College of Surgeons of England. *BMJ*,November 29th 1890:1227-1229 (*BMJ* 1890;2:1227); December 6th, 1890:1283-1286 (*BMJ* 1890;2:1283); December 13th 1890:1348-1353 (*BMJ* 1890;2:1348); December 20th 1890:1418-1422 (*BMJ* 1890;2:1418); April 4th 1891:739-743 (*BMJ* 1891;1:739); April 11th 1891:790-795 (*BMJ* 1891;1:790); April 18th 1891:840-844, (*BMJ* 1891;1:840); April 25th 1891:896-901 (*BMJ* 1891;1:896); June 18th 1892:1290-1293 (*BMJ* 1892;1:1290); June 25th 1892:1352-1356 (*BMJ* 1892;1:1352); July 2nd 1892:11-16 (*BMJ* 1892;2:11).
27. Cheyne, W.W., Extracts from Three Lectures on Tubercular Diseases of Bones and Joints, Delivered at the Royal College of Surgeons of England. *BMJ*,November 29th 1890:1227. (*BMJ* 1890;2:1227).
28. Cheyne, W.W., Extracts from Three Lectures on Tubercular Diseases of Bones and Joints, Delivered at the Royal College of Surgeons of England. *BMJ*,June 18th, 1892:1292. (*BMJ* 1892;1:1290).
29. Ibid.
30. Cheyne, W.W., *Tuberculous Disease of Bones and Joints: its pathology, symptoms and treatment*, Edinburgh and London (Young J. Pentland), 1895: Preface:vi.
31. Ibid:v.
32. Cheyne, W.W., The Harveian Lectures on the Surgical Treatment of Tuberculous Diseases, *BMJ*,16th December 1899:1659-1663 (*BMJ* 1899;2:1659); 23rd December 1899:1721-1726 (*BMJ* 1899;2:1721) 30th December 1889:1779-1784 (*BMJ* 1899;2:1779).

33. The Prevention of Tuberculosis, Chelsea Clinical Society, *BMJ*, April 29, 1899:1038. (*BMJ* 1899;1:1038).
34. Association Intelligence, South Midland Branch [British Medical Association] The Natural History of Tubercle, *BMJ*, July 8th 1899:111. (*BMJ* 1899;2:107).
35. Obituary, Sir William Watson Cheyne, F.R.C.S., F.R.S., *The Lancet*, April 30th, 1932:964.
36. Flügge, C., Cheyne, W.W. (trans.), Micro Organisms, with Special Reference to the Etiology of the Infective Diseases, London (New Sydenham Society), 1890.
37. Micro-Organisms, with Special Reference to the Etiology of the Infectious Diseases. By Dr C. Flügge. Translated by W. Watson Cheyne, *BMJ*, April 12th 1890:844. (*BMJ* 1890;1:842).
38. Obituary, Sir William Watson Cheyne, F.R.C.S., F.R.S., *The Lancet*, April 30th, 1932:964.
39. Cheyne, insert in letter to Albert Carless, June 11th 1891, Albert Carless, Letters and Testimonials, Royal College of Surgeons of England, London. MS 0064/6.
40. Worboys, M., *op. cit.* :213.
41. Ibid.
42. Lord Lister's notes on iodide gauze, 1882-1887, Royal College of Surgeons of England, London, MS0021/4/4, folder 86, item 1:22nd December 1882.
43. Ibid.
44. Cheyne, W.W., Corfield, W.H., Cassell, C.E., *Public Health Laboratory Work*, London (William Clowes and Sons), 1884. Printed and published for the Executive Council of the International Health Exhibition, and for the Council of the Society of Arts.
45. Flügge, C., Cheyne, W.W. (trans.), *Micro Organisms, with Special Reference to the Etiology of the Infective Diseases*, London (New Sydenham Society), 1890: Preface vii.
46. Cheyne, W.W. (ed.), *Recent Essays by Various Authors on Bacteria in Relation to Disease*, London (New Sydenham Society), 1886: Prefatory Note.
47. The Pathology of Tuberculosis, Notes, Letters, etc., *BMJ*, April 9th 1892:795. Response of Editor of *BMJ* to a letter by Dr George F. Crooke. (*BMJ* 1892;1:794).
48. Ibid.
49. Reviews, Tuberculosis of Bones and Joints, by N. Senn, M.C., Professor of Surgery, Rush Medical College, Chicago, Philadelphia and London: F.A. Davison. 1892. Roy. 8vo. 504pp. 22s.6d. *BMJ*, Feb 18th 1893:358. (*BMJ* 1893;1:358).
50. Koch, R., A Further Communication on a Remedy for Tuberculosis, *BMJ*, 22nd November 1890:1193.(*BMJ* 1890;2:1193).
51. (No author given) Professor Koch's Remedy for Tuberculosis: The Rash Caused by Injection of Koch's Fluid, *BMJ*, Dec 6th 1890:1330. (*BMJ* 1890;2:1321).
52. Action taken by English Hospitals, *BMJ*, 22nd November 1890:1197. (*BMJ* 1890;2:1193).
53. *BMJ*, Nov 29th 1890:1265. (*BMJ* 1890;2:1254).
54. Action taken by English Hospitals, *BMJ*, 22nd November 1890:1197. (*BMJ* 1890;2:1193).
55. Worboys, M., *op. cit.* :226.
56. Ibid.
57. Brookhouse, J.O., Letter to the Editor, *BMJ*, 22nd November 1890:1199. (*BMJ* 1890;2:1193).
58. "Special Correspondent" of *BMJ* in Berlin, General Notes from Berlin, *BMJ*, 22nd November 1890:1198. (*BMJ* 1890;2:1193).
59. "F.P.C.P.", The Hegira to Berlin, *BMJ*, 22nd November 1890:1199. (*BMJ* 1890;2:1193).
60. Heron, G.A., Cheyne, W.W., Demonstration to be Given in London, *BMJ*, 22nd Nov 1890:1198. (*BMJ* 1890;2:1193).

61. Professor Koch's Remedy for Tuberculosis, Demonstrations in London, *BMJ*,Dec 6th 1890:1321. (*BMJ* 1890;2:1321).
62. Ibid.
63. "Special Correspondent" of *BMJ*, Notes from Berlin, *BMJ*,December 6th 1890:1327. (*BMJ* 1890;2:1321).
64. Ibid.
65. Ibid.
66. The Ethics of Research (no author stated), *BMJ*,November 7th 1891:999. *BMJ* 1891; 2:999. (*BMJ* 1891;2:999).
67. Professor Koch's Remedy for Tuberculosis, Some General Pathological Conditions, *BMJ*,November 29th 1890:1262. (*BMJ* 1890;2:1254).
68. From an Occasional Correspondent, *BMJ*,Nov. 29th 1890:1264. (*BMJ* 1890;2:1254).
69. Ibid.
70. Professor Koch's Remedy for Tuberculosis, Some General Pathological Conditions, *BMJ*,November 29th 1890:1262. (*BMJ* 1890;2:1254).
71. Honours and Congratulations, *BMJ*,Nov. 29th 1890:1264. (*BMJ* 1890;2:1254).
72. Ibid.
73. *Pall Mall Gazette*, November 18th, 1890:5, British Library, accessed through British Newspaper Archive, 20/01/2014.
74. Dr Koch's Discovery, The Manchester Courier and Lancashire General Advertiser, November 21st, 1890:3, British Library, accessed through British Newspaper Archive, 20/01/2014.
75. Ibid.
76. Ibid.
77. Demonstration at Paddington Green Children's Hospital, *BMJ*,December 13th 1890:1382. (*BMJ* 1890;2:1378).
78. Professor Koch's Remedy for Tuberculosis, King's College Hospital, *BMJ*,December 6th 1890:1322. (*BMJ* 1890;2:1321).
79. Ibid:1321.
80. Demonstration at Paddington Green Children's Hospital, *British Medical Journal* Dec 13th 1890:1382. (*BMJ* 1890;2:1378).
81. Demonstrations of Dr Koch's Method of Treating Tuberculosis, *BMJ*,November 29th, 1890:1265.
82. Professor Koch's Remedy for Tuberculosis, King's College Hospital, *BMJ*,December 6th 1890:1321. (*BMJ* 1890;2:1321).
83. Ibid.
84. Abstract of an Address on the Value of Tuberculin in the Treatment of Surgical Tuberculosis ... by W. Watson Cheyne, M.B., F.R.C.S., *BMJ*,May 2nd 1891:951 (*BMJ* 1891;1:951).
85. Notice in *BMJ*,December 20th 1890:1441. (*BMJ* 1890;2:1438).
86. *Bath Chronicle and Weekly Gazette,* 25th December, 1890, British Library, accessed through the British Newspaper Archive, 20/01/14.
87. Our London Letter, *Exeter and Plymouth Gazette*, 17th December, 1890:5, accessed through the British Newspaper Archive, 20/01/2014.
88. Williams, C. Theodore, Professor Koch's Treatment for Tuberculosis, *BMJ*,December 20th 1890:1434. (*BMJ* 1890;2:1434).
89. Letter from Cheyne to an unknown correspondent, December 16th 1890, Wellcome Library, London. MS7796/1.
90. Cheyne, W.W., Abstract of an Address on the Value of Tuberculin in the Treatment of Surgical Tuberculosis, *BMJ*,May 2nd 1891:951-961. (*BMJ* 1891;1:951).

91. Reports of Societies, Royal Medical and Chirurgical Society, Thursday, May 7th, 1891, *BMJ*,May 16th 1891:1069. (*BMJ* 1891;1:1068).
92. Ibid.
93. What is the Practical Value of Tuberculin? *BMJ*,May 2nd 1891:973. (*BMJ* 1891;1:971).
94. Reports of Societies, Royal Medical and Chirurgical Society, Thursday, May 7th, 1891, *BMJ*,May 16th 1891:1069. (*BMJ* 1891;1:1068).
95. What is the Practical Value of Tuberculin? II. *BMJ*,May 2nd 1891:973. (*BMJ* 1891;1:971).
96. Ibid.
97. The Treatment of Surgical Tuberculoses, *BMJ*,August 8th 1891:327. (*BMJ* 1891;2:326).
98. Sir D'Arcy Power, Le Fanu, W.R., *Lives of the Fellows of the Royal College of Surgeons of England 1930-1951*, London (Royal College of Surgeons), 1953:145.
99. Reviews: Tuberculous Disease of Bones and Joints, its Pathology, Symptoms, and Treatment. By W. Watson Cheyne. *BMJ*,December 7th 1895:1429. (*BMJ* 1895;2:1428).
100. What is the Practical Value of Tuberculin? *BMJ*,May 2nd 1891:973. (*BMJ* 1891;1:971).
101. Worboys, M., *op. cit.* :226.
102. Ibid:227.
103. The Treatment of Surgical Tuberculoses, *BMJ*,August 8th 1891:328. (*BMJ* 1891;2:326).
104. Mail and Ship News, *Leeds Mercury*, 7th April 1888:7, British Library, accessed through the British Newspaper Archive, 20/01/2014.
105. Cheyne, Sir William Watson, King's College website, http://www.kingscollections.org/catalogues/kclca/collection/c/10ch30-1 accessed 21/01/2014.
106. Cheyne, W.W., List of Cheyne's works, sent to E.S. Reid Tait, 1928-29, Shetland Archives, Lerwick. D6/262/9/5, no.21.
107. Letter to Cheyne, dated 19/8/90, from Herbert G. Briston, Honorary Secretary, in possession of Cheyne family.
108. In the 1891 census, he is listed as resident at 59 Welbeck Street, and by 1893, his letters are headed 75 Harley Street. There are, nevertheless ambiguous references placing him in Harley Street by 1888, which could simply mean he was only working from there.
109. Registry of births, National Archives, Kew, accessed via www.ancestry.com 19/09/2013.
110. Registry of deaths, National Archives, Kew, accessed via www.ancestry.com 19/09/2013.
111. Letter from Lord Lister to Mary Watson Cheyne, Royal College of Surgeons (Edin.), Edinburgh, RCS/GD5/139.
112. British Medical Association, Sixtieth Annual Meeting, *BMJ*,May14th 1892:1028. (*BMJ* 1892;1:1028).
113. Registry of deaths, National Archives, Kew, accessed via www.ancestry.co.uk 19/09/2013.
114. Letter to Cheyne from Lionel F. Hill, dated 2/12/1893, in possession of Cheyne family.
115. Godlee, R.J., *op. cit.* :523.
116. Cheyne, W.W., An Address on the Cancer of the Breast, Delivered before the Harveian Society of London, *BMJ*,February 10th, 1894, 289-291. (*BMJ* 1894;1:289).
117. Fellowship of the Royal Society, the Royal Society website https://docs.google.com/spreadsheet/ccc?key=0AmIblj8F2r_GdG9aaFZRMjNrYUZXVHkzeXRzdmhFTmc#gid=0 Accessed 17/09/2013.

Notes and References

118. 1891 England Census, National Archives, Kew, accessed through www.ancestry.co.uk 23/08/2013.
119. Ibid.
120. It is unclear whether Margaret Smith was employed by Cheyne, but family tradition suggests that she became his housekeeper around this point.
121. 1861 Scottish Census for Lerwick, National Archives, Kew, accessed via www.ancestry.co.uk 25/08/2012.
122. Ibid.
123. Ibid.
124. Ibid.
125. Ibid.
126. Ibid.
127. 1851 Scottish Census, National Archives, Kew, accessed via www.ancestry.co.uk 25/08/2012.
128. 1881 Scottish Census, National Archives, Kew, accessed via www.ancestry.co.uk 25/08/2013.
129. Literary Notes, *BMJ*,February 17th 1894:378 (*BMJ* 1894;1:377).
130. Treves, F. (ed.), *A System of Surgery*, Vol. II, London (Cassell), 1896.
131. Cheyne, W.W. *The Treatment of Wounds, Ulcers and Abscesses*, London and Edinburgh (Young J. Pentland), 1894.
132. Opening of the Medical Schools: Metropolitan, *BMJ*,Sept. 1st, 1894:497. (*BMJ* 1894;2:497).
133. Opening of the Winter Session in London: King's College Hospital, *BMJ*,Oct 6th 1894:782. (*BMJ* 1894;2:781)
134. Worboys, M., *op. cit.* :151.
135. Godlee, R.J., *op. cit.* :523.
136. Association Intelligence, West Somerset Branch, *BMJ*,Oct. 13th 1894:842. (*BMJ* 1894;2:842).
137. Marriages,The London Standard, 17th December, 1894:1, accessed through the British Newspaper Archive, 20/01/2014.
138. Notes on books, *BMJ*,March 30th 1895:706.
139. British Medical Association, Sixty-Third Annual Meeting: Surgery, *British Medical Journal* June 1st 1895:1235. (*BMJ* 1895;1:1232).
140. Letter to Cheyne from Major General A.L.Playfair, Haymarket Theatre, dated September 30th, 1895. In possession of the Cheyne family.
141. Ibid.
142. Letter to Cheyne from the Medical Society of London, dated June 20th 1895, in possession of the Cheyne family.
143. Reviews: The Objects and Limits of Operations for Cancer. By W. Watson Cheyne, M.B., F.R.S., F.R.C.S., Professor of Surgery at King's College, London, *BMJ*,December 12th, 1896:1722. (*BMJ* 1896;2:1722).
144. Ibid.
145. Ibid.
146. Letter in possession of the Cheyne family, dated 27th June 1896, from J.W. Clark, Registrar, Cambridge.
147. Obituary, Sir William Watson Cheyne, F.R.C.S., F.R.C., *The Lancet*, April 30th, 1932:964.
148. Ibid.
149. The Council of the Royal College of Surgeons, *BMJ*,June 20th 1896:1520. (*BMJ* 1896;1:1517).
150. The Council of the Royal College of Surgeons, *BMJ*,June 26th, 1897:1673; Elections at the College of Surgeons, *BMJ*,July 10th 1897:98. (*BMJ* 1897;2:95).

151. Royal College of Surgeons of England Council Election, *British Medical Journal* July 3rd 1897:41. (*BMJ* 1897;2:41).
152. Letter to Cheyne from Robert Saundby, British Medical Association, dated January 22nd, 1897. In possession of the Cheyne family.
153. Official Guide and Souvenir of the Sixty-Fifth Annual Meeting of the British Medical Association, Montreal, 1897: http://archive.org/stream/cihm_01515#page/n5/mode/2up Accessed 19/09/2013.
154. Ibid.
155. UK incoming passenger lists, 1897, accessed through www.ancestry.co.uk, 22/12/2013.
156. Cheyne, W.W., An Address Delivered at the Opening of the Section of Pathology at the Annual Meeting of the British Medical Association in Montreal, September,1897, *BMJ,*September 4th 1897 (586-589):589. (*BMJ* 1897;2:586)
157. Letter to Cheyne from Edward Trimmer, Secretary of the Royal College of Surgeons of England, dated 1st July, 1897. In possession of the Cheyne family.
158. Letter to Cheyne from D.M. Forbes, Honorary Surgeon, Scottish Hospital, dated 27th November 1897. In possession of Cheyne family.
159. Letter to Cheyne from d'Arcy Power, Hon. Sec. Pathological Society of London, 10th April, 1899. In possession of Cheyne family.
160. Obituary, Sir William Watson Cheyne, F.R.C.S., F.R.C., *The Lancet*, April 30th, 1932:965.
161. Cheyne, W.W., Objects and Limits of Operations for Cancer, being the Lettsomian Lectures for 1896, New York (William Wood), 1896.
162. Carpenter, M.W., Lister's Relationship with Patients: "A successful case."", Notes and Records of the Royal Society, Notes Rec. R. Soc. published online May 29, 2013.
163. Cheyne, W.W., The Harveian Lectures on the Surgical Treatment of Tuberculous Diseases, *BMJ,*December 30th 1899 (1779-1784):1781. (*BMJ* 1899;2:1779).
164. Cheyne, W.W., Lettsomian Lectures on the Objects and Limits of Operations for Cancer, Delivered before the Medical Society of London, February, 1896, *BMJ,*February 15th 1896, (385-390):385-6. (*BMJ* 1896;1:385).
165. Lettsomian Lectures on the Objects and Limits of Operations for Cancer, Delivered before the Medical Society of London, February, 1896, *BMJ,*March 7th, 1896, 577-582:582. (*BMJ* 1896;1:577).
166. Review: The Objects and Limits of Operations for Cancer, by W. Watson Cheyne, M.B., F.R.S., F.R.C.S., *BMJ,*December 12th 1896:1722. (*BMJ* 1896;2:1722).
167. Ibid.
168. Ibid.
169. Cheyne, W.W., Two Cases of Oöphorectomy for Inoperable Breast Cancer, *BMJ,*May 7th 1898, (1194-1195):1195. (*BMJ* 1898;1:1194).
170. Sir D'Arcy Power, Le Fanu, W.R., *Lives of the Fellows of the Royal College of Surgeons of England 1930-1951*, London (Royal College of Surgeons), 1953:145.
171. Cheyne, W.W., Objects and Limits of Operations for Cancer, being the Lettsomian Lectures for 1896, New York (William Wood), 1896.
172. Sir D'Arcy Power, Le Fanu, W.R., *op. cit.* :145.
173. Repair of the Bridge of the Nose by Rabbit Bone, Clinical Society of London, report in *BMJ,*November 5th 1898:1431. (*BMJ* 1898;2:1430).
174. Ibid.
175. Cheyne, W.W., The Treatment of Incompletely Descended Testicle, *BMJ,*February 15th 1890:351-352. (*BMJ* 1890;1:351).
176. Down Brothers, The Annual Museum, *BMJ,*August 25th 1894:417. (*BMJ* 1894;2:414).

177. Messrs. Down Bros., Annual Museum, *BMJ*,Aug. 10th 1895:371. (*BMJ* 1895;2:369).
178. Ibid.
179. Lister, J., An Address on Catgut Ligature, *BMJ*,Feb 12th 1881, (219-221):220. (*BMJ* 1881;1:219).
180. Watson, Caroline C., Griessenauer, Christoph J., Loukas, Marios, Blount, Jeffrey P., Shane Tubbs, R., William Watson Cheyne (1852–1932): a life in medicine and his innovative surgical treatment of congenital hydrocephalus, Springer, 2013, Childs Nerv Syst. DOI 10.1007/s00381-013-2220-7.
181. Cheyne, W.W., Sutherland, G.A., The Treatment of Hydrocephalus by Intracranial Drainage, *BMJ*,October 15th, 1898:1155-1157. (*BMJ* 1898;2:1127).
182. Watson, Caroline C., Griessenauer, Christoph J., Loukas, Marios, Blount, Jeffrey P., Shane Tubbs, R., William Watson Cheyne (1852–1932): a life in medicine and his innovative surgical treatment of congenital hydrocephalus, Springer, 2013, Childs Nerv Syst. DOI 10.1007/s00381-013-2220-7.
183. Ibid.
184. Cheyne, W.W., Sutherland, G.A., The Treatment of Hydrocephalus by Intracranial Drainage, *BMJ*,October 15th, 1898, (1155-1157):1158. (*BMJ* 1898;2:1127).
185. British Medical Association, Sixty-Sixth Annual Meeting, Section of Diseases of Children, report in *BMJ*,Aug 6th, 1898:366. (*BMJ* 1898;2:361).
186. Cheyne, W.W., Sutherland, G.A., The Treatment of Hydrocephalus by Intracranial Drainage, *BMJ*,October 15th, 1898:1155-1157. (*BMJ* 1898;2:1127).
187. Ibid.
188. Watson, Caroline C., Griessenauer, Christoph J., Loukas, Marios, Blount, Jeffrey P., Shane Tubbs, R., William Watson Cheyne (1852–1932): a life in medicine and his innovative surgical treatment of congenital hydrocephalus, Springer, 2013, Childs Nerv Syst. DOI 10.1007/s00381-013-2220-7.
189. Literary Notes, *BMJ*,October 29th 1898:1367. (*BMJ* 1898;2:1367).
190. Notes on Books, King's College Hospital Reports, Vol. III, *BMJ*,1898:1272 (*BMJ* 1898;1:1272).
191. Cheyne, W.W., List of Cheyne's works, sent to E.S. Reid Tait, 1928-29, Shetland Archives, D6/262/9/5, no.23.
192. Ibid.
193. Ibid.
194. Grey Turner, G., Correspondence: Thyroid Surgery, *BMJ*,April 8th 1950:842. (*BMJ* 1950; 1:841).
195. Ibid.
196. Burghard, F.F., Cheyne, W.W., *A Manual of Surgical Treatment*, Volume IV, Philadelphia/New York (Lea Brothers), 1901: Author's Preface; General Preface:xvi.
197. Reviews: Some Recent Surgical Textbooks, *BMJ*,April 6th 1901:832.(*BMJ* 1901;1:832).
198. Reviews: A Manual of Surgical Treatment. By W. Watson Cheyne, M.B., F.R.C.S., F.R.S., Professor of Surgery in King's College, London, Surgeon to King's College Hospital, etc., and F.F. Burghard, M.B. and M.S.Lond., F.R.C.S., Teacher of Practical Surgery in King's College, London, Surgeon to King's College Hospital, etc. In Six Parts. Part III. *BMJ*,July 21st 1900:172. (*BMJ* 1900, 2:173).
199. Burghard, F.F. Cheyne, W.W., *A Manual of Surgical Treatment*, Volume IV, Philadelphia/New York (Lea Brothers), 1901: Author's Preface; Preface to Part IV.
200. Reviews: Practical Surgery, *British Medical Journal* Sept 6th 1902:705. (*BMJ* 1902, 2:704).
201. Burghard, F.F. Cheyne, W.W., *A Manual of Surgical Treatment*, Volume IV, Philadelphia/New York (Lea Brothers), 1901: Author's Preface; Preface to Part VI.

202. Ibid.
203. Literary Notes, *BMJ*, Dec 14th 1901:1767. (*BMJ* 1901, 2:1767).
204. Reviews: Some Recent Surgical Textbooks, *BMJ*, April 6th 1901:832. (*BMJ* 1901, 1:832).
205. Registry of births, National Archives, Kew, accessed via www.ancestry.co.uk 25/08/2013.
206. 1901 England Census, National Archives, Kew, accessed via www.ancestry.co.uk 25/08/2013.
207. 1891 Scottish Census (for Lerwick), accessed through www.ancestry.co.uk 25/08/2013.
208. 1911 England Census, National Archives, Kew, accessed via www.ancestry.co.uk 25/08/2013.
209. Annual Meeting [of the British Medical Association], The Sections, Pathology and Bacteriology, *BMJ*, September 18th 1897:719 (*BMJ* 1897; 2:719).

Chapter 9
1. Reports of Societies: Harveian Society of London, Thursday, October 16th, 1890, *BMJ*, October 25th 1890, 958-9:959. (*BMJ* 1890; 2:957).
2. Reviews and Notices: The Science and Practice of Surgery. By Frederick James Gant, F.R.C.S., Senior Surgeon to the Royal Free Hospital. Third Edition. London: Ballière, Tindall, and Cox, 1886, *BMJ*, Aug 28th 1886:414. (*BMJ* 1886; 2:414).
3. Reviews and Notices: The International Encyclopaedia of Surgery ... Edited by John Ashhurst ... Vol. II. London: Macmillan, 1882, *BMJ*, Nov 25th 1882:1044. (*BMJ* 1882; 2:1044).
4. Handwritten version of Cheyne's Boylston Prize submission (Volume II). There is no date, but he submitted his final version of the essay for the 1880 prize. The original is in possession of the Cheyne family, and a transcript by the author is available at http://www.watson-cheyne.com/handwritten-manuscript-volume2.pdf.
5. Worboys, M., *op. cit.* :162.
6. Burdon Sanderson, J., 1878 in Worboys, M., *op. cit.* :158.
7. Lister, J., Mr Spence on Surgical Statistics, *BMJ*, February 21st 1880:276. (*BMJ* 1880; 1:276).
8. Ibid.
9. Ibid.
10. Cheyne, W.W., Statistical Report of All Operations Performed on Healthy Joints in Hospital Practice, by Mr Lister, from September 1871 to the present time, together with such accidental wounds of joints as occurred during the same period, *BMJ*, November 29th, 1879:859-864. (*BMJ* 1879; 2:859).
11. Spence, J., Surgical Statistics, *BMJ*, January 24th 1880:119-121. (*BMJ* 1880; 1:119).
12. Ibid:119.
13. Surgical Statistics (no author given), *BMJ*, April 10th, 1880:559. (*BMJ* 1880; 1:556).
14. Lister, J., Mr Spence on Surgical Statistics, *BMJ*, February 14th 1880, 237-239:237. (*BMJ* 1880; 1:237).
15. Cheyne, W.W., Professor Spence on Surgical Statistics, *BMJ*, 14th February 1880, 239-241:241. (*BMJ* 1880; 1:239).
16. Ibid:240.
17. Surgical Statistics (no author given), *BMJ*, April 10th, 1880:559. (*BMJ* 1880; 1:556).
18. Reference from Professor Spence in favour of Cheyne, dated 10th May, 1878, Royal College of Surgeons of England, MSS552(1).

19. Cheyne, W.W., Professor Spence on Surgical Statistics, *BMJ*, 14th February 1880:240-241. (*BMJ* 1880; 1:239).
20. Spence, J., Surgical Statistics, *BMJ*, May 29th 1880:831. (*BMJ* 1880; 1:831).
21. Ibid:832.
22. Cheyne, W.W., Remarks on the Treatment of Wounds in connexion with the Recent Results obtained at St. George's Hospital, *The Lancet*, Volume 175, Issue 4505:15-18, January 1st 1910:15.
23. Spence, J., Surgical Statistics, *BMJ*, May 29th 1880:831. (*BMJ* 1880; 1:831).
24. Cheyne, W.W., Statistical Report of All Operations Performed on Healthy Joints in Hospital Practice, by Mr Lister, from September 1871 to the present time, together with such accidental wounds of joints as occurred during the same period, *BMJ*, November 29th, 1879,859-864:860. (*BMJ* 1879; 2:859).
25. Cheyne, W.W., *Lister and His Achievement*, London (Longmans, Green & Co.), 1925:89.
26. Ibid:92.
27. Ibid.
28. H R Ranke, 'Die Bakterien-vegetation unter 19: 393-401. dem Lister'schen Verbande', Centralblatt für Chirurgie, 1874, 1: 193-4. in Pennington, H., Listerism, its Decline and its Persistence, Medical History, 1995:36.
29. Pennington, H., Listerism, its decline and its persistence: the introduction of aseptic surgical techniques in three British teaching hospitals, 1890-99, Medical History, 02/1995; 39(1):35-60:37.
30. Ibid:38.
31. Ibid:37.
32. Ibid.
33. Cheyne, W.W., *Lister and His Achievement*, London (Longmans, Green & Co.), 1925:88.
34. Mayo Robson, A.W., A Proposed Substitute for Carbolic Spray in Antiseptic Surgery, *BMJ*, September 2nd 1882, 420-422:420. (*BMJ* 1882; 2:420).
35. Obituary Notice, Sir William Watson Cheyne, Baronet - 1852-1932, Royal Society 10.1098/rsbm, 1932 0007 Obit. Not. Fell. R. Soc., December 1932, Vol.1, No.1, 27.
36. Cheyne, W.W., *Lister and His Achievement*, London (Longmans, Green & Co.), 1925:90-91.
37. Cheyne, W.W., The After-Effects of Chloroform, *The Lancet*, February 10th 1894, Vol. 143, Issue 3676:370.
38. Cheyne, W.W., *Lister and His Achievement*, London (Longmans, Green & Co.), 1925:93.
39. Ibid:90.
40. Worboys, M., *op. cit.* :189.
41. Pennington, H., Listerism, its decline and its persistence: the introduction of aseptic surgical techniques in three British teaching hospitals, 1890-99, Medical History, 02/1995; 39(1):35-60:38.
42. Ibid:35.
43. Cheyne, W.W. 'Antiseptic Surgery', in Heath, C. (ed.), *Dictionary of Practical Surgery*, London, Smith, Elder, 1889:71.
44. Ibid.
45. Pennington, H., Listerism, its decline and its persistence: the introduction of aseptic surgical techniques in three British teaching hospitals, 1890-99, *Medical History*, 02/1995; 39(1):35-60:35.
46. Ibid.
47. Ibid:55.

48. Ibid:39.
49. J Lister, On the principles of antiseptic surgery, Virchow-Festschrift, 1891, vol. 3; reprinted in *The collected papers of Joseph, Baron Lister*, 2 vols, Oxford, Clarendon Press, 1909, vol. 2, pp. 340-8. In Pennington, H., Listerism, its decline and its persistence: the introduction of aseptic surgical techniques in three British teaching hospitals, 1890-99, Medical History, 02/1995; 39(1):35-60:39.
50. Antiseptic and Aseptic Surgery, (author not specified) *BMJ*, January 26th, 1895:210; Stewart, J., in Obituary, Sir William Watson Cheyne, F.R.C.S., F.R.S., *The Lancet*, April 30th, 1932:964-965.
51. Cheyne, W.W., Remarks on the Treatment of Wounds in connexion with the Recent Results obtained at St. George's Hospital, *The Lancet*, Volume 175, Issue 4505:15-18, January 1st 1910:15.
52. Cheyne, W.W., *The Treatment of Wounds, Ulcers and Abscesses*, Edinburgh and London, 1894:56.
53. Burghard, F.F., A Discussion on the Present Position of the Aseptic Treatment of Wounds, *BMJ*, Oct 1st 1904, 793-798:797.
54. Cheyne, W.W., Remarks on the Treatment of Wounds in connexion with the Recent Results obtained at St. George's Hospital, *The Lancet*, Volume 175, Issue 4505:15-18, January 1st 1910:15.
55. McCrae, Morrice, *Simpson, the turbulent life of a medical pioneer*, Edinburgh (Birlinn), 2010:120.
56. Ibid.
57. Stiles, Sir H., in Obituary, Sir William Watson Cheyne, F.R.C.S., F.R.S., *The Lancet*, April 30th, 1932:964-965.
58. Ibid:965.
59. Cheyne, W.W., An Address on the Treatment of Wounds in War, Journal of Comparative Pathology and Therapeutics, Vol. 27, 1914:pp.304-323, reprinted from *The Lancet*, November 21st, 1914.
60. Cheyne, W.W., *Three Orations: The Lister Centenary*, London (John Bale, Sons and Danielsson), 1927.
61. Antiseptic and Aseptic Surgery (no author given), *BMJ*, January 26th, 1895, 210-211:210. (*BMJ* 1895; 1:209).
62. Cheyne, W.W., *The Treatment of Wounds, Ulcers and Abscesses*, Edinburgh and London, 1894:Preface, xviii.
63. Worboys, M., *op. cit.* :187.
64. Cheyne, W.W., The Disinfection of Hands and Instruments, *BMJ*, Aug 7th 1897, 373-374:374 (*BMJ* 1897; 2:373) and Lockwood, C.B., The Disinfection of Hands and Instruments, *British Medical Journal* July 31st 1897:314. (*BMJ* 1897; 2:314).
65. Cheyne, W.W., An Address on the Treatment of Wounds in War, Journal of Comparative Pathology and Therapeutics, Vol. 27, 1914, pp:304-323, reprinted from *The Lancet*, November 21st, 1914.
66. King's College Hospital. An Exploratory Laparotomy, *BMJ*, July 1st 1893:14. (*BMJ* 1893; 2:14).
67. Ibid.
68. Ibid.
69. Cheyne, W.W., *The Treatment of Wounds, Ulcers and Abscesses*, Edinburgh and London (Young J. Pentland), 1894:57.
70. Ibid:Preface, viii.
71. Worboys, M., *op. cit.* :189.
72. Ibid:186.
73. Ibid:188.
74. Rose, W. and Carless, A., *A Manual of Surgery*, London (Baillière, Tyndall and Cox), 1898, in Worboys, M., *op. cit.* :190.

75. Stiles, Sir H., in Obituary, Sir William Watson Cheyne, F.R.C.S., F.R.S., *The Lancet*, April 30th, 1932:964.
76. Ibid:965.
77. Ibid.
78. Edmunds, A., The Late Sir Watson Cheyne. To the Editor of *The Lancet*, *The Lancet*, May 7, 1932:1013.
79. Ibid.
80. Letter from Lord Lister to Cheyne, dated January 5th, 1910, Royal College of Surgeons of Edinburgh RCS/GD5/148 (65/8/6).
81. Letter from Lord Lister to Cheyne, dated March 2nd, 1908, Royal College of Surgeons of Edinburgh RCS/GD5/145 (65/8/h)
82. Cheyne, W.W., Remarks on the Treatment of Wounds in connexion with the Recent Results obtained at St. George's Hospital, *The Lancet*, Volume 175, Issue 4505:15-18, January 1st 1910:15.
83. Pendlebury, H.S., Back, I., A Comparison between the Antiseptic and Aseptic Methods of Operation, with Special Reference to the Occurrence of Suppuration, based upon the Results obtained at St. George's Hospital during the years 1906 and 1908 respectively, *The Lancet*, Volume 174, Issue 4500, 27 November 1909, Pages 1578-1580.
84. Cheyne, W.W., Remarks on the Treatment of Wounds in connexion with the Recent Results obtained at St. George's Hospital, *The Lancet*, Volume 175, Issue 4505:15-18, January 1st 1910:15.
85. Ibid.
86. Ibid.
87. Obituary, Sir William Watson Cheyne, F.R.C.S., F.R.S., *The Lancet*, April 30th, 1932:964.
88. Asepsis and Antisepsis, *The Lancet*, January 29th 1910:329.
89. Ibid.
90. Sir d'Arcy Power, le Fanu, W.R., *op. cit.* :145.
91. Stiles, Sir H., in Obituary, Sir William Watson Cheyne, F.R.C.S., F.R.S., *The Lancet*, April 30th, 1932:965.
92. Ibid.
93. Ibid:963.
94. Ibid:964.
95. Ibid:965.
96. Ibid:963.
97. King's College Hospital Reports, 1895, being the Annual Report of King's College Hospital, Vol.1 (Oct. 1st, 1893 - Sept. 30th 1894), London (Adlard and Son), 1895:400.
98. Ibid:398.
99. 1911 Census, National Archives, Kew accessed through ancestry.co.uk 25th August 2013.
100. Ibid.
101. 1911 Scotland Census, accessed through www.ancestry.co.uk 25th August 2013, and Mr K. Hughson, Yell, Shetland (formerly Fetlar), personal communication to the author, 2nd December 2013: the 1901 census lists Grace Watson as resident at Hillside Cottage. Mr Hughson confirmed that this was the house known locally as "The Chapel" and that it had formerly been a Methodist chapel.

Chapter 10

1. Smuts, J., 1906 in Pakenham, T., *The Boer War*, London (BCA) 1999:9.
2. Pakenham, T., *op. cit.* :581.

3. Conan Doyle, A., *The Great Boer War*, London (Smith, Elder and Co.), 1900.
4. Pakenham, T., *op. cit.* :109.
5. Grey, E.F., *Women in Journalism at the Fin de Siècle: Making a Name for Herself*, London (Palgrave MacMillan), 2012: Chapter 7: Flora Shaw and the Times: Becoming a Journalist, Advocating Empire.
6. Morgan, K.O., The Boer War and the Media, *Twentieth Century British History*, 13, No.1 (2002):1-16 (4).
7. Ibid.
8. Ibid.
9. Ibid:6.
10. Crowther, M.A. & Dupree, M.W., *op. cit.* :328.
11. The War, *BMJ*, 28th October 1899:1210. (*BMJ* 1899;2:1209).
12. Civilian Medical Aid in the War, *BMJ*, October 28th, 1899:1217. (*BMJ* 1899;2:1217).
13. Ibid.
14. The War, *BMJ*, October 28th 1899:1209. (*BMJ* 1899;2:1209).
15. Ibid:1210.
16. The War, *BMJ*, November 4th 1899:1303-4. (*BMJ* 1899;2:1301).
17. Ibid.
18. Consulting Surgeons with the Army in South Africa, *BMJ*, December 30th 1899:1803. (*BMJ* 1899;2:1803).
19. Pay of Great War Surgeons, Isle of Man Times, 21st July 1900:9, British Library, accessed through British Newspaper Archive, 20/01/2014.
20. A New Shrine of Hygiea in Sicily, *BMJ*, February 17th 1900:417. (*BMJ* 1900;1:417).
21. People of To-Day, *The Gloucester Citizen*, Friday 11th May 1900:3, British Library, accessed through the British Newspaper Archive, 20/01/2014.
22. *The Morning Post*, 28th May 1898:5, British Library, accessed through British Newspaper Archive, 20/01/2014.
23. People of To-Day, *The Gloucester Citizen*, Friday 11th May 1900:3, British Library, accessed through the British Newspaper Archive, 20/01/2014.
24. Complimentary Dinner to Mr Watson Cheyne and Mr G. Lenthal Cheatle, *The Lancet*, Vol.115, Issue 3985, January 13th, 1900:114-115.
25. Consulting Surgeons with the Forces in South Africa, *BMJ*, January 13th, 1900:93. (*BMJ* 1900;1:91).
26. Ibid.
27. Complimentary Dinner to Mr Watson Cheyne and Mr G. Lenthal Cheatle, *The Lancet*, Vol.115, Issue 3985, January 13th, 1900:115.
28. Ibid.
29. Original handwritten poem in possession of the Cheyne family.
30. Ibid.
31. Ibid.
32. Ibid.
33. Cheyne, W.W., *Lister and his Achievement*, London (Longmans, Green and Co.), 1925:13.
34. Original handwritten poem in possession of the Cheyne family.
35. Cutting from an unknown publication, in an album belonging to the Cheyne family.
36. Daily Telegraph, Jan. 8th, 1900, untitled newspaper cutting in an album belonging to the Cheyne family.
37. Surgeons for the Front, *Lloyd's News*, Jan. 7th, 1900, newspaper cutting in an album belonging to the Cheyne family.

Notes and References

38. Enthusiastic Medical Students, *Westminster Gazette*, Jan. 7th, 1900, newspaper cutting in an album belonging to the Cheyne family.
39. Departure of Mr Watson Cheyne as Consulting Surgeon to the Forces, Shetland Times, January 13th, 1900, newspaper cutting in an album belonging to the Cheyne family.
40. Ibid.
41. Ibid.
42. Two More Surgeons Leave, *Daily Mail*, January 8th, 1900, newspaper cutting in an album belonging to the Cheyne family.
43. Pakenham, T., The Boer War, London (BCA), 1999:581.
44. Cheyne, W.W., The March from Modder River to Bloemfontein, *BMJ*, May 5th 1900, 1093-1096:1093. (*BMJ* 1900;1:1093).
45. Ibid.
46. Ibid.
47. Ibid.
48. Ibid.
49. Makins, G.H., *Surgical Experiences in South Africa, 1899-1900. Being Mainly a Clinical Study of the Nature and Effects of Injuries Produced by Bullets of Small Calibre*, London (Henry Frowde and Hodder & Stoughton), 1913:30.
50. Villiers, J.C., The Medical Aspect of the Anglo-Boer War, 1899-1902, Part 2, *Military History Journal*, Vol. 6, no. 3, June 1984. South African Military History Society, http://samilitaryhistory.org/vol063jc.html/ Accessed 10/11/2012.
51. Makins, G.H., *op. cit.*
52. Ibid.
53. Ibid.
54. Ibid.
55. Makins, G.H., *op. cit.* :7.
56. Ibid.
57. Ibid.
58. Cheyne, W.W., The March from Modder River to Bloemfontein, *BMJ*, May 5th, 1900, (1093-1096):1093. (*BMJ* 1900;1:1093).
59. Pakenham, T., The Boer War, London (BCA), 1999:312.
60. Ibid.
61. Ibid:321.
62. Grey, E.F., *op. cit.*
63. Pakenham, T., *op. cit.* :314.
64. Ibid:311.
65. Cheyne, W.W., The March from Modder River to Bloemfontein, *BMJ*, May 5th, 1900, (1093-1096):1094. (*BMJ* 1900;1:1093).
66. Ibid:1093.
67. Ibid.
68. Ibid:1094.
69. Ibid:1093.
70. Ibid:1094.
71. Ibid.
72. Pakenham, T., *op. cit.* :319.
73. Ibid.
74. Cheyne, W.W., The March from Modder River to Bloemfontein, *BMJ*, May 5th, 1900, (1093-1096):1094. (*BMJ* 1900;1:1093).
75. Ibid.
76. Pakenham, T., *op. cit.* :320.
77. Ibid.

78. Ibid.
79. Ibid:312.
80. Cheyne, W.W., The March from Modder River to Bloemfontein, *BMJ*, May 5th, 1900, (1093-1096):1093. (*BMJ* 1900;1:1093).
81. *The Times*, July 23rd, 1900, untitled newspaper cutting in an album belonging to the Cheyne family.
82. Cheyne, W.W., The March from Modder River to Bloemfontein, *BMJ*, May 5th, 1900, (1093-1096):1094. (*BMJ* 1900;1:1093).
83. Cheyne, W.W., *Antiseptic Surgery, its principles, practice, history and results*, London (Smith, Elder & Co.), 1882:125.
84. Cheyne, W.W., The March from Modder River to Bloemfontein, *BMJ*, May 5th, 1900, (1093-1096):1094. (*BMJ* 1900;1:1093).
85. Ibid.
86. Ibid.
87. Ibid.
88. Ibid.
89. Ibid:1095.
90. Ibid:1093.
91. Cheyne, W.W., On the Treatment of Wounds in War, *BMJ*, November 30th 1901:1592. (*BMJ* 1901; 2:1591).
92. Cheyne, W.W., The March from Modder River to Bloemfontein, *BMJ*, May 5th, 1900, (1093-1096):1094. (*BMJ* 1900;1:1093).
93. Ibid.
94. Ibid:1093.
95. Ibid.
96. Ibid.
97. Ibid:1094.
98. Makins, G.H., *op. cit.* :35-36.
99. Ibid.
100. Ibid:36
101. Ibid.
102. Ibid:36-37
103. Cheyne, W.W., On the Treatment of Wounds in War, *BMJ*, November 30th 1901:1594. (*BMJ* 1901; 2:1591).
104. Conan Doyle, Sir Arthur, *Memories and Adventures*, Cambridge University Press, 2012:186.
105. Ibid. Dutch Attaché Killed, *Daily Telegraph*, Issue 9740, 14th April 1900:5. Accessed 01/05/2015 through Papers Past, http://paperspast.natlib.govt.nz/cgi-bin/paperspast?a=d&d=DTN19000414.2.16.15
106. Treves, F., *The Tale of a Field Hospital*, London (Cassell), 1900:2.
107. Ibid.
108. Ibid:20-21.
109. Ibid:21.
110. Ibid:22.
111. Ibid:23.
112. Makins, G.H., *op. cit.*
113. Ibid.
114. Treves, F., *op. cit.* :24.
115. Cheyne, W.W., The March from Modder River to Bloemfontein, *BMJ*, May 5th, 1900, (1093-1096):1093. (*BMJ* 1900;1:1093).

Notes and References

116. Ibid:1094.
117. Ibid:1095.
118. Pakenham, T., *op. cit.* :320:341.
119. Ibid:242.
120. Ibid.
121. Personal observation by author, who is familiar with Leagarth House in Fetlar. The photographs belong to the Cheyne family.
122. Cheyne, W.W., The March from Modder River to Bloemfontein, *BMJ*, May 5th, 1900, (1093-1096):1094. (*BMJ* 1900;1:1093).
123. Ibid:1094-1095.
124. Ibid:1095.
125. Ibid.
126. Ibid.
127. Ibid.
128. Pakenham, T., *op. cit.* :320:341.
129. Cheyne, W.W., The March from Modder River to Bloemfontein, *BMJ*, May 5th, 1900, (1093-1096):1095. (*BMJ* 1900; 1:1093).
130. Ibid.
131. Ibid.
132. Ibid.
133. Ibid.
134. Personal observation by author, who is familiar with Leagarth House in Fetlar. The photographs belong to the Cheyne family.
135. Cheyne, W.W., The March from Modder River to Bloemfontein, *BMJ*, May 5th, 1900, (1093-1096):1095. (*BMJ* 1900; 1:1093).
136. Pakenham, T., *op. cit.* :320:373.
137. Makins, G.H., *op. cit.*
138. Pakenham, T., *op. cit.* :320:375.
139. Cheyne, W.W., The March from Modder River to Bloemfontein, *BMJ*, May 5th, 1900, (1093-1096):1093. (*BMJ* 1900; 1:1093).
140. Pakenham, T., *op. cit.* :374-375.
141. Ibid:375.
142. Cheyne, W.W., The March from Modder River to Bloemfontein, *BMJ*, May 5th, 1900, (1093-1096):1096. (*BMJ* 1900; 1:1093).
143. Ibid.
144. Ibid.
145. Ibid.
146. Ibid.
147. Ibid.
148. Ibid.
149. Ibid.
150. Ibid.
151. Ibid.
152. Ibid.
153. Cheyne, W.W., The March from Modder River to Bloemfontein, *BMJ*, May 5th, 1900, (1093-1096):1094. (*BMJ* 1900; 1:1093).
154. Conan Doyle, A., *The Great Boer War,* London (Smith, Elder & Co.), 1900: 371. "... there can be no doubt that this severe outbreak had its origin in the Paardeberg water."
155. South African Hospitals Commission, *BMJ*, August 4th 1900:302. (*BMJ* 1900;2:301).
156. Makins, G.H., *op. cit.* :7.
157. Treves, F., *op. cit.* :21.

158. Makins, G.H., *op. cit.* :37.
159. Ibid.
160. Ibid:37.
161. Ibid:8.
162. Ibid.
163. Ibid.
164. Ibid.
165. Cheyne, W.W., The March from Modder River to Bloemfontein, *BMJ*,May 5th, 1900, (1093-1096):1093. (*BMJ* 1900; 1:1093).
166. Ibid.
167. Ibid.
168. Ibid.
169. Ibid:1094.
170. Ibid:1095.
171. Villiers, J.C., The Medical Aspect of the Anglo-Boer War, 1899-1902, Part 2, *Military History Journal*, Vol. 6, no. 3, June 1984. South African Military History Society http://samilitaryhistory.org/vol063jc.html Accessed 10/11/2012.
172. Invalid Soldiers in South Africa, *Daily Telegraph*, June 30th, 1900, newspaper cutting in an album belonging to the Cheyne family.
173. Medical Arrangements in South Africa, Mr Burdett-Coutts's Allegations, *BMJ*,June 30th 1900, 1610-1611:1611. (*BMJ* 1900; 1:1610).
174. Ibid.
175. Cheyne, W.W., The March from Modder River to Bloemfontein, *BMJ*,May 5th, 1900, (1093-1096):1096. (*BMJ* 1900;1:1093).
176. Ibid.
177. Gildea, J., *For King and Country, 1889-1902; Being a record of funds and philanthropic work in connection with the South African War in, London*, (Eyre and Spottiswoode), 1902; reviewed in *Poverty Bay Herald*, "For King and Country", 22nd August, 1902.
178. Ibid.
179. Conan Doyle, A., The Epidemic of Enteric Fever at Bloemfontein, *BMJ*,July 7th, 1900:49. (*BMJ* 1900; 2:49).
180. Makins, G.H., *op. cit.*
181. Cheyne, W.W., The March from Modder River to Bloemfontein, *BMJ*,May 5th, 1900: 1093-1096. (*BMJ* 1900;1:1093).
182. Cheyne, W.W., The Wounded from the Actions between Modder and Driefontein, *BMJ*,May 12th, 1900:1193-1195. (*BMJ* 1900;1:1193).
183. Photographs now at Fetlar Museum Trust, Fetlar, Shetland.
184. Cheyne, W.W., The Wounded from the Actions between Modder and Driefontein, *BMJ*,May 12th, 1900, (1193-1195):1194. (*BMJ* 1900;1:1193).
185. Ibid.
186. Ibid.
187. South African Hospitals Commission, *BMJ*,August 4th 1900:302. (*BMJ* 1900;2:301).
188. Ibid.
189. Ibid.
190. Cheyne, W.W., On the Treatment of Wounds in War, *BMJ*,Nov. 30th 1901:1593. (*BMJ* 1901; 2:1591).
191. Pakenham, T., *op. cit.* :422.
192. Ibid.
193. *The Star*, 16th June, 1900, untitled newspaper cutting in an album belonging to the Cheyne family.
194. Ibid.

Chapter 11
1. Burdett Coutts, W. *The Sick and Wounded in South Africa*, London (Cassell), 1900
2. Ibid.
3. Ibid:3.
4. Ibid:10.
5. Ibid:6.
6. Ibid:193.
7. Ibid:5.
8. Stevens, F.T., *Complete History of the South African War*, London (Nicholson and Sons), 1902:260.
9. Burdett Coutts, W. *op. cit.* :57.
10. Ibid:49-95.
11. The Royal Commission on South African Hospitals, *BMJ*, January 26th 1901:236. (*BMJ* 1901;1236).
12. Burdett Coutts, W. *op. cit.* :2:x.
13. Pakenham, T., T *op. cit.* :382.
14. Burdett Coutts, W. T *op. cit.* :19.
15. Ibid:27.
16. Ibid:29.
17. Stevens, F.T., *op. cit.* :258.
18. Cheyne, W.W. The Organisation of Medical Aid in a Great War, *BMJ*, June 22nd, 1901, 1558-1560:1559. (*BMJ* 1901;1:1558).
19. Stevens, F.T., *op. cit.* :258.
20. Ibid:255.
21. Invalid Soldiers in South Africa, *Daily Telegraph*, June 30th, 1900. Newspaper cutting in album belonging to the Cheyne family.
22. Treves, F., Letter from Mr Treves, *BMJ*, June 30th:1611. (*BMJ* 1900;1:1610).
23. Bowlby, A.A., The Portland Hospital in South Africa, *BMJ*, June 30th 1900:1610. (*BMJ* 1900;1:1610).
24. Cheyne, W.W., *The Times*, July 23rd, 1900. Noted in untitled newspaper cutting belonging to the Cheyne family: "From our Special Correspondent, Cape Town, Saturday, June 30 (4.45p.m.)."
25. The Transvaal Campaign, Boers Repulsed Near Heidelberg, Praise for the Sydney Field Hospital, The South Australian Register, Tuesday, July 24th, 1900:5, Trove Digitised Newspapers, http://trove.nla.gov.au/ndp/del/page/4101659?zoomLevel=1 accessed 12/01/2013.)
26. Army Medical Department: arrangements condemned, Bendigo Advertiser, Wednesday, 25th July 1900:2. Accessed 02/10/2013 through Trove Digitised Newspapers, http://trove.nla.gov.au/ndp/del/article/89614812?searchTerm=Willi am%20Watson%20Cheyne&searchLimits=
27. South African Hospitals Commission, *BMJ*, 4th August, 1900:302. (*BMJ* 1900;2:301).
28. Cheyne, W.W., The March from Modder River to Bloemfontein, *BMJ*, May 5th, 1900, 1093-1096:1096. (*BMJ* 1900;1:1093).
29. South African Hospitals Commission, *BMJ*, 4th August, 1900:302. (*BMJ* 1900;2:301).
30. Ibid.
31. Ibid.
32. Ibid.
33. Ibid.
34. Crowther, M.A. & Dupree, M.W., *op. cit.* :335.
35. Ibid

36. Ibid.
37. Press cutting from unknown newspaper, in album belonging to the Cheyne family.
38. Hospitals in South Africa, Daily Telegraph, July 16th, 1900, newspaper cutting in an album belonging to the Cheyne family.
39. The Royal Commission on South African Hospitals, *BMJ*, January 26th 1901, 236-240:236. (*BMJ* 1901;1:236).
40. Ibid:237.
41. Ibid:236.
42. Ibid:237.
43. South African Hospitals Commission, *BMJ*, January 26th 1901:238. (*BMJ* 1900;2:301).
44. Ibid.
45. Registry of Births, National Archives, Kew, accessed through www.ancestry.co.uk 25/08/2013.
46. 1901 England census, National Archives, Kew, accessed through www.ancestry.co.uk 25/08/2013.
47. Ibid.
48. South African Civil Surgeons' Dinner, *BMJ*, June 8th, 1901:1420. (*BMJ* 1901; 1:1420).
49. Ibid.
50. Ibid.
51. Worboys, M. *op. cit.* :7.
52. South African Civil Surgeons' Dinner, *BMJ*, June 8th, 1901:1420. (*BMJ* 1901; 1:1420).
53. Cheyne, The Organisation of Medical Aid in a Great War, *BMJ*, June 22nd, 1901, 1558-1560:1558. (*BMJ* 1901;1:1558).
54. Ibid.
55. Ibid.
56. Ibid:1559.
57. Ibid:1558.
58. Ibid:1559.
59. Ibid.
60. Ibid.
61. Ibid.
62. Ibid.
63. Cheyne, W.W., On the Treatment of Wounds in War. Being a portion of an address delivered before the Midland Medical Society, *BMJ*, Nov. 30th, 1901:1591-1594. (*BMJ* 1901;2:1591).
64. Ibid:1591.
65. Ibid:1591-1594.
66. Crowther, M.A. & Dupree, M.W., *op. cit.* :333.
67. Cheyne, W.W., *Antiseptic Surgery, its principles, practice, history and results*, London (Smith, Elder & Co.), 1882:125.
68. Ibid.
69. Cheyne, W.W., On the Treatment of Wounds in War. Being a portion of an address delivered before the Midland Medical Society, *BMJ*, Nov. 30th, 1901, 1591-1594:1591. (*BMJ* 1901;2:1591).
70. Cheyne, W.W., *Antiseptic Surgery, its principles, practice, history and results*, London (Smith, Elder & Co.), 1882:125-6.
71. Ogston, A., Remarks on the Influence of Lister upon Military Surgery, *BMJ*, December 13th 1902:1837. (*BMJ* 1902;2:1837).

72. Cheyne, W.W., On the Treatment of Wounds in War. Being a portion of an address delivered before the Midland Medical Society, *BMJ*,Nov. 30th, 1901, 1591-1594:1591. (*BMJ* 1901;2:1591).
73. Ibid:1593.
74. Ibid.
75. Makins, G.H., *op. cit.* :461-2.
76. Cheyne, W.W., The Wounded from the Actions between Modder and Driefontein, *BMJ*,May 12th 1900:1194-1195. (*BMJ* 1900;1:1193).
77. Cheyne, W.W., On the Treatment of Wounds in War. Being a portion of an address delivered before the Midland Medical Society, *BMJ*,Nov. 30th, 1901, 1591-1594:1592. (*BMJ* 1901;2:1591).
78. Ibid.
79. Ibid.
80. Ogston, A., Remarks on the Influence of Lister upon Military Surgery, *BMJ*,December 13th 1902:1837 (footnote). (*BMJ* 1902;2:1837).
81. Cheyne, W.W., On the Treatment of Wounds in War. Being a portion of an address delivered before the Midland Medical Society, *BMJ*,Nov. 30th, 1901, 1591-1594:1593. (*BMJ* 1901;2:1591).
82. Ibid.
83. Ibid:1594.
84. For example, "At Home" at the Hospital, *Folkestone Herald*, January, 1901, newspaper cutting in an album belonging to the Cheyne family.
85. Cheyne, W.W., On the Treatment of Wounds in War. Being a portion of an address delivered before the Midland Medical Society, *BMJ*,Nov. 30th, 1901, 1591-1594:1594. (*BMJ* 1901;2:1591).
86. Ibid.
87. Crowther, M.A. & Dupree, M.W., *op. cit.* :330-331.
88. Ibid:348.
89. Worboys, M., *op. cit.* :270.
90. Ibid:271.
91. Conan Doyle, A., The Epidemic of Enteric Fever at Bloemfontein, *BMJ*, July 7th, 1900:49.
92. Miscellaneous Information: Mentions in Despatches – Army, AngloBoerWar.com http://www.angloboerwar.com/other-information/16-other-information/1843-mentions-in-despatches-army Accessed 12/01/2013.
93. Letter from Lord Lister to Mrs Cheyne, 11th February 1901, Royal College of Surgeons (Edin.), Edinburgh, RCS/GD5/141 (65/8/d).
94. Notice of "Presentation of South African Medals", 12th June 1901, issued by Buckingham Palace, in possession of the Cheyne family. Also: British and Colonial – Order of the Bath, AngloBoerWar.com http://www.angloboerwar.com/medals-and-awards/british/1879-order-of-the-bath Accessed 12/01/2013.
95. Letter from Lord Lister to William Watson Cheyne, written from Lyme Regis, 1901, Royal College of Surgeons (Edin.), Edinburgh. RCS/GD5/142 (65/8/e).
96. "At Home" at the Hospital, *Folkestone Herald*, January, 1901, newspaper cutting in an album belonging to the Cheyne family.
97. Ibid.
98. Ibid.
99. *The Western Mercury* (Plymouth), September 18th, 1900, untitled newspaper cutting in an album belonging to the Cheyne family.
100. Obituary, Sir William Watson Cheyne, *The Lancet*, April 30th 1932:963.
101. Slides at Fetlar Museum Trust, Fetlar Interpretive Centre, Fetlar, Shetland. They have been digitised by the author.

102. "At Home" at the Hospital, *Folkestone Herald*, January, 1901, newspaper cutting in an album belonging to the Cheyne family.
103. The Red Cross Flag, *Derby Daily Telegraph*, 22nd January, 1901:4, British Library, accessed through the British Newspaper Archive, 20/01/2014.
104. Tommy's Tin Patch, *Cheltenham Chronicle and Gloucester Graphic*, Sat 5th January, 1901:8, British Library, accessed through the British Newspaper Archive, 20/01/2014.
105. Ibid.
106. Ibid.
107. Register of Deaths, National Archives, Kew, accessed through www.ancestry.co.uk 25/08/2013.
108. Death of Sir William MacCormac, *BMJ*, December 7th 1901:1708. (*BMJ* 1901;2:1704).
109. Obituary: Alfred William Hughes, F.R.C.S., F.R.S., M.B., C.M. Edin., *BMJ*, November 10th 1900:1409-10. (*BMJ* 1900;2:1409).

Chapter 12

1. Letter from Mary Cheyne to Lord Lister, dated September 16th 1887, Royal College of Surgeons of England, MS0021/4/4 (10).
2. Sir D'Arcy Power, Le Fanu, W.R., *op. cit.* :143.
3. The Viking Club, *Dundee Evening Telegraph*, Saturday 20th January, 1894:2, accessed through the British Newspaper Archive, 05/01/2013.
4. Ibid.
5. Notices, *Aberdeen Journal*, Friday 18th April 1913:5. Accessed through the British Newspaper Archive, 05/01/2013.
6. The Viking Club, *Dundee Evening Telegraph*, Saturday 20th January, 1894:2, accessed through the British Newspaper Archive, 05/01/2013.
7. Ibid.
8. Additional inventory of the personal estate of the late Reverend David Webster, minister of Fetlar and North Yell, died at Manse, Fetlar, on 13th May 1881, Shetland Archives, Lerwick. SC12/36/7.
9. Maugham, W.C. *op. cit.*, Chapter III: Parochial Records; Church Minutes ...,
10. Ibid.
11. Website of the National Trust for Scotland http://unst.org/web/unstpartnership/files/2012/09/HH_Panels.pdf Accessed 07/01/2014.
12. Saxby, J., Shetland's Greatest Men, Shetland Archives, Lerwick. D11/21/14.
13. Register of deaths, accessed through www.ancestry.co.uk 07/01/2014.
14. 1991 census, accessed through www.ancestry.co.uk 07/01/2014.
15. Mr K. Hughson, Yell, Shetland, personal communication to the author, December 1st, 2014.
16. Diary of the Reverend William Watson, Shetland Archives, Lerwick. D1/542/10. Entry for March 12th 1854 said, "Methodists preaching in their Chapel. After sermon in Church ... the young folk ran up to Meth Chap leaving no schol(ars)."
17. Photograph in possession of the Cheyne family, copy at Fetlar Interpretive Centre, F-2591a.
18. See photograph from Fetlar Museum Trust, which is probably Grace Watson sitting in a chair with rugs.
19. 1891 census. The 1891 Census for Scotland was taken on the night of 5/6 April 1891. In the 1871 and 1881 censuses, Mary Brown is given as a servant at the Manse.
20. Shooting Mishap in Shetland, Dundee Evening Telegraph, Tuesday 19th Sept 1899:2, accessed 07/01/2014 via the British Newspaper Archive, British Library.

Notes and References

21. Personal communication to author by John Coutts, who grew up in Fetlar, and was told the story by his mother. The Catherine Hughson (Coutts) in the tale was his grandmother on his father's side.
22. Letter from W. Scott Burn to Arthur Nicolson of Lochend, 4th August 1818, Shetland Archives, Lerwick. D24/49/42.
23. Correspondence between David Webster and Sir Arthur Nicolson, Shetland Archives, Lerwick. D24, Box 38, Item 7.
24. Letter by Laurence Williamson, under the pen-name 'Stakaberg' (the name of a hill in Fetlar), to the *Shetland Times*, dated 6th October 1886, Shetland Archives, Lerwick, D7/48/12. The letter notes that Lady Nicolson's factor, Colin Arthurson (Sir Arthur Nicolson had passed away by this time and the estate had passed to his wife for the duration of her own lifetime), had introduced a pony with the disease, and it had decimated the Fetlar stock. Ponies were important working animals in terms of agriculture and transporting peats. The main peat banks, where peats were cut for fuel, were located quite some distance from where most of the population lived, and people found themselves having to carry the peats themselves. Sale of ponies also made an important contribution to the economy. The incident caused particular hardship on the island. People had petitioned Lady Nicolson for the use of land on which to isolate affected ponies, but she never replied.
25. Shetland in Statistics, Shetland Islands Council, 2001:10.
26. Mrs M. Hughson, Yell, Shetland, personal communication to author. Her mother was one of Eardley's nurse-maids at Brough.
27. Coutts, J., Fetlar and Waltzing Matilda, *The New Shetlander*, Issue 251, Voar 2010.
28. Diaries of Lady Annie Nicolson, Brough Lodge, Fetlar, 25th January 1913. Shetland Archives, Lerwick. D24/Box82.
29. Entry for Thursday 26th June, 1902, Diaries of Lady Annie Nicolson, Shetland Archives, Lerwick. D24/Box82.
30. This information occurs regularly in the diaries of Lady Annie Nicolson, Shetland Archives, Lerwick. D24/Box82.
31. Scottish Statistical Account, 1834-45: United Parishes of Fetlar and North Yell, Volume 15:25.
32. Nicolson Estate papers, Shetland Archives, Lerwick. D24/Box86.
33. Letters to Dolores Elaine Cubbon (the bride), from friends, Shetland Archives, Lerwick. D6/294/2/p.55.
34. Lord Herschell, personal communication to author, 2006.
35. Shetland Archives, Lerwick. D24, Box 38. Also, letter from Hay's to Mr George Garster, Bealagord, Fetlar, Lerwick 29th October 1901, Hay and Co wet book re WW Cheyne and Leagarth Estate, Shetland Archives, Lerwick. D31/18/12.
36. Sir Joseph Cheyne, personal communication to author, 2000.
37. Hay and Co wet book re WW Cheyne and Leagarth Estate, Shetland Archives, Lerwick. D31/18/12.
38. Letter from Hay's to Mr George Garster, Bealagord, Fetlar, Lerwick 29th October 1901, Hay and Co wet book re WW Cheyne and Leagarth Estate, Shetland Archives, Lerwick. D31/18/12.
39. Ibid.
40. Ibid.
41. Ibid:8.
42. Ibid.
43. Ibid
44. Ibid
45. www.scottisharchitects.org.uk accessed 19/12/2013.
46. Mr K. Hughson, Yell, Shetland, personal communication to author, 01/12/2013.

47. Letter from Hay's to Mr Watson Cheyne, 24th June 1901, Hay and Co wet book re WW Cheyne and Leagarth Estate, Shetland Archives, Lerwick. D31/18/12.
48. Ibid: letters dated 1st and 13th July 1901.
49. Photograph at Fetlar Interpretive Centre, F-1027, showing a car being transported on a boat.
50. Diaries of Lady Annie Nicolson, 1907, Friday, July 12th, Shetland Archives, Lerwick. D24/Box82.
51. Mrs M. Hughson, personal communication to author, 2000.
52. John Coutts, personal communication to author.
53. Letter dated 23rd May 1928 from Wellcome Museum to Cheyne, London (Wellcome Library), MS7796.
54. Mrs M. Hughson, personal communication to the author: her mother, Agnes Smith, came from Yell, and was employed to take care of Lady Nicolson's disabled son Eardley. She ultimately married a Fetlar man. Annie Clark also originally came from Yell, and worked as a maid at Brough prior to her marriage to George Bain in Fetlar.
55. Letter from Laurence Williamson to his cousin Andrew, 18th April 1921, Shetland Archives, Lerwick. D1/302.
56. John Coutts, personal communication to the author.
57. John Coutts, personal communication to the author. His mother, who was ultimately one of the people working with the peat each summer when Cheyne's guests arrived, told him this story.
58. I am indebted to Davie Clark, Yell, Shetland, for this information on Cheyne's boats, and to his sister, Andrina Tulloch, for relaying it to me.
59. Collection of Fetlar Museum Trust, Fetlar Interpretive Centre, Fetlar, Shetland. O-86.
60. Water Dianas of the Shetland (sic): The Story of a Great Hunting Expedition, Dundee Evening Telegraph, Saturday, 20th August, 1904:3, accessed via the British Newspaper Archive, 07/01/2014.
61. Notes kept by Leagarth guests in 1907, belonging to the Cheyne family.
62. Ibid.
63. Ibid.
64. Ibid.
65. Ibid.
66. Ibid.
67. Ibid.
68. Ibid.
69. Stiles, Sir H., in Obituary, Sir William Watson Cheyne, F.R.C.S., F.R.S., *The Lancet*, April 30th, 1932:965.
70. Notes kept by Leagarth guests in 1907, belonging to the Cheyne family.
71. Ibid.
72. Ibid.
73. Ibid.
74. Ibid.
75. Ibid.
76. Ibid.
77. Ibid.
78. Ibid.
79. Diaries of Lady Annie Nicolson, Brough Lodge, Fetlar, 2nd August 1907, Shetland Archives, Lerwick. D24/Box82.
80. Notes kept by Leagarth guests in 1907, belonging to the Cheyne family.
81. Ibid.

Notes and References

82. Entry for 31st July 1907, Diaries of Lady Annie Nicolson, Brough Lodge, Fetlar, Shetland Archives, Lerwick. D24/Box82.
83. Notes kept by Leagarth guests in 1907, belonging to the Cheyne family.
84. Ibid.
85. Ibid.
86. Entry for September 1st 1907, Diaries of Lady Annie Nicolson, Brough Lodge, Fetlar, Shetland Archives, Lerwick. D24/Box82.
87. Notes kept by Leagarth guests in 1907, belonging to the Cheyne family.
88. Ibid.
89. Ibid.
90. Entry for 15th August 1907, Diaries of Lady Annie Nicolson, Brough Lodge, Fetlar, Shetland Archives, Lerwick. D24/Box82.
91. Notes kept by Leagarth guests in 1907, belonging to the Cheyne family.
92. Entry for 17th August 1907, Diaries of Lady Annie Nicolson, Brough Lodge, Fetlar, Shetland Archives, Lerwick. D24/Box82.
93. The Shetland Times, Saturday, August 24, 1907, Shetland Archives, Lerwick. SA1/11/22/34/p.8.
94. Ibid.
95. Ibid.
96. Ibid.
97. Entry for 20th September 1907, Diaries of Lady Annie Nicolson, Brough Lodge, Fetlar, Shetland Archives, Lerwick. D24/Box82.
98. Entry for 6th September 1907, Diaries of Lady Annie Nicolson, Brough Lodge, Fetlar, Shetland Archives, Lerwick. D24/Box82.
99. Scott, H., *op. cit.*,
100. Entry for September 21st 1907, Diaries of Lady Annie Nicolson, Brough Lodge, Fetlar, Shetland Archives, Lerwick. D24/Box82.
101. Manson, T.M.Y., Profiles from the Past, No. XXXVI, Sir Watson Cheyne, a world role in surgery, extract from New Shetlander, Shetland Archives, Lerwick. SA4/591/5/83.
102. John Coutts, personal communication to the author. The story is well known in Fetlar.
103. Letter from Laurence Williamson, Mid Yell, to Cheyne, dated 16th August 1901, Shetland Archives, Lerwick. D7/20/45.
104. Ibid.

Chapter 13

1. Letter from Lord Lister to Cheyne, from Lyme Regis, 21st April 1901. Royal College of Surgeons (Edin.), Edinburgh. RCS/GD5/142 (65/8/e).
2. Reports of Societies: Torquay Medical Society, *BMJ*, October 17th 1903:989. (*BMJ* 1903;2:987)
3. Case notes of cases under the care of Watson Cheyne, King's Collage Hospital. King's College Archive, London. KH/CN1/449 1895.
4. Case notes of cases under the care of Watson Cheyne, King's Collage Hospital. King's College Archive, London. KH/CN1/489 1912.
5. The peacock feather dress is on display at Kedlestone Hall, the former residence of Lord and Lady Curzon, and the anecdote about the dress and its significance was told me by Sir Joseph Cheyne, grandson of Sir William Watson Cheyne, around 2000, when I asked about a photograph, given to Cheyne, of Lady Curzon wearing the dress.
6. Letter from Lord Curzon to "Dr Cheyne", from Walmer Castle, dated November 7th, 1904, Royal College of Surgeons of England, London. MSS 552(2).

7. Seventy-Second Annual Meeting of the British Medical Association, Oxford, 1904: Section of Surgery, *BMJ*,July 23rd 1904: Supplement 27:90. (*BMJ* 1904;2S65)
8. The Present Position of Aseptic Surgery, *BMJ*,October 1st 1904:847 (*BMJ* 1904;2:847) and Seventy-Second Annual Meeting of the British Medical Association, Oxford, 1904: Section of Surgery, *BMJ*,July 23rd 1904: Supplement 27:90. (*BMJ* 1904;2S65).
9. Lady Curzon. A slight improvement. The Advertiser, Adelaide, September 26th 1904:5. http://trove.nla.gov.au/ndp/del/article/5009413 Viewed 28/12/2013.
10. Opening of the Winter Session in the Medical Schools: King's College Hospital, *BMJ*,October 8th 1904:944. (*BMJ* 1904;2:943).
11. Lady Curzon. Condition very critical. Sydney Morning Herald, Thursday, 29th September 1904:7. http://trove.nla.gov.au/ndp/del/article/14644113 accessed 28/12/2013.
12. Letter from Lord Curzon to "Dr Cheyne", from Walmer Castle, dated November 7th, 1904, Royal College of Surgeons of England, London. MSS 552(2).
13. Lady Curzon, Aberdeen Journal, Saturday 8th October, 1904:5, accessed through British Newspaper Archive, British Library, 05/01/2013.
14. Letter from Lord Curzon to "Dr Cheyne", from Walmer Castle, dated November 7th, 1904, Royal College of Surgeons of England, London. MSS 552(2).
15. Ibid.
16. Ibid.
17. Lady Curzon's Illness, Derby Daily Telegraph, Friday 14th October 1904:2, British Library, accessed through British Newspaper Archive, 05/01/2014.
18. Ibid.
19. Website of Kedlestone Hall, National Trust, accessed 08/01/2014 http://www.nationaltrustcollections.org.uk/object/107881
20. Premier's Daughter Undergoes Operation for Appendicitis, Dundee Courier, Monday, December 1st, 1913:4, accessed through British Newspaper Archive, British Library, 05/01/2014.
21. Mr A. Cheyne, London, personal communication to the author, December 2013, and Sir Joseph Cheyne, Yell, Shetland, personal communication to the author, 2000.
22. The Proposed Cancer Investigation, *BMJ*,22nd February 1902:480. (*BMJ* 1902;1:480).
23. Ibid.
24. Ibid.
25. Cancer Research, *BMJ*,August 2nd 1902:336. (*BMJ* 1902;2:336).
26. Imperial Cancer Research Fund, *BMJ,BMJ*, July 20th 1912:129. (*BMJ*; 1912;2:129).
27. Cheyne, W.W., Pathological Society of London, *BMJ*,February 9th 1901:343. (*BMJ* 1901;3:343).
28. Harveian Society of London, *BMJ*,May 3rd 1902:1088. (*BMJ* 1902;1:1084).
29. The Council of the Royal College of Surgeons of England, *BMJ*,June 15th 1901:1509. (*BMJ* 1901;1:1509).
30. Notice in *British Medical Journal* on Committee on the Egyptian Medical Congress, July 26th 1902:274. (*BMJ* 1902;2:267).
31. King's College Hospital Removal Fund, *BMJ*,March 19th 1904:701. (*BMJ* 1904;1:701).
32. King's College Hospital Removal Fund, *BMJ*,November 7th 1903:1244. (*BMJ* 1902;2:1243).
33. The New Sydenham Society, *BMJ*,Nov 10th 1906:1315. (*BMJ* 1906;2:1315).
34. Literary Notes, *BMJ*,July 5th 1913:33. (*BMJ* 1913;2:26).
35. Lord Lister's Birthday, *BMJ*,April 13th, 1907:877. (*BMJ* 1907;1:877).

36. Ibid.
37. Ibid.
38. Godlee, R.J., *op. cit.,*:583.
39. Letter from Lord Lister to Sir Hector Cameron, 19th February 1907. Royal College of Surgeons (Edin.), Edinburgh. RCS/GD5/067.
40. Ibid.
41. Ibid.
42. Godlee, R.J., *op. cit.,*:583.
43. Letter from Lord Lister to Sir Hector Cameron, 19th February 1907. Royal College of Surgeons (Edin.), Edinburgh. RCS/GD5/067.
44. Godlee, R.J., *op. cit.,*:583.
45. Ibid.
46. Lord Lister's Birthday, *British Medical Journal* April 13th 1907:877. (*BMJ* 1907;1:877).
47. Godlee, R.J., *op. cit.,*:583.
48. Lord Lister Honoured by City, *BMJ*,July 6th 1907:30. (*BMJ* 1907;2:30).
49. Harveian Society of London, Thursday, October 16th, 1890, Treatment of Tuberculous Joint Disease, *BMJ*,October 25th 1890, (958-9):959. (*BMJ* 1890;2:957).
50. Semon, F., The Celebration of Rudolph Virchow's 80th Birthday. A Personal Impression, *BMJ*,October 19th, 1901:1180. (*BMJ* 1901;2:1180).
51. Letter from Lord Lister to Cheyne, 27th December 1902, Royal College of Surgeons (Edin.), Edinburgh. RCS/GD5/144 (65/8/9).
52. Ibid.
53. Obituary. Robert Koch, M.D., *BMJ*,June 4th, 1910, 1384-:1388. (*BMJ* 1910;1:1384).
54. Ibid: 1389.
55. Societies, *BMJ*,January 19 1907:144. (*BMJ* 1907;1:141).
56. Ibid.
57. Correspondence between Mrs H.G. Campbell, Lerwick, Dr Guthrie and Mr J.N.J. Hartley, Conservator, RCS (Edin.), July 19th 1953 and June 23rd 1954, Royal College of Surgeons (Edin), RCS/GD7/123.
58. Edinburgh Degrees, *Nottingham Evening Post*, Friday 7th April 1905:4. Accessed through British Newspaper Archive, British Library, 05/01/2014.
59. Notice in *British Medical Journal* June 29th 1907: 1562. (*BMJ* 1907;1:1562).
60. Cheyne, Sir William Watson, (1852-1932), Plarr's Live of the Fellows Online, Royal College of Surgeons of England http://livesonline.rcseng.ac.uk/biogs/E000222b.htm Accessed 08/01/2014.
61. The Royal Colleges and the University of London, *BMJ*,March 6th 1909:607. (*BMJ* 1909;1:607).
62. The Birthday Honours, *BMJ*,July 4th 1908:46. (*BMJ* 1908;2:44).
63. Clippingdale, S.D., Medical Baronets, 1645-1911. *BMJ*,May 25th, 1912, (1188-1190):1190.(*BMJ* 1912;1:1188).
64. Lord Lister (1827-1912), Plarr's Lives of the Fellows Online, Royal College of Surgeons of England, http://livesonline.rcseng.ac.uk/biogs/E000500b.htm accessed 08/01/2014.
65. Letter from Lord Lister to Cheyne, Royal College of Surgeons (Edin.), Edinburgh. RCS/GD5/146 (65/8/j).
66. Day by Day, *Dundee Evening Telegraph*, Friday, 1st July 1927:3. Accessed through British Newspaper Archive, British Library, 05/01/2014.
67. Letter from W. Cheyne (Cheyne's grandson) on the occasion of his father's funeral in Fetlar. He said, "The committal was straightforward and the casket was put next door to Nono." Shetland Archives, Lerwick. D1542/13/17.

68. Royal College of Surgeons of England: Congratulations, *BMJ*, July 18th 1908:154. (*BMJ* 1908;2:154).
69. The Birthday Honours, *BMJ*, July 4th, 1908:46. (*BMJ* 1908;2:44).
70. Scotland. Edinburgh Royal Infirmary Residents' Club, *BMJ*, 4th July 1908:53. (*BMJ* 1908;2:53).
71. Complimentary Dinner to Sir W. Watson Cheyne, *The Lancet*, July 11th 1908:107.
72. The King's Household, *BMJ*, 18th June 1910:1509. (*BMJ* 1910;1:1509).
73. Edinburgh Royal Infirmary Residents' Club, *BMJ*, July 2nd 1910:43. (*BMJ* 1910;2:43).
74. Edinburgh Royal Infirmary Residents' Club, *BMJ*, June 24th 1911:1484. (*BMJ* 1911;1:1484).
75. Godlee, R.J., *op. cit.*,:579.
76. Obituary. Lord Lister, *BMJ*, February 17th 1912, 397-402:400. (*BMJ* 1912;1:397).
77. Godlee, R.J., *op. cit.*,:393.
78. Obituary. Lord Lister, *BMJ*, Feb 17th 1912, (397-402):400. (*BMJ* 1912;1:397).
79. Letter from A.W. Howlam, King's College, London, to Cheyne, February 12th, 1912 Royal College of Surgeons of England, London. MSS 552(3).
80. Obituary. Lord Lister, *BMJ*, Feb 17th 1912, (397-402):400. (*BMJ* 1912;1:397).
81. Ibid.
82. Ibid.
83. Cheyne's invitation to the funeral of Lord Lister, Royal College of Surgeons (Edin.), Edinburgh. RCS/GD5/202/1.
84. Lord Lister, *BMJ*, February 17th 1912, 397-402:401. (*BMJ* 1912;1:397).
85. A National Memorial to Lord Lister, *BMJ*, October 26th 1912:1149. (*BMJ* 1912;2:1149).
86. The Lister Memorial, *BMJ*, November 23rd 1912:1484. (*BMJ* 1912;2:1484).
87. The Medical Society of London, *BMJ*, October 12th 1912:994. (*BMJ* 1912;2:994).
88. Memorial to Lord Lister, *BMJ*, March 15th 1913:574. (*BMJ* 1913;1:571).
89. Ibid.
90. Notice in the *BMJ*, March 7th 1914:557. (*BMJ* 1914;1:557).
91. Cambridge Hospital for Special Diseases, *BMJ*, April 1st 1911:764. (*BMJ* 1911;1:764).
92. Ibid.
93. Ibid.
94. Cambridge Hospital for Special Diseases, *British Medical Journal* June 1st 1912:1246. (*BMJ* 1912;1:1246).
95. State Registration of Nurses, *BMJ*, April 2nd 1904:803. (*BMJ* 1904;1:803).
96. Ibid.
97. The History of Nursing and Midwifery Regulation, website of the Nursing and Midwifery Council, http://www.nmc-uk.org/about-us/the-history-of-nursing-and-midwifery-regulation/ accessed 31/12/2013.
98. Gröschel, H.M. and Hornick, Richard B., Who Introduced Typhoid Vaccine: Almroth Wright or Richard Pfeiffer? *Reviews of Infectious Diseases*, Volume 3, No. 6, University of Chicago, November-December 1981:1251.
99. Cheyne, W.W., Professor A.E. Wright's Method of Treating Tuberculosis, *The Lancet*, Jan 13th 1906:80.
100. Ibid.
101. Ibid:79.
102. Ibid.
103. Ibid.
104. Ibid.
105. Cheyne, W.W., An Address on Moveable Kidney with Details of an Operation for Fixing the Kidney, *The Lancet*, April 24th, 1909:1155-1158.

106. Notes for Cases under Watson Cheyne, King's College Hospital, King's College Archive, London. KH/CN1/452 1896.
107. Ibid: KH/CN1/470 1903.
108. Ibid.
109. Dr Arthur Dean Bevan, Chicago, The Surgical treatment of Stone, Tuberculosis and Tumors of the Kidneys. The Canadian Journal of Medicine and Surgery, 1909:5 http://www.forgottenbooks.org/readbook/The_Canadian_Journal_of_Medicine_and_Surgery_1909_v26_1000030151#7 accessed 31/12/2013).
110. Cheyne, W.W., Bradshaw Lecture on the Treatment of Wounds, delivered before the Royal College of Surgeons of England on December 4th 1908, London (J. Bales, Sons and Danielsson Ltd.), 1908:1.
111. Ibid.
112. Ibid:2.
113. Ibid.
114. "What's What", *BMJ*,February 15th, 1902:428. (*BMJ* 1902;1:427).
115. Ibid.
116. Quilter, H., in Allingham, Herbert W., "What's What", *BMJ*,February 22nd 1902:485. (*BMJ* 1902;1:485).
117. "What's What", *BMJ*,18th January 1902:184. (*BMJ* 1902;1:183).
118. 'The Editor of the 'Medical Times', "What's What", *BMJ*,February 22nd 1902:485. (*BMJ* 1902;1:485).
119. Ibid.
120. Mount Vernon Hospital, Hampstead, *BMJ*,July12th, 1913:77. (*BMJ* 1913;2:77).
121. Seventeenth International Congress of Medicine, Section of Surgery: The Social Side of the Section, *BMJ*,May 3rd 1913:949. (*BMJ* 1913;1:949).
122. Ibid.
123. Garden Parties, *BMJ*,August 16th, 1913:426. (*BMJ* 1913;2:417).
124. On 28th July 1911 Mr W.W. Cheyne left Liverpool on the Empress of Britain (owned by Canadian Pacific Railway) accompanied by Mr W.H.W. Cheyne. Incoming Passenger Lists for 1911, www.ancestry.co.uk, accessed 08/01/2014.
125. 1911 Census, England and Wales, National Archives, Kew, accessed through www.ancestry.co.uk 10/01/2014.
126. Bridge, Sir Frederick (ed.), The Form and Order of the Service that is to be Performed and of the Ceremonies that are to be Observed in the Coronation of their Majesties King George V and Queen Mary in the Abbey Church of S. Peter, Westminster, on Thursday the 22nd Day of June 1911, London (Novello and Company, Limited), 1911. Copy belonging to the Cheyne family.

Chapter 14
1. Cheyne, Sir William Watson (1852-1932), Plarr's Lives of the Fellows Online, Royal College of Surgeons of England, http://livesonline.rcseng.ac.uk/biogs/E000222b.htm Accessed 09/01/2014.
2. Naval and Military Appointment, Royal Naval Medical Service, *BMJ*,July 10th 1915:S31. (*BMJ* 1915;2:S25).
3. Biographical Memoirs of Fellows of the Royal Society, Sir William Watson Cheyne, http://rsbm.royalsocietypublishing.org/content/obits/1/1/26.full.pdf+html Accessed 08/01/2014.
4. Royal Navy Medical Club, *BMJ*,May 23rd 1914:1160. (*BMJ* 1914;1:1160).
5. Cheyne, Sir William Watson (1852-1932), Plarr's Lives of the Fellows Online, Royal College of Surgeons of England, http://livesonline.rcseng.ac.uk/biogs/E000222b.htm Accessed 09/01/2014.
6. German "Culture" and Science, *BMJ*,Oct 31st 1914:762. (*BMJ* 1914;2:762).

7. Cheyne, W.W., The Hunterian Oration on the Treatment of Wounds in War, London, delivered before the Royal College of Surgeons of England on February 15th, 1915: Bearing of Results on Treatment of Wounds, *The Lancet*, Vol.185, Issue 4775, 6th March 1915:422.
8. Ibid.
9. Ibid.
10. Ibid.
11. Paul Ehrlich - Biographical, Nobel Prize website, http://www.nobelprize.org/nobel_prizes/medicine/laureates/1908/ehrlich-bio.html accessed 11/02/2105.
12. Jacobs, Professor C., Our Belgian Colleagues at Home and Abroad, The Position of Belgian Doctors and Pharmacists, *BMJ*,Nov 21st 1914:890. (*BMJ* 1914;2:890).
13. Ibid.
14. *The London Gazette*, 27th October 1914.
15 His Majesty's Hospital Ship "Soudan" (Surgeons), Hansard, Official Reports of Debates in Parliament, http://hansard.millbanksystems.com/commons/1920/nov/09/his-majestys-hospital-ship-soudan Accessed 08/01/2014.
16. Notes from cases under the care of Watson Cheyne, King's College Hospital, e.g. KH/CN1/473 1904 and KH/CN1/470 1903, King's College Archive, London.
17. Cheyne, W.W., An Address on the Treatment of Wounds in War, Journal of Comparative Pathology and Therapeutics, Vol. 27, 1914, (304-323):304, reprinted from *The Lancet*, November 21st, 1914.
18. Vincent J Cirillo, Arthur Conan Doyle (1859–1930): Physician during the typhoid epidemic in the Anglo-Boer War (1899–1902), J Med Biogr 2014 22: 2 originally published online 12 July 2013.
19. Meynell, E.W., M.D., J R Army Med Corps 1996;412:43-47. http://www.ramcjournal.com/content/142/1/43.full.pdf Accessed 06/01/14.
20. Hardy, A., "Straight Back to Barbarism": Antityphoid Inoculation and the Great War, 1914:288. Bull. Hist. Med. 2000, 74:265-290.
21. Cheyne, W.W., An Address on the Treatment of Wounds in War, Journal of Comparative Pathology and Therapeutics, Vol. 27, 1914:304-323, reprinted from *The Lancet*, November 21st, 1914.
22. Ibid.
23. Makins, G.H., *op. cit.*,:461-2.
24. Cheyne, W.W., On the Treatment of Wounds in War. Being a portion of an address delivered before the Midland Medical Society, *BMJ*,Nov. 30th, 1901, (1591-1594):1592. (*BMJ* 1901;2:1591).
25. Cheyne, W.W., Remarks on the Treatment of Wounds in War, *BMJ*,November 21st, 1914,(865-871):866. (*BMJ* 1914;2:865).
26. Ibid
27. Ibid.
28. Ibid.
29. Ibid:867
30. Ibid:870.
31. Ibid.
32. Ibid.
33. Ibid:871.
34. Godlee, R.J., in The War, Surgical Experiences of the Present War, Discussion at the Medical Society of London, *BMJ*,21st November 1914:891. (*BMJ* 1914;2:891).
35. Ibid.
36. Ibid.
37. Ibid.
38. Wilson, Albert, The Treatment of Wounds in the Present War, *BMJ*,December 5th 1914:998. (*BMJ* 1914;2:998).

39. Ibid.
40. Ibid.
41. Cheyne, Sir W.W., Observations on the Treatment of Wounds in War, *The Lancet*, July 31st, 1915, Vol.186, Issue 4796, 213-219:213.
42. Cheyne, W.W., The Hunterian Oration on the Treatment of Wounds in War, delivered before the Royal College of Surgeons of England on February 15th, 1915, *The Lancet*, Vol.185, Issue 4775, 6th March 1915:423.
43. Ibid.
44. Arthur Edmunds, Plarr's Lives of the Fellows Online, Royal College of Surgeons of England, http://livesonline.rcseng.ac.uk/biogs/E004013b.htm Accessed 09/01/2014.
45. Cheyne, W.W., The Hunterian Oration on the Treatment of Wounds in War, delivered before the Royal College of Surgeons of England on February 15th, 1915, *The Lancet*, Vol.185, Issue 4775, 6th March 1915:425.
46. Ibid:430.
47. Ibid:424.
48. Ibid.
49. Ibid.
50. Cheyne, W.W., The Recommendations of the Naval Medical Committee, *BMJ*, May 22nd, 1915:912. (*BMJ* 1915;1:912).
51. Cheyne, W.W., The Hunterian Oration on the Treatment of Wounds in War, London, delivered before the Royal College of Surgeons of England on February 15th, 1915, *The Lancet*, Vol.185, Issue 4775, 6th March 1915:425:430.
52. Cheyne, Sir W.W., Observations on the Treatment of Wounds in War, *The Lancet*, July 31st, 1915:214.
53. Ibid.
54. Ibid:215.
55. Ibid.
56. Ibid.
57. Ibid:216.
58. Ibid.
59. Ibid.
60. Ibid:217.
61. Ibid:218.
62. Ibid:217-18.
63. Ibid:219.
64. Ibid.
65. Cheyne, W.W., Professor A.E. Wright's Method of Treating Tuberculosis, *The Lancet*, January 13th 1906:78.
66. Meynell, E.W., Some Account of the British Military Hospitals of World War I at Etaples, in the orbit of Sir Almroth Wright, RAMC Journal, J R Army Med Corps 1996;142:43-47, http://www.ramcjournal.com/content/142/1/43.full.pdf Accessed 06/01/14.
67. Report signed by W. Watson Cheyne and dated January 13th 1916, Wellcome Library, London, Rare Materials Room, MS1591.
68. Ibid.
69. Wright, Sir A., The Question as to How Septic War Wounds should be Treated, *The Lancet*, September 16th 1916:506.
70. Ibid.
71. Ibid:507.
72. Ibid:512.
73. Ibid:507.

74. Ibid.
75. Ibid.
76. Ibid.
77. Ibid:508.
78. Ibid:510.
79. Ibid.
80. Cheyne, W.W., The Treatment of Septic War Wounds: To the Editor of *The Lancet*, letter dated September 18th 1916, *The Lancet*, September 23rd 1916:580.
81. Meynell, E.W., Some Account of the British Military Hospitals of World War I at Etaples, in the orbit of Sir Almroth Wright, J R Army Med Corps 1996; 142: 43-47. http://www.ramcjournal.com/content/142/1/43.full.pdf Accessed 06/01/14.

Chapter 15

1. Parliamentary Representation of Edinburgh and St. Andrews, *BMJ*, August 4th, 1917:164. (*BMJ* 1917;2:164)
2. Dons in the House, University of Cambridge website, http://www.cam.ac.uk/for-staff/features/dons-in-the-house accessed 11.02.2015.
3. Parliamentary Representation of Edinburgh and St. Andrews, *BMJ*, August 4th, 1917:164. (*BMJ* 1917;2:164)
4. Sir Watson Cheyne, Bt., M.P., *BMJ*, August 18th 1917:235. (*BMJ* 1917;2:235).
5. Conservative Banquet to the Lord-Advocate. Speech by the Prime Minister, Aberdeen Journal, 4th July 1885, British Library, accessed through British Newspaper Archive, 22/01/2014.
6. Parliamentary Representation of Edinburgh and St. Andrews, *BMJ*, August 4th, 1917:164. (*BMJ* 1917;2:164).
7. Ibid.
8. Ibid.
9. National Insurance: Meetings of the Profession to be Held: London, *BMJ*, Dec 16th 1911:S619. (*BMJ* 1911;2:S609).
10. National Insurance: London, *BMJ*, December 23rd 1911:S661. (*BMJ* 1911;2:S657).
11. Ibid:S161-162.
12. Ibid:S162.
13. Ibid.
14. Ibid.
15. Ibid.
16. Ibid.
17. Ibid.
18. Ibid.
19. Ibid.
20. Cheyne, Sir William Watson (1852-1932), Plarr's Lives of the Fellows Online, Royal College of Surgeons of England, http://livesonline.rcseng.ac.uk/biogs/E000222b.htm Accessed 10/01/2014.
21. National Insurance: London, *BMJ*, December 23rd 1911:S663. (*BMJ* 1911;2:S657).
22. Ibid.
23. The Profession and the Politicians, *BMJ*, March 2nd 1912:519. (*BMJ* 1912;1:519).
24. National Insurance: London, *BMJ*, December 23rd 1911:S663. (*BMJ* 1911;2:S657).
25. Carre-Smith Dr Herbert L., Letter to *BMJ*, Dec 30th 1911:705. (*BMJ* 1911;2:S697).
26. The Profession and the Politicians, *BMJ*, March 2nd 1912:519. (*BMJ* 1912;1:519).

27. Cox, A., Letter to the Members of the House of Lords, *BMJ*,December 16th 1911:S609. (*BMJ* 1911;2:S609).
28. Domestics and Insurance Act, *Derby Daily Telegraph*, Friday, 28th June 1912:2, British Library, accessed through the British Newspaper Archive, 15/01/2104.
29. Ibid.
30. News of the Week, *The Spectator*, 29th June 1912:18. Last accessed through The Spectator Archive online, 01/03/2015. http://archive.spectator.co.uk/article/29th-june-1912/18/a-crowded-meeting-summoned-to-protest-against-the-
31. Domestics and Insurance Act, *Derby Daily Telegraph*, Friday, 28th June 1912:2, British Library, accessed through the British Newspaper Archive, 15/01/2104.
32. News of the Week, *The Spectator*, 29th June 1912:18. Last accessed through The Spectator Archive online, 01/03/2015. http://archive.spectator.co.uk/article/29th-june-1912/18/a-crowded-meeting-summoned-to-protest-against-the-
33. Domestics and Insurance Act, *Derby Daily Telegraph*, Friday, 28th June 1912:2, British Library, accessed through the British Newspaper Archive, 15/01/2104.
34. Ibid.
35. Ibid.
36. Letter from Cheyne to Mr Reid Tait, from Leagarth, Fetlar, Shetland, dated September 22nd, 1928. Shetland Archives, Lerwick. SA4/2/43.
37. Sir Watson Cheyne, Bt., M.P., *BMJ*,August 18th 1917:235. (*BMJ* 1917;2:235).
38. Ibid:236.
39. Cheyne, Sir William Watson (1852-1932), Plarr's Lives of the Fellows Online, Royal College of Surgeons of England, http://livesonline.rcseng.ac.uk/biogs/E000222b.htm accessed 10/01/2014.
40. Sir Watson Cheyne, Bt., M.P., *BMJ*,August 18th 1917:236. (*BMJ* 1917;2:235).
41. Letter from Cheyne to Mr Reid Tait, from Leagarth, Fetlar, Shetland, dated September 22nd, 1928. Shetland Archives, Lerwick. SA4/2/43.
42. Medical Notes in Parliament. Army Medical Vote: Manipulative Treatment, *BMJ*,August 18th, 1917:228-229. (*BMJ* 1917;2:228).
43. Ibid.
44. Ibid.
45. Ibid.
46. Ibid.
47. Ibid.
48. Ibid.
49. Letter from Cheyne to Mr Reid Tait, from Leagarth, Fetlar, Shetland, dated September 22nd, 1928. Shetland Archives, Lerwick. SA4/2/43.
50. Hospital Treatment of Discharged Disabled Men, *BMJ*,August 18th 1917:S47 and Godlee, R.J., Letter sent on behalf of members of the Medical Advisory Committee to the Ministry of Pensions, *BMJ*,August 18th 1917:S48. (*BMJ* 1917;2:S47).
51. Medical Officers for the Army: Proposed Committee of Inquiry, *BMJ*,August 18th 1917:S48. (*BMJ* 1917;2:S47).
52. Ibid.
53. The story was told me by my husband, John Coutts, who, as a boy, had been told by Mr Laurenson that he had found the papers. It correlates with the note in the Wellcome Library material, which says the "Fragments of a War Diary ... were obtained by a Mr Laurenson of Leagarth, Fetlar, Shetland, having been recovered by him from the beach, when he was attending the sale of Sir W. W. Cheyne's belongings on July 26th, 1933." They erroneously gave his address as Leagarth, when he lived at Aithbank, but Leagarth must be where he said he had found them. Wellcome Library, MS.1591.

54. Hunterian Oration on the treatment of wounds in war 1915: typescript with holograph corrections and additions. (Incomplete), and On the use of hypochlorite solutions in the treatment of war-wounds, with notes of cases in the Dardanelles Campaign, September-October 1915, Wellcome Library, London. MS.1591.
55. Fragment of a War-diary, 11-21 September 1917, Wellcome Library, London. MS.1591.
56. Ibid.
57. Ibid.
58. Ibid.
59. Medical Notes in Parliament: The New Session, *BMJ*,February 16th 1918:212. (*BMJ* 1918;1:212).
60. Ibid.
61. Ibid.
62. Forms and Reforms, *BMJ*,February 23rd 1918:235. (*BMJ* 1918;1:235).
63. Ibid.
64. Ibid:236.
65. Ibid.
66. The Air Force Medical Service, *BMJ*,Nov 9th 1918:524. (*BMJ* 1918;2:520).
67. Ibid.
68. Ibid.
69. Military Service Bill: The Extended Age for Medical Men, *BMJ*,April 20th 1918:462. (*BMJ* 1918;1:462).
70. Ibid.
71. Lundie, R.A., The Military Service Act, 1918, *BMJ*,May 4th 1918:520. (*BMJ* 1918;1:519).
72. Military Service Bill: The Extended Age for Medical Men, *BMJ*,April 20th 1918:463. (*BMJ* 1918;1:462).
73. Ibid.
74. Correspondence. Medical Men in Parliament: Medical Reconstruction, *The Lancet*, October 12th 1918:504.
75. Correspondence: War Emergency Fund of the Royal Medical Benevolent Fund, *The Lancet*, June 23rd 1917:964.
76. War Pensions Bill: Penalty for Refusing to Submit to Treatment, October 26th 1918:474. (*BMJ* 1918;2:474).
77. Eighty-Sixth Annual Meeting of the British Medical Association: Medico-Political Committee, The Education Bill, *BMJ*,August 17th 1918:S34. (*BMJ* 1918;2:S33).
78. Universities Constituency, Aberdeen Journal, Monday, 18th November 1918. British Library, accessed through the British Newspaper Archive, 15/01/2104.
79. Medicine in the General Election, *BMJ*,December 7th 1918:632. (*BMJ* 1918;2:632).
80. N. Bishop Harman, Letter to the *BMJ*,Oct 19th 1918:449. (*BMJ* 1918;2:249).
81. Representation of the Medical Profession. "Mass Meeting" in London, *BMJ*,February 8th 1919:160. (*BMJ* 1919;1:159).
82. Ibid:159.
83. A Commons Medical Committee, *BMJ*,February 22nd 1919:226. (*BMJ* 1919;1:226).
84. Parliamentary Representation of the Medical Profession, *BMJ*,May 10th 1919:579. (*BMJ* 1919;1:579).
85. Ibid.
86. Ibid.
87. National Archives website, http://webarchive.nationalarchives.gov.uk/+/www.dh.gov.uk/en/Aboutus/HowDHworks/DH_074813 accessed 02/01/2014

88. The Ministry of Health Bill, *BMJ*,March 1st 1919:258. (*BMJ* 1919;1:258).
89. Ibid.
90. Future of Hospitals, *Dundee Evening Telegraph*, Wednesday 15th August 1923:2, British Library, accessed through the British Newspaper Archive, 15/01/2104.
91. Ibid.
92. Ibid.
93. State Grants for Scientific Investigation, *BMJ*,March 6th 1920:346. (*BMJ* 1920; 1:346).
94. Cutting from an unidentified newspaper entitled Shetlander Lord Lieutenant of the County, under a column called "Day to Day", Cheyne family album, accessed by author, 1999.
95. Ibid.
96. Ibid.
97. Hansard, Dogs' Protection Bill, HC Deb 27 June 1919 vol 117 cc511-53 http://hansard.millbanksystems.com/commons/1919/jun/27/dogs-protection-bill#S5CV0117P0_19190627_HOC_21 Accessed 14/10/12.
98. Hansard, Dogs' Protection Bill, HC Deb 27 June 1919 vol 117 cc511-53 http://hansard.millbanksystems.com/commons/1919/jun/27/dogs-protection-bill#S5CV0117P0_19190627_HOC_21 Accessed 10/01/2014.
99. Ibid.
100. Leeson, J.R., *op. cit.,*:69.
101. Hansard, Dogs' Protection Bill, HC Deb 27 June 1919 vol 117 cc511-53 http://hansard.millbanksystems.com/commons/1919/jun/27/dogs-protection-bill#S5CV0117P0_19190627_HOC_21 Accessed 10/01/2014.
102. Anaesthetics for Operations on Animals, *BMJ*,May 10th 1919:590. (*BMJ* 1919;1:588).
103. Ibid.
104. The Dogs' Protection Bill. Deputation to the Home Office, *BMJ*,May 17th 1919:613-4. (*BMJ* 1919;1:613).
105. Ibid.
106. The Dogs' Protection Bill, *BMJ*,February 28th 1920:302. (*BMJ* 1920;1:301).
107. "The most cost-effective strategy for preventing rabies in people is by eliminating rabies in dogs through vaccination", Website of the World Health Organisation, http://www.who.int/mediacentre/factsheets/fs099/en/index.html Accessed 10/01/2014.
108. Dogs' Protection Bill. *BMJ*,May 31st 1919, 684-687:686. (*BMJ* 1919;1:683).
109. The Dogs' Protection Bill. The Measure Rejected, *BMJ*,July 5th 1919:23. (*BMJ* 1919;2:23).
110. Ibid:24.
111. Ibid:24.
112. Ministry of Health Bill, Clause 4: Consultative Councils, *BMJ*,March 29th 1919:388. (*BMJ* 1919;1:388).
113. Ibid.
114. Ibid:389.
115. Letter from Cheyne to Mr Reid Tait, from Leagarth, Fetlar, Shetland, dated September 22nd, 1928. Shetland Archives, Lerwick. SA4/2/43.

Chapter 16

1. Letter from Cheyne to Mr Reid Tait, from Leagarth, Fetlar, Shetland, dated September 22nd, 1928. Shetland Archives, Lerwick. SA4/2/43.
2. Estate inventory of Sir William Watson Cheyne, Shetland Archives, Lerwick, Shetland. SC12/36/23/page 1.

3. Sydenham and Forest Hill Local History website http://sydenhamforesthillhistory. blogspot.com.es/2011/05/history-of-beechgrove-sydenham-hil.html Accessed 10/01/2014.
4. University Students' Greek Drama, *Aberdeen Journal*, Sat 15th March 1919:4, British Library, accessed through the British Newspaper Archive, 11/01/2104.
5. Scheme to Encourage British Music, *Aberdeen Journal*, Saturday, 7th October, 1922:6 British Library, accessed through the British Newspaper Archive, 11/01/2104.
6. 1911 Census England and Wales, National Archives, Kew, accessed through www.ancestry.co.uk 10/01/2014.
7. 1881 Census, accessed through www.ancestry.co.uk 10/01/2014.
8. 1911 Census, England and Wales, National Archives, Kew, accessed through www.ancestry.co.uk 10/01/2014.
9. Small silver tray in collection of Fetlar Museum Trust, Fetlar Interpretive Centre, Fetlar, Shetland. O-1375.
10. Last Will and Testament of William Watson Cheyne, Shetland Archives, Lerwick. SC12/37/14/page1.
11. Day by Day: Shetlander Lord Lieutenant of the County, cutting from newspaper of unknown provenance, 7th June, 1919, Shetland Archives, Lerwick, Shetland. D21/4/p.91.
12. The New Lord Lieutenant of the County: Sir W. Watson Cheyne, newspaper cutting of unknown provenance, Shetland Archives, Lerwick.
13. Local Intelligence: Sir William Watson Cheyne, Shetland News, April 1st, 1920, Shetland Archives, Lerwick. SA1/14/36/14/p.4.
14. Newspaper cutting of unknown provenance, showing photographs and article, Shetland Archives, Lerwick. D21/4/p.110.
15. *Aberdeen Journal*, 23rd December 1926:4, British Library, accessed through the British Newspaper Archive, 11/01/2104.
16. Website of the Dulwich Society, http://www.dulwichsociety.com/newsletters/42-summer-2006/237-villagers-notebook. Last accessed 15/02/2015.
17. Ibid.
18. Exemption from Firearms certificate, signed by Gifford Gray, Superintendant, Zetland County Police, 29th December, 1920, Wellcome Library, London. MS7796.
19. Personal communication to the author by Sir Joseph Cheyne, 2000.
20. Saxby, J., Shetland's Greatest Men, Shetland Archives, Lerwick. D11/21/14.
21. Personal communication to author by Sir Joseph Cheyne, 2000.
22. Johnston, Colonel H. H., "Additions to the Flora of Shetland, papers 2, 3 & 4", Edinburgh, 1st May 1919. Copy viewed in collection of Fetlar Museum Trust, Fetlar, Shetland. D-209.
23. Mr A. Cheyne, London, personal communication to author, December 2012.
24. Saxby, J., Shetland's Greatest Men, Shetland Archives, Lerwick. D11/21/14.
25. Ibid.
26. Manson, T.M.Y., Profiles from the Past, No. XXXVI, Sir Watson Cheyne, a world role in surgery, New Shetlander, Shetland Archives, Lerwick. SA4/591/5/83.
27. Robertson, J., Sir William Watson Cheyne, Letters to the Editor, Shetland Life, May 1987.
28. Manson, T.M.Y., Profiles from the Past, No. XXXVI, Sir Watson Cheyne, a world role in surgery, New Shetlander, Shetland Archives, Lerwick. SA4/591/5/83.
29. The prescription is part of the collection of Fetlar Museum Trust, Fetlar, Shetland. D-246.
30. John Coutts, personal communication to author, 2014.
31. Ibid.

32. Letter to Sir William Watson Cheyne from Buckingham Palace, dated 16th August, 1920, in possession of the Cheyne family.
33. Personal comment made to author during the process of compiling the Fetlar Community Action Plan for the Shetland Biodiversity Action Plan, December 2002-February 2003. Document available at http://www.livingshetland.org.uk/ Last accessed 06/02/2014.
34. Catherine Coutts, New Zealand, through her daughters and son-in-law, personal communication to the author, 2013.
35. John Coutts, personal communication to the author, 2014.
36. Cheyne, W.W., An Address on the Treatment of Wounds in War, Journal of Comparative Pathology and Therapeutics, Vol. 27, 1914, pp:304-323, reprinted from *The Lancet*, November 21st, 1914.
37. Cheyne, W.W., Atkinson, F.P., Milk in Connexion with Septic Disease, *The Lancet*, Vol. 137, Issue 3516, 17 January 1891:166.
38. John Coutts, personal communication to author, 2014.
39. Gurney, D.F., God's Garden, verse 4, 1913. http://www.theotherpages.org/poems/gurney01.html Accessed 08/03/2015.
40. I collected this information in Fetlar, for Fetlar Interpretive Centre, at various points between the early 1990s and 2007.
41. Cheyne, W.W., Abstracts of Lectures on the Treatment of Surgical Tuberculous Diseases, Delivered at the Royal College of Surgeons of England, *BMJ*, June 18th, 1892, (1290-1293):1292. (*BMJ* 1892;1:1290).
42. Laurence Williamson noted in a letter to his cousin on 18th April 1921 that there were overall 236 people resident in the island, though 30 of these were away at the time, Shetland Archives, Lerwick. D1/302.
43. One of these is an illuminated address from his tenants, and is in the possession of the Cheyne family.
44. Letter from tenants of Leagarth Estate, signed by James Campbell, Parish Ministers and Ronald McAffer, United Free Church Minister, to J. L. Cheyne, dated January 1909. Shetland Archives, Lerwick. 131/542/15/15.
45. Fetlar Rejoicings, *Aberdeen Journal*, 19th February 1938:3, British Library, accessed through the British Newspaper Archive, 22/01/2014.
46. John Coutts, personal communication to the author, 2014.
47. A Shetland Concert, *Aberdeen Journal*, Wednesday January 19th, 1921:6, accessed through the British Newspaper Archive, 22/01/2014.
48. Ibid.
49. Ibid.
50. Births, Marriages and Deaths: Death: Watson Cheyne, *BMJ*, April 29th 1922;124. (*BMJ* 1922;1:S113).
51. Plaque in Fetlar Kirk, Fetlar, Shetland.
52. Cheyne, W.W., *Three Orations: The Lister Centenary*, London (John Bale, Sons and Danielsson), 1927:2.
53. Ibid.
54. Outgoing passenger lists, January 1924, accessed through www.ancestry.co.uk 11/01/2014.
55. Incoming passenger lists, 1924, accessed through www.ancestry.co.uk 11/01/2014.
56. Outgoing passenger lists, December 1924, accessed through www.ancestry.co.uk 11/01/2014.
57. Shipping News, *Western Daily Press*, 23rd December 1924:9, British Library, accessed through the British Newspaper Archive, 15/01/2014.
58. Sir Joseph Cheyne, personal communication to the author, 2000.
59. Ibid.

60. Outgoing passenger lists, December 1925, accessed through www.ancestry.co.uk 11/01/2014.
61. National Probate Calendar, 1931, Grace Ella Margaret Watson Browne, National Archives, Kew, accessed through www.ancestry.co.uk, 23/02/2014.
62. Inventory of the Personal Estate of the late Sir William Watson Cheyne, including certified copy of Trust Disposition and Settlement, Shetland Archives, Lerwick. SC12/36/23/page1.
63. Marriage index England and Wales, accessed through www.ancestry.co.uk 22/12/2013.
64. Outgoing passenger lists, November 1926, accessed through www.ancestry.co.uk 11/01/2014.
65. Ibid.
66. Lister Memorial Fund, *BMJ*,July 24th 1920:132. (*BMJ* 1920;2:130).
67. Letter from Cheyne to Reid Tait, containing a list of his works, Shetland Archives. Lerwick. D6/262/9/6.
68. Ibid.
69. Notes on Books, *BMJ*,Nov 21st 1925:957. (*BMJ* 1925;2:956).
70. Lister Memorial Lecture, *BMJ*,April 25th 1925:796. (*BMJ* 1925;1:791).
71. Notes on books, *BMJ*,November 21st 1925:956-7. (*BMJ* 1925;2:956).
72. Cheyne, W.W., *Lister and His Achievement*, London (Longman, Green and Co.), 1925.
73. In a letter to the Wellcome Museum, London, dated February 1927, Wellcome Library, London. WA/HMM/EX/A.11.
74. Letter from Mr Malcolm, Wellcome Museum, to Cheyne, 16th May 1928. Wellcome Library, London. WA/HMM/EX/A.11.
75. Telegram from Mr Malcolm to Cheyne, 24th January 1927, Wellcome Library, London. WA/HMM/EX/A.11.
76. Letter from P. Johnston-Saint at the Wellcome Museum, dated 7th December 1926, to Sir William Watson Cheyne, Bart., c/o R.H. Pringle, Tuloa Farm, Molo, Kenya Colony. Wellcome Library, London. WA/HMM/EX/A.11.
77. Mr A. Cheyne, London, personal communication to the author, December 2012.
78. Telegram from Mr Malcolm to Cheyne, 24th January 1927, Wellcome Library, London. WA/HMM/EX/A.11.
79. Letter from Cheyne to Wellcome Museum, dated February 1927, Wellcome Library, London. WA/HMM/EX/A.11.
80. Letter from Cheyne to Mr Malcolm, Wellcome Museum, March 12th 1927, Wellcome Library, London. WA/HMM/EX/A.11.
81. Note from Mr Malcolm, Wellcome Museum, to Cheyne at the Langham Hotel, dated 6th April 1927, Wellcome Library, London. WA/HMM/EX/A.11.
82. Ibid.
83. The Lister Centenary Celebrations. Reception at the Wellcome Historical Museum, The Canadian Medical Association Journal, 1927:608. (Can Med Assoc J. 1927 May; 17(5): 608).
84. Ibid.
85. Lister Centenary Celebration in Edinburgh, *BMJ*,February 12th 1927:299. (*BMJ* 1927;1:299).
86. *Aberdeen Journal*, Saturday 22nd July 1927, accessed through British Newspaper Archive, 11/01/2014.
87. Letter from Mr Malcolm, Wellcome Museum, to Cheyne, dated April 14th 1927, Wellcome Library, London. WA/HMM/EX/A.11.
88. Letter from Cheyne to Mr Malcolm, Wellcome Museum, dated June 14th 1927, Wellcome Library, London. WA/HMM/EX/A.11.

89. Ibid.
90. Ibid.
91. Ibid.
92. Letter from Mr Johnston-Saint, Wellcome Museum, to Cheyne, dated 29th June 1927. Wellcome Library, London. WA/HMM/EX/A.11.
93. Letter from Cheyne to Mr Johnston-Saint, Wellcome Museum, dated 6th July 1927. Wellcome Library, London. WA/HMM/EX/A.11.
94. Ibid.
95. Letter from Mr Malcolm, Wellcome Museum, to Cheyne, dated 1st November 1927. Wellcome Library, London. WA/HMM/EX/A.11.
96. Letter from Cheyne to Mr Malcolm, Wellcome Museum, dated 13th May 1928, Wellcome Library, London. WA/HMM/EX/A.11.
97. Letter from Mr Malcolm, Wellcome Museum, to Cheyne, dated May 16th 1928. Wellcome Library, London. WA/HMM/EX/A.11.
98. Letter from Cheyne to Mr Malcolm, Wellcome Museum, dated May 19th 1928. Wellcome Library, London. WA/HMM/EX/A.11.
99. Letter from Mr Malcolm, Wellcome Museum, to Cheyne, dated May 23rd 1928. Wellcome Library, London. WA/HMM/EX/A.11.
100. Letter from Cheyne to Mr Malcolm, Wellcome Museum, dated 26th May 1928, Wellcome Library, London. WA/HMM/EX/A.11.
101. Letter from Cheyne to Mr Malcolm, Wellcome Museum, dated 19th May 1928, Wellcome Library, London. WA/HMM/EX/A.11.
102. Letter from Mr Malcolm, Wellcome Museum, to Cheyne, dated 23rd May 1928, Wellcome Library, London. M.1343.28.
103. Letter from Cheyne to Mr Malcolm, Wellcome Museum, dated 26th May 1928, Wellcome Library, London. WA/HMM/EX/A.11.
104. Ibid.
105. Letter from Mr Malcolm, Wellcome Museum, to Cheyne, dated May 16th 1928. Wellcome Library, London. WA/HMM/EX/A.11.
106. Sir Joseph Cheyne, personal communication to author, 2000.
107. A Noted Surgeon, Sir Watson Cheyne, Visit to Wellington, Wellington Evening Post, Volume CV1, Issue 132, 12th December 1928:12. Viewed at Papers Past: http://paperspast.natlib.govt.nz/cgi-bin/paperspast?a=d&cl=search&d=EP19281212.2.87 11/02/2015.
108. Outgoing Passenger Lists, November 1928, accessed through www.ancestry.co.uk, 11/01/2014.
109. Manson, T.M.Y., Profiles from the Past, No. XXXVI, Sir Watson Cheyne, a world role in surgery, *New Shetlander*, Shetland Archives, Lerwick, Shetland. SA4/591/5/83.
110. A Noted Surgeon, Sir Watson Cheyne, Visit to Wellington, *Wellington Evening Post*, Volume CV1, Issue 132, 12th December 1928:12. Viewed at Papers Past: http://paperspast.natlib.govt.nz/cgi-bin/paperspast?a=d&cl=search&d=EP19281212.2.87 11/02/2015.
111. *Aberdeen Journal*, Tuesday 17th December, 1909, accessed through British Newspaper Archive, 11/01/2014.
112. A Noted Surgeon, Sir Watson Cheyne, Visit to Wellington, *Wellington Evening Post*, Volume CV1, Issue 132, 12th December 1928:12. Viewed at Papers Past: http://paperspast.natlib.govt.nz/cgi-bin/paperspast?a=d&cl=search&d=EP19281212.2.87 11/02/2015.
113. Sir Watson Cheyne: A Great Welcome, honoured by Shetlanders, *Wellington Evening Post*, Volume CVI, Issue 132, 17 December 1928, Page 17, accessed 11/01/2014 through Papers Past, http://paperspast.natlib.govt.nz/cgi-bin/paperspast?a=d&cl=search&d=EP19281217.2.134

114. Ibid.
115. Ibid.
116. Ibid.
117. Ibid.
118. The folder belongs to the Cheyne family, and was on display at Fetlar Interpretive Centre in December 2013, when I last visited.
119. Sir Watson Cheyne: A Great Welcome, honoured by Shetlanders, *Wellington Evening Post*, Volume CVI, Issue 132, 17 December 1928, Page 17, accessed 11/01/2014 through Papers Past, http://paperspast.natlib.govt.nz/cgi-bin/paperspast?a=d&cl=search&d=EP19281217.2.134
120. Ibid.
121. Ibid.
122. Eminent Surgeon Sir W. Cheyne's Visit, *Wellington Evening Post*, Volume CVII, Issue 1, 2 January 1929, Page 8, accessed 11/01/2014 through Papers Past website http://paperspast.natlib.govt.nz
123. Sir Watson Cheyne: A Great Welcome, honoured by Shetlanders, *Wellington Evening Post*, Volume CVI, Issue 132, 17 December 1928, Page 17, accessed 11/01/2014 through Papers Past, http://paperspast.natlib.govt.nz/cgi-bin/paperspast?a=d&cl=search&d=EP19281217.2.134
124. Glass slide from Leagarth House, digitised by Fetlar Museum Trust, Fetlar, Shetland, F-3453.
125. Manson, T.M.Y., Profiles from the Past, No. XXXVI, Sir Watson Cheyne, a world role in surgery, New Shetlander, Shetland Archives, Lerwick. SA4/591/5/83.
126. Letter from Cheyne to Mr Reid Tait, from Leagarth, April 1929. Shetland Archives, Lerwick. D6/262/9/3.
127. Letter from Mr Carson to Mr Reid Tait, from Leagarth, 10th November, 1929, Shetland Archives, Lerwick. D6/262/9.
128. Plaque inside the Kirk (Church of Scotland), Fetlar, Shetland.
129. National Probate Calendar, accessed through www.ancestry.co.uk, 24/02/2014.
130. Sir Joseph Cheyne, personal communication to the author, 2000.
131. Inventory of the Estate of Sir William Watson Cheyne, Shetland Archives, Lerwick. SC12/36/23/ page 1.
132. Cutting from unknown newspaper, Shetland Surgeon leaves £58,000, Shetland Archives, Lerwick. 21D/4a/15.
133. Inventory of the Estate of Sir William Watson Cheyne, Shetland Archives, Lerwick. SC12/36/23/ page 1.
134. Ibid.
135. Mr A. Cheyne, Sir Watson's grandson, personal communication to the author, December 2013.
136. Programme from Memorial Service for W. Watson Cheyne, St. Columba's Church, Lerwick, 1st May 1932, Shetland Archives, Lerwick. SA4/3000/18/60/33.
137. Obituary, Sir William Watson Cheyne, *The Lancet*, April 30th 1932:965.
138. Ibid.
139. Ibid.
140. Obituary, The Royal Society, December 1932, Vol. 1, No.1:30.
141. Ibid.
142. Ibid.
143. Ibid:28.
144. Obituary, Sir William Watson Cheyne, *The Lancet*, April 30th 1932:965.
145. Ibid.
146. Obituary, Sir William Watson Cheyne, Bt., K.C.M.G., C.B., F.R.S., F.R.C.S., *BMJ*, April 30th 1932:822. (*BMJ* 1932;1:821).

... *with his ubiquitous cigar*

Select bibliography

Printed sources

Abrams, L., *Myth and Materiality in a Woman's World: Shetland 1800-2000*, Manchester University Press, 2005.
Brunton, D. (ed.), *Medicine Transformed: Health, Disease and Society in Europe, 1800-1930*, Manchester University Press, 2004.
Brunton, D., *The Politics of Vaccination*, University of Rochester Press, 2008.
Burdett Coutts, W., *The Sick and Wounded in South Africa*, London (Cassell), 1900.
Burghard, F.F. and Cheyne, W.W., *A Manual of Surgical Treatment*, Vols 1-6, 1899-1903.
Cheyne, Captain A., *Sailing Directions from New South Wales to China and Japan, Including the Whole Islands and Dangers in the Western Pacific Ocean and Coasts of New Guinea and Safest Route through Torres Strait*, London (J.D. Potter), 1855.
Cheyne, W.W., *Antiseptic Surgery, its Principles, Practice, History and Results*, London (Smith, Elder and Co.), 1882.
Cheyne, W.W, *Die antiseptische Chirurgie; ihre Grundsätze, Ausübung, Geschichte und Resultate*, Leipzig, 1883.
Cheyne, W.W., Corfield, W.H., Cassell, C.E., *Public Health Laboratory Work*, London (William Clowes and Sons), 1884.
Cheyne, W.W., *Manual of the Antiseptic Treatment of Wounds for Students and Practitioners*. London (Smith and Elder), 1885.
Cheyne, W.W., "Antiseptic Surgery" in Heath, C. (ed.) *Dictionary of Practical Surgery*, Volume 1, London (Smith and Elder) 1886.
Cheyne, W.W. (ed.), *Essays by Various Authors on Bacteria in Relation to Disease*, London (New Sydenham Society), 1886.
Cheyne, W.W. *The Treatment of Wounds, Ulcers and Abscesses*, London and Edinburgh (Young J. Pentland), 1894.
Cheyne, W.W., *Tuberculous Disease of Bones and Joints: its pathology, symptoms and treatment*, Edinburgh and London (Young J. Pentland), 1895.
Cheyne, W.W., *Objects and Limits of Operations for Cancer*, being the Lettsomian Lectures for 1896, New York (William Wood), 1896.
Cheyne, W.W., *Bradshaw Lecture on the Treatment of Wounds*, delivered before the Royal College of Surgeons of England on December 4th 1908, London (J. Bale, Sons and Danielsson Ltd.), 1908.
Cheyne, W.W., *Lister and His Achievement*, London (Longman, Green and Co.), 1925.
Cheyne, W.W., *Three Orations. The Lister Centenary*, London (John Bale, Sons & John Danielsson, Ltd.), 1927.
Collier, A., *The Crofting Problem*, Cambridge University Press, 1953:30.
Conan Doyle, A., *The Great Boer War*, London (Smith, Elder and Co.), 1900.
Conan Doyle, A. *Memories and Adventures*, Cambridge University Press, 2012:186.
Cowie, R., Shetland: Descriptive and Historical, 1874: "Memoir" section, digital transcript viewed on http://www.electricscotland.com/History//index.htm 20/11/2014.
Crookshank, E.M., *Photography of Bacteria*, London (H.K. Lewis), 1887.
Crowther, M.A. & Dupree, M.W., *Medical Lives in the Age of Surgical Revolution*, Cambridge University Press, 2010.
Darwin, Charles, *The Voyage of the Beagle*, 1845 second edition, Hertfordshire (Wordsworth Editions Limited), 1997.

Select Bibliography

Edmondston, A., A *View of the Ancient and Present State of the Zetland Islands*, London (Longman, Hurst, Rees and Orm) and Edinburgh (John Ballantyne and Co.), Volume II, 1809:88-89.
Fenton, A., *The Northern Isles: Orkney and Shetland*, East Lothian (Tuckwell Press), 1997.
Flügge, C., Cheyne, W.W. (trans.), *Micro-organisms, with special reference to the etiology of the infective diseases*, London (New Sydenham Society), 1890.
Gildea, J., *For King and Country, 1889-1902; Being a record of funds and philanthropic work in connection with the South African War*, London, (Eyre and Spottiswoode), 1902; reviewed in *Poverty Bay Herald*, "For King and Country", 22nd August, 1902.
Godlee, R.J., *Lord Lister*, London (MacMillan & Co.), 1918.
Goldman, M., *Lister Ward*, Bristol and Boston (Adam Hilger), 1987.
Johnson, L.G., *Laurence Williamson of Mid Yell*, Shetland Times, 1971.
Grey, E.F., *Women in Journalism at the Fin de Siècle: Making a Name for Herself*, London (Palgrave MacMillan), 2012.
Gurney, D.F., God's Garden, verse 4, 1913, http://www.theotherpages.org/poems/gurney01.html Accessed 08/03/2015.
Hamilton, D., *The Healers: a history of medicine in Scotland*, Canongate Books, 1981:174.
Hezel, Francis X., *The First Taint of Civilisation, a history of the Caroline and Marshall Islands*, University of Hawaii Press, 1994:307.
Howden Smith, Arthur D. (ed.), *The Narrative of Samuel Hancock, 1845-1860*, New York, 1927.
Hutt, P., Heath, I., Neighbour, R., Confronting an Ill Society, Oxford (Radcliffe), 2005:102.
King's College Hospital Reports, 1895, being the Annual Report of King's College Hospital, Vol.1 (Oct. 1st, 1893 - Sept. 30th 1894), London (Adlard and Son), 1895:400.
Koch, R., (trans. Cheyne, W.W.) *Investigations into the Etiology of Traumatic Infective Diseases*. Translated by W. Watson Cheyne, F.R.C.S., Assistant Surgeon to King's College Hospital, London. The New Sydenham Society, 1880.
Leeson, J.R., *Lister As I Knew Him*, London, 1927.
MacCarthy, F., *Byron: Life and Legend*, Faber and Faber Ltd., 2005:11.
Makins, G.H., *Surgical Experiences in South Africa, 1899-1900. Being Mainly a Clinical Study of the Nature and Effects of Injuries Produced by Bullets of Small Calibre*, London (Henry Frowde and Hodder & Stoughton), 1913.
McCrae, Morrice, *Simpson: the turbulent life of a medical pioneer*, Edinburgh (Birlinn), 2011.
Nightingale, F., *Notes on Nursing*, London (Harrison), 1860: 120-121.
Pakenham, T., *The Boer War*, London (BCA) 1999.
Power, Sir D'Arcy and Le Fanu, W.R., *Lives of the Fellows of the Royal College of Surgeons of England 1930-1951*, London, 1953.
Robertson, T.A., Graham, John J.J., *Grammar and Usage of the Shetland dialect*, Lerwick (The Shetland Times), 1991.
Rose, W. and Carless, A., *A Manual of Surgery*, London (Bailliere, Tyndall and Cox), 1898.
Scott, H., *Fasti Ecclesiae Scoticanae, The Succession of Ministers in the Church of Scotland from the Reformation*, Edinburgh (Oliver and Boyd), 1928.
Second Statistical Account of Scotland, 1834-45: United Parishes of Fetlar and North Yell, Volume 15.
Shetland Islands Council, Shetland in Statistics, 2001.
Shineberg, D., *They Came for Sandalwood: a study of the sandalwood trade in the south-west Pacific, 1830-1865*. Melbourne University Press, 1967.
Shineberg, D. (ed.), *The Trading Voyages of Andrew Cheyne 1841-1844*, Canberra (Australian National University Press), 1971.

Stephenson, G.S. (under the pseudonym Alisma), *Reminiscences of a Student's Life at Edinburgh in the Seventies*, Edinburgh (Oliver and Boyd), 1918.
Stevens, F.T., Complete History of the South African War, London (Nicholson and Sons), 1902.
Stepansky, Paul E., *Freud, Surgery and the Surgeons*, New Jersey, (Analytic Press), 1999:38.
Tetens, A., *Among the Savages of the South Seas: Memoirs of Micronesia*, Oxford University Press and Stanford University Press, 1958.
The Third King's College Hospital and the Medical School 1913-1918, Surgeon-Rear-Admiral Sir William Watson Cheyne Bt.:339.
Tirard, N., Watson Cheyne, W., Phillips, J., Halliburton, W.D. (eds.), King's College Hospital Reports, Volume I, London (Adlard and Son), 1895:398.
Treves, F. (ed.), *A System of Surgery*, Vol. II, London (Cassell), 1896.
Treves, F., *The Tale of a Field Hospital*, London (Cassell), 1900.
Waddington, I., *The Medical Profession in the Industrial Revolution*, Dublin (Gill and Macmillan, Humanities Press), 1984:138-143. In Brunton, D. (ed.), *op.cit.*:90-93.
Woodward, J., *To Do the Sick no Harm, A Study of the British Hospital System to 1875*, London and Boston (Routledge Kegan Paul), 1974. In Brunton, D. (ed.), *op.cit.*:39-40.
Worboys, M., *Spreading Disease: Disease Theories and Medical Practice in Britain, 1865-1900*, Cambridge University Press, 2000.
Wylie, W.G., Hospitals: their history, organization and construction, Boylston Prize essay for 1876. Accessed 02/09/2013 via https://ia801408.us.archive.org/8/items/hospitalstheirhi00wyli/hospitalstheirhi00wyli.pdf

Parliamentary Papers
Loch, John D., Immigration Agent, Hobart Town, letter dated 31st January 1853, as part of Despatches from Lieutenant-Governor Sir W. Denison:Van Diemen's Land, sent 10th January 1853, Hansard, Parliamentary Papers, 1854:56 (436) (436-I) Emigration (Australia).
His Majesty's Hospital Ship "Soudan" (Surgeons), Hansard, Official Reports of Debates in Parliament, http://hansard.millbanksystems.com/commons/1920/nov/09/his-majestys-hospital-ship-soudan
Hansard, Dogs' Protection Bill, HC Deb 27 June 1919 vol 117 cc511-53
Cruelty to Animals Act 1876 (Amendment), House of Commons, Hansard, HC Deb 30 January 1968 vol. 757 cc1093-4

Journals
Atalić, B., Drenjančević-Perić, I., Fatovic-Ferenčić, S., Emanuel Edward Klein, a diligent and industrious plodder or the father of British microbiology, Medicinski Glasnik, Volumen 7, Number 2, August 2010:112.
Baxter, Thomas C., The Antiseptic Trio; Pasteur, Lister and Stewart – Their Legacy for our Times, Proceedings of the 17th Annual History of Medicine Days, University of Calgary, 2008, http://dspace.ucalgary.ca/bitstream/1880/47474/1/2008_HMD_Baxter.pdf Accessed 31/10/12.
Bevan, Arthur Dean Chicago, The Surgical Treatment of Stone, Tuberculosis and Tumors of the Kidneys. The Canadian Journal of Medicine and Surgery, 1909:5
Bishop Harman, N., Letter to the *BMJ*, Oct 19th 1918:449. (*BMJ* 1918;2:249).
Bowlby, A.A., The Portland Hospital in South Africa, *BMJ*, June 30th 1900:1610. (*BMJ* 1900;1:1610).
Bradbury, S., Galbraith, W. and Lyster, M.E., Cytological Laboratory, Department of Zoology, University Museum, Oxford. A Microscopical Finder Slide, first published online: 11th November, 2011. DOI: 10.1111/j.1365-2818.1872.tb02107.x http://jcs.

Select Bibliography

biologists.org/content/s3-97/38/197.full.pdf Accessed 09/08/2013.
Brody, H., Rip, M.R., Vinten-Johansen, P., Paneth, N., Rachman, S., Mapmaking and myth-making in Broad Street: the London cholera epidemic, 1854, *The Lancet* 356: 64 - 68, 2000.
Brookhouse, J.O., Letter to the Editor, *BMJ*, 22nd November 1890:1199. (*BMJ* 1890;2:1193).
Burghard, F.F., A Discussion on the Present Position of the Aseptic Treatment of Wounds, *BMJ*, Oct 1st 1904, 793-798.
Carpenter, M.W., 2012, The Patient's Pain in her Own Words: Margaret Mathewson's "Sketch of Eight Months a Patient, in the Royal Infirmary of Edinburgh, A.D. 1877", *Interdisciplinary Studies in the Long 19th Century*, no. 15 (2012):I(2).
Carpenter, M.W., Lister's relationship with patients: a successful case. Notes and Records of the Royal Society, August 2013:4 Notes Rec. R. Soc. Published online May 29, 2013.
Carre-Smith Dr Herbert L., Letter to *BMJ*, Dec 30th 1911:705. (*BMJ* 1911;2:S697).
Cartwright, F.F., Robert Bentley Todd's contributions to medicine, *Proceedings of the Royal Society of Medicine*, September 1974:895, Proc R Soc Med. 1974 September; 67(9): 893–897.
Cheyne, Captain Andrew, *Nautical Magazine and Naval Chronicle*, 24 (September 1855), Cambridge University Press, 2013.
Cheyne, W.W., Statistical Report of all Operations performed on Healthy Joints in Hospital Practice by Mr Lister from September 1871 to the Present Time, together with such accidental wounds of joints as occurred during the same period, *BMJ*, November 29th, 1879:859-864. (*BMJ* 1879;2:859).
Cheyne, W.W., Professor Spence on Surgical Statistics, *BMJ*, 14th February 1880, 239-241. (*BMJ* 1880; 1:239).
Cheyne, W.W., Report to the Association for the Advancement of Medicine by Research on the Relation of Micro-Organisms to Tuberculosis, Handed in to the Association for the Advancement of Medicine by Research on February 1, 1883 by W. Watson Cheyne, M.B., F.R.C.S., Assistant Surgeon to King's College Hospital. Published in *The Practitioner*, April 1883, No. 178, (Vol.XXX. No. 4).
Cheyne, W.W., Report on Micrococci in Relation to Wounds, Abscesses, and Septic Diseases, Reports to the Scientific Grants Committee of the British Medical Association, *BMJ*, September 27th 1884, (599-605) (*BMJ* 1884;2:599).
Cheyne W.W., The Bacillus of Tubercle, A Speech made during the discussion on Dr Kidd's paper at the Royal Medical and Chirurgical Society, London, on January 13th, 1885, *BMJ*, Jan 24th, 1885:169. (*BMJ* 1885;1:169).
Cheyne, W.W., The Cholera-Bacillus of Koch, *BMJ*, April 11th 1885:756-7. (*BMJ* 1885;1:756).
Cheyne, W.W., Report on the Cholera Bacillus, Reports to the Scientific Grants Committee of the British Medical Association, *BMJ*, April 25th, 1885:821-823 (*BMJ* 1885;1:821); May 2nd, 1885:877-879 (*BMJ* 1885;1:877); May 9th, 1885:931-934 (*BMJ* 1885;1:931); May 23rd 1885:1027-1031. (*BMJ* 1885;1:1027).
Cheyne, W.W., The Comma-Bacillus: Proposal for a Commission of Inquiry, *BMJ*, May 30th 1885:1127. (*BMJ* 1885;1:1126).
Cheyne, W.W., On Foulbrood, *BMJ*, Oct 10th 1885:697. (*BMJ* 1885;2:697).
Cheyne, W.W., Reports to the Scientific Grants Committee of the British Medical Association, Report on a Study of Certain of the Conditions of Infection *BMJ*, July 31st 1886:197-207. (*BMJ* 1886;2:197).
Cheyne, W.W., On Early Tracheotomy in Diphtheria, *BMJ*, March 5th 1887:504. (*BMJ* 1887;1:502).
Cheyne, W.W. Lectures on Suppuration and Septic Diseases, delivered at the Royal

College of Surgeons, February 1888, *BMJ*, February 25th, 1888:404-409 (*BMJ* 1888;1:404), March 3rd 1888:452-458 (*BMJ* 1888;1:452) and March 10th 1888:524-529 (*BMJ* 1888;1:524).

Cheyne, W.W., The Treatment of Incompletely Descended Testicle, *BMJ*, February 15th 1890:351-352. (*BMJ* 1890;1:351).

Cheyne, W.W., Extracts from Three Lectures on Tubercular Diseases of Bones and Joints, Delivered at the Royal College of Surgeons of England. *BMJ*, November 29th 1890:1227-1229 (*BMJ* 1890;2:1227); December 6th, 1890:1283-1286 (*BMJ* 1890;2:1283); December 13th 1890:1348-1353 (*BMJ* 1890;2:1348); December 20th 1890:1418-1422 (*BMJ* 1890;2:1418); April 4th 1891:739-743 (*BMJ* 1891;1:739); April 11th 1891:790-795 (*BMJ* 1891;1:790); April 18th 1891:840-844, (*BMJ* 1891;1:840); April 25th 1891:896-901 (*BMJ* 1891;1:896); June 18th 1892:1290-1293 (*BMJ* 1892;1:1290); June 25th 1892:1352-1356 (*BMJ* 1892;1:1352); July 2nd 1892:11-16 (*BMJ* 1892;2:11).

Cheyne, W.W., Atkinson, F.P., Milk in Connexion with Septic Disease, *The Lancet*, Vol. 137, Issue 3516, 17 January 1891:166.

Cheyne, W.W., Abstract of an Address on the Value of Tuberculin in the Treatment of Surgical Tuberculosis, *BMJ*, May 2nd 1891:951 (*BMJ* 1891;1:951).

Cheyne, W.W., Abstracts of Lectures on the Treatment of Surgical Tuberculous Diseases, Delivered at the Royal College of Surgeons of England, *BMJ*, June 18th, 1892, (1290-1293). (*BMJ* 1892;1:1290).

Cheyne, W.W., An Address on the Cancer of the Breast, Delivered before the Harveian Society of London, *BMJ*, February 10th, 1894, 289-291. (*BMJ* 1894;1:289).

Cheyne, W.W., The After-Effects of Chloroform, *The Lancet*, February 10th 1894, Vol. 143, Issue 3676:370.

Cheyne, W.W., Lettsomian Lectures on the Objects and Limits of Operations for Cancer, Delivered before the Medical Society of London, February, 1896, *BMJ*, February 15th 1896, (385-390). (*BMJ* 1896;1:385).

Cheyne, W.W., The Disinfection of Hands and Instruments, *BMJ*, Aug 7th 1897, 373-374 (*BMJ* 1897; 2:373).

Cheyne, W.W., An Address Delivered at the Opening of the Section of Pathology at the Annual Meeting of the British Medical Association in Montreal, September, 1897, *BMJ*, September 4th 1897 (586-589). (*BMJ* 1897;2:586).

Cheyne, W.W., Two Cases of Oöphorectomy for Inoperable Breast Cancer, *BMJ*, May 7th 1898, (1194-1195). (*BMJ* 1898;1:1194).

Cheyne, W.W., Sutherland, G.A., The Treatment of Hydrocephalus by Intracranial Drainage, *BMJ*, October 15th, 1898:1155-1157. (*BMJ* 1898;2:1127).

Cheyne, W.W., The Harveian Lectures on the Surgical Treatment of Tuberculous Diseases, *BMJ*, 16th December 1899:1659-1663 (*BMJ* 1899;2:1659); 23rd December 1899:1721-1726 (*BMJ* 1899;2:1721) 30th 476 December 1889:1779-1784 (*BMJ* 1899;2:1779).

Cheyne, W.W., The March from Modder River to Bloemfontein, *BMJ*, May 5th 1900, 1093-1096. (*BMJ* 1900;1:1093).

Cheyne, W.W., The Wounded from the Actions between Modder and Driefontein, *BMJ*, May 12th, 1900, (1193-1195). (*BMJ* 1900;1:1193).

Cheyne, W.W., Pathological Society of London, *BMJ*, February 9th 1901:343. (*BMJ* 1901;3:343).

Cheyne, W.W. The Organisation of Medical Aid in a Great War, *BMJ*, June 22nd, 1901, 1558-1560. (*BMJ* 1901;1:1558).

Cheyne, W.W., On the Treatment of Wounds in War. Being a portion of an address delivered before the Midland Medical Society, *BMJ*, Nov. 30th, 1901:1591-1594. (*BMJ* 1901;2:1591).

Select Bibliography

Cheyne, W.W., Listerism and the Development of Operative Surgery, *BMJ*, December 13th 1902:1851. (*BMJ* 1902;2:1851).
Cheyne, W.W., Professor A.E. Wright's Method of Treating Tuberculosis, *The Lancet*, Jan 13th 1906:78-80.
Cheyne, W.W., An Address on Moveable Kidney with Details of an Operation for Fixing the Kidney, *The Lancet*, April 24th, 1909:1155-1158.
Cheyne, W.W., Remarks on the Treatment of Wounds in connexion with the Recent Results obtained at St. George's Hospital, *The Lancet*, Volume 175, Issue 4505:15-18, January 1st 1910:15.
Cheyne, W.W., Remarks on the Treatment of Wounds in War, *BMJ*, November 21st, 1914,(865-871). (*BMJ* 1914;2:865).
Cheyne, W.W., An Address on the Treatment of Wounds in War, *Journal of Comparative Pathology and Therapeutics*, Vol. 27, 1914:304-323, reprinted from *The Lancet*, November 21st, 1914.
Cheyne, W.W., The Hunterian Oration on the Treatment of Wounds in War, London, delivered before the Royal College of Surgeons of England on February 15th, 1915: Bearing of Results on Treatment of Wounds, *The Lancet*, Vol.185, Issue 4775, 6th March 1915:422-430.
Cheyne, W.W., The Recommendations of the Naval Medical Committee, *BMJ*, May 22nd, 1915:912. (*BMJ* 1915;1:912).
Cheyne, W.W., Observations on the Treatment of Wounds in War, *The Lancet*, July 31st, 1915, Vol.186, Issue 4796, 213-219.
Cheyne, W.W., The Treatment of Septic War Wounds: To the Editor of *The Lancet*, letter dated September 18th 1916, *The Lancet*, September 23rd 1916:580.
Cirillo, Vincent J., Arthur Conan Doyle (1859–1930): Physician during the typhoid epidemic in the Anglo-Boer War (1899–1902), J Med Biogr 2014 22: 2 originally published online 12 July 2013.
Clippingdale, S.D., Medical Baronets, 1645-1911. *BMJ*, May 25th, 1912, (1188-1190):1190.(*BMJ* 1912;1:1188).
Conan Doyle, A., The Epidemic of Enteric Fever at Bloemfontein, *BMJ*, July 7th, 1900:49. (*BMJ* 1900; 2:49).
Coutts, J., Fetlar and Waltzing Matilda, *The New Shetlander*, Issue 251, Voar 2010.
Cox, A., Letter to the Members of the House of Lords, *BMJ*, December 16th 1911:S609. (*BMJ* 1911;2:S609).
Edmunds, A., The Late Sir Watson Cheyne. To the Editor of *The Lancet*, *The Lancet*, May 7, 1932:1013.
Edwardes, E.J., Assistant Physician to the N.W. London Hospital, April 30th, 1887, *BMJ*, *BMJ*, April 30th, 1887:961. (*BMJ* 1887;1:961).
"F.P.C.P.", The Hegira to Berlin, *BMJ*, 22nd November 1890:1199. (*BMJ* 1890;2:1193).
Godlee, R.J., in The War, Surgical Experiences of the Present War, Discussion at the Medical Society of London, *BMJ*, 21st November 1914:891. (*BMJ* 1914;2:891).
Godlee, R.J., Letter sent on behalf of members of the Medical Advisory Committee to the Ministry of Pensions, *BMJ*, August 18th 1917:S48. (*BMJ* 1917;2:S47).
Green, Henry T., An Address given at the Opening of the Section of Pathology at the Annual Meeting of the British Medical Association, in Liverpool, August, 1883, *BMJ*, August 4th 1883, (229-232):231. (*BMJ* 1883;2:229).
Grey Turner, G., Correspondence: Thyroid Surgery, *BMJ*, April 8th 1950:842. (*BMJ* 1950; 1:841).
Gröschel, H.M. and Hornick, Richard B., Who Introduced Typhoid Vaccine: Almroth Wright or Richard Pfeiffer? *Reviews of Infectious Diseases*, Volume 3, No. 6, University of Chicago, November-December 1981:1251.
Hardy, A., Tracheotomy versus intubation: Surgical intervention in diphtheria in

Europe and the United States, 1825-1930, Bull. Hist. Med., 1992, 66: 536-559. http://discovery.ucl.ac.uk/3441/1/3441.pdf - accessed 05/08/2013.

Hardy, A., "Straight Back to Barbarism": Antityphoid Inoculation and the Great War, 1914:288. Bull. Hist. Med. 2000, 74:265-290.

Hart, E., Abstract of a Lecture on the International Health Exhibition of 1884, its influence and possible sequels, delivered before the Society of Arts, Wednesday, November 26th, 1884, *BMJ*, Dec. 6th, 1884,(1115-1122):1120. (*BMJ* 1884;2:1115).

Heron, G.A., Cheyne, W.W., Demonstration to be Given in London, *BMJ*, 22nd Nov 1890:1198. (*BMJ* 1890;2:1193).

Hime, T.W., The Cholera-Bacillus from a Public Health Point of View, *BMJ*, May 23rd 1885:1080. (*BMJ* 1885;1:1079).

Hime, T.W., The Cholera-Bacillus: Proposal for a Commission of Inquiry, *BMJ*, June 20th 1885:1270. (*BMJ* 1885;1:1268).

Illingworth, C.R., Cholera and Comma-Bacilli, *BMJ*, June 13th, 1885:1231. (*BMJ* 1885;1:1231).

Johnson, G., On the Biology, Pathology and Treatment of Cholera, Royal Medical and Chirurgical Society, Reports of Societies, *BMJ*, March 28th, 1885:654-6. (*BMJ* 1885;1:654).

Johnson, Robert L., The deserted homesteads of Fetlar, *Shetland Life*, Vol. 13, November 1981:26.

Johnson, R.L., From a Minister's Diary, Lerwick, *Shetland Life*, April 1987:20-22.

Johnson, R.L., From a Minister's Diary, Lerwick, *Shetland Life*, May 1987:21-23.

Johnston, Colonel H. H., "Additions to the Flora of Shetland, papers 2, 3 & 4", Edinburgh, 1st May 1919.

Klein, E., Remarks on the Etiology of Asiatic Cholera, *BMJ*, March 28th 1885, (650-652). (*BMJ* 1885;1:650).

Klein, E., Some Remarks on the Present State of our Knowledge of the Comma-Bacilli of Koch, *BMJ*, April 4th 1885:693.(*BMJ* 1885;1:693)

Klein, E., The Comma-Bacillus, Proposal for a Commission of Inquiry, *BMJ*, 13th June 1885:1224. (*BMJ* 1885;1;1224).

Koch, R., Further Researches on Cholera, *BMJ*, January 2nd 1886, 6-8:7. (*BMJ* 1886;1:6)

Koch, R., A Further Communication on a Remedy for Tuberculosis, *BMJ*, 22nd November 1890:1193.(*BMJ* 1890;2:1193).

Lister, J., Admission of Ladies to the Meetings of the British Medical Association, *BMJ*, 9th February 1878:212-213. (*BMJ* 1878;1:211).

Lister, J., On the lactic fermentation and its bearing on pathology, *Transactions of the Pathological Society of London*, Vol. 29, London, October 1878 (425-467).

Lister, J., Mr Spence on Surgical Statistics, *BMJ*, February 21st 1880:276. (*BMJ* 1880; 1:276).

Lister, J., An Address on Catgut Ligature, *BMJ*, Feb 12th 1881, (219-221):220. (*BMJ* 1881;1:219).

Lockwood, C.B., The Disinfection of Hands and Instruments, *BMJ* July 31st 1897:314. (*BMJ* 1897; 2:314).

Lundie, R.A., The Military Service Act, 1918, *BMJ*, May 4th 1918:520. (*BMJ* 1918;1:519).

Mack, J., Thomas Irvine of Midbrake (II), *Shetland Life*, June 1988, Volume 92.

Maglen, Krista, 'The First Line of Defence': British Quarantine and the Port Sanitary Authorities in the Nineteenth Century," 0951-631X *Social History of Medicine* Vol. 15 No. 3:413-428.

Manson, T.M.Y., Profiles from the Past, No.XXXVI: Sir Watson Cheyne, A World Role in Surgery, The New Shetlander, Shetland Archives, SA4/591/5/83.

Morgan, K.O., The Boer War and the Media, *Twentieth Century British History*, 13, No.1 (2002):1-16 (4).
Mayo Robson, A.W., A Proposed Substitute for Carbolic Spray in Antiseptic Surgery, *BMJ*, September 2nd 1882, 420-422:420. (*BMJ* 1882;2:420).
Meynell, E.W., Some Account of the British Military Hospitals of World War I at Étaples, in the orbit of Sir Almroth Wright, J R Army Med Corps 1996; 142: 43-47. http://www.ramcjournal.com/content/142/1/43.full.pdf Accessed 06/01/14.
Ogston, A., Remarks on the Influence of Lister upon Military Surgery, *BMJ*, December 13th 1902:1837. (*BMJ* 1902;2:1837).
Pendlebury, H.S., Back, I., A Comparison between the Antiseptic and Aseptic Methods of Operation, with Special Reference to the Occurrence of Suppuration, based upon the Results obtained at St. George's Hospital during the years 1906 and 1908 respectively, *The Lancet*, Volume 174, Issue 4500, 27 November 1909:1578-1580.
Pennington, T.H., Listerism, its Decline and its Persistence: the Introduction of Aseptic Surgical Techniques in three British Teaching Hospitals, 1890-99, *Medical History*, 1995.
Quilter, H., in Allingham, Herbert W., "What's What", *BMJ*, February 22nd 1902:485. (*BMJ* 1902;1:485).
Richardson, R., Inflammation, Suppuration, Putrefaction, Fermentation: Joseph Lister's Microbiology, Notes and Records of the Royal Society, May 29th, 2013.
Richardson, R., *Imperial Contagions. Henry Vandyke Carter and Indian Leprosy: Anatomy, Geography, Morphology, Medicine, Hygiene and Cultures of Planning in Asia*, Hong Kong University Press, 2013.
Santer, M., Joseph Lister: first use of a bacterium as a 'model organism' to illustrate the cause of infectious disease of humans, Notes and Records of the Royal Society, London, Notes Rec R Soc Lond. 2010 Mar 20;64(1).
Semon, F., The Celebration of Rudolph Virchow's 80th Birthday. A Personal Impression, *BMJ*, October 19th, 1901:1180. (*BMJ* 1901;2:1180).
Smith, B., The Development of the spoken and written Shetland dialect: a historian's view, in Waugh, Doreen J. (ed.), Shetland's Northern Links: language and history, Edinburgh, Scottish Society for Northern Studies:30-43.
Spence, J., Surgical Statistics, *BMJ*, January 24th 1880:119-121. (*BMJ* 1880;1:119).
Spence, J., Surgical Statistics, *BMJ*, May 29th 1880:831. (*BMJ* 1880; 1:831).
Spencer Wells, T., The Hunterian Oration, delivered February 14th, at the Royal College of Surgeons of England, *BMJ*, February 24th, 1883, (341-343):341. (*BMJ* 1883;1:341).
Treves, F., Letter from Mr Treves, *BMJ*, June 30th:1611. (*BMJ* 1900;1:1610).
'The Editor of the 'Medical Times', "What's What", *BMJ*, February 22nd 1902:485. (*BMJ* 1902;1:485).
Villiers, J.C., The Medical Aspect of the Anglo-Boer War, 1899-1902, Part 2, *Military History Journal*, Vol. 6, no. 3, June 1984. South African Military History Society, http://samilitaryhistory.org/vol063jc.html Accessed 10/11/2012.
Watson, Caroline C., Griessenauer, Christoph J., Loukas, Marios, Blount, Jeffrey P., Shane Tubbs, R., William Watson Cheyne (1852–1932): a life in medicine and his innovative surgical treatment of congenital hydrocephalus, Springer, 2013, Childs Nerv Syst. DOI 10.1007/s00381-013-2220-7.181.
Williams, C. Theodore, Professor Koch's Treatment for Tuberculosis, *BMJ*, December 20th 1890:1434. (*BMJ* 1890;2:1434).
Wilson, Albert, The Treatment of Wounds in the Present War, *BMJ*, December 5th 1914:998. (*BMJ* 1914;2:998).
Winter, A., Ethereal Epidemic: Mesmerism and the Introduction of inhalation anaesthesia to early Victorian London, *Social History of Medicine*, 4, 1991. In

Brunton, D. (ed.) *op.cit.*:104.
Wright, A., The Question as to How Septic War Wounds should be Treated, *The Lancet*, September 16th 1916:506.
Wright, William, M.D., D.Sc., F.R.C.S., The Fellowship of the Royal College of Surgeons of England, The Canadian Medical Association Journal, May 1932:600. Accessed 04/11/12 via http://www.ncbi.nlm.nih.gov/pmc/articles/PMC402358/pdf/canmedaj00118-0083.pdf

Journals articles where no author given (in chronological order)
Pathological Society of London, Tuesday, May 6th 1879, Micrococci in Antiseptic Dressings, *BMJ*, May 24th 1879:781.(*BMJ* 1879;1:776).
Surgical Statistics, *BMJ*, April 10th, 1880:559. (*BMJ* 1880; 1:556).
Antiseptic Methods, *BMJ*, June 26th 1880:984. (*BMJ* 1880;1:977).
The Boylston Prize, *BMJ*, June 18th 1881:974. (*BMJ* 1881;1:972).
International Medical Congress: Discussion on the Relations of Minute Organisms to Unhealthy Processes occurring in Wounds, *BMJ*, October 1st 1881:546. (*BMJ* 1881;2:545).
The Bacilli of Tubercle, *BMJ*, May 13th 1882:709. (*BMJ* 1882;1:706).
Micro-organisms, *BMJ*, May 27th, 1882:787. (*BMJ* 1882;1:785).
Reports of Societies, Royal Medical and Chirurgical Society, Tuesday May 23rd 1882, Micro-organisms in Disease, *BMJ*, May 27th 1882: 778. (*BMJ* 1882;1:777).
Reviews and Notices: The International Encyclopaedia of Surgery ... Edited by John Ashhurst ... Vol. II. London: Macmillan, 1882, *BMJ*, Nov 25th 1882:1044. (*BMJ* 1882; 2:1044).
Mr Watson Cheyne, on the Bacillus of Tubercle, *BMJ*, March 17th 1883:527. (*BMJ* 1883;1:520).
Research Scholarships of the British Medical Association, Notice in the *BMJ*, October 27th, 1883:841. (*BMJ* 1883;2:828).
International Health Exhibition, 1884, South Kensington, *BMJ*, May 17th 1884:968. (*BMJ* 1884;1:959).
The Biological Library at the International Health Exhibition, *BMJ*, July 5th, 1884:27. (*BMJ* 1884;2:24)
Working-Health Laboratories at the International Health Exhibition, *BMJ*, July 12th 1884:82. (*BMJ* 1884;2:71).
Reports of Societies, Royal Medical and Chirurgical Society, Tuesday, December 9th 1884, *BMJ*, Dec. 13th, 1884: 1193. (*BMJ* 1884;2:1193).
Preliminary Report of the India Cholera Commission, *BMJ*, 3rd January 1885:43. (*BMJ* 1885;1:33).
The Organisms of Cholera, *BMJ*, Jan 17th 1885:141. (*BMJ* 1885;1:140-141).
The Discussion on Cholera at the Royal Medical and Chirurgical Society, *BMJ*, March 21st, 1885:613 (*BMJ* 1885;1:606).
Obituary: Frances Helen Prideaux, *BMJ*, December 5th 1885:1089. (*BMJ* 1885;2:1089).
Reviews and Notices: The Science and Practice of Surgery. By Frederick James Gant, F.R.C.S., Senior Surgeon to the Royal Free Hospital. Third Edition. London: Balliere, Tindall, and Cox, 1886, *BMJ*, Aug 28th 1886:414. (*BMJ* 1886; 2:414).
The Astley-Cooper Prize, *BMJ*, June 28th, 1890:1495. (*BMJ* 1890;1:1490).
The Annual Meeting [of the British Medical Association], Section of Pathology, *BMJ*, August 9th, 1890, (347-354):352. (*BMJ* 1890;2:347).
Harveian Society of London, Thursday, October 16th, 1890, Treatment of Tuberculous Joint Disease, *BMJ*, October 25th 1890, (958-9):959. (*BMJ* 1890;2:957).
Reports of Societies: Harveian Society of London, Thursday, October 16th, 1890, *BMJ*, October 25th 1890, 958-9:959. (*BMJ* 1890; 2:957).

Select Bibliography

"Special Correspondent" of *BMJ* in Berlin, General Notes from Berlin, *BMJ*, 22nd November 1890:1198. (*BMJ* 1890;2:1193).
From an Occasional Correspondent, *BMJ*, Nov. 29th 1890:1264. (*BMJ* 1890;2:1254).
Honours and Congratulations, *BMJ*, Nov. 29th 1890:1264. (*BMJ* 1890;2:1254).
Professor Koch's Remedy for Tuberculosis, Some General Pathological Conditions, *BMJ*, November 29th 1890:1262. (*BMJ* 1890;2:1254).
Demonstrations of Dr Koch's Method of Treating Tuberculosis, *BMJ*, November 29th, 1890:1265.
Professor Koch's Remedy for Tuberculosis, Demonstrations in London, *BMJ*, Dec 6th 1890:1321. (*BMJ* 1890;2:1321).
Professor Koch's Remedy for Tuberculosis, King's College Hospital, *BMJ*, December 6th 1890:1321-2. (*BMJ* 1890;2:1321).
"Special Correspondent" of *BMJ*, Notes from Berlin, *BMJ*, December 6th 1890:1327. (*BMJ* 1890;2:1321).
Professor Koch's Remedy for Tuberculosis: The Rash Caused by Injection of Koch's Fluid, *BMJ*, Dec 6th 1890:1330. (*BMJ* 1890;2:1321).
Demonstration at Paddington Green Children's Hospital, *BMJ*, December 13th 1890:1382. (*BMJ* 1890;2:1378).
Action taken by English Hospitals, *BMJ*, 22nd November 1890:1197. (*BMJ* 1890;2:1193).
What is the Practical Value of Tuberculin? *BMJ*, May 2nd 1891:973. (*BMJ* 1891;1:971).
Reports of Societies, Royal Medical and Chirurgical Society, Thursday, May 7th, 1891, *BMJ*, May 16th 1891:1069. (*BMJ* 1891;1:1068).
The Treatment of Surgical Tuberculoses, *BMJ*, August 8th 1891:327. (*BMJ* 1891;2:326).
The Ethics of Research (no author stated), *BMJ*, November 7th 1891:999. *BMJ* 1891;2:999. (*BMJ* 1891;2:999).
The Pathology of Tuberculosis, Notes, Letters, etc., *BMJ*, April 9th 1892:795. Response of Editor of *BMJ* to a letter by Dr George F. Crooke. (*BMJ* 1892;1:794).
British Medical Association, Sixtieth Annual Meeting, *BMJ*, May 14th 1892:1028. (*BMJ* 1892;1:1028).
Reviews, Tuberculosis of Bones and Joints, by N. Senn, M.C., Professor of Surgery, Rush Medical College, Chicago, Philadelphia and London: F.A. Davison. 1892. *BMJ*, Feb 18th 1893:358. (*BMJ* 1893;1:358).
King's College Hospital. An Exploratory Laparotomy, *BMJ*, July 1st 1893:14. (*BMJ* 1893; 2:14).
Literary Notes, *BMJ*, February 17th 1894:378 (*BMJ* 1894;1:377).
Down Brothers, The Annual Museum, *BMJ*, August 25th 1894:417. (*BMJ* 1894;2:414).
Opening of the Medical Schools: Metropolitan, *BMJ*, Sept. 1st, 1894:497. (*BMJ* 1894;2:497).
Opening of the Winter Session in London: King's College Hospital, *BMJ*, Oct 6th 1894:782. (*BMJ* 1894;2:781)
Association Intelligence, West Somerset Branch, *BMJ*, Oct. 13th 1894:842. (*BMJ* 1894;2:842).
Antiseptic and Aseptic Surgery, *BMJ*, January 26th, 1895:210.
Notes on books, *BMJ*, March 30th 1895:706.
British Medical Association, Sixty-Third Annual Meeting: Surgery, *BMJ* June 1st 1895:1235. (*BMJ* 1895;1:1232).
Messrs. Down Bros., Annual Museum, *BMJ*, Aug. 10th 1895:371. (*BMJ* 1895;2:369).
Reviews: Tuberculous Disease of Bones and Joints, its Pathology, Symptoms, and Treatment. By W. Watson Cheyne. *BMJ*, December 7th 1895:1429. (*BMJ* 1895;2:1428).
Lettsomian Lectures on the Objects and Limits of Operations for Cancer, Delivered

before the Medical Society of London, February, 1896, *BMJ*, March 7th, 1896, 577-582:582. (*BMJ* 1896;1:577).
The Council of the Royal College of Surgeons, *BMJ*, June 20th 1896:1520. (*BMJ* 1896;1:1517).
Reviews: The Objects and Limits of Operations for Cancer. By W. Watson Cheyne, M.B., F.R.S., F.R.C.S., Professor of Surgery at King's College, London, *BMJ*, December 12th, 1896:1722. (*BMJ* 1896;2:1722).
The Council of the Royal College of Surgeons, *BMJ*, June 26th, 1897:1673. Elections at the College of Surgeons, *BMJ*, July 10th 1897:98. (*BMJ* 1897;2:95).
Royal College of Surgeons of England Council Election, *BMJ* July 3rd 1897:41. (*BMJ* 1897;2:41).
Annual Meeting [of the British Medical Association], The Sections, Pathology and Bacteriology, *BMJ*, September 18th 1897:719 (*BMJ* 1897; 2:719).
Notes on Books, King's College Hospital Reports, Vol. III, *BMJ*, 1898:1272 (*BMJ* 1898;1:1272).
British Medical Association, Sixty-Sixth Annual Meeting, Section of Diseases of Children, report in *BMJ*, Aug 6th, 1898:366. (*BMJ* 1898;2:361)
Literary Notes, *BMJ*, October 29th 1898:1367. (*BMJ* 1898;2:1367).
Repair of the Bridge of the Nose by Rabbit Bone, Clinical Society of London, report in *BMJ*, November 5th 1898:1431. (*BMJ* 1898;2:1430).
The Jacksonian Prize – a Restrospect, *BMJ*, Vol. 1, No.1996, April 1st, 1899:818. (*BMJ* 1899;1:818).
The Prevention of Tuberculosis, Chelsea Clinical Society, *BMJ*, April 29, 1899:1038. (*BMJ* 1899;1:1038).
Association Intelligence, South Midland Branch [British Medical Association] The Natural History of Tubercle, *BMJ*, July 8th 1899:111. (*BMJ* 1899;2:107).
The War, *BMJ*, 28th October 1899:1210. (*BMJ* 1899;2:1209).
Civilian Medical Aid in the War, *BMJ*, October 28th, 1899:1217. (*BMJ* 1899; 2:1217).
Consulting Surgeons with the Army in South Africa, *BMJ*, December 30th 1899:1803. (*BMJ* 1899;2:1803).
Complimentary Dinner to Mr Watson Cheyne and Mr G. Lenthal Cheatle, *The Lancet*, Vol.115, Issue 3985, January 13th, 1900:114-115.
Consulting Surgeons with the Forces in South Africa, *BMJ*, January 13th, 1900:93. (*BMJ* 1900;1:91
A New Shrine of Hygiea in Sicily, *BMJ*, February 17th 1900:417. (*BMJ* 1900;1:417).
Medical Arrangements in South Africa, Mr Burdett-Coutts's Allegations, *BMJ*, June 30th 1900, 1610-1611:1611. (*BMJ* 1900; 1:1610).
Reviews: A Manual of Surgical Treatment. By W. Watson Cheyne, M.B., F.R.C.S., F.R.S., Professor of Surgery in King's College, London, Surgeon to King's College Hospital, etc., and F.F. Burghard, M.B. and M.S.Lond., F.R.C.S., Teacher of Practical Surgery in King's College, London, Surgeon to King's College Hospital, etc. In Six Parts. Part III. *BMJ*, July 21st 1900:172. (*BMJ* 1900, 2:173).
South African Hospitals Commission, *BMJ*, August 4th 1900:302. (*BMJ* 1900;2:301).
Obituary: Alfred William Hughes, F.R.C.S., F.R.S., M.B., C.M. Edin., *BMJ*, November 10th 1900:1409-10. (*BMJ* 1900;2:1409).
The Royal Commission on South African Hospitals, *BMJ*, January 26th 1901:236. (*BMJ* 1901;1236).
South African Civil Surgeons' Dinner, *BMJ*, June 8th, 1901:1420. (*BMJ* 1901; 1:1420).
Reviews: Some Recent Surgical Textbooks, *BMJ*, April 6th 1901:832.(*BMJ* 1901;1:832).
The Council of the Royal College of Surgeons of England, *BMJ*, June 15th 1901:1509. (*BMJ* 1901;1:1509).
Death of Sir William MacCormac, *BMJ*, December 7th 1901:1708. (*BMJ* 1901;2:1704).

Select Bibliography

Literary Notes, *BMJ*, Dec 14th 1901:1767. (*BMJ* 1901, 2:1767).
"What's What", *BMJ*, 18th January 1902:184. (*BMJ* 1902;1:183).
"What's What", *BMJ*, February 15th, 1902:428. (*BMJ* 1902;1:427).
Harveian Society of London, *BMJ*, May 3rd 1902:1088. (*BMJ* 1902;1:1084)
Notice in *BMJ* on Committee on the Egyptian Medical Congress, July 26th 1902:274. (*BMJ* 1902;2:267).
Cancer Research, *BMJ*, August 2nd 1902:336. (*BMJ* 1902;2:336).
Reviews: Practical Surgery, *BMJ* Sept 6th 1902:705. (*BMJ* 1902, 2:704).
Some Open Scholarships and Prizes, *BMJ*, September 19th, 1903:667. (*BMJ* 1903;2:663).
Reports of Societies: Torquay Medical Society, *BMJ*, October 17th 1903:989. (*BMJ* 1903;2:987)
King's College Hospital Removal Fund, *BMJ*, November 7th 1903:1244. (*BMJ* 1902;2:1243).
King's College Hospital Removal Fund, *BMJ*, March 19th 1904:701. (*BMJ* 1904;1:701).
State Registration of Nurses, *BMJ*, April 2nd 1904:803. (*BMJ* 1904;1:803).
Seventy-Second Annual Meeting of the British Medical Association, Oxford, 1904: Section of Surgery, *BMJ*, July 23rd 1904: Supplement 27:90. (*BMJ* 1904;2S65)
The Present Position of Aseptic Surgery, *BMJ*, October 1st 1904:847 (*BMJ* 1904;2:847).
Opening of the Winter Session in the Medical Schools: King's College Hospital, *BMJ*, October 8th 1904:944. (*BMJ* 1904;2:943).
The New Sydenham Society, *BMJ*, Nov 10th 1906:1315. (*BMJ* 1906;2:1315).
Lord Lister's Birthday, *BMJ*, April 13th, 1907:877. (*BMJ*1907;1:877).
Lord Lister Honoured by City, *BMJ*, July 6th 1907:30. (*BMJ* 1907;2:30).
The Birthday Honours, *BMJ*, July 4th, 1908:46. (*BMJ* 1908;2:44).
Scotland. Edinburgh Royal Infirmary Residents' Club, *BMJ*, 4th July 1908:53. (*BMJ* 1908;2:53).
Complimentary Dinner to Sir W. Watson Cheyne, *The Lancet*, July 11th 1908:107.
Royal College of Surgeons of England: Congratulations, *BMJ*, July 18th 1908:154. (*BMJ* 1908;2:154).
The Royal Colleges and the University of London, *BMJ*, March 6th 1909:607. (*BMJ* 1909;1:607).
Asepsis and Antisepsis, *The Lancet*, January 29th 1910:329.
Obituary. Robert Koch, M.D., *BMJ*, June 4th, 1910, 1384-1388. (*BMJ* 1910;1:1384).
The King's Household, *BMJ*, 18th June 1910:1509. (*BMJ* 1910;1:1509).
Edinburgh Royal Infirmary Residents' Club, *BMJ*, July 2nd 1910:43. (*BMJ* 1910;2:43).
Cambridge Hospital for Special Diseases, *BMJ*, April 1st 1911:764. (*BMJ* 1911;1:764).
Edinburgh Royal Infirmary Residents' Club, *BMJ*, June 24th 1911:1484. (*BMJ* 1911;1:1484).
National Insurance: Meetings of the Profession to be Held: London, *BMJ*, Dec 16th 1911:S619. (*BMJ* 1911;2:S609).
National Insurance: London, *BMJ*, December 23rd 1911:S661-163. (*BMJ* 1911;2:S657).
Obituary. Lord Lister, *BMJ*, Feb 17th 1912, (397-402):400. (*BMJ* 1912;1:397).
The Profession and the Politicians, *BMJ*, March 2nd 1912:519. (*BMJ* 1912;1:519).
Cambridge Hospital for Special Diseases, *BMJ* June 1st 1912:1246. (*BMJ* 1912;1:1246).
The Proposed Cancer Investigation, *BMJ*, 22nd February Imperial Cancer Research Fund, *BMJ*, *BMJ*, July 20th 1912:129. (*BMJ*; 1912;2:129).
The Medical Society of London, *BMJ*, October 12th 1912:994. (*BMJ* 1912;2:994).
A National Memorial to Lord Lister, *BMJ*, October 26th 1912:1149. (*BMJ* 1912;2:1149).
The Lister Memorial, *BMJ*, November 23rd 1912:1484. (*BMJ* 1912;2:1484).
Memorial to Lord Lister, *BMJ*, March 15th 1913:574. (*BMJ* 1913;1:571).

Seventeenth International Congress of Medicine, Section of Surgery: The Social Side of the Section, *BMJ*, May 3rd 1913:949. (*BMJ* 1913;1:949).
Literary Notes, *BMJ*, July 5th 1913:33. (*BMJ* 1913;2:26).
Mount Vernon Hospital, Hampstead, *BMJ*, July 12th, 1913:77. (*BMJ* 1913;2:77).
Garden Parties, *BMJ*, August 16th, 1913:426. (*BMJ* 1913;2:417).
Royal Navy Medical Club, *BMJ*, May 23rd 1914:1160. (*BMJ* 1914;1:1160).
German "Culture" and Science, *BMJ*, Oct 31st 1914:762. (*BMJ* 1914;2:762).
Jacobs, Professor C., Our Belgian Colleagues at Home and Abroad, The Position of Belgian Doctors and Pharmacists, *BMJ*, Nov 21st 1914:890. (*BMJ* 1914;2:890).
Naval and Military Appointment, Royal Naval Medical Service, *BMJ*, July 10th 1915:S31. (*BMJ* 1915;2:S25).
Correspondence: War Emergency Fund of the Royal Medical Benevolent Fund, *The Lancet*, June 23rd 1917:964.
Parliamentary Representation of Edinburgh and St. Andrews, *BMJ*, August 4th, 1917:164. (*BMJ* 1917;2:164).
Sir Watson Cheyne, Bt., M.P., *BMJ*, August 18th 1917:235-236. (*BMJ* 1917;2:235).
Medical Notes in Parliament. Army Medical Vote: Manipulative Treatment, *BMJ*, August 18th, 1917:228-229. (*BMJ* 1917;2:228).
Hospital Treatment of Discharged Disabled Men, *BMJ*, August 18th 1917:S47
Medical Officers for the Army: Proposed Committee of Inquiry, *BMJ*, August 18th 1917:S48. (*BMJ* 1917;2:S47).
Medical Notes in Parliament: The New Session, *BMJ*, February 16th 1918:212. (*BMJ* 1918;1:212).
Forms and Reforms, *BMJ*, February 23rd 1918:235. (*BMJ* 1918;1:235).
Military Service Bill: The Extended Age for Medical Men, *BMJ*, April 20th 1918:462. (*BMJ* 1918;1:462).
Eighty-Sixth Annual Meeting of the British Medical Association: Medico-Political Committee, The Education Bill, *BMJ*, August 17th 1918:S34. (*BMJ* 1918;2:S33).
Correspondence. Medical Men in Parliament: Medical Reconstruction, *The Lancet*, October 12th 1918:504.
War Pensions Bill: Penalty for Refusing to Submit to Treatment, October 26th 1918:474. (*BMJ* 1918;2:474).
The Air Force Medical Service, *BMJ*, Nov 9th 1918:524. (*BMJ* 1918;2:520).
Medicine in the General Election, *BMJ*, December 7th 1918:632. (*BMJ* 1918;2:632).
Representation of the Medical Profession. "Mass Meeting" in London, *BMJ*, February 8th 1919:160. (*BMJ* 1919;1:159).
A Commons Medical Committee, *BMJ*, February 22nd 1919:226. (*BMJ* 1919;1:226).
The Ministry of Health Bill, *BMJ*, March 1st 1919:258. (*BMJ* 1919;1:258).
Ministry of Health Bill, Clause 4: Consultative Councils, *BMJ*, March 29th 1919:388. (*BMJ* 1919;1:388).
Parliamentary Representation of the Medical Profession, *BMJ*, May 10th 1919:579. (*BMJ* 1919;1:579).
Anaesthetics for Operations on Animals, *BMJ*, May 10th 1919:590. (*BMJ* 1919;1:588).
The Dogs' Protection Bill. Deputation to the Home Office, *BMJ*, May 17th 1919:613-4. (*BMJ* 1919;1:613).
Dogs' Protection Bill. *BMJ*, May 31st 1919, 684-687:686. (*BMJ* 1919;1:683).
The Dogs' Protection Bill. The Measure Rejected, *BMJ*, July 5th 1919:23. (*BMJ* 1919;2:23).
The Dogs' Protection Bill, *BMJ*, February 28th 1920:302. (*BMJ* 1920;1:301).
State Grants for Scientific Investigation, *BMJ*, March 6th 1920:346. (*BMJ* 1920; 1:346).
Lister Memorial Fund, *BMJ*, July 24th 1920:132. (*BMJ* 1920;2:130).
Lister Memorial Lecture, *BMJ*, April 25th 1925:796. (*BMJ* 1925;1:791).

Notes on Books, *BMJ*, Nov 21st 1925:956-7. (*BMJ* 1925;2:956).
Lister Centenary Celebration in Edinburgh, *BMJ*, February 12th 1927:299. (*BMJ* 1927;1:299).
The Lister Centenary Celebrations. Reception at the Wellcome Historical Museum, The Canadian Medical Association Journal, 1927:608. (Can Med Assoc J. 1927 May; 17(5): 608).
Obituary: Waldemar Haffkine, C.I.E., *BMJ*, November 8th, 1930:801. (*BMJ* 1930;2:801).
Obituary, Sir William Watson Cheyne, Bt., K.C.M.G., C.B., F.R.S., F.R.C.S., *BMJ*, April 30th, 1932. 1932:821. (*BMJ* 1932;1:821)
Obituary, Sir William Watson Cheyne, F.R.C.S., F.R.S., *The Lancet,* April 30th, 1932:963.
Obituary Notices, the Royal Society, 10.1098/rsbm.1932.0007 Obit. Not. Fell. R. Soc. 1 December 1932 vol. 1 no. 1 26-30.
Nova et Vetera: Alexander Ogston - Bacteriologist, *BMJ*, August 7th 1954:356. (*BMJ* 1954;2:355).

Newspaper articles (chronological)
Mortality on Board the Ship Lady Montague, *The Courier*, Hobart, Wednesday, 3rd November 1852, http://trove.nla.gov.au/ndp/del/article/2958264 Accessed 17/03/2013.
Shipping Intelligence, Port of Hobart Town: Departures, *Colonial Times*, Hobart, Tasmania, Friday, 21st January 1853, http://trove.nla.gov.au/ndp/del/article/8772820?searchTerm=Captain%20Andrew%20Cheyne&searchLimits Accessed 23/03/2013.
Law Report: U.S. District Court, *Daily Alta*, California, 6th October, 1853, Volume 4, No.267. http://cdnc.ucr.edu/cdnc/cgi-bin/cdnc?a=d&cl=search&d=DAC18531006.2.16&srpos=1&e=06-10-1853-06-10-1853--en--20-DAC-1--txt-IN-Cheyne Accessed 17/03/2013.
Headlam, Charles, Letter to The Right Hon. Sir John S. Pakington, Bart., printed in *Launceston Examiner*, Tasmania, February 25th, 1854:2. http://trove.nla.gov.au/ndp/del/article/36288467 Accessed 16/03/2013.
Marriage notices, *Surrey Mirror*, Saturday, 12th July 1884:1, accessed through the British Newspaper Archive, 13/01/2014.
Conservative Banquet to the Lord-Advocate. Speech by the Prime Minister, *Aberdeen Journal*, 4th July 1885, British Library, accessed through British Newspaper Archive, 22/01/2014.
Carlisle Poisoning Case, *Tamworth Herald*, Saturday, 18th September, 1886:6, accessed through British Newspaper Archive, 19/01/2014.
Mail and Ship News, *Leeds Mercury*, 7th April 1888:7, British Library, accessed through the British Newspaper Archive, 20/01/2014.
Pall Mall Gazette, November 18th, 1890:5, British Library, accessed through British Newspaper Archive, 20/01/2014.
Dr Koch's Discovery, *The Manchester Courier and Lancashire General Advertiser*, November 21st, 1890:3, British Library, accessed through British Newspaper Archive, 20/01/2014.
Our London Letter, *Exeter and Plymouth Gazette*, 17th December, 1890:5, accessed through the British Newspaper Archive, 20/01/2014.
Bath Chronicle and Weekly Gazette, 25th December, 1890, British Library, accessed through the British Newspaper Archive, 20/01/14.
Vivisection, *Aberdeen Journal*, Friday, 1st May 1891:5, accessed through British Newspaper Archive, 13/01/2104.
The Viking Club, *Dundee Evening Telegraph*, Saturday 20th January, 1894:2, accessed

through the British Newspaper Archive, 05/01/2013.
Marriages, *The London Standard*, 17th December, 1894:1, accessed through the British Newspaper Archive, 20/01/2014.
The Morning Post, 28th May 1898:5, British Library, accessed through British Newspaper Archive, 20/01/2014.
Shooting Mishap in Shetland, *Dundee Evening Telegraph*, Tuesday 19th Sept 1899:2, accessed 07/01/2014 via the British Newspaper Archive, British Library.
Surgeons for the Front, *Lloyd's News*, Jan. 7th, 1900.
Enthusiastic Medical Students, *Westminster Gazette*, Jan. 7th, 1900.
Two More Surgeons Leave, *Daily Mail*, January 8th, 1900.
Departure of Mr Watson Cheyne as Consulting Surgeon to the Forces, *Shetland Times*, January 13th, 1900
Dutch Attaché Killed, *Daily Telegraph*, Issue 9740, 14th April 1900:5. Accessed 01/05/2015 through Papers Past, http://paperspast.natlib.govt.nz/cgi-bin/paperspast?a=d&d=DTN19000414.2.16.15
People of To-Day, *The Gloucester Citizen*, Friday 11th May 1900:3, British Library, accessed through the British Newspaper Archive, 20/01/2014.
Invalid Soldiers in South Africa, *Daily Telegraph*, June 30th, 1900.
Hospitals in South Africa, *Daily Telegraph*, July 16th, 1900.
Pay of Great War Surgeons, *Isle of Man Times*, 21st July 1900:9, British Library, accessed through British Newspaper Archive, 20/01/2014.
The Transvaal Campaign, Boers Repulsed Near Heidelberg, Praise for the Sydney Field Hospital, *The South Australian Register*, Tuesday, July 24th, 1900:5, Trove Digitised Newspapers, http://trove.nla.gov.au/ndp/del/page/4101659?zoomLevel=1 Accessed 12/01/2013.
Army Medical Department: arrangements condemned, *Bendigo Advertiser*, Wednesday, 25th July 1900:2. Accessed 02/10/2013 through Trove Digitised Newspapers, http://trove.nla.gov.au/ndp/del/article/89614812?searchTerm=William%20Watson%20Cheyne&searchLimits=
"At Home" at the Hospital, *Folkestone Herald*, January, 1901.
Tommy's Tin Patch, *Cheltenham Chronicle and Gloucester Graphic*, Sat 5th January, 1901:8, British Library, accessed through the British Newspaper Archive, 20/01/2014.
The Red Cross Flag, *Derby Daily Telegraph*, 22nd January, 1901:4, British Library, accessed through the British Newspaper Archive, 20/01/2014.
Water Dianas of the Shetland (sic): The Story of a Great Hunting Expedition, *Dundee Evening Telegraph*, Saturday, 20th August, 1904:3, accessed via the British Newspaper Archive, 07/01/2014.
Lady Curzon. A slight improvement. The Advertiser, Adelaide, September 26th 1904:5. http://trove.nla.gov.au/ndp/del/article/5009413 Viewed 28/12/2013.
Lady Curzon. Condition very critical. *Sydney Morning Herald*, Thursday, 29th September 1904:7. http://trove.nla.gov.au/ndp/del/article/14644113 Accessed 28/12/2013.
Lady Curzon, *Aberdeen Journal*, Saturday 8th October, 1904:5, accessed through British Newspaper Archive, British Library, 05/01/2013.
Lady Curzon's Illness, *Derby Daily Telegraph*, 14th October 1904:2, British Library, accessed through British Newspaper Archive, 05/01/2014.
Edinburgh Degrees, *Nottingham Evening Post*, Friday 7th April 1905:4. Accessed through British Newspaper Archive, British Library, 05/01/2014.
Aberdeen Journal, Tuesday 17th December, 1909, accessed through British Newspaper Archive, 11/01/2014.
Domestics and Insurance Act, *Derby Daily Telegraph*, Friday, 28th June 1912:2, British Library, accessed through the British Newspaper Archive, 15/01/2104.

Select Bibliography

News of the Week, *The Spectator*, 29th June 1912:18. The Spectator Archive online, 01/03/2015. http://archive.spectator.co.uk/article/29th-june-1912/18/a-crowded-meeting-summoned-to-protest-againstthe-
Notices, *Aberdeen Journal*, Friday 18th April 1913:5. Accessed through the British Newspaper Archive, 05/01/2013.
Premier's Daughter Undergoes Operation for Appendicitis, *Dundee Courier*, Monday, Dec. 1st, 1913:4, accessed through British Newspaper Archive, British Library, 05/01/2014.
Obituary, Harry Cheyne, *The Scotsman*, Edinburgh, Monday, February 29th, 1915.
Woman Doctors Justified, *Aberdeen Evening Press*, 3rd October 1918, British Library, accessed through British Newspaper Archive, 19/01/2014.
Universities Constituency, *Aberdeen Journal*, Monday, 18th November 1918. British Library, accessed through the British Newspaper Archive, 15/01/2104.
University Students' Greek Drama, *Aberdeen Journal*, Sat 15th March 1919:4, British Library, accessed through the British Newspaper Archive, 11/01/2104.
Day by Day: Shetlander Lord Lieutenant of the County, cutting from newspaper of unknown provenance, 7th June, 1919, Shetland Archives, Lerwick, Shetland. D21/4/p.91
Local Intelligence: Sir William Watson Cheyne, *Shetland News*, April 1st, 1920, Shetland Archives, Lerwick. SA1/14/36/14/p.4.
Little Reward for Brain Work, *Dundee Evening Telegraph*, August 26th, 1920:3, British Library, accessed through the British Newspaper Archive, 22/01/14.
Cutting from an unidentified newspaper entitled Shetlander Lord Lieutenant of the County, under a column called "Day to Day", Cheyne family album, accessed by author, 1999.
A Shetland Concert, *Aberdeen Journal*, Wednesday January 19th, 1921:6, accessed through the British Newspaper Archive, 22/01/2014.
Scheme to Encourage British Music, *Aberdeen Journal*, Saturday, 7th October, 1922:6 British Library, accessed through the British Newspaper Archive, 11/01/2104.
Future of Hospitals, *Dundee Evening Telegraph*, Wednesday 15th August 1923:2, British Library, accessed through the British Newspaper Archive, 15/01/2104.
Shipping News, *Western Daily Press*, 23rd December 1924:9, British Library, accessed through the British Newspaper Archive, 15/01/2014.
Aberdeen Journal, 23rd December 1926:4, British Library, accessed through the British Newspaper Archive, 11/01/2104.
Day by Day, *Dundee Evening Telegraph*, Friday, 1st July 1927:3. Accessed through British Newspaper Archive, British Library, 05/01/2014.
Aberdeen Journal, Saturday 22nd July 1927, accessed through British Newspaper Archive, 11/01/2014.
A Noted Surgeon, Sir Watson Cheyne, Visit to Wellington, *Wellington Evening Post*, Volume CV1, Issue 132, 12th December 1928:12. Viewed at Papers Past: http://paperspast.natlib.govt.nz/cgi-bin/paperspast?a=d&cl=search&d=EP19281212.2.87 11/02/2015.
Sir Watson Cheyne: A Great Welcome, honoured by Shetlanders, *Wellington Evening Post*, Volume CVI, Issue 132, 17 December 1928:17, accessed 11/01/2014 through Papers Past, http://paperspast.natlib.govt.nz/cgi-bin/paperspast?a=d&cl=search&d=EP19281217.2.134
Eminent Surgeon Sir W. Cheyne's Visit, *Wellington Evening Post*, Volume CVII, Issue 1, 2 January 1929:8, accessed 11/01/2014 through Papers Past http://paperspast.natlib.govt.nz
Cutting from unknown newspaper, Shetland Surgeon leaves £58,000, Shetland Archives, Lerwick. 21D/4a/15.

Saxby, Jessie., Shetland's Greatest Men, Shetland Archives, Lerwick. D11/21/14.
Fetlar Rejoicings, *Aberdeen Journal,* 19th February 1938:3, British Library, accessed through the British Newspaper Archive, 22/01/2014.
Robertson, J., Sir William Watson Cheyne, Letters to the Editor, Shetland Life, May 1987.
Squires, N., The Man who lost a "coral kingdom", BBC news, June 7th, 2007. http://news.bbc.co.uk/2/hi/programmes/from_our_own_correspondent/6730047.stm Accessed 26/8/2013.
Times Past: 100 Years Ago, *Shetland Times,* October 28th, 2011. http://www.shetlandtimes.co.uk/2011/10/28/times-past-135 Viewed 06/03/2015

Archival sources
Addington, Foreign Office, letter dated September 8th, 1853, National Archives, Kew, HO45/4725.
Additional inventory of the personal estate of the late Reverend David Webster, minister of Fetlar and North Yell, died at Manse, Fetlar, on 13th May 1881, Shetland Archives, Lerwick. SC12/36/7.
Anderson, Bruford, Henderson Tobar an Dualchais, School of Scottish Studies, University of Edinburgh, SA1970.240, http://www.tobarandualchais.co.uk/en/fullrecord/65344/3;jsessionid=1AFAADA422BFFF7EFACDC8230C99C628 Accessed 0/08/2013.
Bridge, Sir Frederick (ed.), The Form and Order of the Service that is to be Performed and of the Ceremonies that are to be Observed in the Coronation of their Majesties King George V and Queen Mary in the Abbey Church of S. Peter, Westminster, on Thursday the 22nd Day of June 1911, London (Novello and Company, Limited), 1911. Copy belonging to the Cheyne family.
Briston, Herbert G., letter to Cheyne, dated 19/8/90, in possession of Cheyne family.
Buckingham Palace, letter to Sir William Watson Cheyne from, dated 16th August, 1920, in possession of the Cheyne family.
Cheyne's invitation to the funeral of Lord Lister, Royal College of Surgeons (Edin.), Edinburgh. RCS/GD5/202/1.
Campbell, James and McAffer, Ronald, letter on behalf of tenants of Leagarth Estate, to J. L. Cheyne, dated January 1909. Shetland Archives, Lerwick. 131/542/15/15.
Campbell, Mrs. H.G., Lerwick, correspondence with Dr Guthrie and Mr. J.N.J. Hartley, Conservator, RCS (Edin.), July 19th 1953 and June 23rd 1954, Royal College of Surgeons (Edin), RCS/GD7/123.
Carson, Mr., letter to E.S. Reid Tait, from Leagarth, 10th November, 1929, Shetland Archives, Lerwick. D6/262/9.
Catalogue to the Manse Library at Hillswick, Shetland Archives, Lerwick. D25/32.
Catalogue of the Parish Library of Fetlar, 1831, Shetland Archives, Lerwick. Box 37, Item 1.
Chapman, T.D. and 8 other signatories, Southern Tasmanian Council of the League Protest Against the Convicts per *Lady Montague,* 11th December, 1852, Colonial Times, Hobart, Tasmania, Tuesday 14th December 1852.
Accessed: 17/03/2013 via http://trove.nla.gov.au/ndp/del/article/8772544
Cheyne, Captain Andrew, contract to transport 290 named convicts to Van Diemen's Land, National Archives, Kew, TS18/498.
Cheyne, Eliza, letter to Christian Watson, April 8th 1892, Shetland Archives, Lerwick. D1/542/13/11.
Cheyne, Mary, letter to Lord Lister, dated September 16th 1887, Royal College of Surgeons of England, MS0021/4/4 (10).
Cheyne W. (Cheyne's grandson), letter on the occasion of his father's funeral in Fetlar.

Select Bibliography

Shetland Archives, Lerwick. D1542/13/17.
Cheyne's certificates from the University of Edinburgh, 1871-75, in possession of the Cheyne family.
Cheyne, W.W., Natural History Notebook, C. Wyville Thomson Prof., Summer Session 1871, Edinburgh University Special Collections: MS2708.
Cheyne, W.W., Lectures on Clinical Surgery by Joseph Lister. Winter Session. 1972-73, Royal College of Surgeons of England, London, MS0021/4/1.
Cheyne's registration certificate and lecture timetable from the University of Vienna, 1875, in possession of the Cheyne family.
Cheyne, W.W. Handwritten draft of Boylston Prize submission (Volumes I and II), 1880, in possession of the Cheyne family: transcript by the author is available at http://www.watson-cheyne.com/handwritten-manuscript-volume2.pdf
Cheyne, W.W., letters to the Committee of the Association for the Advancement of Medicine by Research, 7th November 1882 and February 1st, 1883, Wellcome Library, London, MS5310.
Cheyne, W.W., Astley Cooper Prize submission, King's College Archives, London, G/PP5/7/1-2.
Letter from Cheyne to an unknown correspondent, December 16th 1890, Wellcome Library, London. MS7796/1.
Cheyne, W.W, insert in letter to Albert Carless, June 11th 1891, Albert Carless, Letters and Testimonials, Royal College of Surgeons of England, London. MS0064/6.
Cheyne, W.W., Case notes, King's College Hospital. King's College Archive, London. KH/CN1/449 1895, KH/CN1/452 1896, KH/CN1/470 1903, KH/CN1/473 1904, KH/CN1/489 1912.
Cheyne, W.W., letter to E.S. Reid Tait, containing a list of his works, Shetland Archives. Lerwick. D6/262/9/6.
Cheyne, W.W., letter to E.S. Reid Tait, from Leagarth, Fetlar, Shetland, dated September 22nd, 1928. Shetland Archives, Lerwick. SA4/2/43.
Cheyne, W.W., letter to E.S. Reid Tait, from Leagarth, April 1929. Shetland Archives, Lerwick. D6/262/9/3.
Cheyne, W.W., letter to L.W.G. Malcolm, Conservator at the Wellcome Museum, dated October 6th, 1926, Wellcome Library, London, WA/HMM/CO/AP/123.
Cheyne, W.W. correspondence with L.W.G. Malcolm, Wellcome Museum, 24th January 1927-26th May 1928, Wellcome Library, London. WA/HMM/EX/A.11.
Cheyne, W.W., report dated January 13th 1916, Wellcome Library, London, Rare Materials Room, MS1591.
Cheyne, W.W., Last Will and Testament, Shetland Archives, Lerwick. SC12/37/14/page1.
Clark, J.W., Registrar, Cambridge, letter to Cheyne dated 27th June 1896, in possession of the Cheyne family.
Curzon, Lord, letter to W.W. Cheyne from Walmer Castle, dated November 7th, 1904, Royal College of Surgeons of England, London. MSS552(2).
Papers of Captain Andrew Cheyne, Shetland Archives, Lerwick, Shetland. D/542. Copy deeds relating to 24 merks land in Bouster, Yell, Shetland.
Cubbon, Dolores, letters from friends, Shetland Archives, Lerwick. D6/294/2/p.55.
D'Arcy Power, Hon. Sec. Pathological Society of London, letter to Cheyne, 10th April, 1899. In possession of Cheyne family.
Davey, John, Letter, loose page inside the log books of the schooner *Acis*, 1863-66, Shetland Archives, Lerwick, Shetland. D1/542.
Dundas, W.P., Notice to registrars with requirements for compulsory vaccination, and sample copies of schedules A-C, 1863, Shetland Archives, Lerwick. D23/156/1A.
Donnelly, Samuel, Surgeon on the *Lady Montague*, Journal, National Archives, Kew:

ADM 101/254, 1A-1H.
Watson, Reverend William, Diary, 1849-1854, Shetland Archives, Lerwick. D1/542.
Exemption from Firearms certificate, signed by Gifford Gray, Superintendant, Zetland County Police, 29th December, 1920, Wellcome Library, London. MS7796.
Forbes, D.M., Honorary Surgeon, Scottish Hospital, letter to Cheyne dated 27th November 1897. In possession of Cheyne family.
Hill, Lionel F., letter to Cheyne dated 2/12/1893, in possession of Cheyne family.
Home Office, letter to the Office of the Committee of the Council for Trade, 10th June 1854, National Archives, Kew. HO 45/5814 7721.
Hay and Co., letters to Mr Watson Cheyne, 24th June, 1st July and 13th July 1901, Hay and Co. wet book re W.W. Cheyne and Leagarth Estate, Shetland Archives, Lerwick. D31/18/12.
Hay and Co., letter to Mr George Garster, Bealagord, Fetlar, Lerwick 29th October 1901, Hay and Co. wet book re W.W. Cheyne and Leagarth Estate, Shetland Archives, Lerwick. D31/18/12.
Howlam, A.W., King's College, London, letter to Cheyne, February 12th, 1912 Royal College of Surgeons of England, London. MSS 552(3).
Inventory of the personal estate and extract trust disposition and settlement of the Reverend William Watson, minister of Fetlar and North Yell, Shetland Archives, Lerwick. SC12/36/4:42.
Inventory of the Personal Estate of the late Sir William Watson Cheyne, including certified copy of Trust Disposition and Settlement, Shetland Archives, Lerwick. SC12/36/23/page1.
Inventory of the Personal Estate of the Late Revd. David Webster, 1881, Shetland Archives, Lerwick, Shetland. SC12/36/7/p.194.
Johnson, Robert L., unpublished manuscript on the life of Sir William Watson Cheyne, dated April 1982, copy belonging to the Cheyne family.
Johnston-Saint P., Wellcome Museum, to Sir William Watson Cheyne, Bart., c/o R.H. Pringle, Tuloa Farm, Molo, Kenya Colony, 7th Dec. 1926. Wellcome Library, London. WA/HMM/EX/A.11.
Journal kept by Leagarth guests in 1907, belonging to the Cheyne family.
Lister, Sir Joseph, letter to Mr. J.A Collinson., 31st May 1870, in private ownership, photocopy at Fetlar Museum Trust, Fetlar, Shetland.
Lister, Sir Joseph, reference in favour of W.W. Cheyne, 10th February, 1880. Royal College of Surgeons, Edinburgh, GD5/138.
Lister, Sir Joseph, Notes and letters on Iodide Gauze, 1882-1888, Royal College of Surgeons of England, London. MS0021/4/4.
Lister, Sir Joseph, letter to Mary Watson Cheyne, Royal College of Surgeons (Edin.), Edinburgh, RCS/GD5/139.
Lister, Sir Joseph, letter to Mrs Cheyne, 11th February 1901, Royal College of Surgeons (Edin.), Edinburgh, RCS/GD5/141 (65/8/d).
Lister, Sir Joseph, letter to William Watson Cheyne, written from Lyme Regis, 21st April 1901, Royal College of Surgeons (Edin.), Edinburgh. RCS/GD5/142 (65/8/e).
Lister, Sir Joseph, letter to Cheyne, 27th December 1902, Royal College of Surgeons (Edin.), Edinburgh. RCS/GD5/144 (65/8/9).
Lister, Sir Joseph, letter to Cheyne, Royal College of Surgeons (Edin.), Edinburgh. RCS/GD5/146 (65/8/j).
Lister, Sir Joseph, letter to Sir Hector Cameron, 19th February 1907. Royal College of Surgeons (Edin.), Edinburgh. RCS/GD5/067.
Lister, Sir Joseph, letter to Cheyne, dated March 2nd, 1908, Royal College of Surgeons of Edinburgh RCS/GD5/145 (65/8/h).
Lister, Sir Joseph, letter to Cheyne, dated January 5th, 1910, Royal College of Surgeons

Select Bibliography

of Edinburgh RCS/GD5/148 (65/8/6).
Loeterbagh, Dr., Lerwick, letters to A.D. Mathewson, registrar, Yell, Shetland, dated 7th June and 26th July 1865, Shetland Archives, Lerwick. D23/151/31/724.
Manson, T.M.Y., Transcript of a lecture given to the Northmavine History Group at the Sullom Public Hall on 21st March 1989, copy at Fetlar Museum Trust, Fetlar Interpretive Centre, Fetlar, Shetland.
Mathewson, A.D., letter dated September 8th 1873, Shetland Archives, Lerwick. D23/150/32/23.
Mathewson, A.D., letter to his children, dated 18 October 1880, Shetland Archives, Lerwick. D1/411/3/6.
Mathewson, Margaret C., letter to her brother Arthur, April 30th, 1876, Shetland Archives, Lerwick. D23/193/17/27.
Mathewson, Margaret, letter to her brother Arthur, dated October 22nd, 1876, Shetland Archives, Lerwick. D23/193/17/72.
Mathewson, Margaret, letter to her father, dated 22nd February, 1877, Shetland Archives, Lerwick. D23/151/43/4.
Mathewson, Margaret in a letter to her brother Arthur, dated February 24th 1877, cited in Goldman, M., *op.cit.*:37
Mathewson, Margaret, letter to her family, February 27th, 1877. Shetland Archives, Lerwick. D23/151/43/5.
Mathewson Margaret, letter to her father and brother, April 30th 1877, Shetland Archives, Lerwick. D.23/151/43/12.
Mathewson, Margaret, letter to her father, May 10th 1877, Shetland Archives, Lerwick. D24/151/43/15.
Mathewson, Margaret, letter to her father, May 24th 1877, Shetland Archives, Lerwick. D24/151/43/18.
Mathewson, Margaret, letter to her father, June 11th 1877, Shetland Archives, Lerwick. D24/151/43/22.
Mathewson, Margaret, letter to her father, July 2nd 1877, Shetland Archives, Lerwick. D24/151/43/27.
Mathewson, Margaret, letter to her father, July 16th, 1877, Shetland Archives, Lerwick. D23/151/43/30.
Mathewson, Margaret, letter to her father, October 18th, 1877, Shetland Archives, Lerwick. D23/151/43/39.
Mathewson, Margaret, letter to her father, November 6th, 1877, Shetland Archives, Lerwick. D23/151/43/42.
Mathewson, Margaret C., Holograph manuscript of part of Sketch of her treatment in Edinburgh Royal Infirmary in 1877, Edinburgh, 15th December, 1878, Shetland Archives, Lerwick, D1/557/4:5-76.
Mathewson, Margaret C., Holograph "Sketch of Eight Months a Patient in the Royal Infirmary of Edinburgh, 1877," July 26th 1879, Shetland Archives, Lerwick, D1/557/1.
Letter from Margaret Mathewson to Dr Cheyne on Fetlar, dated 4th August 1879, Shetland Archives, Lerwick. D/23/193/20/13.
Mathewson, Margaret C., *Sketch of Eight Months a Patient in the Royal Infirmary, Edinburgh, August 8th 1879*, copy in Shetland Archives, SA2/340.
Medical Society of London, letter to Cheyne, June 20th 1895, in possession of the Cheyne family.
Minutes of the first meeting of the Council of the Association for the Advancement of Medicine by Research, April 20th 1882, Wellcome Library, London, MS5310:3.
Nicolson, Lady Annie, Brough Lodge, Fetlar, Diaries, 1902-1920s, Shetland Archives, Lerwick. D24/Box82.

Poem written by Cheyne's students on his departure for the Boer War, 1899. In possession of the Cheyne family.
Playfair, Major General A.L., Haymarket Theatre, letter to Cheyne, September 30th, 1895. In possession of the Cheyne family.
"Presentation of South African Medals", 12th June 1901, issued by Buckingham Palace, in possession of the Cheyne family.
Rickets (British Consul in Manila), letter to Henry Cheyne, 3rd June, 1867, responding to an enquiry made by Henry Cheyne in April that year. Shetland Archives, Lerwick. D1/542/13/12.
Programme from Memorial Service for W. Watson Cheyne, St. Columba's Church, Lerwick, 1st May 1932, Shetland Archives, Lerwick. SA4/3000/18/60/33.
Rules for the Management of the Fetlar Parochial Library, September 18th, 1828, Shetland Archives, Lerwick. D24, Box 37, Item 1.
Saundby Robert, British Medical Association, letter to Cheyne, January 22nd, 1897. In possession of the Cheyne family.
Scott Burn, W., letter to Arthur Nicolson of Lochend, 4th August 1818, Shetland Archives, Lerwick. D24/49/42.
Shineberg, D., letter to Sir Joseph Cheyne, 24th April 1967, Shetland Archives, Lerwick. D1/542/15/1.
Shineberg, D., letter to Sir Joseph Cheyne, 29th December, 1985. Shetland Archives, Lerwick, Shetland. D1/542/15/14.
"Small-Pox": public notice urging the residents of Lerwick to seek vaccination against smallpox and offering free vaccination, January 1877, Shetland Archives, Lerwick. 50.10.19.
Spence, James, reference in favour of William Watson Cheyne, 10th May 1878, London, Royal College of Surgeons of England, MSS552(1).
Stevens, Charles E., 1867, Report of Proceedings at the Pelew islands in the Matter of the Murder of Andrew Cheyne, National Archives, Kew, FO72/1155.
Trimmer, Edward, Secretary of the Royal College of Surgeons of England, letter to Cheyne dated 1st July, 1897. In possession of the Cheyne family.
Watson, Reverend William, letter to Sir Arthur Nicolson, 7th August 1839, copy in possession of Cheyne family.
Watson, Reverend William, letter to a friend, 27th November 1848, Shetland Archives, Lerwick. D1/542/13/3.
Watson, Reverend William, Diary, 1849-1854, Shetland Archives, Lerwick. D1/542.
Watson, Reverend William, letter to a friend, dated 21st February 1852, Shetland Archives, Lerwick. D1/542/13/10
Webster, Reverend David, Correspondence with Sir Arthur Nicolson, 1856-1863, Shetland Archives, Lerwick. D.24, Box 38. Item 7.
Wellcome Museum, letter to Cheyne from Wellcome Museum, London, 23rd May 1928 Wellcome Library, London. MS7796.
Williamson, Laurence under the pen-name 'Stakaberg' (the name of a hill in Fetlar), to the *Shetland Times*, dated 6th October 1886, Shetland Archives, Lerwick, D7/48/12.
Williamson, Laurence, letter to Cheyne, dated 16th August 1901, Shetland Archives, Lerwick. D7/20/45.
Williamson Laurence, letter to his cousin Andrew, 18th April 1921, Shetland Archives, Lerwick. D1/302.

Websites (not mentioned elsewhere)
Aberdeen Grammar School http://grammar.org.uk/visitors/history.php Accessed 26/03/2013.
Anderson High School, Lerwick: http://www.anderson.shetland.sch.uk/wordpress/ahs/

Select Bibliography

history-of-the-school Accessed 26/03/2013.
Ancestry: www.ancestry.co.uk
Australian Government website http://australia.gov.au/about-australia/australian-story/convicts-and-the-british-colonies Accessed 26/08/2013.
The Dulwich Society, http://www.dulwichsociety.com/newsletters/42-summer-2006/237-villagers-notebook Accessed 15/02/2015.
Kedlestone Hall, National Trust, accessed 08/01/2014 http://www.nationaltrustcollections.org.uk/object/107881
King's College Collections, London. http://kingscollections.org
Lost Hospitals of London website http://ezitis.myzen.co.uk/paddingtongreen.html Accessed 14/09/2013.
http://www.measuringworth.com Accessed 16/06/2015.
National Records of Scotland, http://www.nrscotland.gov.uk/research/guides/poor-relief-records Accessed 16/06/2015.
Natural History Museum, accessed 16/06/2015 http://www.nhm.ac.uk/nature-online/science-of-natural-history/expeditions-collecting/hms-challenger-expedition/index.html
National Trust for Scotland http://unst.org/web/unstpartnership/files/2012/09/HH_Panels.pdf Accessed 07/01/2014.
Nobel Prize: Robert Koch: http://www.nobelprize.org/nobel_prizes/medicine/laureates/1905/koch-bio.html Accessed 21/04/2013 and
Nobel prize: Paul Ehrlich http://www.nobelprize.org/nobel_prizes/medicine/laureates/1908/ehrlich-bio.html Accessed 16/06/2015.
Official Guide and Souvenir of the Sixty-Fifth Annual Meeting of the British Medical Association, Montreal, 1897: http://archive.org/stream/cihm_01515#page/n5/mode/2up Accessed 19/09/2013.
Plarr's Lives of the Fellows Online, http://livesonline.rcseng.ac.uk/biogs/E000222b.htm
ScotlandsPeople: http://www.scotlandspeople.gov.uk
Scottish Architects, www.scottisharchitects.org.uk accessed 19/12/2013.
University of Cambridge website, Dons in the House, http://www.cam.ac.uk/for-staff/features/dons-in-the-house Accessed 11.02.2015.
National Archives, Kew, http://webarchive.nationalarchives.gov.uk Accessed 02/01/2014.
World Health Organisation, http://www.who.int/mediacentre/factsheets/fs099/en/index.html Accessed 10/01/2014.

Photograph credits

Sir William Watson Cheyne, page ii: by permission of the Cheyne family and Fetlar Museum Trust.
Sir Watson with his ubiquitous cigar, page vi: by permission of the Cheyne family and Fetlar Museum Trust.
Leagarth House with verandah in the 1920s, page xii: by permission of the Cheyne family and Fetlar Museum Trust.
Sir William Watson Cheyne, cover and page xvi: copyright: The Royal College of Surgeons of Edinburgh.
A garden party in the grounds of Leagarth House, early 1900s, page xviii: by permission of the Cheyne family and Fetlar Museum Trust.
Tangwick Haa, Shetland, childhood home of Captain Andrew Cheyne, page 11: by permission of Eileen Mullay, Shetland.
Tangwick Haa, page 12: by permission of Susan Thompson, Shetland.
This image may be Andrew Cheyne, page 13: courtesy of Shetland Archives.
This may be Grace Watson, Cheyne's aunt, as an old lady, page 38: by permission of Fetlar Museum Trust.
The West Manse in Fetlar, page 43: by permission of Fetlar Museum Trust.
The West Manse, Fetlar, in 2014, page 46: Photograph by Jane Coutts.
One of the last houses in Fetlar with a fire in the centre, page 48: by permission of the Cheyne family and Fetlar Museum Trust.
Certificate for Natural History from the University of Edinburgh, page 71: by permission of the Cheyne family.
Certificate for Chemistry from the University of Edinburgh, page 73: by permission of the Cheyne family.
Cheyne's certificate for Anatomy, page 74: by permission of the Cheyne family.
Cheyne's certificate for Junior Surgery, page 75: by permission of the Cheyne family.
Cheyne's certificate for Senior Surgery, page 76: by permission of the Cheyne family.
Lister's carbolic hand spray, page 90: from Cheyne's *Antiseptic Surgery*, 1880, with the permission of the Cheyne family.
Vienna registration document and class attendance certificate, page 96: by permission of the Cheyne family.
Margaret Mathewson, Lord Lister's Shetland patient, page 103: The Old Haa, Shetland.
A Shetland woman walking down a road knitting, page 105: by permission of the Cheyne family and Fetlar Museum Trust.
Illustration from Cheyne's *Antiseptic Surgery (1880)*, page 133: by permission of the Cheyne family.
Drawing of Bacillus alvei, found inside one of Cheyne's books, page 162: by permission of the Cheyne family.
Page from Cheyne's own collection of his reports, page 168: reproduced by permission of the Cheyne family.
Pages from Cheyne's own amendments to *The Treatment of Wounds...*, page 207: by permission of the Cheyne family.
A 'Watson Cheyne' probe, page 213: by permission of Fetlar Museum Trust.
Basil Cheyne, Sir Watson's youngest son, page 218: by permission of the Cheyne family and Fetlar Museum Trust.

Photograph Credits

Lister's carbolic steam spray, page 222: from Cheyne's *Antiseptic Surgery*, 1880, with the permission of the Cheyne family.
Signed photograph of Sir William Watson Cheyne, page 234: by permission of the Cheyne family.
Probably on the march to Bloemfontein, page 247: by permission of the Cheyne family and Fetlar Museum Trust.
What appears to be a destroyed railway bridge, page 251: by permission of the Cheyne family and Fetlar Museum Trust.
Attending to bodies during the Boer War, page 252: by permission of the Cheyne family and Fetlar Museum Trust.
Erecting a temporary bridge, page 254: by permission of the Cheyne family and Fetlar Museum Trust.
Men clearly in some distress, and may be suffering from fever, page 255: by permission of the Cheyne family and Fetlar Museum Trust.
Building probably used as a temporary hospital, page 256: by permission of the Cheyne family and Fetlar Museum Trust.
Difficulties medical teams faced with transport, page 258: by permission of the Cheyne family and Fetlar Museum Trust.
Lord Roberts and Boer General Cronje, page 261: by permission of the Cheyne family and Fetlar Museum Trust.
In talks after General Cronje's surrender, page 262: by permission of the Cheyne family and Fetlar Museum Trust.
Lord Roberts leaving Cronje and his secretary, page 263: by permission of the Cheyne family and Fetlar Museum Trust.
Cronje's Secretary speaking to Lord Roberts, page 264: by permission of the Cheyne family and Fetlar Museum Trust.
Scenes of devastation in the Boer laager, page 265: by permission of the Cheyne family and Fetlar Museum Trust.
Devastation, probably in Cronje's laager after his surrender, page 267: by permission of the Cheyne family and Fetlar Museum Trust.
Showing the difficulties medical teams faced with transport, page 270: by permission of the Cheyne family and Fetlar Museum Trust.
Royal Army Medical Corps group in South Africa, page 277: by permission of the Cheyne family and Fetlar Museum Trust.
Sir Watson with his camera outside the Chapel, page 300: by permission of the Cheyne family and Fetlar Museum Trust.
Brough Lodge, Fetlar, the home of the Nicolson family, page 302: by permission of Fetlar Museum Trust.
The tower at Brough Lodge, Fetlar, a mid-19th century folly, page 302: by permission of Fetlar Museum Trust.
Lady Annie Nicolson, page 304: by permission of Fetlar Museum Trust.
Sir Arthur T.B.R. Nicolson of Brough Lodge, page 305: by permission of Fetlar Museum Trust.
Flit boats coming into Brough, Fetlar. 306: by permission of Fetlar Museum Trust.
The flit boat transport to (or from) the Earl of Zetland, page 307: by permission of Fetlar Museum Trust.
Golf at Brough Lodge, page 308: by permission of Fetlar Museum Trust.
Leagarth House in its original state, page 311: by permission of the Cheyne family and Fetlar Museum Trust.
Car being brought ashore by flit boat, page 313: by permission of Fetlar Museum Trust.
Vera Nicolson of Brough Lodge, page 314: by permission of Fetlar Museum Trust.

Microbes and the Fetlar Man

The Kirk on Sunday, pages 316, 317: by permission of the Cheyne family and Fetlar Museum Trust.
Flittin peats by boat from Lambhoga, Fetlar, page 318: by permission of Fetlar Museum Trust.
Flittin peats by pony in Lambhoga, Fetlar, page 319: by permission of Fetlar Museum Trust.
One of Sir Watson's boats, page 320: by permission of the Cheyne family and Fetlar Museum Trust.
Sir Watson in one of his boats, page 321: by permission of the Cheyne family and Fetlar Museum Trust.
Standing by the flit boat, page 322: by permission of the Cheyne family and Fetlar Museum Trust.
A Leagarth party disembarking into the flit boat, page 323: by permission of the Cheyne family and Fetlar Museum Trust.
A Leagarth party, early 1900s, page 324: by permission of the Cheyne family and Fetlar Museum Trust.
Images from the guest book at Leagarth, page 326: by permission of the Cheyne family.
Sir Watson (left) in his motor boat, page 327: by permission of the Cheyne family and Fetlar Museum Trust.
Meta Cheyne, Basil Cheyne. Outside Leagarth House, early 1900s, page 328: by permission of the Cheyne family and Fetlar Museum Trust.
Framed photograph of Lady Curzon, page 334: by permission of the Cheyne family.
Sir Lister Cheyne, Sir Watson's eldest son, page 335: by permission of the Cheyne family and Fetlar Museum Trust.
Complimentary dinner given for Cheyne, page 343: by permission of the Cheyne family.
The notice contains signatures of the guests, page 344: by permission of the Cheyne family.
The Order of St. Michael and St. George (KCMG), page 373, by permission of the Cheyne family.
J.J. ('Jeemsie') Laurenson, page 383: by permission of John Coutts
Leagarth House and gardens from the sea, page 396: by permission of the John Hughson Collection.
The silver tray engraved "Beechgrove, 1920", page 397: by permission of Fetlar Museum Trust.
The Leagarth staff, page 399: by permission of the Cheyne family and Fetlar Museum Trust.
Leagarth staff, 1920s, page 400: by permission of the Cheyne family and Fetlar Museum Trust.
The entrance to the laboratory next to Cheyne's apartments, page 401: by permission of the Cheyne family and Fetlar Museum Trust.
Prescription written by Sir Watson for Christian Anderson, page 402: by permission of Fetlar Museum Trust.
Frank Coutts, manager of Cheyne's Home Farm, page 404: by permission of Frank Coutts.
Overview of Leagarth gardens, early 1920s, page 406: by permission of the Cheyne family and Fetlar Museum Trust.
Sir Watson in the completed gardens at Leagarth, page 407: by permission of the Cheyne family and Fetlar Museum Trust.
In the sunken alpine garden at Leagarth, page 408: Reproduced by permission of the Cheyne family and Fetlar Museum Trust.

Photograph Credits

The head gardener at Leagarth, James Young Robertson ("Jimmy Young"), and his assistant, Willie Garriock, page 409: by permission of the Cheyne family and Fetlar Museum Trust.
Sir Watson in the verandah at Leagarth, page 410: by permission of the Cheyne family and Fetlar Museum Trust.
The Fetlar community, page 411: by permission of Fetlar Museum Trust.
Fetlar people dressed for a concert party, probably 1920s, page 411: by permission of Fetlar Museum Trust.
One of the many Leagarth concert parties, 1920s, page 412: by permission of the Cheyne family and Fetlar Museum Trust.
Sports on the lawn at Leagarth, page 412: by permission of the Cheyne family and Fetlar Museum Trust.
A picnic on the lawn outside the verandah at Leagarth, 1920s, page 413: by permission of the Cheyne family and Fetlar Museum Trust.
Sir Watson teaching his son Basil to fish in the Houbie Burn, Fetlar, page 415: by permission of the Cheyne family and Fetlar Museum Trust.
The trout ladder Sir Watson built on the Houbie Burn, page 416: by permission of the Cheyne family and Fetlar Museum Trust.
Sir Watson on one of his many journeys, page 418: by permission of the Cheyne family and Fetlar Museum Trust.
Possibly Ella Davis, on Sir Watson's New Zealand trip in 1928, page 425: by permission of the Cheyne family and Fetlar Museum Trust.
The illuminated copy of the welcome speech, pages 427 and 428: by permission of the Cheyne family and Fetlar Museum Trust.
Probably Ella Davies holding a pickaxe outside a hut, page 430: by permission of the Cheyne family and Fetlar Museum Trust.
Hunter Cheyne, who followed his father into medicine, page 431: by permission of the Cheyne family and Fetlar Museum Trust.
Sir Watson's grandchildren on Tresta Beach, Fetlar, page 433: by permission of the Cheyne family and Fetlar Museum Trust.
Sir Watson's final resting place in the Fetlar Kirkyard, page 434: photograph by Jane Coutts.
Sir Watson and his dog, outside the verandah at Leagarth, page 436: by permission of the Cheyne family and Fetlar Museum Trust.
… … with his ubiquitous cigar, page 519: by permission of the Cheyne family and Fetlar Museum Trust.

Index

99th Regiment 28, 29

A

Aberdeen 49, 56-60, 70, 93, 110, 119, 135, 151, 227, 320, 321, 337, 387, 388, 395, 397, 400, 414, 445, 447, 449
Aberdeen Grammar School 49, 56, 57-59, 447
Aberdeen, University of 49, 59, 60, 70, 397, 445, 449
Aberdeen Evening Press 473
Aberdeen Journal 467, 497, 501, 508, 511, 513, 514, 516, 517
Acis 33, 69, 442
Aithsting 12
Aitken, John Morgan 310
Altham, James 139
Amakirrima 32
anaesthesia (anaesthetics) 62, 64-65, 100, 144, 158, 216, 368, 393, 448, 512
Anderson, Arthur, MP 23, 56
Anderson, Christian (Sir Watson's staff) 219, 397
Anderson, Christian xv, 402-403
Anderson, Janet 13
Anderson, Thomas, Professor of Chemistry 80
Anderson Educational Institute 56
Aneityum 19-21, 42
Annandale, Thomas 139, 140
Ansell, Florence 204
anthrax 150, 156, 157, 185, 227
anticontagionism 66, 67
antiseptic(s) xvii, 1, 3, 4, 6, 9, 62, 68, 77, 80, 81, 83-86, 89-92, 95-97 100, 101, 109, 127-129, 132-134, 137, 138, 142, 148, 149, 152-154, 157, 166, 167, 169, 179, 181-183, 185, 192, 212, 220, 221-237, 244, 245, 253, 285, 288-291, 296, 331, 338, 339, 342, 345, 348, 350, 352, 355, 358, 360-372, 435, 461, 485
antiseptic surgery xiii, xviii, 133, 152-154, 163, 169, 183, 220, 221, 228, 229, 250, 466, 470, 474, 484, 485, 490, 495

Antiseptic Surgery, its Principles, Practice, History and Results 152, 466, 474, 490, 495
Aquitania 370
Ariel 43
Army Medical Nursing Reserve 242
aseptic 133, 148, 213, 227-238, 244, 245, 287, 288, 290, 296, 319, 331, 339, 348, 360-362, 372, 417, 432, 461, 484-, 486, 501
Asquith, Herbert Henry 334-336, 375
Asquith, Elizabeth 332, 334-336
Association for the Advancement of Medicine by Research 158, 159, 197, 467, 468
Astley Cooper Prize 186, 189-191, 474, 475
Auckland 19
Australia 15, 20, 21, 23, 32, 36, 110, 247, 303, 420, 429, 439, 440-442, 444

B

bacilli xi, 134, 135, 150, 156, 157, 159-161, 163, 164, 166, 167, 171-177, 179, 186-189, 193, 194, 196-199, 285, 357, 358, 405, 473
Bacillus alvei xiii, 162, 166
Bacillus anthracis 149
Back, Ivor 236, 486
bacteria 1, 2, 4, 82-84, 89, 101, 129, 130-135, 138, 142, 143, 145, 148-154, 157, 160, 166, 169-175, 177, 181, 182, 187, 190, 192, 193, 219, 222, 223, 227-229, 233-236, 244, 252, 285, 288, 291, 350, 361, 363, 364, 368, 369, 371, 405, 463-465
bacteriology xi, xvii, 1, 3, 4, 6, 30, 85, 98, 101, 132, 133, 135, 136, 150, 152, 155, 156, 160, 164, 167, 169, 172, 174, 177, 178, 186, 191, 201, 210, 219, 231, 238, 239, 341, 348, 364
Bacterium lactis 148, 151, 170
Bain, Christina 397
Bain, George 499
Balfour, Professor John Hutton 71

536

Balfour, Arthur, Earl of Balfour 336, 390
Balfour Hospital, Orkney, 389
Banbury, Sir Frederick 391
Banks, Sir Joseph 15
Bannister, Rose 204
Barclay, Reverend James 105, 107, 110-112, 121, 123
Barlow, Sir Thomas 333
basilar meningitis 215
Bassett-Smith, Fleet Surgeon 363, 364
Bastian, Henry Charlton 221
Bay of Islands 10
Beagle 20, 33, 439, 442
Beala, Fetlar 309
Beale, Keyton 341
bêche-de-mer 18, 21, 31, 54, 55
Becher Island 32
Bee 10, 15
Beechgrove xiv, 396-398, 403
Behring, Emil von 179
Bell, Joseph 61
Bell, Nurse 345
Bell and Robertson Ltd (Aberdeen) 414
Belloc, J. Hilaire P.R. 379
Bengal 10
Bennett, John Hughes 85
Bentley Todd, Robert 147, 465
Bergmann, Ernst von 196
Berlin 95, 151, 156, 160, 161, 165, 176, 181, 182, 194, 196-199, 340, 477
Billroth, Theodor 95, 97-98
Bird, Golding 68
Bloemfontein xiii, xix, 5, 247, 249, 253, 257, 260, 267-269, 271-275, 277-282, 284, 287, 292, 358, 489-492, 494, 496
Boers 240, 247, 248, 250, 257, 268, 274, 294, 493
Boer War xi, xiii, xix, 5, 238-240, 252, 257, 277, 290, 292, 301, 331, 336, 351, 357-359, 369, 384, 399, 432, 487, 489, 491, 492, 506
boiled water 80, 274, 360, 361
bonesetters 380-382
borsal 363-365, 367
Boulogne 358, 369
Bouster 96, 454
Boyd, Stanley 181
Boylston Prize xviii, 140, 145, 146, 153, 183, 221, 462, 464, 473, 483
Bradshaw Lecture 352, 504
Brandy (Cheyne's dog) 395, 400, 435
Brennand, John Chaffer 351

British East India Company 9
British Expeditionary Force 358
British Journal of Surgery 337
British Medical Association 81, 84, 164, 167, 173, 180, 182, 189, 191, 201, 203, 205, 208, 209, 215, 219, 230, 332, 376-378, 387-389, 390, 426, 468-473, 475, 476, 479, 480, 482, 483, 501, 511
British Medical Journal (BMJ) 56, 72, 146, 152, 157, 161, 164-167, 169, 172, 173, 175, 181, 189, 193-197, 211, 215-217, 220, 223, 224, 233, 243, 249, 271-273, 281, 285, 287, 290, 292, 340, 342, 345, 346, 348, 353, 362, 377, 378, 385, 420, 433, 446, 447, 449, 450, 452, 453, 461-486, 488-496, 500-516, 518
Brough Lodge viii, xiv, 107, 301-306, 308, 309, 311, 312, 314, 315, 323, 325, 327, 330 456, 457, 498-500
Brown, Alexander C. 72, 73
Brown Animal Sanitary Institute 158, 172
Brown Dog 391
Brown, Mary 300, 497
Browne, Herbert M.T. 419, 420
Brown, John 399
Brücke, Ernst von 97
Bruns, Victor von 227, 250
Brunton, Deborah 99, 184, 341, 448, 449, 455
Buller, General Sir Redvers H. 241, 242, 274
Buness 60
Burdett-Coutts, William 275-279, 281, 282, 284, 492
Burdon Sanderson, John 135, 149, 160, 163, 172, 175, 184, 210, 222, 483
Burgess, Haldane 297
Burghard, Frédéric François 215, 230, 332, 339, 482, 483, 485
Burravoe (Yell) 322
Burroughs, Wellcome and Co. 356

C

Callao 28, 33, 34, 418, 419
Cambridge Hospital for Special Diseases 347, 390
Cameron, Sir Hector 337-340, 345, 422, 423, 502
Camito 419
Campbell (Minister of Fetlar and North Yell) 43, 298, 322, 325, 328, 413, 414
Canadian Journal of Surgery 352

carbolic xiii, 82-86, 89, 90, 98, 114, 127-129, 132, 142, 145, 166, 169, 185, 192, 222, 226-229, 231, 233, 236, 244, 252, 285, 291, 360, 367
Carisbrooke Castle 245
Carless, Albert 191, 476, 486
Carlisle 83, 167, 209, 470
catgut 214, 215
Chapman, Rose 219
Cheatle, Sir George Lenthal 208, 242-245, 291, 488
Cheshire, Frank R. 166
Cheyne, Andrew (Sandy) viii, 433,
Cheyne, Captain Andrew viii, ix, xi, xiii, 3, 7, 9-37, 39-42, 44, 45, 51, 54, 55, 57, 58, 69, 429, 437, 440, 444, 447, 454
Cheyne, Eliza (Watson) 21-23, 26-28, 30-39, 41-44, 48, 52, 54, 440
Cheyne, Elizabeth (Betsy) 13, 14, 26, 69
Cheyne, Gerald xix
Cheyne, George Basil xiii, xiv, xv, 218, 219, 285, 323, 324, 328, 354, 407, 414, 415, 417
Cheyne, Grace Ella Margaret Watson (Meta) xiv, 208, 285, 323-325, 328, 354, 397, 415, 417, 419-421, 429, 515
Cheyne, Henry 11, 14, 36, 69, 438, 443, 447, 449
Cheyne, James 10, 11, 21
Cheyne, John 11
Cheyne, Sir Joseph Lister (Sir Lister Cheyne) xiv, 185, 204, 285, 300, 319, 327, 328, 335, 336, 341, 357, 413, 419, 422, 431, 432
Cheyne, Sir Joseph ix, xviii, 44, 433, 438, 443, 444, 449, 498, 500, 501, 513, 515, 517, 518
Cheyne, Julia Millicent 284, 295
Cheyne. Mary Emma (Servanté) 184, 185, 202-208, 210, 296, 348, 479, 496
Cheyne, Mary Frances 202, 203, 295
Cheyne, Lady Margaret (Margaret Smith) 204-209, 219, 239, 245, 284-285, 293, 300, 323, 353-354, 379, 397-399, 415, 417, 479
Cheyne, William Hunter Watson xv, 185, 195, 204, 239, 285, 300, 319, 328, 354, 357, 366, 367, 396, 431, 505
Cheyne, Sir William Watson *passim*
Chiene, Professor John 112, 130
China 9, 10, 15, 16, 18, 19, 21, 27, 30-32, 36, 37, 54, 442, 443
chlorine 66

chloroform 61, 65, 89, 100, 120, 158, 159, 233
cholera 67, 170-173, 175-179
Church of Scotland 2, 10, 14, 27, 46, 104, 105, 111, 298, 322, 439, 449, 518
Clearances 25, 301
Clunies-Ross, John 20, 21, 54
Cocos Islands 20, 21, 54
Cohn, Ferdinand Julius 150
Cohnheim, Julius Friedrich 160
Colenso 259, 269
Colonial Land and Emigration Commissions 23
Colonial Times 29, 440, 441
comma bacillus 172, 173, 175, 177
Commons Medical Committee 388, 511
Conan Doyle, Sir Arthur 61, 75, 194, 241, 257, 269, 272, 292, 358, 451, 487, 490-492, 496, 506
Condy's Fluid 80
contagionism 67
Corfield, W.H. 165, 167, 476
Coutts, Catherine 301, 497
Coutts, Frank xv, 403-405
Cowie, Dr. Robert 99, 100, 107, 109, 455, 457
Crewkerne (Somerset) 396
Cronje, General Piet xiv, xix, 248-250, 253, 261-267, 294, 432
Crooke, Dr. George F. 193, 476
Crookshank, Edgar M. 172, 178, 188, 191, 195, 475
Cruelty to Animals Act (1876) 158, 391, 467
Curragh Incident 336
Curzon, George, Marquess Curzon of Kedleston 332-334, 500, 501
Curzon, Mary Victoria, Baroness Curzon of Kedleston xiv, 332-334, 399, 500
Cynosure 33

D

Dacre, Ranulph 15, 17
Daily Alta (California) 32, 442
Dakin, Henry Drysdale 369, 383
Dakin's Solution 369
Dardanelles xi, 153, 355, 365-369, 510
Darwin, Charles 20, 33, 61, 158, 439, 442
Davey, John 33, 42, 57, 58, 442
Davis, Ella xv, 425, 426, 429
de Beers (mining company) 247

Index

De Kiel's Drift 248
Delting 204
Dessouter, Marcel 332
De Wet, Christiaan 249, 261, 267
Diana 15
diarrhoea 29, 270, 281
Dobie, W.H. 139
Donnelly, Samuel (ship's surgeon) 28, 29, 440, 441
Dow, Barry 380
Down Bros. (surgical instrument manufacturer) 213, 481
dresser 79, 89-91, 93, 113, 115, 123-127
dressership 88
Driefontein 268, 282, 492, 495
Duncan, John 227
Dundee 25, 454, 496, 497, 499, 501, 503, 511
Dundee Evening Telegraph 454, 496, 497, 499, 503, 511
dysentery 29, 177, 369

E

Earl of Zetland (ship) xiv, 307, 322, 323
Earls of Zetland 308, 309, 398
East Booth, Hillswick 12
Edinburgh iv, vii, xi, xiii, xix, 3, 14, 21, 26, 27, 49, 52, 60-62, 64, 65, 68, 70-76, 78-80, 85, 86, 88, 91, 92, 94, 95, 98, 100-102, 105, 107-112, 119, 123, 125, 130, 134, 135, 137-142, 147-149, 153, 157, 180, 181, 183, 191, 199, 204, 208, 223, 224, 231, 239, 257, 296, 300, 301, 309, 341-343, 345, 374, 380, 386, 387, 420, 422-424, 438-440, 443, 447-451, 453, 454, 456, 457, 459, 460, 462, 463, 475, 479, 485, 486, 496, 500, 502-504, 508, 513, 516
Edmondston family (Unst, Shetland) 26, 60, 100, 106, 299, 456
Edmunds, Arthur 235, 363-370, 486, 507
Ehrlich, Paul 157, 187, 192, 356, 506
Elizabeth 21, 22, 26
Enslin 247
Esmarch, Johann Friedrich August von 154, 250, 288, 289

F

Feal (Fetlar) 315, 403, 404
Fetlar i, iii, v, vii-ix, xi, xiii-xv, xvii, xix, 2, 5, 7, 14, 22-26, 31, 34-37, 41-53, 56, 57, 60, 77, 94, 99, 102-104, 106, 107, 110, 111, 123, 126, 130, 184, 219, 236, 239, 262, 294, 296-304, 306, 308-310, 312, 315, 317, 318, 319, 322, 323, 325, 329, 330, 341-343, 362, 383, 395-397, 399, 402, 403, 407, 408, 410, 411, 413-415, 417, 419, 423, 424, 429, 430, 433-435, 437-439, 441-447, 449, 456, 458, 460, 461, 474, 487, 491, 492, 496-500, 503, 509, 510, 512-514, 517, 518
Fetlar Manse xiii, 2, 22, 35-37, 39, 41-44, 46, 48-55, 57, 69, 130, 184, 298-301, 307-309, 407, 430, 497
Fiji 10
Finkler (bacteriologist) 172, 173
Flinders, Captain Matthew 15, 438
flit boat xiv, 306, 307, 310, 312, 313, 322, 323
Flügge, Carl G.F.W. 170, 191, 192, 470, 476
Fox, Wilson 160
Fraenkel (bacteriologist) 160
France 1, 79-81, 159, 166, 345, 349, 357-359, 365-369, 371, 382-384
Franks, Kendall 242, 293
Free Church of Scotland 46, 325, 413, 414
French, John (Field Marshall) 28-30, 32, 35, 44, 57, 79, 91, 95, 171, 191, 193, 247-249
Freud, Sigmund 97, 455

G

Gangee, Sampson 222
garden party xiii, xviii, xix
Garster, George 309, 498
gauze 113, 185, 192, 244, 285, 288-291, 361, 362, 476
General Medical Council 68, 73
German(y) 4, 40, 42, 57, 70, 79, 83, 90, 91, 95, 98, 100, 101, 132, 149-152, 153, 155-157, 159, 160, 165, 166, 169-174, 182, 187, 189, 191, 193, 196, 197, 201, 211, 227, 228, 230, 231, 249, 250, 266, 288, 292, 334, 345, 354-357, 380, 506
Gilbert Bain Hospital (Lerwick) xix, 402
Glasgow 61, 80, 85, 110, 339, 345, 387, 420, 449, 451
Godlee, Sir Rickman 97, 98, 102, 137, 138, 144, 149, 180, 184, 337-339, 342, 345, 356, 362, 382, 447, 451, 454, 455, 462, 465, 473, 479, 480, 502, 503, 507, 510

539

Godwin, George 67
Goltdammer (Robert Koch's assistant) 156
Gossler, Gustav von 195
Grove House School 79
Guam 31

H

Haffkine, Waldemar 178, 472
Haig, Douglas 268
Harley Street 5, 144, 191, 202, 219, 313, 329, 396-398, 403, 479
Hart (surgeon) 123-125, 165, 166
Harveian Lectures 191, 203, 211, 220, 475, 481
Harveian Society 189, 336, 475, 479, 483, 501, 502
Hay and Company 309, 312, 498
Hebb (Captain, 107th Field Ambulance) 384
Henle, Jakob 143, 150
Henley, William 109, 114
Henry, Elizabeth, 43
Henry, Andrew 42
heritors 10, 51, 52, 53
Heron, George A. 172, 195-199, 477
Herschell, 2nd Baron 314, 414
Herschell, 3rd Baron 315, 498
Hillside Cottage 184, 299, 487
Hillswick 15, 26, 44, 50, 52, 438
Hime, Thomas 175, 176, 472
HMS Challenger 72
HMS Perseus 57
Hobart 24, 27-31, 429, 439-443
Hogaland 12
Home Farm, Leagarth vii, xv, 403-405
Hong Kong 31, 32, 472
Horrell (Methodist Minister) 108-109
Horsley, Sir Victor 377, 378
Hôtel Métropole 243
Houbie (Hubie) xv, 107, 311, 312, 322, 327, 330, 415, 416
House of Commons 24, 30, 35, 242, 392, 395, 432, 467
Howlam, A.W. 345, 503
Hughes, A.W. 295, 496
Hughson, John 312
Hunter, John 363
Hunter, William 201
Hunterian Oration 163, 356, 363, 383, 468, 506, 507, 510
hydrocephalus 214, 215, 482
hypochlorites 369, 370, 372, 383, 510

I

Ibedul of Koror 54, 57-58
idiopathic purpura haemorrhagica 167
International Medical Congress 152, 227, 228, 466
Irvine, Thomas (of Midbrake) 11, 437
Isle of Pines 10, 15-17, 32, 35

J

Jacobsdal 248-250, 265, 266
Jameson, Dr. L.S. 240
Jameson Raid 240, 247
Japan 30, 32, 36, 442, 443
Jellicoe, John, 1st Earl Jellicoe 355
Jenner, Edward 106
Jenner, Sir William 158
Jex-Blake, Sophia 181
Johnson and Johnson (manufacturers) 214
Johnston, Colonel H.H. 400, 513
Johnston-Saint, P. 422, 515, 516

K

Karee Siding 273, 274, 290
Kekewich, Major General Robert George 247
Kenwood, Henry R. 167
Kenya 420, 421, 423, 424, 515
Keogh, General Sir Alfred 358
Kidd, Percy 163, 193, 469
Kimberley 240, 241, 247-249, 266-268
King Edward's Hospital Fund 398
King's College vii, xix, 2, 4, 137, 139, 140, 147, 152, 178, 180, 184, 197, 198, 200, 201, 205, 208, 209, 215, 242, 243, 245, 296, 337, 341, 343, 345, 351, 397, 422, 432, 437, 448, 463-465, 467, 468, 474, 475, 477, 478, 480, 482, 486, 487, 500-504, 506
Kirkwall 110, 389
Kitchener, Horatio Herbert, 1st Earl 261, 262, 267, 269, 358
Klebs, Edwin 162, 179
Klein, Emanuel Edward 135, 165, 171-177, 471, 472
Klip Kraal Drift 250, 261
Knight Commander of the Order of St. Michael and St. George (K.C.M.G.) 372, 446, 447, 449, 450, 453, 462-465, 518

Index

Koch, Robert xvii, 4, 79, 101, 143, 149, 150, 151, 155-157, 159-165, 167, 170-178, 181, 182, 188, 192-201, 207, 227, 228, 340, 341, 345, 349, 350, 465, 468, 471, 472, 475-478, 502
Koror 19, 40, 54, 55, 57, 58
Kroonstad 274, 280
Kruger, Paul 241, 267

L

laboratory methodology 3, 154, 157, 174
Lady Montague 27-30, 32-34, 36, 440, 441
Ladysmith 240, 259
Laing, Malcolm Alfred 398
Laing, A.L. (Lerwick chemist) 403
Lambhoga xiv, 311, 317-319, 322, 343
Lansdowne, Henry, 5th Marquess 292
lantern slides xix, 189, 215, 236, 294
Laurenson, James J. (Jeemsie) ix, xiv, 383, 414, 435, 444, 445, 510
Leagarth House vii, viii, xi-xv, xviii, xix, 6, 162, 296, 304, 310, 311, 313, 314, 317-319, 323-330, 334, 341, 343, 344, 362, 383, 395, 396, 398-403, 405-415, 417, 421, 429, 431, 432, 435, 436, 491, 498-500, 509, 510, 512, 514, 517, 518
Leagarth gardens xiv, xv, 396, 406, 407, 432
Leeson, John Rudd 64, 81, 83, 87-91, 102, 114, 115, 117, 120, 124, 136, 138, 141, 147, 159, 184, 339, 392, 448, 452-454, 456, 458, 460, 462, 463, 465, 467, 474, 512
Leith 26, 110
Lerwick xix, 25-27, 30, 53, 56, 97, 99, 108, 204, 219, 301, 309, 310, 312, 321, 342, 397, 398, 402, 403, 426, 432, 437-449, 454-462, 464, 466, 470, 479, 483, 497, 498-500, 502, 503, 509, 510, 512-515, 517, 518
Lerwick Seaman's Mission 398
Lettsomian Lectures 208, 210, 212, 481
Libbertz, A. 195, 199
Lillie, Reverend John 30
Lister, Agnes 79, 85, 180, 184, 185, 202, 205, 342, 345
Lister, Dr. A.H. 337, 338
Lister, Joseph, 1st Baron vii, xi, xiii, xvii, xix, 1-4, 9, 30, 61, 62, 77-95, 97-104, 108, 109, 111-118, 120, 122-125, 127-130, 132-154, 158, 159, 161, 169, 170, 172, 175, 180-182, 184-186, 188, 191, 192, 194, 198, 203-205, 209, 210, 212, 214, 216, 221-229, 231, 233, 234, 236-239, 244, 253, 283, 289, 291-293, 296, 314, 329, 331, 336-348, 353, 359, 362, 363, 367, 371, 392, 395, 420-424, 435, 437, 447-458, 460-465, 467, 470, 473, 474, 476, 479, 481, 483-486, 488, 495, 496, 500, 502-504, 512, 515, 516
Lister and His Achievement 421, 437, 448-456, 460-465, 473, 484, 485, 515
Listerian 1, 84, 86-88, 90, 92, 97-99, 101, 104, 109, 133, 134, 136, 142, 145, 146, 148, 149, 151-155, 169, 181, 220-224, 226-238, 288, 289, 331, 332, 341, 346, 348, 362, 366, 372, 461
Listerism 4, 97, 101, 134, 136, 144, 154, 169, 190, 203, 222, 223, 230, 234-236, 288, 452, 461, 463, 484, 485
Lister Memorial 77, 346, 420, 448, 450, 451, 504, 515
Liston, Robert 62, 64, 65, 79
Littlejohn, Harvey 380
Lloyd George, David 375, 379
Loch, John D. 24-26, 30, 31, 439-442
Lockwood, Charles Barrett 232, 486
Loeffler, Friedrich 151, 179
Loeterbagh, Dr. Petrus 100, 455
Logan, Mary 113-115, 120, 127, 140
Logue, Lionel 398
London vii, xi, xix, 4, 5, 17, 21-28, 30, 32-34, 36, 56, 62, 64, 66, 79-81, 89, 98, 99, 109, 114, 119, 122, 125, 132, 137-143, 145-147, 149, 162, 164, 167, 170, 171, 178, 181-184, 186, 195, 197-199, 202, 203, 205, 206, 208, 210, 216, 220, 224, 242, 243, 245, 257, 259, 275, 281, 283, 285, 286, 296-298, 300, 308, 312, 314, 317, 323, 333, 336, 338-340, 342, 343, 345, 346, 348, 353, 358, 375, 383, 385, 391, 396-398, 414, 419, 420-424, 426, 432, 435, 437, 442, 443, 447-456, 458, 460-470, 472-493, 495, 499, 500-511, 513, 515-517
London Orkney and Shetland Association 296
London School of Medicine for Women 181
Lord Lieutenant for Orkney and Shetland 398
Lücke, Georg Albert 98, 101
Lundie, R.A. 386, 511
lupus 199

541

Mac

MacCormac, Sir William 242, 295, 496
MacDonald, Ramsey 374, 419
MacFarlan, J.F. and Co. 214
Mackenzie Davidson, Sir James 292
MacNamara, C. 200

M

macrophages 358
Mafeking 240
Magersfontein 265
Makins, Sir George 242, 246, 247, 253, 256, 260, 267, 269, 273, 290, 369, 372, 489, 490-492, 495, 506
Malcolm, Mr. (Curator of Wellcome Museum) 421-424, 458, 515-517
Malta 367, 369
Maltwood's finder 187, 188
Manila 17, 21, 32, 55, 69, 447, 449
Manson, T.M.Y. 402, 403, 437, 438, 441-447, 459, 500, 513, 514, 517
Manual of Surgical Treatment 215, 332, 339, 363, 482, 483
Maori 426-428
Marquesas 10
Martin, C.J. 337-339
Mathewson, Andrew Dishington 49, 100, 455-457, 461
Mathewson, Margaret ix, xiii, 3, 100, 102-131, 139-140, 183, 186, 329, 448, 455-462, 473-474
Mauritius 26
Mayo Robson, Arthur William 227, 484
Medical Act (1858) 68
Medical Consultative Board to the Royal Navy 355
Medical Parliamentary Committee 388
Medical Society of London 208, 210, 358, 480, 481, 504, 507
Medical Times and Gazette 223
Medical Times and Hospital Gazette 353
Mercantile Marine Act 28
Merson, George 214
Mesmerism 65, 448
Metchnikoff, Elie 349
Methodists 46, 299, 497
miasms 66, 67, 81, 150, 170, 171
microbes xi, 1, 3, 4, 86, 136, 167, 244, 364
micrococci 132, 134-136, 145, 152, 161, 162, 167, 227, 228

microorganisms xi, 80, 81, 83-85, 97, 99, 101, 132, 134, 150, 152, 153, 156, 157, 169, 179, 182, 186, 220, 227, 228, 232, 340
milk 80, 82, 142, 143, 279, 405
Milles, W. Jennings 162
Ministry of Health 156, 388, 389, 511, 512
Minos 28
Modder xix, 5, 241, 245-251, 253, 267, 269, 271, 277, 287, 489-492, 494, 495
Moor 275
Moowick 324
Morris, Sir Henry 389
Mount Vernon Hospital for Consumption and Diseases of the Chest 353
Mudros 369, 370

N

Naiad 19, 20
National Insurance Act 375, 378, 379
Nautical Magazine 17, 36, 443
Neesik 318, 324
Nelson, E.M. 156, 157
New Caledonia 10, 15
New Hebrides 10, 20, 31
New South Wales 23, 32, 36, 249, 281, 282, 438, 442, 443
New South Wales Ambulance Corps 249
New Sydenham Society 151, 169, 170, 174, 191, 193, 337, 465, 470, 475, 476, 502
New Zealand vii, xv, 10, 15, 247, 424-426, 428-430, 514
Nicolson, Eardley 303, 315, 498, 499
Nicolson, Lady Annie, xiv, 107, 303, 304, 307, 312, 316, 323, 325, 327, 456, 457, 498-500
Nicolson, Lady Eliza 497, 498
Nicolson, Lionel 414
Nicolson, Sir Arthur 44-47, 51-54, 301, 303, 309, 444-446, 497
Nicolson, Sir Arthur T.B.R. xiv, 106, 107, 303, 305-307, 309, 311, 322, 323, 325
Nicolson, Sir Arthur J. 307, 308, 327, 328, 414
Nicolson, Sir Harold Stanley 414
Nicolson, Vera (Lady Herschell) xiv, 314, 315, 325, 414
Nightingale, Florence 66, 67, 114, 147, 148, 185, 276, 348, 448
Norna 75
Northmavine 11, 14, 21, 41, 50, 54, 69, 106, 437, 438, 441-447

Index

Northmavine Manse 15, 41, 44, 52, 438
Nounsbrough (Nounsburgh) 12, 21, 58, 69
Nova Scotia 139
Numidian 209
Nyasaland 420

O

Ogston, Sir Alexander 135, 151, 152, 167, 227, 289, 292, 466, 495
Ollaberry 69
opium 10, 21, 28
opsonins 349, 369
Orange River 246, 256, 269, 270
Orkney, Shetland and Northern Society 296
Ormuz 420
Ortega 417
Orwell 15, 16
Ossfontein 267
Owen, Edmund 189

P

Paardeberg 7, 250, 253, 260, 261, 266, 267, 269, 271, 274, 281, 282, 491
Pacific Ocean 2, 9, 10, 15, 17, 19, 21-23, 26, 30-36, 41, 42, 44, 54, 299, 417, 437-439, 441-444, 446, 505
Pacini, Filippo 171, 175
Paddington Green 178, 179, 181, 184, 198, 201, 214-216, 477
Paddon. James 19-21, 42
Palau 19, 40, 42, 54, 55, 57, 58, 69, 429
Palmerston, H.J., 3rd Viscount 28, 29
Parliament 6, 44, 158, 159, 272, 276, 278, 280, 347, 348, 372, 375, 377, 380, 382, 384, 386, 387-389, 391-393, 395, 396, 417, 506, 510, 511
Passenger Act (1855) 94
Pasteur, Louis 1, 79-82, 84, 150, 151, 165, 170, 178, 194, 197, 198, 340, 345, 357, 462
Pathological Society of London 143, 210, 336, 340, 461, 463, 481, 501
Penn, W.C. 341
Pennington, T.H. 228, 229, 461, 484, 485
Perseus 57, 58
Petrie, James 324
Pfeiffer, Richard 349, 357, 504
phagocytes 349, 369
photography 188
physiology 74, 77, 87, 97, 145, 186
Playfair, Major General A.L. 208

Plymouth 28, 29, 294, 369, 478, 496
Pollock, Elizabeth 61, 76
Ponape 18, 31
Poplar Grove 260, 267, 268
Port Resolution 20
Post Office 47
Potter, J.D. 17, 442, 443
Prideaux, Frances Helen 180, 181, 473
Pringle, Andrew 188, 189, 191
Pringle, Angelique Lucille 114, 115, 130, 458
Pringle, Robert H. 421, 515
puerperal fever 66, 81

Q

Quakers 47, 79, 185
quarantine 177, 472
Quarff 204

R

rabies 170, 393, 394, 512
Ramdam 247, 248
Ranke, H.R. 227, 484
Ravensbourne 415
Recklinghausen, Friedrich Daniel von 98
Redding, Emily 204
Reid Tait, E.S. 136, 166, 167, 169, 201, 429, 461, 464, 466, 469, 470, 479, 482, 509, 510, 512, 515, 518
Retractor for use in intestinal operations 213
Rewa 367, 369, 370
Reyher, Carl von 288, 289, 363
Rhodes, Cecil 240, 241, 247
Riet 248, 249
Roberts, Frederick, 1st Earl xiv, xix, 5, 149, 153, 241, 242, 245, 246-249, 260-264, 266-269, 272-274, 276-278, 280, 281, 283, 284, 293, 294, 357
Robertson, Elizabeth 10-12, 21
Robertson, James Young xv, 324, 408, 409, 429
Roentgen (X-)rays 351, 352
Röntgen, Wilhelm 292
Rothschild, Lord 346
Roxburgh, Dr. (Surgeon, Edinburgh Royal Infirmary) 124, 129, 140
Royal Air Force 385
Royal Army Medical Corps xiv, 242, 243, 246, 249, 259-261, 263, 266, 268, 272-274 277-279, 282-287, 292, 380, 382, 385, 388, 389, 508

543

Microbes and the Fetlar Man

Royal College of Physicians 68, 105, 336, 341, 422
Royal College of Surgeons vii, xix, 5, 49, 79, 144-145, 152, 182, 189, 208, 210, 237, 242, 336, 337, 341, 342, 345, 352, 355, 357, 363, 372, 380, 420, 447-453, 456, 460-464, 467, 468, 473-476, 478-481, 484, 486, 496, 500-507, 509, 510, 514
Royal Medical and Chirurgical Society 157, 164, 172, 199, 349, 467, 469, 471, 478
Royal Medical Society 76, 210
Royal Society 72, 80, 119, 177, 204, 227, 345, 346, 393, 420, 432, 456, 459, 460, 463-465, 470, 472, 479, 481, 485, 505, 518

S

Sailing Directions from New South Wales to China and Japan 32, 36, 442, 443
St. Andrews 374, 380, 387, 508
St Bartholomew's 184, 229, 348
St. Petersburg 198, 199
St. Thomas' Hospital 64, 66, 89, 114, 147, 213, 242
Sambrooke Surgical Registrar 143, 363
sandalwood 9, 10, 15, 16, 18, 19, 31, 35, 43, 55, 437-439, 441-444, 446
sandalwooders 19, 31, 40, 42
San Francisco 32, 33, 442
Saxby, Jessie 60, 297, 299, 399, 401, 447, 497, 513
Scalloway 318
Schüller, Max 161
Scollay, Daniel 110
Scott, Thomas Walter 318
Semmelweis, Ignaz 65, 66
Senn, N. 193, 476
sepsis 1, 63, 64, 65, 68, 80, 81, 83-85, 91, 132-134, 135, 137, 138, 141, 145, 146, 148, 149, 154, 167, 169, 182, 200, 224, 228-230, 232, 236, 265, 266, 273, 290, 346, 355, 357-361, 364, 365, 367, 417, 461
Servanté, Mary Emma (see Cheyne)
Servanté, William (Vicar of Plumstead) 184, 185
Servants' Tax Registers' Defence Association 379
Sharpey, William 79
Shaw, Flora 241, 247, 487
Shetland vii- ix, xiii, xvii, 2, 3, 7, 10-12, 20-27, 30, 33, 35, 37, 39, 41, 44, 45, 50, 51, 53, 55-58, 60, 69, 75, 99, 100, 102-107, 109-112, 118, 119, 122-124, 130, 131, 139, 183, 186, 204, 205, 207, 219, 239, 294, 296-299, 301, 309-311, 315, 317, 318, 321, 322, 325, 329, 330, 341, 343, 395, 398, 400, 402, 406, 407, 414, 415, 420, 426-428, 432, 433, 435, 437-449, 454-462, 464, 466, 470, 479, 482, 487, 488, 492, 496-501, 503, 509, 510, 512, 513-515, 517, 518
Shetland Mainland 7, 11, 22, 49, 56, 69, 106, 107, 110, 204, 296, 318
Shetland Society 426, 428
Shewan, Walter viii, 318, 414
Shineberg, Dorothy ix, 17, 19, 30, 32, 40, 54, 437-444, 446, 447, 449
Siddons House Private Hospital 208
Simpson, James Young 61, 65, 66, 100, 142, 147, 447, 448, 463, 485
Sir Lister Hall 413, 414
sixereen 50
Skua 324
Slater, Andrina 219, 397
Slater, Catherine, 219
smallpox 33, 79, 99, 106-108, 455
Smith, George 204
Smith, Gilbert 46, 48, 52,
Smith, James 204
Smith, John Scott 205, 397
Smith, Julia 285
Smith, Margaret (see Cheyne, Lady Margaret)
Smugga 318
Smuts, Jan 240, 487
Society for the Propagation of Christian Knowledge 49
Soudan 357, 366, 367, 506
South Africa xiv, 153, 216, 217, 239-243, 245, 246, 259, 272, 276-278, 281, 282, 285, 287-289, 291, 292, 294, 295, 359, 389, 399, 488, 489, 492-494
South Seas 14, 15, 36, 41, 42, 444
Spence, James 74, 75, 140, 142, 148, 223-226, 462, 463, 465, 483, 484
Spence, Basil 26,
Spencer, W.G. 200
Spencer Wells, Thomas 81, 163, 468
SS St. Giles 320
staphylococcus 135
Stephenson, George Skelton 70-72, 75, 76, 83, 86, 95, 180, 184, 339, 449-451,

453, 454, 471, 473, 474
Stevens, Captain Charles E. 40, 41, 57, 58, 444, 447
Stevenson, Colonel (Royal Army Medical Corps) 246, 269, 272
Stewart, John 139, 140, 463, 464, 485
Stiles, Sir Harold 191, 210, 231, 235, 236, 238, 319, 382, 486, 487, 499
Still (Fetlar) 49, 318
Stokes, Sir William 242
Strasbourg 98, 100, 101, 132, 145
Sunderland 26, 27
Sunflower 312
surgery xi, 1, 2, 4, 9, 62-64, 67, 74, 77-79, 82, 85, 87, 91, 95, 97, 100, 101, 109, 111, 127, 135, 136, 142-145, 153, 157, 186, 189, 202, 205, 210-213, 221, 225, 228, 229, 234, 237-239, 251, 257, 269, 282, 287-289, 292, 293, 319, 331, 338, 339, 341, 342, 347-349, 352, 354, 362, 363, 384, 435, 485, 500, 513, 514, 517, 518
Sutherland, G.A. 214, 215, 482
Sydenham Hill 396
Sydney 10, 15-17, 33, 332, 493, 501
Syme, James 61, 79, 136, 142, 231
Syme, Lucy 342

T

Tamaroa 426
Tanganyika (Tanzania) 419, 429
Tangwick viii, xiii, 11, 12, 14, 22, 55, 438
Tasmania 23, 24, 27, 440, 441
Taylor, Dr. Harry P. 107, 314, 328-330, 403
tea xix, 9, 10, 21, 104, 107, 115, 325, 406
Tetens, Captain Alfred 40, 41, 55, 58, 444, 446, 447
The Chapel 184, 299, 300, 487
The Lancet 177, 236, 349, 370, 405, 433, 454, 464, 470, 472, 476, 480, 481, 484-488, 496, 499, 503, 504, 506-508, 511, 514, 518
The Science and Practice of Surgery 220, 483
The Spectator 379, 509
The Treatment of Wounds, Ulcers and Abscesses 207, 232, 234, 479, 485, 486
Thomas, Lizzie 140
Thomason, Agnes 315
Toulouse 160, 161
Toussaint, M. 159-163, 468
Towns, Robert 42
Transvaal 240, 493

Treaty of Commerce and Constitution of the Pelew Islands 54
Tresta (Fetlar) xv, 330, 397, 407, 430, 433
Treves, Sir Frederick 205, 242, 259, 260, 269, 281, 285, 345, 397, 479, 490, 492, 493
Truck 21, 45, 104
tubercle xi, 156, 157, 159, 161, 163, 164, 166, 186-189, 191, 193, 194, 196, 197, 199, 405, 466, 467, 469, 476
tuberculin xi, 186, 193, 194, 197-201, 340, 349, 350, 405
tuberculosis xi, xvii, 4, 29, 35, 43, 51-53, 70, 104, 108, 123, 128, 131, 144, 156, 159, 160-162, 164, 172, 186, 189, 190, 191, 194, 195, 197, 199, 201, 211, 238, 296, 349, 354, 375, 405, 407, 415, 417
Tuloa Farm 421, 515
Turner, G. Grey 216-217, 482
Turner, William 73-74, 136,
Tyndall, John 157, 175
typhoid 5, 241, 269, 271, 274, 275, 277, 278, 280-282, 292, 295, 349, 357, 358, 506

U

Uitlanders 240
University of Edinburgh vii, xiii, 49, 60-62, 68, 70-80, 86, 88, 91-93
University of Glasgow 61, 80, 345, 451, 387, 420
University of Vienna 95-98
Unst 26, 47, 60, 100, 107, 110, 184, 297, 299, 300, 399
Up Helly Aa 297

V

Van Diemen's Land 24, 27-29, 439-442
Vatsie (Yell) 330
Vaux, John 27-28, 36
Victoria Park Hospital for Diseases of the Chest 197
Vienna xi, xiii, 65, 86, 95-97, 455
Viking Club 296, 297, 496, 497
Viking Society for Northern Research 296
Virchow, Rudolf 98, 340, 485, 502
vivisection 158, 177, 391, 392, 395
Volcano Island 32

W

Walewski, Count 28
Walker, Dr. (doctor in Newcastle) 167
Walmer Castle 332, 333, 500, 501
Wassau 20
Waterval Drift 248, 267
Watson, Grace xiii, 22, 26, 32, 37, 38, 52, 96, 184, 208, 239, 298-300, 487, 497, 515
Watson, Reverend William Watson, 14, 15, 21, 22, 25, 27, 30, 34-39, 44-48, 50-52, 56, 68, 69, 104, 105, 111, 298, 299, 307, 432, 438-46, 457, 497
Webster, Christian (Watson) 23, 26, 27, 35-39, 41, 45, 48, 56-59, 69, 70, 184, 298, 440
Webster, David 22-26, 35, 36, 37, 39, 41, 44, 45, 47-54, 56-59, 69, 70, 96, 184, 298, 299, 302, 444-446, 454, 497
Webster, John 48, 56, 61, 70, 76, 77, 299, 445
Wegdraai 248
Wellcome Museum viii, 83, 314, 383, 422, 452, 454, 458, 467-468, 478, 499, 508, 510, 513, 515-517
Wellington Evening Post 517
whales 319
Whalsay 204, 322
White, Sir George 243
whitefish 10, 22, 45
Whitla, William 391, 393, 394
Wild Wave 36
Williams, George 32
Williams, C.T. 199, 478
Williams, Dawson 337-339
Williamson, Laurence ix, 125, 315, 329, 330, 460, 497, 499, 500, 514
Williamson, John (Johnnie Notions) 106
Wilson, Albert 362, 507
Wilson Carpenter, Mary vii, 125, 456, 457
Woodhouse, Colonel Percy 384
Woodin, Edward 40
World War I xi, 5, 238, 341, 348, 350, 355, 358, 385, 386, 414, 508
Wright, Sir Almroth 6, 292, 341, 349, 350, 357, 358, 361, 368-372, 504, 507, 508
Wyville-Thomson, Sir Charles 72, 77

X

x- (Roentgen) rays 292, 351, 352

Y

Yell viii, ix, 11, 22, 47-49, 96, 99, 100, 102, 104, 107, 110, 111, 121, 125, 130, 131, 297, 300, 314, 315, 319, 322, 325, 328-330, 403, 443-446, 449, 454, 455, 460, 474, 487, 497-501
Yeo, Gerald 152

Z

Zetland xiv, 204, 307-309, 312, 322, 323, 398, 456, 513
Zetland Territorial Force Association 398
Zoflora 107

www.ingramcontent.com/pod-product-compliance
Lightning Source LLC
Chambersburg PA
CBHW051106230426
43667CB00014B/2457